INSEAD

COMMITTED TO
IMPROVING THE STATE
OF THE WORLD

The Global Information Technology Report 2006–2007

Connecting to the Networked Economy

Soumitra Dutta

Irene Mia

*The Global Information Technology Report
2006–2007* is a special project within the framework
of the Global Competitiveness Network. *The Global
Information Technology Report* is the result of a
collaboration between the World Economic Forum
and INSEAD.

Professor Klaus Schwab, Executive Chairman,
World Economic Forum

EDITORS

Soumitra Dutta, Roland Berger Professor of
Business and Technology and Dean of External
Relations, INSEAD

Irene Mia, Senior Economist,
World Economic Forum

GLOBAL COMPETITIVENESS NETWORK

Fiona Paua, Senior Adviser

Jennifer Blanke, Senior Economist
Ciara Browne, Senior Community Manager
Margareta Drzeniek, Senior Economist
Thierry Geiger, Economist
Aviva Rajczyk, Coordinator

**INFORMATION TECHNOLOGY AND
TELECOMMUNICATIONS INDUSTRIES TEAM**

Alex Wong, Head of Centre for Global Industries (Geneva)

Sandra Betemps, Coordinator
Simon Mulcahy, Head of IT Industries
James Tee, Global Leadership Fellow
Silvia Von Gunten, Head of Telecommunications Industry

A special thank you to Hope Steele for her superb
editing work and Ha Nguyen for her excellent
graphic design and layout. We are very grateful to
Nathali Glanzmann, Tamara Gomes, and Pearl
Samandari for their invaluable research assistance.

The terms *country* and *nation* as used in this
report do not in all cases refer to a territorial entity
that is a state as understood by international law
and practice. The terms cover well-defined, geographically self-contained economic areas that may
not be states but for which statistical data are
maintained on a separate and independent basis.

First published 2007 by
PALGRAVE MACMILLAN
Houndmills, Basingstoke, Hampshire RG21 6XS and
175 Fifth Avenue, New York, N. Y. 10010
Companies and representatives throughout the world

PALGRAVE MACMILLAN is the global academic imprint of the
Palgrave Macmillan division of St. Martin's Press, LLC and of
Palgrave Macmillan Ltd. Macmillan® is a registered trademark
in the United States, United Kingdom and other countries.
Palgrave is a registered trademark in the European Union and
other countries.

ISBN-13: 978–1–4039–9931–3
ISBN-10: 1–4039–9931–7

This book is printed on paper suitable for recycling and made
from fully managed and sustained forest sources.

A catalogue record for this book is available from the British Library.
A catalog record for this book is available from the Library of
Congress.

10   9   8   7   6   5   4   3   2   1
16   15   14   13   12   11   10   09   08   07

Printed and bound in Great Britain by
Hobbs the Printers Ltd, Totton, Hampshire

Contents

Preface

KLAUS SCHWAB

Executive Chairman, World Economic Forum

In recent years, the world has witnessed the power of information and communication technologies (ICT) in revolutionizing the business and economic landscape. With the advent of Web 2.0 and related technologies, the world is also seeing how ICT empowers individuals while fostering social networks and virtual communities, with profound impact on business, politics, and society.

What is equally evident is the increasing role that technology plays in accelerating economic growth and promoting development. The diffusion of mobile telephones, for example, has made a huge difference for remote rural communities, providing relatively cheap and easy access to new markets. Likewise, greater Internet access and more afforadable computers are radically changing the way education is provided in many countries, offering students virtual and enhanced resources while facilitating cooperation and exchange.

Now, more than ever, policymakers and business leaders recognize the need to create an enabling environment to support the adoption of technologies and spread their benefits across all sectors of society. The importance of networked readiness, especially at the national level, has achieved prominence on the public policy agenda, with the realization that the tools provided by ICT can help countries fulfill their national potential and enable a better quality of life for their citizens.

Recognizing the key role of ICT as a driver of growth and prosperity, each year since 2001—and jointly with INSEAD since 2002—the World Economic Forum has produced *The Global Information Technology Report*. This *Report* series assesses the progress of networked readiness in countries, revealing the obstacles that prevent governments, businesses, and individuals from fully capturing the benefits of ICT. Beyond providing a yearly snapshot of networked readiness, the *Report* has evolved into an authoritative benchmarking instrument and an invaluable tool for facilitating public-private dialogue, whereby policymakers, business leaders, and other stakeholders can evaluate progress on a continual basis.

The Global Information Technology Report 2006–2007, the sixth in the series, presents the latest findings of our research and highlights the best policies and practices for promoting networked readiness. In line with past editions, the *Report* offers a comprehensive snapshot of the state of networked readiness of the world, this year covering a record number of 122 economies. Also included in the *Report* is an extensive section of data tables with global rankings covering over 60 indicators as well as a number of essay contributions looking at different aspects of networked readiness.

We commend the contributors to this *Report* for their thought leadership and for helping this volume to serve, once again, as a platform for dialogue and reflection for both government and business. We especially wish to thank the editors of the *Report*, Soumitra Dutta at INSEAD and Irene Mia at the World Economic Forum, for their energy and their commitment to the project. Appreciation also goes to Fiona Paua, who leads the Global Competitiveness Network, and the other members of the team: Jennifer Blanke, Ciara Browne, Margareta Drzeniek, Thierry Geiger, and Aviva Rajczyk. Finally, we would like to express our appreciation to our network of Partner Institutes around the world, without whose dedication and hard work the 2007 Executive Opinion Survey and this *Report* would not have been possible.

Foreword

JOHN CHAMBERS

Chairman and CEO, Cisco Systems, Inc.

The theme of *The Global Information Technology Report 2006–2007* is "connecting to the networked economy." It is an ideal time to focus on this topic, in my opinion. Whether or not the global economy will become networked is no longer debatable—the vast majority of industries are increasingly adopting networked business processes—and the discussion now focuses not on *if* but on *how* we get connected to maximize the benefits to business and society.

Clearly, some parts of the world are more connected than others. In developing nations, our ability to get connected will have a dramatic impact not only on the global economy but also on the quality of life for millions as we begin to share widely the economic and social benefits of connectedness. Even in the most connected places—Asia, Europe, North America—I believe we are only at the beginning of what's possible using the network.

Different types of organizations are in various stages of connectedness. The world's leading enterprises and service providers are at the forefront, investing in information and communication technologies (ICT) to create and maintain competitive advantage. Their efforts to deploy networked processes and improve communications pave the way for mainstream small- and medium-sized businesses as well as government organizations of all sizes. But, ultimately, organizations are made up of individuals, and connecting to the networked economy means creating a human network that connects people with other people in meaningful ways.

We are truly in the midst of a market transition that will have a profound impact on the ways in which people communicate with each other, interact with companies, and access resources and information. Broadband adoption has gained significant momentum and the ICT infrastructure allows easy access to video, voice, and data communications any time, anywhere, from any device. Soon the ability to have your content—files, music, pictures, email, and so on—follow you from the office to your vehicle to your handheld device and into the home will be widespread.

But this transition is not inevitable. Networking technology is complex; enterprises have large teams of trained and credentialed engineers to manage it. However, to connect individuals effectively using a human network that empowers people to personalize and enhance their life experiences, the challenge becomes making something that is inherently complex so easy almost anyone can use it.

We believe the answer lies in an increasingly intelligent network that absorbs much of the complexity, making the end user's experience fast and seamless in everything from adding new devices to the network and sharing resources between devices to accessing content any time, anywhere, from any device.

In order for this to occur, five network characteristics must be in place:

- **Convergence** of disparate forms of communication, such as voice, video, and data, onto an Internet protocol (IP) network platform must occur. This convergence is already well underway.

- The development of the human network requires **virtualization**, which allows content to be seamlessly replicated on multiple devices hosted on multiple networks. This allows the content to "follow" you, so it is accessible wherever you go.

- The human network must be **simple**. By working together to use the network's intelligence to mask its complexity, technology vendors, application developers, and service providers can eliminate this challenge.

- In order to achieve the simplicity necessary for end users, the collective ICT community must agree on a common set of **open standards**. We believe IP provides the ideal framework upon which to build the human network.

- And finally, the human network needs to be **safe**. It requires strong security, as well as the means to effectively protect your family with features such as parental controls.

With these five characteristics in place, I firmly believe that we can effectively create the "human network," connecting not only organizations but individuals to the networked economy. Cisco is pleased to sponsor *The Global Information Technology Report 2006–2007*, including the Networked Readiness Index. The future lies in using the network to create meaningful, personal experiences for the people using the technology. This *Report* will provide us all with greater insight and help us to shape the experience.

Executive Summary

SOUMITRA DUTTA, INSEAD

IRENE MIA, World Economic Forum

The pace of technological innovation throughout the global community is relentlessly pushing forward. Ray Kurzweil, the American futurist, has observed that "In the nineteenth century, we saw more technological change than in the nine centuries preceding it. Then in the first twenty years of the twentieth century, we saw more advancement than in all of the nineteenth century. Now, paradigm shifts occur in only a few years time.... So we won't experience 100 years of progress in the 21st century —it will be more like 20,000 years of progress (at today's rate)."[1] Such a stunning assertion, assuming it proves true, has significant implications for how we lead our day-to-day lives and pursue our daily work. It means the capacity to adjust to unprecedented change at all levels of human experience will be the new paradigm not only for success but for survival. It means becoming constantly innovative individuals, workers, and citizens.

Technology has long had an important impact on innovation and the development of societies and economies. This impact can be visualized as occurring in three stages. In the first stage of "substitution," new technology substitutes for the old. For example, with the invention of the mobile telephone, consumers start substituting their fixed telephone lines with mobile telephones. The second stage of impact is "diffusion," which occurs when the new technology is adopted widely across society because it is cheaper, better, and in general more effective than the previous technology. Continuing with the example of the mobile telephone, there are many countries in the world today where the number of mobile telephones exceeds the population of the country. The final stage of impact is "transformation," which occurs when new ways of living and working start emerging because the new technology is diffused so widely in society. For example, the widespread adoption of the mobile telephone has led to interesting innovations in the communication patterns of individuals —such as executives conducting business while waiting in airport lounges or traveling in trains.

Keeping the above three-stage model in mind, one can argue that we are at a critical stage of transformation in society and business with new (Internet-based) information and communication technologies (ICT). For example, mobile telephones are not only creating innovative patterns of social and business communication but are also becoming sources of entrepreneurial business generation. Grameen phone is contributing to poverty reduction in Bangladesh by giving mobile telephones (with adequate training and backup system support) to women who turn into entrepreneurs by providing communications services to their respective village societies. The penetration of the Internet has reached a critical threshold, and even poor countries in Africa are moving actively to leverage the transformational potential of the global information architecture created by the Internet. Ethiopia, despite being one of the continent's poorest countries, is spending nearly one tenth of its GDP on information technology every year. Hundreds of government offices and schools have already been equipped with broadband Internet connections, and more are yet to come. Both government and private sector leaders in Ethiopia—in an astonishing example of forging headlong into the "global village"— have committed huge resources to seeing that by 2007 all of Ethiopia's 74 million people live no more than a few kilometers from a broadband connection. Mozambique is using information technology to improve governance and public administration while guarantying its citizens greater access to the benefits of a global knowledge base.

The Global Information Technology Report 2006–2007 makes its appearance at a critical juncture in the impact of ICT on the world economy. There is growing evidence that ICT is driving innovation by allowing creative thinking and responsive problem-solving to provide the promise of never-before-seen opportunities for all. Access to the global networked economy is becoming an important cornerstone of the development of economies and societies. It is against this background of optimism about the innovations and transformations induced by connecting to the global networked economy that this *Report* is being published. It builds on the work of the five previous editions, and thus it may be seen as part of a long-term commitment at both the World Economic Forum and INSEAD to the dissemination of business-relevant research on information technology issues with a strong practical focus.

The *Report* is composed of four thematic parts. Part 1 features the results of the Networked Readiness Index 2006–2007 and related analysis, together with a number of thoughtful essays on selected issues of networked readiness of particular relevance for today's world, written by eminent academics and knowledgeable industry experts and practitioners. These essays cover topics ranging from the opportunities and challenges brought about by next-generation telecommunications networks, to the several ways in which networks impact our every day life, the e-readiness of cities, and the moral dilemma multilateral corporations experience when dealing with Internet filtering regimes. Part 2 groups together a series of insightful country or regional case-studies under the common theme of access to ICT, considering sub-Saharan Africa, Estonia, Japan, and China. The distinguished authors relate the ICT policies and practices implemented in the country/region covered and present the challenges ahead. Part 3 provides detailed profiles for each of the 122 economies covered in the *Report*, displaying valuable background information on each economy's current networked readiness status and allowing for international and historical comparison on specific variables or components of the NRI.

Last but not least, Part 4 provides detailed data tables for each of the 67 variables composing the NRI this year, with global rankings.

Part 1: Selected Issues of Networked Readiness

Each year, *The Global Information Technology Report* selects a few issues that have with a particular relevance for countries' networked readiness. This year we look closely at four special areas: (1) networks and changes in every day life, (2) generation networks in telecommunications, (3) cities' e-government and global competition, and (4) filtered Internet and the moral dilemma for multinational corporations.

The Networked Readiness Index

In Chapter 1.1, "Connecting the World to the Networked Economy: A Progress Report Based on the Findings of the Networked Readiness Index 2006–2007," we present the latest findings of a research project undertaken by the World Economic Forum in 2001—and jointly with INSEAD since 2002—aimed at gauging countries' capacity to leverage ICT for growth and development. This year's *Report* assesses a record number of 122 economies all over the world, 7 countries more than last year and almost the double the number of countries covered in the very first edition in 2000.

Based on a mixture of hard data, collected from well-respected international organizations such as the International Telecommunication Union (ITU) and the World Bank, and survey data coming from the Executive Opinion Survey conducted annually by the World Economic Forum in each of the economies covered by the *Report*, the Networked Readiness Index (NRI)

measures the level of ICT development of nations, looking at a large number of relevant variables—67 this year.

The networked readiness framework used for the analysis is constant from one year to the next. It rests on three main subindexes, capturing:

- the presence of an ICT-conducive environment in a given country by assessing a number of features of the broad business environment, some regulatory aspects, and the soft and hard infrastructure for ICT;

- the level of ICT readiness and propensity of the three main national stakeholders—individuals, the business sector, and the government; and

- the actual use of ICT by the above three stakeholders.

The NRI rankings for 2006–2007 are broadly in line with those of last year, published in *The Global Information Technology Report 2005–2006*. Denmark is emerging as the world's leader in networked readiness this year, culminating an upward trend observed since 2003. The same upward trend can be seen for all other Nordic countries but Iceland, with Finland, Sweden, and Norway all gaining positions this year.

Among the top 20, the United States loses its top position and drops to 7th place, overtaken not only by Denmark but also by Sweden, which comes in at 2nd place this year, with Singapore at 3rd place, Finland at 4th, Switzerland at 5th, and the Netherlands at 6th. The latter, in particular, realizes the greatest improvement from last year (6 positions up). Also Estonia, at 20th, for the first time enters the top-20 league.

With respect to the largest Asian emerging markets, China (59th) and India (44th) are both losing ground from previous year, with a 9- and 4-position drop respectively. Most Latin American and Caribbean countries register encouraging improvements this year, with Jamaica (45th), Mexico (49th), Costa Rica (56th), Uruguay (60th), Argentina (63rd), the Dominican Republic (66th), Peru (78th), and Guatemala (79th) all gaining several positions. Africa is unfortunately experiencing an opposite trend, with all countries but Nigeria, Tunisia, and Algeria going down in the rankings, while the Middle East, led by Israel (18th) posts a rather stable showing.

The chapter provides some analysis of the trends in networked readiness at the global level, comparing regions in terms of aggregate ICT performance. It also explores the links between networked readiness and overall competitiveness, looking at countries by stages of development —the principal methodological tool used by the Forum to gauge national competitiveness.

Networks and changes in everyday life

In their paper "Networks Changing the Way We Work, Live, Play, and Learn," authors Roger Farnsworth, Lionel Gibbons, Tracey Lewis, and Marsha Powell (all at Cisco Systems, Inc.) relate the dramatic advancements in ICT that have created great opportunities for people around the world to change the way they work, live, play, and learn.

Each era of technological advancement since the Industrial Revolution—from steam engines and railways to steel and electricity, and oil and automobiles—has enabled businesses to expand their commerce and production globally. Today's innovators, however, are using ICT to completely rethink how they use information, and they are designing new business models and new capabilities that integrate dispersed partners and coworkers who network closely with each other daily.

In this new business environment, teams are formed not by traditional hierarchical organization or geography, but rather on an ad hoc footing based on the skills and expertise of individuals. Employees become members of fluid teams, empowered by collaborative tools and corresponding processes.

The impact on the individual is profound. Work is no longer a place to go but something people do. Network connectivity and collaboration tools give people the flexibility to work anytime from anywhere, for a better balance between their work and personal lives.

Governments worldwide are employing ICT to deliver more services to citizens at less cost and to improve their networking capabilities. The major challenges they face, especially in emerging countries, are finding investment capital, setting regulations for security and interoperability, and providing Internet access to and educating citizens. Citizens, at the same time, are seeking the same types of electronic entertainment they've sought for years—television, movies, games. Today's technologies, such as IPTV and mobile TV, personalize that entertainment in ways that let consumers pick and choose what they want, when they want it. These new capabilities also make good business sense, providing fresh revenue opportunities for service providers, cheaper ways to distribute movies for theaters, and so on.

In education, IP networks are turning the traditional classroom into virtual schoolhouses that deliver education to remote students, life-long learners, and others. Essentially, education has been transformed from a teacher-led class to a student-centric experience accentuated by self-learning; peer-to-peer teaching; rich, readily available content; greater accessibility; and discovery-based learning.

The authors point out that this ICT revolution is still young. We can look forward to technology advances to change the way we work, live, play, and learn in exciting new ways far into the future.

Generation networks in telecommunications

In "Opportunities and Challenges of Next-Generation Networks in Telecommunications," authors Scott Beardsley, Luis Enriquez, Mehmet Guvendi, Duarte Braga, Wim Torfs, and Sergio Sandoval (all at McKinsey & Company Inc.) discuss the opportunities and challenges of deploying next-generation networks (NGNs) for telecommunications providers.

Many telecommunications operators around the world are keen to capture the benefits of building NGNs—benefits such as reducing operational costs and increasing the speed of product development. However, NGNs also pose significant risks for operators, as they will require huge investments in infrastructure and could potentially erode fixed-voice revenues—on which the industry is still highly dependent—by changing the traditional way of pricing these services.

The urgency of capturing these benefits while at the same time weighing the risks has caused some of the regulatory and business battles we have seen over the conditions under which industry investments should be made. The authors argue that the prominence of this debate has obscured the fact that overall capital expenditure (CAPEX) investment in the telecommunications industry today is significantly higher than NGN investments alone. However, if the debate is not handled properly, it could affect the overall CAPEX investment of the industry.

Regulators will play a key role in mitigating or increasing the risks operators face. The authors argue that regulators should look for an alternative policy framework—one that recognizes the impact of NGN and regulation on the overall structure of the industry—instead of simply looking at the new issues through the same old regulatory lens. Operators, regulators, and other industry stakeholders must work together to manage the transition to NGNs successfully. This might be the only way that investors will be encouraged to invest, and that industry stakeholders and society can capture the full benefits NGNs have to offer.

Cities' e-government and global competition

By 2050 some 6 billion people (two-thirds of the world population) will live in cities. Most of the growth will take place in developing countries, where the urban population will double in 30 years, from 2 billion in 2000 to 4 billion in 2030. In less than 10 years from now, most of the "mega-cities" emerging from that process will be located in developing countries.

In their paper "The Next Frontier of E-Government: Local Governments May Hold the Keys to Global Competition," Bruno Lanvin and Anat Lewin (both at the World Bank) illustrate some of the main challenges brought about by the increasingly important role of cities, as opposed to central governments, all over the world.

Indeed, the above projections raise questions regarding the ability of the cities of the future to sustain extremely high growth rates while maintaining adequate levels of production and delivery of key public services such as water, transport, electricity, sanitation, education, and containment of crime and pollution. There is, however, another side to this equation, often overlooked. It relates to the emerging role of cities (and subnational entities generally) that are becoming global players—as attractors of foreign investment, as competitiveness hubs, and as platforms for the combination of local and international components of global production and supply chains.

The next few years will see the convergence of three major trends: (1) the continuing delegation of functions and responsibilities from central to local governments; (2) the maturation of outsourcing strategies, both domestically (from the public to the private sector, including through PPPs) and internationally (through off-shoring and near-shoring); and (3) the emergence of local global players (LGPs), such as major cities and economic centers, as global competitors and magnets for talents and investments. The chapter outlines some of the foreseeable consequences of such a convergence, and their implications for decision makers, public and private, central and local.

Based on some of the most recent data available, it offers a world mapping of e-government performance, comparing the performances of LGPs with those of their respective countries. Some unexpected results emerge from that analysis, calling for a radically new way to consider, analyze, and build competitiveness at the local and central levels. However, the results obtained also point to the need for a significant effort (locally, centrally, and at the international level) to collect relevant data and build relevant indicators. A strong call is made by the authors to build internationally comparable indicators of networked readiness and e-readiness at the local level, in particular for cities.

Filtered Internet and the moral dilemma for multinational corporations

In his paper "Reluctant Gatekeepers: Corporate Ethics on a Filtered Internet," Harvard Law School professor John G. Palfrey, Jr., deals with the issue of Internet filtering and the ethical issues multinational corporations doing business related to the Internet face.

As multinational corporations enter new markets, they come across states that practice sophisticated forms of online censorship and surveillance. The number of states with such regimes in place has risen sharply over the past five years, from a small handful to roughly three dozen as of 2007. The job of online censorship and surveillance is difficult for the state to manage itself, if not altogether impossible. In order to carry out these regimes, then, states turn to private firms that can provide the tools necessary

to effect the censorship and surveillance. These private firms—which should not be lumped together in terms of their ethical obligations, but rather disaggregated—include hardware manufacturers, software firms, online service providers, and local access providers, among others. The ethical problem arises when the corporation is asked to do something at odds with the ethical framework of the corporation's home state. Should a search engine firm agree to censor its search results as a condition of doing business in a new place? Should an email service provider turn over the names of its subscribers to the government of a foreign state without knowing what that government is looking for? Should a blog service provider code its application so as to disallow someone from typing a "banned" term into a subject line?

These questions—prompted by the hard cases that lie between simple acts of law enforcement and clear violations of international norms—are not easily answered through legislation or international treaty. Law is most likely not the primary answer. Traditional legal mechanisms will take so long to put in place that the contours of the problem will have changed beyond recognition by the time of enactment. Changes to the statute or treaty may be equally hard-won. Laws fashioned in this fast-moving environment will function as a hopelessly trailing indicator. The corporations themselves, as an industry, are best placed to work together to resolve this tension by adopting a code of conduct to govern their activities in these increasingly common situations. Palfrey argues that the corporations should call upon the knowledge and goodwill of NGOs, academics, states, and others to help to frame this code of conduct and to make it a meaningful, flexible, and lasting solution.

Part 2: Access to ICT: Selected Case Studies

This year's *Report* presents four case studies of the impact of ICT in different parts of the world. Estonia, sub-Saharan Africa, Japan, and China were chosen because each represents a clear instance of the dramatic effect of and access to ICT, and each can provide guidelines or lessons learned that can prove helpful to both developed and emerging economies.

E-ready Estonia and the challenges ahead

In his chapter "Estonia: A Sustainable Success in Networked Readiness?" Soumitra Dutta provides a compelling account of the accomplishments of Estonia's e-leadership strategy in recent times—which propelled the country to find itself among the top 20 performers in the NRI this year—considering the factors for its success as well as the challenges that it faces in sustaining its success over the next years. Indeed, an astonishing amount of innovation has emerged from Estonia, a country of 1.4 million inhabitants that has regained independence less than 20 years ago.

E-leadership has proven to be instrumental in helping Estonia through the painful transition from centralized state planning to the model of modern governance it is today. Estonia has pioneered and developed unique solutions and systems that have become an integral part of the life of most Estonians. Estonia's clear vision and leadership in ICT have led to results that often surpass those achieved by the older democracies of Western Europe. This is especially remarkable when one notes that the nation was ruled by foreign powers—including Denmark, Germany, Sweden, and Russia—for centuries. The merger of e-leadership and political vision has been one of the critical factors in its economic growth, its embrace of democracy, and its resulting accession to the European Union.

Leadership from the top has been vital for the success of Estonia. Maart Laar, the Prime Minister of Estonia from 1992 to 2002, and his key advisers spearheaded the development of ICT in the country. Not only has the government created a supportive policy environment for ICT, it has also been a pioneer in using ICT for its own processes. In 2000, the Estonian parliament approved a proposal to guarantee Internet access to each of its citizens, just like any other constitutional right. People all over the country can access the Internet free of charge from hundreds of public access points.

Nevertheless, Dutta notices that all is not perfect in the land of e-cabinets, Skype, mobile payments, and electronic ID cards. Despite many major and very visible successes, Estonia still faces a number of challenges and must overcome certain weaknesses. The question that remains open today is whether the country can leverage its knowledge and best practices and turn this advance into a truly a sustainable model. For example, Estonia's business environment is dominated by small- and medium-sized businesses that are often suppliers to international companies and do not invest sufficiently in basic research and development (R&D). The success of everyday mobile applications in Estonia masks the low level of R&D investment (0.8 percent of GDP 2005) in the country. Further, about three-quarters of this R&D funding comes from public sources (compared with about two-thirds of R&D funding from private sources in other European nations).

Communications services in sub-Saharan Africa

The impact of ICT on economic development and growth is well documented in the literature, yet there is the perception that sub-Saharan Africa may have missed the boat vis-à-vis other regions, such as Asia, which have been more successful in reaping the gains from the relocation and globalization of activities enabled by ICT. In his chapter "Access to Communications Services in Sub-Saharan Africa," IMF economist Markus Haacker sheds lights on the real contribution ICT is making to economic development in the region, focusing specifically on the determinants of access to communications services and their implications for doing business in the region and for sub-Saharan Africa's advantageous insertion in the global economy.

Starting with an analysis of what drives investment decisions for mobile telephone services providers in the region, the author observes that there is a role for policy (a license is required for market entry), but that the value of such a license and the decision about entry is driven by market size. He then introduces data on the number of providers and discusses some aspects of market structure (such as the increase of the number of providers since the age of "fixed-lines only" and the presence of international providers).

For the determinants of access to communications services, Haacker suggests that the higher extent of competition introduced by mobile telephone services has resulted in lower prices for these services while also providing improved access to them. He also discusses the viability of mobile telephone services in an environment characterized by less-developed financial services and weak contract enforcement. Mobile telephone technology, which facilitates use of pre-paid services, works where fee-based subscriptions may not work (because of the ease of financial transfers and the absence of contract-enforcement issues, for example); indeed, mobile telephone technologies sometimes enable financial transactions ("e-payments") other than payment for telephone services.

In conclusion, the chapter finds that sub-Saharan Africa has indeed not benefited to the same extent as some Asian countries from the global transformations directly enabled by improved communications technologies (with the exception of some sectors, such as textiles). However, from a microeconomic standpoint, Africa *disproportionally* benefits from advances in ICT, in terms of improved access to these services. Consequently, advances in ICT have substantially contributed to the ease of doing business in Africa, with direct implications for productivity and growth, and have greatly enhanced the potential for growth through enhanced participation in the global economy.

Information and communications policy in Japan

Although ICT usage in Japan lagged behind that of other advanced nations in the 1990s, Japan has recently emerged as one of the world's most advanced broadband communications societies. This is primarily a consequence of the high penetration of broadband technologies such as fiber optic cable for household use. As a result of competition, a number of DSL customers switched to fiber optic cable. In addition, Japan's mobile telephones have the highest

Internet access rate (87 percent) in the world, and over 60 percent of these mobile telephones are 3rd-generation telephones. Mobile telephones in Japan can be used not only for e-mail, downloading music, and playing games, but also for taking high-resolution photos and movies, watching television, using electronic money, and purchasing electronic tickets. Finally, broadband fees in Japan are the lowest in the world (100 kilobytes cost US$0.07 a month).

In their chapter "Information and Communications Technologies Policy in Japan: Meeting the Challenges Ahead," Hideo Shiumizu, Kuniko Ogawa, and Koichi Fujinuma (all at the Ministry of Internal Affairs and Communications, Japan) explore the ICT strategy adopted by the Japanese government in 2000. This strategy enabled and triggered the above advancements, significantly raising the country's level of networked readiness. The authors also consider the direction in which ICT is heading in that country.

Acknowledging the importance of ICT for economic growth, in 2000 the Japanese government adopted a framework IT law setting specific targets with the aim of turning Japan into the most advanced ICT country in the world. This law enabled the adoption of different e-Japan strategies that fostered ICT use and penetration in different areas. In December 2004, building on the achievements of the previous e-strategies, a u-Japan Policy was adopted in order to create an ubiquitous networked society by 2010. The u-Japan Policy is focused on ensuring broadband access for everybody and on making Japan's communications infrastructure totally broadband capable.

As Japan evolves into an increasingly networked society, some challenges—such Internet privacy and security issues and the need to upgrade competition laws, among others—have emerged that will need to be addressed urgently in the near future,

The rise of China as an ICT giant

One of the most remarkable developments observed in recent years has to do with China's extraordinary rise as a world leader in the production and use of ICT. In "Made in China: Information Technologies and the Internet," Graham Vickery and Sacha Wunsch-Vincent (both at the OECD) give an exhaustive account of China's ICT success story, looking also at future trends and main challenges in the years to come.

Starting from the supply side, the authors point out that China has become one of the most important locations in the world for the assembly and production of ICT, a feature mainly driven by foreign firms. Since 2004, China has been the biggest exporter of ICT goods, surpassing Japan and the European Union in 2003 and taking the lead from the United States. Since then, Chinese ICT exports have continued to grow at an astonishing pace, and its strong ICT exports continued until early 2006.

China continues to import electronic components—now increasingly from other Asian countries—while exporting computers and related equipment. The increase in ICT exports can mostly be traced to the transfer to China of foreign companies' often low-value-added assembly and production activities. Recently, even ICT firms from Taiwan and Hong Kong have moved manufacturing to mainland China to reduce costs. OECD countries are benefiting from low-cost ICT assembly in China, which is adding to lower global ICT prices—and thus to increased ICT use and associated productivity gains across industries.

However, there is mounting evidence that ICT-related foreign affiliates are evolving from simple assembly and manufacturing to more complex original design and production and to fulfilling more important roles in global innovation networks. In spite of their relatively limited size and technological know-how, Chinese ICT firms are rapidly developing their production and export capacities (especially in the area of telecommunications equipment). The Chinese ICT industry faces the challenge of making a successful transition from being low-cost manufacturers to becoming global providers of higher-value-added products and services. The Chinese government has recently begun to encourage Chinese companies to invest overseas to gain technology, brands, and distribution channels.

On the demand and use side, China is now the sixth most important global ICT market, although it still lags in the area of ICT services. Personal computer penetration and Internet access and use (including e-commerce) are developing rapidly, albeit from low per capita levels. In general, ICT spending as a percentage of GDP is lower in China (about 4.5 percent of GDP in 2005) than in leading OECD economies (where it was about 9 percent of GDP in 2005), but it is catching up rapidly as Chinese firms increase their IT capital stock— especially in sectors outside manufacturing, and as household consumption increases. The sheer scale of the Chinese ICT market and its potential to serve as a self-supporting base for industrial development are a key difference from other countries that have climbed the ICT value ladder. But, as highlighted by the authors, the production of ICT is not helping China to reap its full benefits. ICT uptake in Chinese firms, its efficient integration in value chains, and complementary innovations (such as organizational restructuring and investment in skills) are lagging. To benefit fully from ICT, its integration in the Chinese economy and society should be high on the Chinese policy agenda.

Parts 3 and 4: Country/Economy Profiles and Data Presentation

Parts 3 and 4 include detailed profiles for each of the 122 economies covered in the *Report* and data tables for each of the 67 variables composing the NRI, with global rankings. Each part is preceded by a description of how to interpret the data provided. Technical notes and sources, included at the end of Part 4, provide details on the characteristics and sources of the individual hard variables included in the *Report*.

Notes

1 See Kurzweil (2001).

References

Kurzweil, R. 2001. "The Law of Accelerating Returns." *KurzweilAI.net*. Available at www.kurzweilai.net/articles/art0134.html?printable=1.

The Networked Readiness Index Rankings

The Networked Readiness Index 2006–2007 rankings

Rank	Country/ Economy	Score		Rank	Country/ Economy	Score
1	Denmark	5.71		62	Indonesia	3.59
2	Sweden	5.66		63	Argentina	3.59
3	Singapore	5.60		64	Colombia	3.59
4	Finland	5.59		65	Panama	3.58
5	Switzerland	5.58		66	Dominican Republic	3.56
6	Netherlands	5.54		67	Botswana	3.56
7	United States	5.54		68	Trinidad and Tobago	3.55
8	Iceland	5.50		69	Philippines	3.55
9	United Kingdom	5.45		70	Russian Federation	3.54
10	Norway	5.42		71	Azerbaijan	3.53
11	Canada	5.35		72	Bulgaria	3.53
12	Hong Kong SAR	5.35		73	Kazakhstan	3.52
13	Taiwan, China	5.28		74	Serbia and Montenegro	3.48
14	Japan	5.27		75	Ukraine	3.46
15	Australia	5.24		76	Morocco	3.45
16	Germany	5.22		77	Egypt	3.44
17	Austria	5.17		78	Peru	3.43
18	Israel	5.14		79	Guatemala	3.41
19	Korea, Rep.	5.14		80	Algeria	3.41
20	Estonia	5.02		81	Macedonia, FYR	3.41
21	Ireland	5.01		82	Vietnam	3.40
22	New Zealand	5.01		83	Venezuela	3.32
23	France	4.99		84	Pakistan	3.31
24	Belgium	4.93		85	Namibia	3.28
25	Luxembourg	4.90		86	Sri Lanka	3.27
26	Malaysia	4.74		87	Mauritania	3.25
27	Malta	4.52		88	Nigeria	3.23
28	Portugal	4.48		89	Bosnia and Herzegovina	3.20
29	United Arab Emirates	4.42		90	Mongolia	3.18
30	Slovenia	4.41		91	Tanzania	3.13
31	Chile	4.36		92	Moldova	3.13
32	Spain	4.35		93	Georgia	3.12
33	Hungary	4.33		94	Honduras	3.09
34	Czech Republic	4.28		95	Kenya	3.07
35	Tunisia	4.24		96	Armenia	3.07
36	Qatar	4.21		97	Ecuador	3.05
37	Thailand	4.21		98	Guyana	3.01
38	Italy	4.19		99	Burkina Faso	2.97
39	Lithuania	4.18		100	Uganda	2.97
40	Barbados	4.18		101	Mali	2.96
41	Slovak Republic	4.15		102	Madagascar	2.95
42	Latvia	4.13		103	Nicaragua	2.95
43	Cyprus	4.12		104	Bolivia	2.93
44	India	4.06		105	Kyrgyz Republic	2.90
45	Jamaica	4.05		106	Cambodia	2.88
46	Croatia	4.00		107	Albania	2.87
47	South Africa	4.00		108	Nepal	2.83
48	Greece	3.98		109	Benin	2.83
49	Mexico	3.91		110	Suriname	2.82
50	Bahrain	3.89		111	Malawi	2.79
51	Mauritius	3.87		112	Zambia	2.75
52	Turkey	3.86		113	Cameroon	2.74
53	Brazil	3.84		114	Paraguay	2.69
54	Kuwait	3.80		115	Mozambique	2.64
55	Romania	3.80		116	Lesotho	2.61
56	Costa Rica	3.77		117	Zimbabwe	2.60
57	Jordan	3.74		118	Bangladesh	2.55
58	Poland	3.69		119	Ethiopia	2.55
59	China	3.68		120	Angola	2.42
60	Uruguay	3.67		121	Burundi	2.40
61	El Salvador	3.66		122	Chad	2.16

(cont'd.)

Part 1
Selected Issues of Networked Readiness

CHAPTER 1.1

Connecting the World to the Networked Economy: A Progress Report Based on the Findings of the Networked Readiness Index 2006–2007

IRENE MIA, World Economic Forum

SOUMITRA DUTTA, INSEAD

In today's interconnected world, the potential for countries' sustained competitiveness and growth depends more and more on the capability and propensity of governments and civil society alike to integrate knowledge into the production processes and into everyday life. In particular, information and telecommunication technologies (ICT) seems to have turned into the steam engine of our time, the "general-purpose technology" able to transform production processes across sectors and industries and boost productivity,[1] hence contributing to a substantial share of countries' overall growth rates. In this sense, leveraging—and benefiting from—ICT becomes an essential tool for countries and national stakeholders to ensure continued levels of prosperity for their people, irrespective of their development levels. This also explains why the digital divide features so prominently in the international debate as one of the biggest impediments to development and as a major challenge for the international community for the near future.

The World Economic Forum has long recognized the importance of technological readiness and innovation in its competitiveness work, notably in its most recent methodological competitiveness model, the Global Competitiveness Index (GCI). Of the nine pillars of growth on which the GCI is based, one addresses factors related to a country's capacity to absorb technology from abroad, the economy's ICT readiness, and other factors related to national endogenous potential for innovation. Although these factors are believed to be important drivers of any country's competitiveness, they become central for nations and companies that, for their stage of development, need efficient production processes and innovation to compete. For a discussion on the crucial role of networked readiness in fostering national competitiveness, see Box 1.

In the same spirit, the Forum and INSEAD have been partnering since 2002 on a wide-ranging research project looking specifically at the capacity of countries to leverage ICT for development and growth. The main outcome of this project has been the production of the *Global Information Technology Report* (GITR) series,[2] published annually. Over the years this annual report has become the most comprehensive and respected international assessment of its kind.

The 2007 edition of the GITR features once again the Networked Readiness Index (NRI) as the main methodological framework used to assess countries' propensity and preparation in benefiting from, and participating in, ICT advancements. This year's NRI includes a record number of 122 economies around the world: 7 more than last year and almost double the 72 economies covered in the original 2001–2002 edition.

The NRI not only maps out the factors that have proven to be key for countries' ability to leverage ICT for improved competitiveness but also offers a valuable

3

Box 1: Networked readiness as a key driver of national competitiveness

Figure 1 plots the Global Competitive Index (GCI) score against the NRI score for 2006–07. The distribution of data points in the figure shows a high degree of correlation between countries that are more competitive (have higher GCI scores) and those that are more ready for leveraging the networked economy (have higher NRI scores). This distribution is aligned with our observations about networked readiness being a fundamental driver of countries' competitiveness.

Although a detailed overview of the GCI is beyond the scope of this chapter, the GCI data of Figure 1 capture an additional dimension that is worth considering and also provides an insightful way to analyze countries' NRI performance.[1] The GCI builds upon the idea that national competitiveness is driven by many and diverse factors (grouped into the nine pillars of growth), each of them mattering to a certain extent for all countries, but with a different relative importance according to a particular country's specific level of development. In this sense,

the model captures the following stages of development, in which the process of economic development evolves:

- **the factor-driven stage** (economies and firms compete in prices, taking advantage of cheap factors),

- **the efficiency-driven stage** (cheap factors are no longer a sufficient condition for sustained economic growth, efficient production practices and efficient markets become key to increasing productivity), and

- **the innovation-driven stage** (productivity increases rest on economies' capability to produce innovative products using sophisticated production methods).

In addition to the above three stages of development, the GCI identifies two transition stages: from stage 1 to 2 and from stage 2 to 3; Appendix B lists the countries covered in this *Report* according to their stage of development. In Figure 1, one observes that the higher the development stage of the economy, the greater the

Figure 1: The Global Competitiveness Index and the Networked Readiness Index for 2006–07

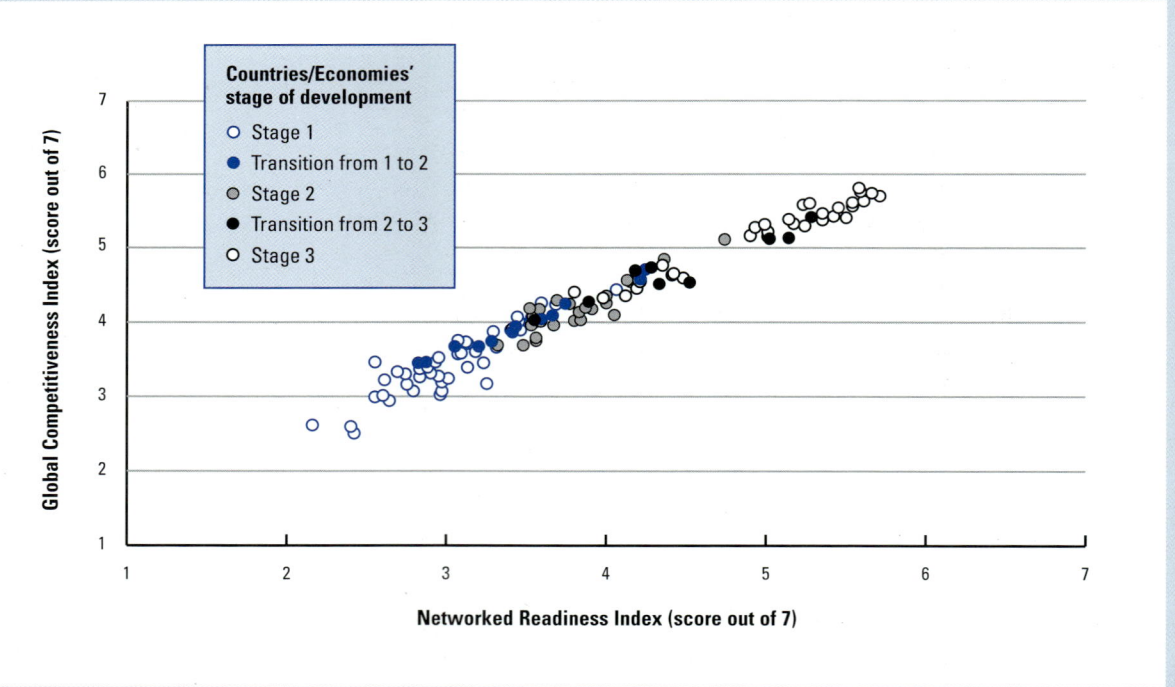

Note: See Appendix B for a list of countries/economies covered by the NRI 2006–07, by stage of development.

Box 1: Networked readiness as a key driver of national competitiveness (cont'd.)

networked readiness of the economy. Although the plot of Figure 1 does not prove causality, it does show a high degree of correlation between the two aspects above. Intuitively this can be understood when one views the use of ICT for innovation—be it the creation of new ICT technologies (for example, Web 2.0), the application of new value-adding ICT-enabled features in everyday products (for example, mobile telephones), or the use of ICT for generating new sources of customer value (for example, customer relationship management systems). Innovation in many, if not most, spheres of business and everyday life depends upon the application and usage of ICT. If the primary actors (government, businesses, and citizens) of an economy are not networked-ready, it will be hard for the economy to transition to the innovation-driven stage of development. For example, if broadband technologies are not widespread in sub-Saharan Africa, it is more difficult for these economies to innovate in global e-commerce.

1 See Lopez-Claros et al. (2005) for a full description of the GCI.

Networked readiness: Defining the framework for 2006–07

As in the past, the networked readiness framework used to compute the NRI this year measures countries' preparedness to use ICT effectively on the basis of three main theoretical assumptions as follows:

1. National actors cannot operate in a vacuum when it comes to networked readiness: an appropriate ICT-friendly and conducive environment must be in place or established as precondition. The term *environment* is used here in a very broad sense, embracing the business environment, the political and regulatory frame for ICT, and the actual ICT infrastructure.

2. Countries' capability to fully leverage ICT depends crucially on the joint effort of the main national actors—notably the government, the business sector, and individuals. They each have a role to play in improving networked readiness. Experience has shown that the countries most successful in ICT have been those in which the government has been able to mobilize business and civil society toward a common ICT development vision and strategy.

3. The actual usage of ICT by the three above actors is greatly influenced and determined by their readiness and propensity to adopt ICT advancements.

In line with the above principles, the framework is structured around the three main dimensions of environment, readiness, and usage. The latter two are broken down along the contributions given respectively by the three primary actors—the government, the business sector, and individuals. Figure 1 provides a graphic description of the framework.

The framework outlined above translates into a nine-pillar index, composed of market environment, political and regulatory environment, infrastructure environment, individual readiness, business readiness, government readiness, individual usage, business usage, and government usage pillars; the entire Index is evaluated for a total of 67 variables. The nine pillars are then regrouped into the three subindexes mentioned above: environment, readiness, and usage. The same weight is given to each pillar in the calculation of the three subindexes, and the overall NRI is an unweighted average of the three subindexes. The underlying assumption is that all index components contribute similarly in determining the overall networked readiness of a country. (For a more detailed description of its composition, please see Appendix A: Technical composition and computation of the Networked Readiness Index 2006–2007.)

benchmarking instrument vis-à-vis other countries as well as with each country's own past performance, thanks to its time series stretching back to 2002. In this sense, the NRI provides a snapshot of countries' competitive advantages and disadvantages with regard to ICT development. The results have been used extensively as a unique platform for private-public debate on national ICT weaknesses and for drawing roadmaps toward improvements in networked readiness. Moreover, the GITR publications have often included country-specific case studies, showcasing best ICT practices and policies and offering sources of inspiration for countries in their efforts to invest in ICT developments.

This 2007 edition of the GITR aims again at presenting a rigorous assessment of countries' ICT progress, raising national stakeholders' awareness of the importance of this progress for sustained growth. It also provides a useful toolkit for policymakers for the design and adoption of policies and actions conducive to ICT growth.

This chapter starts by outlining the main features of the methodological framework adopted for the 2006–2007 NRI. The main results for the 2006–2007 NRI are then discussed and analyzed, looking at the regional trends and comparative performance of economies by region and stage of development.

Figure 1: The framework of the Networked Readiness Index 2006–07

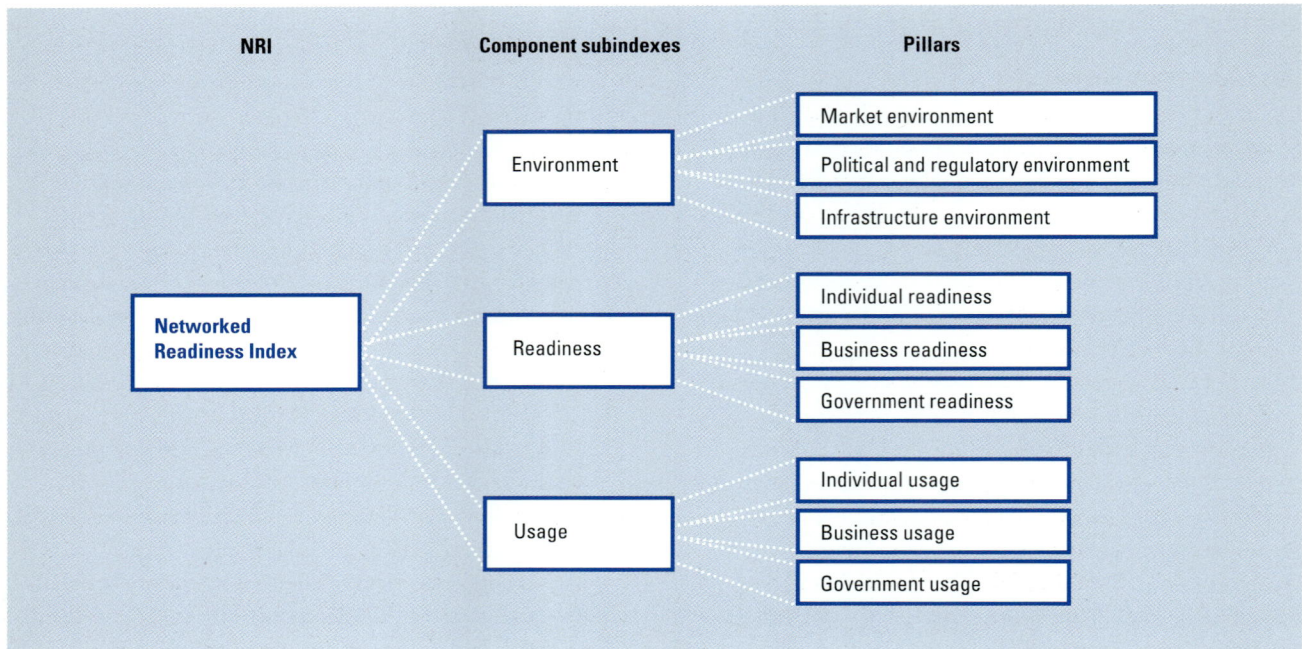

Readers can interpret the NRI at different levels of analysis: while the overall NRI score can give a broad indication on a country's level of networked readiness, the specific subindex and pillar scores provide more detailed insights regarding the areas of relative weakness and strength in national ICT performances. A brief analysis of the structure of the NRI and of its building blocks follows.

Environment

An ICT-conducive environment is an essential element of a country's networked readiness since it is a prerequisite for allowing the main national actors to benefit from and participate in ICT developments. Thus the environment subindex is designed to measure the degree to which the environment of a country is conducive to the development and use of ICT. The environment subindex includes a total of 28 variables, measuring the general market environment, the ICT-friendliness of the regulatory and political environment, and the quality of both hard and soft infrastructure for ICT.

The *market environment pillar* (12 variables) aims at measuring the openness of the general business environment for ICT development, taking into account the presence of appropriate capital sources, the degree of business sophistication and innovation potential, and the ease of doing business as well as the intensity of local competition and the freedom of the press in the country.

The *regulatory and political environment pillar* (9 variables) assesses the general efficiency and fairness of public institutions and of the legal framework, as well as of ICT-specific laws, the extent of protection of property rights and intellectual property, and the quality of competition in the ISP sector.

Last but not least, the *infrastructure environment pillar* (7 variables) measures the existence of ICT-conducive soft infrastructure as well as the state of ICT hard infrastructure in countries. Specifically, the tertiary enrollment rate, the quality of scientific research institutions, and the availability of scientists and engineers together with the degree of ICT penetration (number of telephone lines, secure Internet servers) and electricity production are assessed.

From the above analysis, it is clear that although many of the preconditions for an ICT-friendly environment depend upon the actions of the government, businesses and civil society also need to play a central role, notably by facilitating the establishment of an effective market environment and strong ICT soft and hard infrastructures.

Readiness

The readiness component measures the capability of the principal agents of a given nation's economy (citizens, businesses, and governments) to leverage the potential of ICT. This capability is based on a combination of factors (measured by 24 variables), such as the existence of necessary human skills for using ICT, access to and afford-ability of ICT for companies, and the government's own use of ICT for its services and processes.

The *individual readiness pillar* (10 variables) assesses the extent to which individuals within a country are disposed to use ICT as well as their degree of preparedness to do so, taking into account the presence of appropriate human skills together with the extent of access to ICT and the affordability of telephone and Internet connection and use. Examples of variables assessed in this pillar are the quality of the educational system, with particular emphasis on math and science education; the availability of Internet access in schools; residential telephone connection charges; broadband and telephone subscription charges; and the cost of mobile telephone calls.

The *business readiness pillar* (9 variables) looks at com-panies' readiness to incorporate ICT fully in their operations and businesses, focusing on factors such as the presence of an appropriately trained labor force, the extent of company spending on R&D and collaboration between universities and firms on R&D, the affordability of ICT for business, and levels of ICT imports.

The *government readiness pillar* (5 variables) measures the prioritization of ICT by the government and the extent to which the government has a clear vision on how to promote its use and penetration. The extent of e-government and e-democracy are also examined.

Usage

The usage component aims to measure the degree of ICT usage by the principal actors of the NRI framework: individuals, businesses, and governments. In the absence of reliable data about the specific impact of ICT on the key agents, the usage component provides an indication of the potential gains in efficiency and productivity associated with the adoption of ICT. This third subindex of the NRI assesses the actual degree of ICT usage in each of the countries covered, relying on 15 mainly quantitative variables.

The *individual usage pillar* (5 variables) includes variables relating to ICT penetration among the civil society, namely telephone, personal computer (PC), and Internet penetration.

The *business usage pillar* (6 variables) examines the extent of innovation and technology absorption in the business sector, the availability and usage of fixed lines and mobile telephones for business, and Internet penetration within firms.

The *government usage pillar* (4 variables) deals with government use of ICT as measured by the availability of online services, the improvement of government produc-tivity as a consequence of ICT introduction and use, ICT pervasiveness in governmental offices, and government success in promoting general ICT penetration.

Computation methodology and data

This section outlines the methodology employed to compute the 2006–07 NRI in the context of the networked readiness framework and its building blocks presented earlier.

In line with the Forum's general competitiveness methodology, the NRI builds on a mix of hard, quantita-tive data, collected by international institutions such as the International Telecommunication Union (ITU) and the World Bank, and on qualitative data coming from the Forum's Executive Opinion Survey (Survey), administered to more than 11,000 business leaders across 125 economies in 2006.[3] The Survey data provide a valuable complement to the hard data since they capture key aspects for assessing countries' networked readiness but for which no hard data are available for all countries covered. For example, it is difficult to find precise quantifiable data about the effec-tiveness of law-making bodies in different economies. However, this is an important component of the overall business environment of an economy that affects both the levels of innovation in the ICT industry and the adoption of ICT by businesses across sectors in the economy. Therefore a question capturing this concept is included in the Survey questionnaire, and the resulting data are used for the computation of the NRI (as a contributing variable to the political/regulatory environment in the economy).

Of the 125 economies covered by the Survey this year, three—Gambia, Tajikistan, and Timor Leste—could unfortunately not be included in the NRI computation because of the scarcity of reliable hard data for them. Table 1 shows the evolution of the geographical coverage of *The Global Information Technology Report* since its inception, as compared with *The Global Competitiveness Report*.

Table 1: Evolution of the geographical coverage of *The Global Information Technology Report* compared with *The Global Competitiveness Report*

Years	Global Information Technology Report No. of countries/economies	Global Competitiveness Report No. of countries/economies
2001–02	72	75
2002–03	82	80
2003–04	102	102
2004–05	104	104
2005–06	115	117
2006–07	122	125

Turning to the composition of the NRI, one can notice some differences in the number and in the nature of the variables included from year to year. This is because of the dynamism of the ICT sector and the need to review the appropriate set of variables in good time to capture each of the networked readiness framework's component subindexes. The changes also reflects the need to include recent data related to ICT developments. Many aspects of ICT usage in economies change rapidly, and our research uses the most recent data available. This means that some time-sensitive variables that have been used for computing the NRI in the past but have not been updated in recent years by the relevant collecting agency are no longer used in computing the NRI.

It is therefore important to introduce a note of caution in comparing the NRI rankings over time since, at the variable level, there have been some changes. However, the uniformity of the networked readiness framework ensures the comparability of the data at the global, subindex, and pillar levels.

In previous years, the variables used to compute the various component subindexes of the NRI have been chosen from a larger set of possible variables using factor-analytical techniques. Although factor-analytical techniques are technically rigorous, they often diminish the ability to easily explain the underlying logic for including specific variables and to make strict comparisons over time. With the benefit of our previous experience in computing the NRI, expert opinion has played the primary role in selecting the variables this year. As a result, most of the variables used for computing the NRI last year have been retained. However, 10 new variables have been introduced this year for various reasons—some because new variables have been identified as relevant to the environment for using ICT (such as the freedom of the press), some because the need to capture recent ICT developments has been deemed important (through variables on broadband penetration and usage), and some because older variables were no longer updated by the collecting agencies and had to be dropped. Particular care has been taken to ensure that the total set of variables used for the NRI this year ensures comparability of the results for this year with those for previous years.[4]

The above shift in the selection procedure of the variables also aligns the procedures for the NRI with those used by the Forum for the computation of the GCI. As part of these changes, the NRI has been computed this year on the (increasing) 1–7 scale traditionally used by the Forum in its competitiveness work. This is a change from the scales (positive and negative scores around a standardized mean of 0) used for the computation of NRI scores for the last couple of years.[5]

The current state of networked readiness in the world: Findings from the NRI 2006–07

Tables 2, 3, and 4 show, respectively, the NRI rankings for 2006–07 and their comparison with last year's rankings, the top performers for each of the nine pillars and the evolution of the top 10 NRI rankings from the very first computation in 2001–02 up to now.[6] Tables 5, 6, and 7 display the rankings for each of the three subindexes composing the NRI— environment, readiness, and usage—and the details for each pillar.

As it can be seen from Table 2, **Denmark** leads the way in networked readiness this year, climbing to the top position for the first time as the culmination of an upward trend observed since 2003. Denmark's recipe for networked success has much to do with the country's excellent regulatory and political environment (where it ranks 1st) and with its clear government leadership and vision in developing ICT penetration and usage, promptly followed by the civil society: Denmark ranks 2nd and 5th out of 122 economies for government readiness and usage and 6th and 3rd for individual readiness and usage, respectively. Indeed, an early liberalization of the telecommunications sector in 1996 has given a major boost to the national ICT industry and to ICT penetration in the country. In this sense, Denmark displays impressive levels of Internet and PC usage as well as of government online services, and a very dynamic e-business environment. The well-developed internal market—coupled with a continued emphasis on education and a talent for developing pioneering applications and technologies—laid the basis for the development of a world-class national high-tech industry whose total exports accounted for more than 9 percent of total national exports in 2004 according to ITU.

Among the other best-performing countries, the following trends can be highlighted:

- All **Nordic countries**, except Iceland, share Denmark's upward trend, with Finland, Sweden, and Norway gaining 1, 6, and 3 positions respectively. Iceland is the only country of the group losing some ground, down 4 positions from last year to what remains an impressive 8th rank.

- The **United States** loses its networked readiness pre-eminence to Denmark, Sweden, Singapore, Finland, Switzerland, and the Netherlands with a 6-place drop to 7th position. Nevertheless, the United States continues to lead the world in the efficiency and quality of its market environment and ICT soft and hard infrastructure, confirming itself as a world ICT powerhouse and innovative country.

- *Switzerland* (up 4 positions to 5th place) registers one of the greatest improvements of the top 20 countries, together with the Netherlands and Sweden, which also registered large improvements. In this, the NRI echoes Switzerland's impressive rise to the top position in the GCI and gives credit to the country's efficient market environment (where it ranks 3rd) and high levels of individual and business readiness and usage. Switzerland ranks 3rd and 1st respectively for individual and business readiness and 4th for both individual and business usage.[7]

- Last but not least, *Estonia* gains 3 positions from last year and enters the top 20 league, ranking 20th overall. Estonia must be praised for the progress realized in a very short period and for showing the way, in networked readiness and general competitiveness alike, not only to the rest of the recent European Union (EU) accession countries but also to much of the EU old guard. The remarkable leadership role assumed by the Estonian government in promoting ICT usage and penetration is portrayed in Chapter 2.1 in this *Report*, "Estonia: A Sustainable Success in Networked Readiness?"

Before turning our attention to the analysis of the NRI by region, one must spend a few words on Table 3, highlighting the best performer per pillar. As the table shows, the picture at the pillar level is rather mixed in terms of ICT national showings. Only one country outperforms the rest of the world in more than one pillar: Singapore. In this respect, the NRI highlights the Singaporean government's clear vision for ICT and the subsequent leading role undertaken by the latter in promoting ICT diffusion and penetration, with a 1st position in government readiness and usage. As already pointed out, the United States is outperforming the rest of the world in the quality and efficiency of the market environment. The Nordic countries are each topping a pillar: Denmark tops the regulatory and political environment; Iceland, ICT infrastructure; and Finland, individual readiness. The Netherlands, Switzerland, and Japan rank 1st in individual usage, business readiness, and in business usage respectively.

Europe and North America

Europe continues to figure prominently in the NRI rankings this year, with *Denmark* (1st), *Sweden* (2nd), *Finland* (4th), *Switzerland* (5th), the *Netherlands* (6th), *Iceland* (8th), the *United Kingdom* (9th), *Norway* (10th), *Germany* (16th), *Austria* (17th), and *Estonia* (20th) all among the top 20.

We have already mentioned the outstanding networked readiness performance of the *Nordic countries*, which have featured consistently among the top 10 over the last six years, with high ICT penetration and diffusion rates. Quite predictably, the Nordic countries do exceptionally well also in the GCI's rankings, reflecting the importance of ICT readiness for global national competitiveness. Indeed, these countries have in common a top-class education system, a culture for innovation, a friendly business climate, and a distinct public and private inclination to adopt new technology, all factors that boost their levels of networked readiness.

The networked readiness map for the *EU area* tends to be more diverse: countries such as the Netherlands, the United Kingdom, Germany, Austria, Estonia, *Ireland* (21st), and *France* (23rd) seem to be fully leveraging and benefiting of ICT advances. Some other "old-timers" such as *Italy* (38th) and *Greece* (48th) keep lagging behind, although it must be pointed out that Italy's 4-position recovery from 2005 confirms last year's upward trend after its dismal 17-place drop from 2004 to 2005.

Among the most recent EU accession countries, *Estonia* and *Poland*, at 58th—down 5 places from last year—represent two extremes in the networked readiness continuum. It is worth noticing that Estonia has gained a total of 5 positions in the last two years, displaying an impressive dynamism in networked readiness.

Turkey, at 52nd, is down 4 positions from last year, showing much room for improvement in all dimensions of the NRI, particularly in individual readiness.

Russia gains 2 positions from last year, reversing the 2005–06 major fall and positioning itself at 70th, with rather significant improvements registered in the regulatory environment and in the ICT infrastructure.

As far as *North America* is concerned, the *United States* (7th) and *Canada* (11th) continue to feature among the world's best performers, but they both experienced a significant drop in their NRI rankings, of 6 and 5 places respectively. Although the US performance has been discussed above, it is worth noting that Canada's main weakness is to be found in the levels of usage by all national actors.

Asia and the Pacific

As in previous years, Asia and the Pacific, as a region, shows an extremely varied performance, with economies spread all over the NRI rankings. In this sense, if *Singapore* (3rd), *Hong Kong* (12th), *Taiwan* (13th), *Japan* (14th), *Australia* (15th), *Korea* (19th), and, to a certain extent, *Malaysia* (26th) and *Thailand* (37th) stand out for their world-class levels of networked readiness, *Bangladesh* (118th), *Cambodia* (106th), and *Kyrgyz Republic* (105th) continue to feature in the bottom part of the rankings.

Among the best performers, *Singapore* maintains its dominant position for the fifth consecutive year, thanks to its excellent business environment and the government's savvy early focus on ICT diffusion and on development of synergies with the private sector.

Table 2: The Networked Readiness Index 2006–07 and 2005–06 comparison

2006–07 rank	Country/Economy	Score	2005–06 rank	2006–07 rank	Country/Economy	Score	2005–06 rank
1	Denmark	5.71	3	62	Indonesia	3.59	68
2	Sweden	5.66	8	63	Argentina	3.59	71
3	Singapore	5.60	2	64	Colombia	3.59	62
4	Finland	5.59	5	65	Panama	3.58	66
5	Switzerland	5.58	9	66	Dominican Republic	3.56	89
6	Netherlands	5.54	12	67	Botswana	3.56	56
7	United States	5.54	1	68	Trinidad and Tobago	3.55	74
8	Iceland	5.50	4	69	Philippines	3.55	70
9	United Kingdom	5.45	10	70	Russian Federation	3.54	72
10	Norway	5.42	13	71	Azerbaijan	3.53	73
11	Canada	5.35	6	72	Bulgaria	3.53	64
12	Hong Kong SAR	5.35	11	73	Kazakhstan	3.52	60
13	Taiwan, China	5.28	7	74	Serbia and Montenegro	3.48	80
14	Japan	5.27	16	75	Ukraine	3.46	76
15	Australia	5.24	15	76	Morocco	3.45	77
16	Germany	5.22	17	77	Egypt	3.44	63
17	Austria	5.17	18	78	Peru	3.43	85
18	Israel	5.14	19	79	Guatemala	3.41	98
19	Korea, Rep.	5.14	14	80	Algeria	3.41	87
20	Estonia	5.02	23	81	Macedonia, FYR	3.41	82
21	Ireland	5.01	20	82	Vietnam	3.40	75
22	New Zealand	5.01	21	83	Venezuela	3.32	81
23	France	4.99	22	84	Pakistan	3.31	67
24	Belgium	4.93	25	85	Namibia	3.28	78
25	Luxembourg	4.90	26	86	Sri Lanka	3.27	83
26	Malaysia	4.74	24	87	Mauritania	3.25	—
27	Malta	4.52	30	88	Nigeria	3.23	90
28	Portugal	4.48	27	89	Bosnia and Herzegovina	3.20	97
29	United Arab Emirates	4.42	28	90	Mongolia	3.18	92
30	Slovenia	4.41	35	91	Tanzania	3.13	84
31	Chile	4.36	29	92	Moldova	3.13	94
32	Spain	4.35	31	93	Georgia	3.12	96
33	Hungary	4.33	38	94	Honduras	3.09	100
34	Czech Republic	4.28	32	95	Kenya	3.07	91
35	Tunisia	4.24	36	96	Armenia	3.07	86
36	Qatar	4.21	39	97	Ecuador	3.05	107
37	Thailand	4.21	34	98	Guyana	3.01	111
38	Italy	4.19	42	99	Burkina Faso	2.97	—
39	Lithuania	4.18	44	100	Uganda	2.97	79
40	Barbados	4.18	—	101	Mali	2.96	95
41	Slovak Republic	4.15	41	102	Madagascar	2.95	102
42	Latvia	4.13	51	103	Nicaragua	2.95	112
43	Cyprus	4.12	33	104	Bolivia	2.93	109
44	India	4.06	40	105	Kyrgyz Republic	2.90	103
45	Jamaica	4.05	54	106	Cambodia	2.88	104
46	Croatia	4.00	57	107	Albania	2.87	106
47	South Africa	4.00	37	108	Nepal	2.83	—
48	Greece	3.98	43	109	Benin	2.83	108
49	Mexico	3.91	55	110	Suriname	2.82	—
50	Bahrain	3.89	49	111	Malawi	2.79	—
51	Mauritius	3.87	45	112	Zambia	2.75	—
52	Turkey	3.86	48	113	Cameroon	2.74	99
53	Brazil	3.84	52	114	Paraguay	2.69	113
54	Kuwait	3.80	46	115	Mozambique	2.64	101
55	Romania	3.80	58	116	Lesotho	2.61	—
56	Costa Rica	3.77	69	117	Zimbabwe	2.60	105
57	Jordan	3.74	47	118	Bangladesh	2.55	110
58	Poland	3.69	53	119	Ethiopia	2.55	115
59	China	3.68	50	120	Angola	2.42	—
60	Uruguay	3.67	65	121	Burundi	2.40	—
61	El Salvador	3.66	59	122	Chad	2.16	114

(cont'd.)

Table 3: Top performer on each pillar of the Networked Readiness Index 2006–2007

Country/Economy	Market environment	Regulatory environment	Infrastructure environment	Individual readiness	Business readiness	Government readiness	Individual usage	Business usage	Government usage
United States	**1**	17	2	19	4	5	15	14	22
Denmark	16	**1**	7	6	7	2	3	7	5
Iceland	10	3	**1**	10	34	25	6	8	3
Finland	2	9	4	**1**	2	8	14	6	13
Switzerland	3	8	10	3	**1**	19	4	4	19
Singapore	6	11	15	2	15	**1**	10	13	**1**
Netherlands	12	5	11	15	10	12	**1**	9	18
Japan	7	15	14	14	5	11	22	**1**	35

Table 4: Networked Readiness Index: History of the top 10 rankings

Country/Economy	2006–07	2005–06	2004–05	2003–04	2002–03	2001–02
(Number of countries/economies)	(122)	(115)	(104)	(102)	(82)	(72)
Denmark	1	3	4	5	8	7
Sweden	2	8	6	4	4	4
Singapore	3	2	1	2	3	8
Finland	4	5	3	3	1	3
Switzerland	5	9	9	7	13	16
Netherlands	6	12	16	13	11	6
United States	7	1	5	1	2	1
Iceland	8	4	2	10	5	2
United Kingdom	9	10	12	15	7	10
Norway	10	13	13	8	17	5

11

Despite *Taiwan*'s drop of 6 positions from 2005,[8] it still comes in at an impressive 13th position overall. This confirms the success story of a mostly rural and resource-poor economy that turned into one of the world ICT powerhouses in the space of three decades. Notable enabling factors in this transformation have been the strong leadership exercised by the government in ICT, fostering public-private partnership, investing heavily in education and R&D, and reversing the brain drain of the 1960–1970 period through incentives as well as the access to the large Chinese market.[9]

Korea also moves down 5 places from last year, with a relative worsening of the market environment in particular. However, its 19th position overall reflects its sound ICT fundamentals and the amazing progress realized by the nation in the short span of a few decades, in a way very similar to Taiwan's success story.

India (44th) and *China* (59th) seem to be both losing ground in networked readiness, with a drop of 4 and 9 places respectively from last year. In particular, India's ICT environment and readiness register a relative drop (from 40th to 46th and 29th to 37th respectively), mainly because of the diminishing quality of the regulatory environment

for ICT (from 30th to 48th) and the levels of government readiness (from 28th to 39th).

As for China, all component subindexes see a relative deterioration with respect to last year, with notable drops in the quality of the market environment (from 43rd to 61st), the business readiness level (from 48th to 65th) and individual usage (from 63rd to 80th).

A note of caution must be introduced here—since both countries show very different regional levels of ICT diffusion and development, a difference that is partly hidden by the overall NRI score—India and China's general performance appears to be especially hindered by weak infrastructures, a very low level of individual ICT usage for India and of individual and business readiness and usage for China. Graham Vickery and Sacha Wunsch-Vincent give a detailed account of the state of ICT penetration in China in "Made in China: Information Technologies and the Internet," Chapter 2.4 of this *Report*.

Kazakhstan, at 73rd place, is losing its predominance in Central Asia to Azerbaijan (71st), dropping 13 positions in the rankings. Kazakhstan's capability to wholly leverage ICT seems to be slowed by a lack of individual readiness (where it is ranked 95th) and usage (83rd).

Table 5: Environment component subindex

Rank	Country/Economy	Score	Market environment Rank	Market environment Score	Political and regulatory environment Rank	Political and regulatory environment Score	Infrastructure environment Rank	Infrastructure environment Score
1	Iceland	5.75	10	5.02	3	6.04	1	6.17
2	United States	5.71	1	5.64	17	5.49	2	6.00
3	Finland	5.63	2	5.38	9	5.90	4	5.62
4	Denmark	5.54	16	4.89	1	6.18	7	5.54
5	Sweden	5.52	9	5.05	10	5.85	3	5.65
6	Switzerland	5.50	3	5.35	8	5.91	10	5.24
7	Norway	5.45	18	4.78	4	6.00	6	5.56
8	Australia	5.39	19	4.66	7	5.92	5	5.60
9	Canada	5.36	13	4.96	14	5.66	8	5.46
10	United Kingdom	5.32	11	5.02	2	6.09	13	4.84
11	Netherlands	5.29	12	4.97	5	5.96	11	4.95
12	Japan	5.20	7	5.23	15	5.65	14	4.74
13	Singapore	5.19	6	5.32	11	5.76	16	4.48
14	Germany	5.12	15	4.92	6	5.94	15	4.49
15	New Zealand	5.10	21	4.55	16	5.50	9	5.26
16	Hong Kong SAR	5.03	4	5.34	12	5.75	23	4.02
17	Taiwan, China	4.97	5	5.32	31	4.73	12	4.85
18	Ireland	4.93	14	4.95	18	5.47	18	4.37
19	Israel	4.91	8	5.06	20	5.32	19	4.35
20	Austria	4.89	22	4.53	13	5.68	17	4.46
21	France	4.72	20	4.57	19	5.46	21	4.13
22	Luxembourg	4.62	24	4.46	22	5.31	22	4.10
23	Korea, Rep.	4.55	27	4.34	26	4.96	20	4.34
24	Belgium	4.54	23	4.48	24	5.13	24	4.00
25	Estonia	4.48	25	4.45	21	5.31	26	3.67
26	Malaysia	4.30	17	4.79	23	5.24	50	2.86
27	Portugal	4.20	32	4.14	27	4.92	28	3.54
28	Hungary	4.13	29	4.28	34	4.66	33	3.45
29	Malta	4.10	38	3.99	29	4.80	31	3.49
30	Spain	4.06	39	3.99	35	4.66	29	3.53
31	Barbados	4.02	56	3.65	30	4.80	27	3.62
32	Cyprus	4.00	42	3.96	36	4.60	34	3.44
33	Greece	3.90	71	3.47	41	4.45	25	3.79
34	Chile	3.90	30	4.17	33	4.67	49	2.87
35	United Arab Emirates	3.87	26	4.36	44	4.41	51	2.86
36	Czech Republic	3.86	43	3.93	52	4.16	32	3.48
37	Tunisia	3.84	37	4.00	28	4.90	57	2.63
38	Thailand	3.83	31	4.17	32	4.73	59	2.60
39	Latvia	3.83	41	3.96	42	4.43	38	3.11
40	South Africa	3.83	34	4.07	25	4.98	70	2.42
41	Lithuania	3.82	50	3.78	43	4.42	36	3.26
42	Slovenia	3.78	57	3.65	51	4.18	30	3.52
43	Slovak Republic	3.73	35	4.04	47	4.25	43	2.89
44	Qatar	3.72	53	3.71	37	4.56	45	2.89
45	Kuwait	3.71	33	4.09	53	4.15	48	2.88
46	India	3.68	28	4.28	48	4.23	63	2.54
47	Jamaica	3.61	36	4.03	40	4.48	77	2.32
48	Mauritius	3.58	51	3.75	46	4.40	61	2.58
49	Croatia	3.57	62	3.59	54	4.07	39	3.05
50	Turkey	3.55	49	3.80	50	4.22	56	2.65
51	Italy	3.55	58	3.63	74	3.68	35	3.33
52	Jordan	3.53	63	3.57	39	4.48	62	2.54
53	Costa Rica	3.50	54	3.68	56	4.00	52	2.82
54	Mexico	3.42	44	3.93	60	3.92	69	2.42
55	Uruguay	3.42	80	3.37	57	4.00	47	2.88
56	Indonesia	3.36	40	3.97	68	3.77	76	2.34
57	Russian Federation	3.35	82	3.33	89	3.49	37	3.22
58	Panama	3.32	45	3.91	80	3.60	66	2.46
59	Kazakhstan	3.32	70	3.50	65	3.87	60	2.59
60	China	3.32	61	3.59	55	4.00	74	2.36
61	Philippines	3.31	47	3.85	59	3.95	85	2.14
62	Romania	3.31	52	3.75	87	3.54	55	2.65
63	Botswana	3.31	66	3.53	45	4.40	93	2.01
64	Poland	3.30	69	3.50	95	3.42	40	2.99
65	Bahrain	3.30	60	3.63	82	3.60	54	2.68
66	Ukraine	3.29	75	3.42	88	3.50	41	2.95
67	Namibia	3.26	64	3.54	38	4.49	115	1.74
68	Trinidad and Tobago	3.23	46	3.90	103	3.27	65	2.53
69	Bulgaria	3.20	95	3.12	84	3.57	44	2.89
70	Dominican Republic	3.17	65	3.53	61	3.89	87	2.11
71	Nigeria	3.16	59	3.63	62	3.87	95	1.99
72	Tanzania	3.16	85	3.29	49	4.23	100	1.95
73	El Salvador	3.15	48	3.82	78	3.62	92	2.01
74	Egypt	3.15	73	3.44	77	3.62	72	2.38
75	Georgia	3.13	74	3.42	83	3.58	71	2.40
76	Morocco	3.13	68	3.50	69	3.74	84	2.14
77	Argentina	3.13	91	3.17	101	3.29	42	2.93
78	Macedonia, FYR	3.13	86	3.27	96	3.42	53	2.70
79	Azerbaijan	3.13	89	3.24	64	3.87	78	2.28
80	Colombia	3.13	77	3.40	79	3.60	73	2.38
81	Sri Lanka	3.11	81	3.34	58	3.95	89	2.04
82	Vietnam	3.10	79	3.38	67	3.78	86	2.14
83	Brazil	3.10	109	2.98	73	3.69	58	2.61
84	Serbia and Montenegro	3.09	97	3.12	102	3.28	46	2.89
85	Mongolia	3.08	88	3.24	91	3.46	64	2.53
86	Armenia	3.08	99	3.11	72	3.70	68	2.42
87	Guatemala	3.04	55	3.66	90	3.49	97	1.97
88	Peru	3.01	72	3.46	98	3.37	82	2.19
89	Uganda	2.99	107	3.01	63	3.87	88	2.11
90	Kenya	2.99	100	3.10	66	3.82	90	2.04
91	Nepal	2.95	78	3.39	76	3.64	110	1.81
92	Algeria	2.94	110	2.97	81	3.60	79	2.27
93	Mali	2.93	101	3.09	70	3.73	98	1.96
94	Pakistan	2.92	67	3.52	100	3.32	102	1.93
95	Moldova	2.91	111	2.94	93	3.45	75	2.34
96	Burkina Faso	2.89	93	3.13	75	3.66	104	1.88
97	Madagascar	2.88	94	3.13	86	3.55	99	1.96
98	Honduras	2.85	84	3.32	97	3.41	109	1.82
99	Nicaragua	2.84	76	3.41	104	3.26	105	1.86
100	Malawi	2.83	105	3.03	71	3.72	114	1.75
101	Mauritania	2.82	83	3.32	94	3.44	118	1.70
102	Bosnia and Herzegovina	2.80	103	3.04	107	3.11	80	2.26
103	Kyrgyz Republic	2.80	102	3.06	105	3.14	81	2.21
104	Benin	2.78	90	3.21	99	3.33	111	1.79
105	Zambia	2.73	113	2.87	85	3.57	113	1.75
106	Cambodia	2.70	108	3.01	92	3.45	120	1.65
107	Albania	2.69	92	3.14	108	3.08	108	1.84
108	Guyana	2.68	112	2.89	106	3.13	94	2.00
109	Lesotho	2.67	96	3.12	109	3.04	107	1.84
110	Ecuador	2.66	104	3.03	110	3.01	101	1.93
111	Bolivia	2.62	106	3.01	115	2.82	91	2.02
112	Suriname	2.60	116	2.70	114	2.92	83	2.19
113	Ethiopia	2.59	98	3.11	112	2.95	119	1.70
114	Bangladesh	2.55	87	3.24	121	2.56	106	1.86
115	Venezuela	2.54	121	2.47	118	2.73	67	2.43
116	Paraguay	2.49	114	2.82	120	2.67	96	1.98
117	Zimbabwe	2.48	119	2.55	111	2.97	103	1.90
118	Cameroon	2.41	115	2.73	119	2.72	112	1.79
119	Mozambique	2.38	118	2.58	113	2.94	121	1.62
120	Burundi	2.36	117	2.62	117	2.75	116	1.72
121	Angola	2.24	120	2.52	116	2.76	122	1.43
122	Chad	2.08	122	2.23	122	2.29	117	1.71

(cont'd)

Table 6: Readiness component subindex

Rank	Country/Economy	Score	Individual readiness Rank	Score	Business readiness Rank	Score	Government readiness Rank	Score
1	Singapore	6.09	2	6.51	15	5.47	1	6.30
2	Finland	5.86	1	6.55	2	5.81	8	5.21
3	Denmark	5.84	6	6.29	7	5.60	2	5.64
4	United States	5.77	19	6.05	4	5.70	5	5.55
5	Switzerland	5.74	3	6.45	1	5.97	19	4.79
6	United Kingdom	5.69	27	5.96	14	5.50	4	5.62
7	Taiwan, China	5.66	7	6.21	18	5.40	6	5.36
8	Japan	5.62	14	6.10	5	5.64	11	5.11
9	Korea, Rep.	5.62	23	6.01	21	5.21	3	5.63
10	Hong Kong SAR	5.61	5	6.32	13	5.53	14	4.99
11	Sweden	5.59	21	6.04	11	5.55	9	5.17
12	Netherlands	5.57	15	6.08	10	5.56	12	5.09
13	Canada	5.56	8	6.21	20	5.24	7	5.22
14	Germany	5.53	25	5.97	3	5.75	17	4.87
15	Belgium	5.48	4	6.34	8	5.59	27	4.50
16	Austria	5.47	11	6.19	9	5.56	22	4.67
17	Australia	5.42	13	6.11	22	5.08	13	5.07
18	Malaysia	5.40	17	6.06	16	5.45	21	4.68
19	France	5.40	12	6.15	17	5.42	23	4.62
20	Ireland	5.38	9	6.20	6	5.64	31	4.31
21	Norway	5.37	20	6.04	19	5.28	20	4.79
22	Israel	5.35	22	6.02	12	5.55	29	4.48
23	Estonia	5.35	26	5.97	25	4.92	10	5.15
24	New Zealand	5.32	16	6.06	24	4.96	15	4.94
25	Iceland	5.16	10	6.20	34	4.71	25	4.57
26	Luxembourg	5.05	18	6.05	29	4.82	32	4.29
27	United Arab Emirates	4.99	33	5.67	30	4.82	28	4.48
28	Malta	4.97	32	5.77	53	4.20	16	4.93
29	Tunisia	4.92	24	6.00	33	4.76	38	4.00
30	Portugal	4.91	40	5.50	37	4.66	26	4.56
31	Slovenia	4.90	30	5.82	27	4.83	37	4.04
32	Czech Republic	4.90	28	5.90	23	5.01	46	3.79
33	Chile	4.89	60	5.07	32	4.77	18	4.83
34	Hungary	4.82	35	5.65	41	4.59	33	4.23
35	Thailand	4.81	47	5.40	39	4.63	30	4.41
36	Qatar	4.77	36	5.64	43	4.54	36	4.12
37	India	4.71	52	5.31	28	4.83	39	3.97
38	Spain	4.68	44	5.45	26	4.89	57	3.71
39	Croatia	4.65	38	5.54	40	4.63	50	3.77
40	Slovak Republic	4.63	43	5.47	35	4.69	53	3.74
41	Mexico	4.60	67	4.94	51	4.23	24	4.61
42	Latvia	4.57	34	5.66	45	4.44	63	3.61
43	Lithuania	4.55	37	5.59	47	4.33	55	3.71
44	Romania	4.54	39	5.53	49	4.29	45	3.81
45	South Africa	4.54	70	4.85	31	4.79	40	3.97
46	Italy	4.53	48	5.39	46	4.41	47	3.78
47	Barbados	4.51	29	5.88	59	4.04	62	3.62
48	Mauritius	4.51	49	5.37	54	4.20	41	3.97
49	Cyprus	4.50	31	5.77	52	4.22	71	3.49
50	Brazil	4.49	72	4.75	42	4.57	35	4.15
51	Greece	4.48	45	5.44	48	4.31	58	3.69
52	Indonesia	4.45	42	5.48	38	4.66	92	3.22
53	Costa Rica	4.41	53	5.30	36	4.68	88	3.25
54	Trinidad and Tobago	4.40	46	5.41	44	4.45	84	3.35
55	Jamaica	4.38	58	5.11	50	4.25	48	3.78
56	Poland	4.34	51	5.32	62	4.02	59	3.69
57	Turkey	4.32	64	5.03	56	4.19	52	3.75
58	China	4.29	61	5.06	65	3.97	44	3.85
59	Bahrain	4.29	50	5.35	73	3.81	56	3.71
60	Argentina	4.25	65	5.03	55	4.19	67	3.53
61	Colombia	4.23	81	4.42	58	4.05	34	4.22
62	Kuwait	4.20	41	5.49	57	4.12	101	3.00
63	Botswana	4.19	56	5.18	61	4.03	83	3.37
64	Jordan	4.17	54	5.24	85	3.62	60	3.66
65	Ukraine	4.16	57	5.16	77	3.78	68	3.53
66	Venezuela	4.15	71	4.76	66	3.96	54	3.72
67	Panama	4.14	62	5.03	67	3.96	78	3.43
68	Bulgaria	4.12	59	5.08	78	3.73	66	3.55
69	Serbia and Montenegro	4.11	55	5.24	60	4.04	99	3.07
70	El Salvador	4.11	73	4.75	74	3.81	49	3.77
71	Macedonia, FYR	4.05	66	5.01	70	3.93	91	3.22
72	Uruguay	4.04	69	4.90	69	3.93	86	3.30
73	Algeria	4.04	74	4.68	68	3.94	72	3.49
74	Azerbaijan	4.00	77	4.53	63	4.01	74	3.47
75	Russian Federation	3.98	63	5.03	93	3.48	76	3.44
76	Vietnam	3.96	76	4.54	76	3.79	65	3.55
77	Philippines	3.96	80	4.46	89	3.52	42	3.89
78	Dominican Republic	3.89	79	4.48	86	3.61	64	3.57
79	Peru	3.88	89	4.21	71	3.91	69	3.53
80	Kazakhstan	3.87	95	3.92	75	3.80	43	3.88
81	Morocco	3.81	78	4.48	84	3.62	85	3.32
82	Egypt	3.80	85	4.41	82	3.63	81	3.37
83	Pakistan	3.80	94	4.01	64	4.01	80	3.38
84	Bosnia and Herzegovina	3.77	68	4.91	94	3.48	104	2.92
85	Guatemala	3.72	92	4.13	88	3.53	70	3.49
86	Namibia	3.63	91	4.16	72	3.83	106	2.89
87	Sri Lanka	3.62	86	4.37	101	3.09	79	3.40
88	Guyana	3.61	84	4.42	95	3.43	103	2.98
89	Ecuador	3.59	82	4.42	80	3.65	113	2.70
90	Honduras	3.58	93	4.07	91	3.51	95	3.17
91	Moldova	3.56	90	4.20	90	3.51	102	2.98
92	Suriname	3.46	83	4.42	83	3.63	120	2.33
93	Mongolia	3.46	97	3.72	105	3.00	61	3.64
94	Kyrgyz Republic	3.39	88	4.21	110	2.97	100	3.00
95	Nigeria	3.37	103	3.08	87	3.58	75	3.45
96	Armenia	3.36	98	3.61	97	3.29	94	3.19
97	Albania	3.34	87	4.30	104	3.03	114	2.69
98	Georgia	3.32	75	4.59	117	2.59	109	2.79
99	Bolivia	3.31	96	3.84	100	3.16	105	2.91
100	Kenya	3.23	108	2.88	79	3.65	97	3.14
101	Mauritania	3.21	109	2.83	103	3.04	51	3.76
102	Tanzania	3.20	107	2.89	98	3.27	77	3.44
103	Cambodia	3.20	105	3.05	102	3.06	73	3.48
104	Cameroon	3.17	102	3.35	92	3.50	115	2.64
105	Madagascar	3.10	104	3.07	106	2.99	90	3.25
106	Nicaragua	3.10	99	3.56	119	2.53	93	3.21
107	Paraguay	3.01	100	3.55	109	2.97	117	2.51
108	Zimbabwe	2.97	110	2.82	81	3.64	119	2.46
109	Malawi	2.97	112	2.74	96	3.37	108	2.80
110	Zambia	2.95	101	3.47	115	2.76	116	2.63
111	Burkina Faso	2.92	117	2.42	99	3.18	96	3.15
112	Nepal	2.91	106	2.93	111	2.91	107	2.88
113	Mozambique	2.90	116	2.47	107	2.98	89	3.25
114	Benin	2.83	113	2.57	114	2.78	98	3.12
115	Mali	2.79	119	2.24	112	2.87	87	3.26
116	Uganda	2.70	121	1.93	113	2.81	82	3.37
117	Lesotho	2.66	111	2.78	120	2.41	110	2.78
118	Bangladesh	2.60	114	2.52	118	2.55	112	2.73
119	Angola	2.50	120	2.03	108	2.98	118	2.49
120	Ethiopia	2.42	118	2.26	122	2.24	111	2.76
121	Burundi	2.40	115	2.47	116	2.60	122	2.12
122	Chad	2.09	122	1.81	121	2.31	121	2.17

(cont'd)

The Middle East and at the Gulf region remain quite stable as far as networked readiness is concerned, with the partial exception of **Kuwait** (54th), which loses 8 positions from last year.

Israel, at 18th place, continues to lead the way in the region, with outstanding levels of technological sophistication and innovation, world-class research institutions, and excellent ICT penetration.[12]

The **United Arab Emirates**, down 1 place from last year to position 29, remains the best performer in the Gulf, followed by **Qatar** (36th place) and **Bahrain** (50th). The United Arab Emirates continues to lead the way in the region in networked readiness, with remarkable progress since last year in business (from 57th to 30th) and government (from 36th to 28th) readiness. The government, in particular, has a major role in promoting ICT penetration and usage, as witnessed by the excellent rank registered by the country in government usage (10th), and as detailed in Box 2 (which also provides an aggregate analysis of the levels of regional networked readiness).

Conclusions

Networked readiness is a complex phenomenon, the sum of diverse and interrelated forces. Measuring a country's networked readiness remains a significant challenge, and any framework or model designed to represent it is, by necessity, a simplification. Moreover, limitations in the availability of reliable and current data restrict the measurement of the phenomenon to a subset of countries.

Nevertheless, as has been seen in this chapter, the networked readiness framework and the NRI can be useful tools for a country's policy and decision makers to better understand and benchmark the use of ICT. The

---

Box 2: Networked readiness levels: A regional comparison

An analysis of the results of the NRI 2006–07 shows that there are significant variations in the levels of networked readiness across different regions of the world.

Figure 1 plots the average scores for the NRI and its three component subindexes for different regions of the world. There is a large difference across the NRI and all subindex scores between the OECD countries (with an NRI average score of 5.13) and sub-Saharan Africa (with an NRI average score of 2.95).

Figure 1: The Networked Readiness Index 2006–07 by region

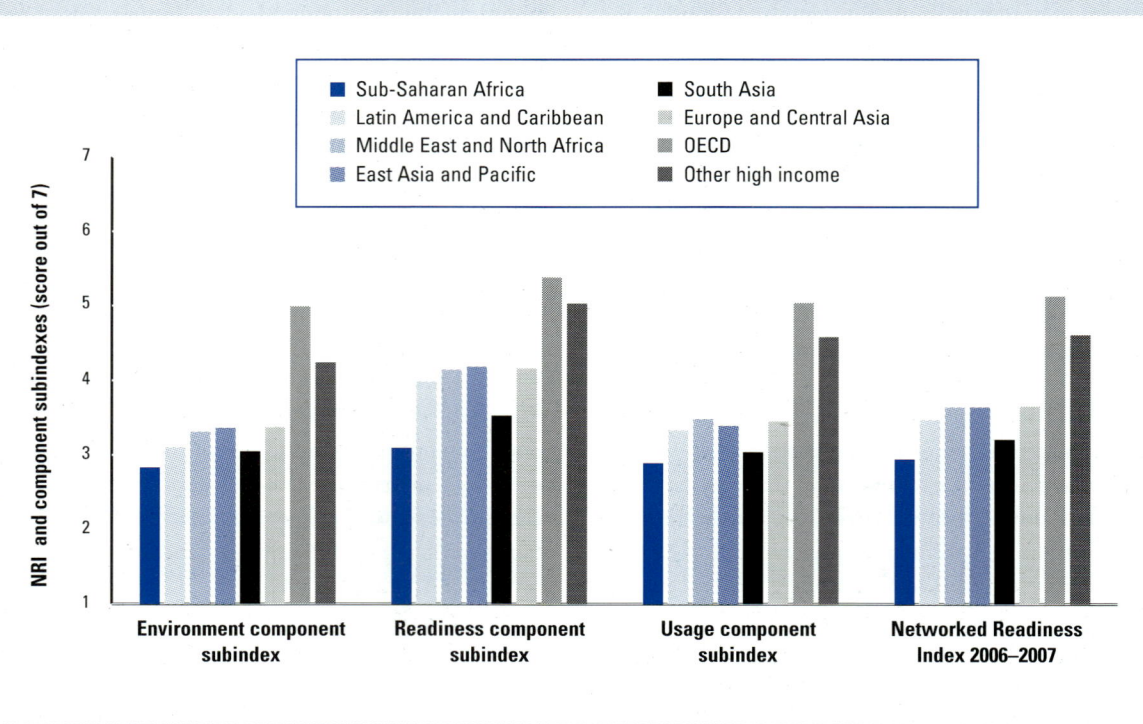

Box 2: Networked readiness levels: A regional comparison *(cont'd.)*

Notwithstanding the low current levels of networked readiness in much of *Africa*, many nations in the region have increased their levels of investment and ambition with respect to ICT. Ethiopia, despite being one of the continent's poorest countries, is spending nearly one tenth of its GDP on IT every year. Hundreds of government offices and schools have already been equipped with broadband Internet connections, and more are yet to come. Both government and private sector leaders in Ethiopia—in an attempt to forge headlong into the "global village"—have committed huge resources to seeing that by 2007 all of Ethiopia's 74 million people live no more than a few kilometers from a broadband connection.

Progress in the *Middle East* is being spearheaded by the Gulf countries. Many countries in the Gulf are now emphasizing the role of ICT for national development. For example, the United Arab Emirates (UAE) has launched a number of ICT initiatives including the Dubai Media City (DMC, which started in November 2000), Dubai Internet City (DIC), and Knowledge Village (KV). The major goal of the multibillion dollar DMC, DIC, and KV complex is to create a cluster of educators, incubators, logistics companies, multimedia businesses, telecommunications companies, remote service providers, software developers, and venture capitalists in one place. The latest addition to the Dubai high-tech corridor is Dubai Silicon Oasis (DSO), which is intended to be one of the world's leading high-technology parks for the semiconductor and microelectronics industry. The centerpiece of ICT implementation in the UAE is the Dubai e-government initiative. Initiated in April 2000, this project has been recognized as a success story by practitioners and researchers alike in the ICT-related and economic fields. Dubai's e-government initiative is an integral component of Dubai Vision 2010, which aims to establish Dubai as a knowledge-based economy by leveraging tourism, IT, media, trade, and services as pivotal industries in an effort to move away from dependence on oil-related products.

Europe presents a mixed picture, with the Nordic countries leading the way and other larger economies following at different speeds. Finland, for example, has successfully transformed itself into a prototype of the future networked society by focusing relentlessly on innovation, education, and information technology. Finland was the first country in the world to introduce the concept of a national innovation system as a basic frame of reference in policy formulation. The leadership for these changes came from the very top of the country. Finland's investment in R&D (3.4 percent of GDP) is one of the the the very highest in the world. Leadership from the top has also been vital for the success of Estonia. Maart Laar—Prime Minister of Estonia from 1992 to 2002—and his key advisers spearheaded the development of ICT in the country. Not only has the government created a supportive policy environment for ICT, it has also been a pioneer in using ICT for its own processes. For example, in August 2000, the government changed its cabinet meetings to paperless sessions using a Web-based document system. In a sight rarely seen in other countries, Estonian ministers peruse draft bills and regulations, make comments and suggestions, and vote entirely online at computer terminals. Citizens have access to nearly all government documents online and are encouraged to participate online in legislative processes.

Regardless of which region one considers, a key message emerges from the regional comparisons. The economies with higher levels of networked readiness are significantly better in ICT usage than others. High levels of ICT usage are supported by private and public leaders in creating environments conducive for doing business, for obtaining high levels of readiness across all actor groups, and for stimulating ICT penetration. High levels of ICT usage are supported by private and public leaders by creating an environment conducive for doing business, for higher levels of readiness across all actor groups and for increasing ICT penetration. All of this leads to a more competitive economy.

framework attempts to interpret the underlying complexity of the development and use of ICT in an intuitive and easy-to-comprehend model. The NRI is a summary measure of a nation's ability to participate in and benefit from ICT developments; it provides guidance to business leaders and policymakers for enhancing the impact of ICT on important actors—individuals, businesses, and governments.

Governments and policymakers can have significant influence on the adoption and usage of ICT. The NRI allows a nation to benchmark its ICT performance and to determine the effectiveness of ICT-related policies. It also permits a country to learn from the policies and performances of other countries with similar profiles, and to identify best practices by highlighting areas of excellent or poor performance.

Typically, countries that are performing well have put ICT on the national agenda and have striven to make it an area of excellence, whereas those nations that are underperforming have not done so. The former set of countries have succeeded in going beyond individual measures of national income, or national ICT spending, in an effort to provide an optimal environment for ICT development, thus promoting high levels of readiness and usage within all three key stakeholders. Denmark, Singapore, and Estonia are some examples of such leaders, and can serve as role models for other nations in their quest for ICT excellence.

Notes

1 Trajtenberg 2005.

2 The first *Global Information Technology Report*, published in 2002 and edited by Kirkman et al., was the result of a cooperation between the World Economic Forum and the Center for International Development at Harvard University. The 2003 and 2004 editions have been prepared by the Forum and INSEAD together with Infodev at the World Bank. Starting in 2005, the Forum and INSEAD have been exclusive partners on the project.

3 Twenty-six out of the total 67 variables used in this year computation are hard data.

4 On a related note, the treatment of missing variables has been aligned to the Forum's methodology: whereas in the past missing variables were estimated using analytical techniques such as regression and clustering, in the current computation they are indicated with "n/a" and not taken in consideration in the calculation of the specific pillar to which they belong. Additional information on the NRI computation is available upon request from gcp@weforum.org.

5 For a more detailed analysis, see Dutta and Jain (2006).

6 Although Table 3 includes the results of the first calculation in 2001–02 to provide the full NRI time series, some caution must be used when comparing the first year's results with those of following years, since the current networked framework was adopted only in the 2002–03 edition of the GITR. For more details on the 2001–02 framework, see Kirkman et al. (2002).

7 In this regard, Switzerland appears as a rather interesting case, since its ICT performance is led largely by the private sector rather than the government, which lags behind in readiness (19th) and usage (19th).

8 This drop is the result of a worsening of selected elements of the market environment and of individual and business usage.

9 For a full account of Taiwan's story, see Dahl and Lopez-Claros (2006).

10 See A. Lopez-Claros et al. (2006a) for a discussion of Latin America's technological readiness and innovation.

11 See ECLAC-CEPAL (2004).

12 Israel's ICT development story was the object of a country case-study in last year's GITR; see Lopez-Claros and Mia (2006).

References

Dutta, S., A. de Meyer, A. Jain, and G. Richter. 2006. *The Information Society in an Enlarged Europe*. Berlin: Springer-Verlag.

Dutta, S. and A. Jain. 2006. "Networked Readiness and the Benchmarking of ICT Competitiveness." *The Global Information Technology Report 2005–2006: Leveraging ICT for Development*. Hampshire: Palgrave Macmillan. 3–24.

Dutta, S. and A. Lopez-Claros, eds. 2005. *The Global Information Technology Report 2004–2005: Efficiency in an Increasingly Connected World*. Hampshire: Palgrave Macmillan.

Dahl, A. and A. Lopez-Claros. 2006. "The Impact of Information and Communication Technologies on the Economic Competitiveness and Social Development of Taiwan." *The Global Information Technology Report 2005–2006: Leveraging ICT for Development*. Hampshire: Palgrave Macmillan. 107–18.

Dutta, S., A. Lopez-Claros, and I. Mia, eds. 2006. *The Global Information Technology Report 2005–2006: Leveraging ICT for Development*. Hampshire: Palgrave Macmillan.

Dutta, S., F. Paua, and B. Lanvin, eds. 2004. *The Global Information Technology Report 2003–2004: Towards an Equitable Information Society*. New York: Oxford University Press

ECLAC-CEPAL (Economic Commission for Latin America and the Caribbean). 2004. *Productive Development in Open Economies*. Thirtieth Session of ECLAC-CEPAL: San Juan.

Kirkman, G., P. Cornelius, J. Sachs, and K. Schwab, eds. 2002. *The Global Information Technology Report 2001–2002: Readiness for the Networked World*. New York: Oxford University Press.

Lopez-Claros, A., L. Altinger, J. Blanke, M. Drzeniek, and I. Mia. 2006a. "Assessing Latin American Competitiveness: Challenges and Opportunities." *The Latin America Competitiveness Review 2006: Paving the Way for Regional Prosperity*. Geneva: World Economic Forum. 1–36

Lopez-Claros, A., L. Altinger, J. Blanke, M. Drzeniek, and I. Mia. 2006b. "The Global Competitiveness Index: Identifying the Key Elements of Sustainable Growth." *The Global Competitiveness Report 2006–2007*. Hampshire: Palgrave Macmillan. 3–50.

Lopez-Claros, A., J. Blanke, M. Drzeniek, I. Mia, and S. Zaidhi. 2005. "Policies and Institutions Underpinning Economic Growth: Results from the Competitiveness Index." *The Global Competitiveness Report 2005–2006*. Hampshire: Palgrave Macmillan. 3–37.

Lopez-Claros, A. and I. Mia. 2006. "Israel: Factor in the Emergence of an ICT Powerhouse." *The Global Information Technology Report 2005–2006: Leveraging ICT for Development*. Hampshire: Palgrave Macmillan. 89–105.

Lopez-Claros, A., M. Porter, X. Sala-i-Martin, and K. Schwab, eds. 2006. *The Global Competitiveness Report 2006–2007*. Hampshire: Palgrave Macmillan

Sala-i-Martin, X. and E. V. Artadi. 2004. "The Global Competitiveness Index." *The Global Competitiveness Report 2004–2005*. Hampshire: Palgrave Macmillan. 51–80.

Trajtenberg, M. 2005. "Innovation Policy for Development: an Overview." Paper prepared for LAEBA, Second Annual Meeting. Tel Aviv University. NBER and CEPR. November.

Appendix A: Technical composition and computation of the Networked Readiness Index 2006–07

The Networked Readiness Index 2006–07 separates environmental factors from ICT readiness and usage, and thus is composed of three component subindexes. Each component subindex is further subdivided into three pillars, with a total of 67 variables.

The Networked Readiness Index is defined as follows:

Networked Readiness
Index = 1/3 Environment component subindex
+ 1/3 Readiness component subindex
+ 1/3 Usage component subindex

I: The Environment Component index is defined as follows:

Environment = 1/3 Market environment
+ 1/3 Political and regulatory environment
+ 1/3 Infrastructure environment

The Market environment pillar is defined by the following variables:
1.01 Venture capital availability
1.02 Financial market sophistication
1.03 Technological readiness
1.04 State of cluster development
1.05 US utility patents (hard data)
1.06 High-tech exports (hard data)
1.07 Burden of government regulation
1.08 Extent and effect of taxation
1.09 Time required to start a business (hard data)
1.10 Number of administrative procedures required to start a business (hard data)
1.11 Intensity of local competition
1.12 Freedom of the press

The Political and regulatory environment pillar is defined by the following variables:
2.01 Effectiveness of law-making bodies
2.02 Laws relating to ICT
2.03 Judicial independence
2.04 Intellectual property protection
2.05 Efficiency of legal framework
2.06 Property rights
2.07 Quality of competition in the ISP sector
2.08 Number of administrative procedures to enforce a contract (hard data)
2.09 Time to enforce a contract (hard data)

The Infrastructure environment pillar is defined by the following variables:
3.01 Telephone lines (hard data)
3.02 Secure Internet servers (hard data)
3.03 Internet hosts (hard data)
3.04 Electricity production (hard data)
3.05 Availability of scientists and engineers
3.06 Quality of scientific research institutions
3.07 Tertiary enrollment (hard data)

II. The Readiness component subindex is defined as follows:

Readiness = 1/3 Individual readiness
+ 1/3 Business readiness
+ 1/3 Government readiness

The Individual readiness pillar is defined by the following variables:
4.01 Quality of math and science education
4.02 Quality of the educational system
4.03 Quality of public schools
4.04 Internet access in schools
4.05 Buyer sophistication
4.06 Residential telephone connection charge (hard data)
4.07 Residential monthly telephone subscription (hard data)
4.08 High-speed monthly broadband subscription charge (hard data)
4.09 Lowest cost of broadband (hard data)
4.10 Cost of cellular phone call (hard data)

The Business readiness pillar is defined by the following variables:
5.01 Extent of staff training
5.02 Local availability of specialized research and training services
5.03 Quality of management schools
5.04 Company spending on research and development
5.05 University-industry research collaboration
5.06 Business telephone connection charge (hard data)
5.07 Business monthly telephone subscription (hard data)
5.08 Local supplier quality
5.09 Computer, communications, and other services imports (hard data)

The Government readiness pillar is defined by the following variables:
6.01 Government prioritization of ICT
6.02 Government procurement of advanced technology products
6.03 Importance of ICT to government's vision of the future
6.04 E-participation index (hard data)
6.05 E-government readiness index (hard data)

(cont'd.)

19

Appendix A: Technical composition and computation of the Networked Readiness Index 2006–07 *(cont'd.)*

III. The Usage component subindex is defined as follows:

$$\text{Usage} = \tfrac{1}{3} \text{ Individual usage} + \tfrac{1}{3} \text{ Business usage} + \tfrac{1}{3} \text{ Government usage}$$

The Individual usage pillar is defined by the following variables:

7.01 Mobile telephone subscribers (hard data)
7.02 Personal computers (hard data)
7.03 Broadband Internet subscribers (hard data)
7.04 Internet users (hard data)
7.05 Internet bandwidth (hard data)

The Business usage pillar is defined by the following variables:

8.01 Prevalence of foreign technology licensing
8.02 Firm-level technology absorption
8.03 Capacity for innovation
8.04 Availability of new telephone lines
8.05 Availability of mobile telephones
8.06 Extent of business Internet use

The Government usage pillar is defined by the following variables:

9.01 Government success in ICT promotion
9.02 Availability of online services
9.03 ICT use and government efficiency
9.04 ICT pervasiveness

Appendix B: Countries/Economies covered by the Networked Readiness Index 2006–07, by stage of development

Stage 1	Transition from 1 to 2	Stage 2	Transition from 2 to 3	Stage 3
GDP per capita, <US$2,000	GDP per capita, US$2,000–3,000	GDP per capita, US$3,000–9,000	GDP per capita, US$9,000–17,000	GDP per capita, >US$17,000
Angola	Albania	Algeria	Bahrain	Australia
Armenia	Bosnia and Herzegovina	Argentina	Barbados	Austria
Azerbaijan	Colombia	Botswana	Czech Republic	Belgium
Bangladesh	Ecuador	Brazil	Estonia	Canada
Benin	El Salvador	Bulgaria	Hungary	Cyprus
Bolivia	Jordan	Chile	Korea, Rep.	Denmark
Burkina Faso	Macedonia, FYR	Costa Rica	Malta	Finland
Burundi	Namibia	Croatia	Taiwan, China	France
Cambodia	Peru	Dominican Republic	Trinidad and Tobago	Germany
Cameroon	Suriname	Jamaica		Greece
Chad	Thailand	Kazakhstan		Hong Kong SAR
China	Tunisia	Latvia		Iceland
Egypt		Lithuania		Ireland
Ethiopia		Malaysia		Israel
Georgia		Mauritius		Italy
Guatemala		Mexico		Japan
Guyana		Panama		Kuwait
Honduras		Poland		Luxembourg
India		Romania		Netherlands
Indonesia		Russian Federation		New Zealand
Kenya		Serbia and Montenegro		Norway
Kyrgyz Republic		Slovak Republic		Portugal
Lesotho		South Africa		Qatar
Madagascar		Turkey		Singapore
Malawi		Uruguay		Slovenia
Mali		Venezuela		Spain
Mauritania				Sweden
Moldova				Switzerland
Mongolia				United Arab Emirates
Morocco				United Kingdom
Mozambique				United States
Nepal				
Nicaragua				
Nigeria				
Pakistan				
Paraguay				
Philippines				
Sri Lanka				
Tanzania				
Uganda				
Ukraine				
Vietnam				
Zambia				
Zimbabwe				

CHAPTER 1.2

Networks Changing the Way We Work, Live, Play, and Learn

ROGER FARNSWORTH, Cisco Systems, Inc.

LIONEL GIBBONS, Cisco Systems, Inc.

TRACEY LEWIS, Cisco Systems, Inc.

MARSHA POWELL, Cisco Systems, Inc.

There is little doubt that the way people live the world over is undergoing profound and rapid change. As well, especially in the developing world, we have seen spectacular growth and capability in an endless variety of technologies, particularly in information and communication technologies (ICT).

ICT and the evolution of work

But what has caused the greatest disruption—and generated the greatest promise for a hopeful future—is the acceleration of lifestyle changes and opportunities as a direct result of dramatic developments in ICT. This convergence of people, their aspirations for a better life, and the network-based tools they use has a positive impact on the way the world works, lives, plays, and learns.

Blurring the line between work and life

For most of recorded history there has been a clear delineation in both the definition of and the time allotted for *work* and *play*. The "workday" has come to mean the typically eight hours per day spent performing the duties for which employees are compensated. Within the hours dedicated to work, there has been a further refinement: there are "blue collar" jobs, consisting of various forms of manual labor or working with one's hands; and "white collar" jobs, which are typically less physical and more analytical or service-oriented. In more developed regions of the world, the balance of white collar versus blue has clearly tipped toward professionals who manipulate information rather than materials. Today, in such industries as financial services, health-care, technology, pharmaceuticals, and media and entertainment, professionals now account for 25 percent or more of the workforce.[1]

But the rise of ICT is causing fundamental changes in these definitions, blurring the line between when we work and when we play, redefining the concept of work itself, and reshaping the skills required to meet the demands of the always-changing world of work.

A brief history of modern commerce

Throughout history, countries have generally created wealth and prosperity by expanding their borders in the search for more cost-effective labor and new markets. Today, businesses seek to "expand their borders" by collaborating—and competing—in a global environment where businesses may never meet their customers and co-workers may never meet in person.

ICT has been a key enabler of this globalization, and is only the latest in a series of technological revolutions that, in the decades since the Industrial Revolution, has changed the face of how businesses and individuals work. According to Carlota Perez, visiting senior research fellow at Cambridge University, the Industrial Revolution led to

Figure 1: Technology revolutions: A historical perspective

	Technology revolution	Irruption	Frenzy		Synergy	Maturity
5	Age of information and telecommunications	1971–87	1987–2001		Platform = Network	
4	Age of oil, automobiles, and mass production	1908–20	1920–29	TURNING POINT	Platform = Distribution system	1960–74
3	Age of steel, electricity, and heavy engineering	1875–84	1884–93		Platform = Energy grid	1908–18
2	Age of steam and railways	1830s	1840s		Platform = Transportation system	1857–73
1	The Industrial Revolution	1770s and early 1780s	Late 1780s and early 1790s		Platform = Machinery/automation	1813–29

Source: Perez, 2002.

the development of machinery and automation that dramatically increased the rate and volume of production of created goods.[2] The age of steam and railways ushered in the capability to locate factories near the resources they used rather than near sources of energy, and laid the foundation for a national transportation system. In the age of steel, electricity, and heavy engineering, industry expanded beyond railways to all manufacturing, both national and international, and led to the development of the energy grid. The age of oil, automobiles, and mass production led ultimately to the ability to cost-effectively produce and distribute goods on a global scale (Figure 1).

Each of these eras profoundly changed the context and capability of businesses globally, but did not fundamentally alter the nature of commerce or production or the boundaries between the participants.

In the current age of information and telecommunications, innovators are re-thinking how to harness and deliver information in completely new ways. Rather than taking existing business models and layering on automation with ICT, they are now able to begin with networked communications and collaboration tools as a "given," then re-design a business model and its processes around this new platform.

Driving this new approach is the fact that businesses and workers are growing more closely networked every day. As of September 2006, almost 1.1 billion people, or 16.7 percent of the global population, had access to the Internet, over either a dedicated or shared connection, with a global average growth rate of 201 percent between 2000 and 2006.[3]

This shifting of focus that places networks at the core of the business model is leading to a shift in the way resources, services, and applications are created and made available to users. It is also fundamentally changing the landscape in which individuals participate in the creation and delivery of these products and services.

Redrawing business borders

World-changing technological developments—such as the rise of the personal computer (PC), the Internet, and workflow software that helps to automate and expedite the flow of processes and content—are enabling businesses to compete and collaborate in ways never possible before. Thomas Friedman, in *The World Is Flat*,[4] describes how businesses are disaggregating their value chains and re-evaluating how to optimize their individual components, driving cost and time out of their core processes. Work is being redistributed to regions where it can be performed more efficiently and cost-effectively, leading to such trends as production off-shoring and information technology (IT) development and service outsourcing (see Figure 2).

Businesses are creating sophisticated global supply chains that optimize parts deliveries to factories, then from

Figure 2: Porter Value Chain

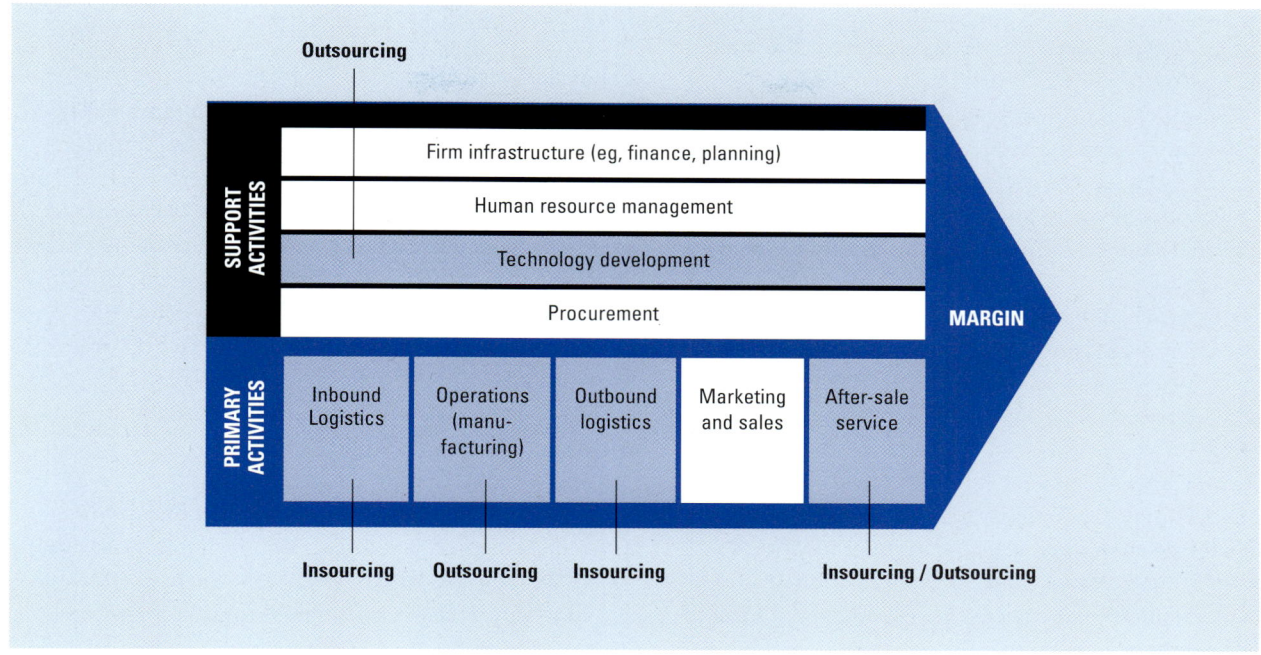

Source: Porter, 1985.

factories to retail outlets at the lowest possible cost. Principles such as collaborative planning, forecasting, and replenishment (CPFR) combine the intelligence of multiple trading partners to achieve a common understanding of customer demand (both quantity and category mix), matching productivity to supply as closely as possible.[5] Suppliers such as Procter & Gamble, Nabisco, Sara Lee, and Hewlett-Packard have used CPFR to create highly accurate joint sales forecasts with retailers such as Wal-Mart and United States–based Wegman's, achieving improvements of 2 to 8 percent for in-stock percentages and inventory reductions of 10 to 40 percent across the entire supply chain.

Taking collaboration further is the concept of insourcing, where businesses quickly and efficiently gain expertise in noncore business functions, such as logistics or warehousing, by integrating third-party-managed capabilities into their infrastructure in a very transparent way to the end customer.

Leading this dramatic re-drawing of corporate boundaries is ICT, which enables businesses to—instantly and securely—share data that optimize the flow of goods and services, from individual components to finished product, around the clock.

Not only do people collaborate with people, but machines collaborate with machines, resulting in a dramatic increase in the volume of instantaneous, ad hoc

communication between organizations. Traditional points of transactional "friction" at the borders of businesses are being all but eradicated. New trends, such as experience-based differentiation, where customers provide preferences and usage patterns *backward* through the supply chain to the start of product design, in near real-time, are increasingly becoming a factor in product life-cycle management. Forrester Research identifies what it calls the "X Internet," which connects information systems (IS) directly to products as they travel through the supply chain, providing near real-time information on potential supply constraints.[6]

A new definition of *team*

The ability to communicate and collaborate in ways previously unheard of has implications for process development and organizational structures as well, especially in large corporate or governmental entities.[7] The complexity and opportunity of the new global economy requires the ability to connect the knowledge and experience of individuals across diverse disciplines and geographies. Traditional, top-down, or heavily matrixed organizations are ill-suited to effective collaboration; increasingly, professionals must interact efficiently and effectively with an expanding community of colleagues, both inside and outside their own organizations, to accomplish tasks. New approaches are required, such as aligning teams according to business process flows or establishing "knowledge" or "talent" marketplaces within organizations.

Organizations that can adapt in these ways are better able to marshal resources quickly across diverse functions and bring together virtual teams that reach their objectives, then move on to their next project.[8] People with common interests, such as similar skills (writing), fields of expertise (computer engineering), or geography (South America) naturally form "social networks." Formalizing those relationships within organizations can lead to new forms of collaboration where participants are motivated by interest or passion instead of simply being compelled to participate by organizational structure. Add collaborative tools and workspaces, powered by ICT, to the mix, and workers are able to accelerate productivity and innovation on a global scale.

United Kingdom–based WPP, the world's second-largest advertising-marketing communications consortium, exemplifies this new approach to creating value for its clients. WPP was created from the consolidation of some of the oldest and most-respected advertising brand names, which joined forces to be able to service more and bigger clients' needs. Shortly after its creation, WPP discovered that the capability for delivering the best solution does not always exist within just one of its companies, or even in the traditional integration of more than one member of the consortium. Rather, the solution increasingly lies in choosing personnel from across its business units, disciplines, and management levels, to assemble the ideal team for each individual client's needs.

WPP removed the boundaries between its member companies and began to view its entire employee base as a single, vast pool of individual specialists who can be assembled horizontally into any number of collaborative teams, based on the unique demands of each individual project or client. What WPP learned, most importantly, is the power inherent in being able to flexibly form and re-form the organization, as business dictates. This "fluid" type of structure is now possible because of collaborative tools and processes that bridge miles and minutes between colleagues, ensuring that teams stay on task, on time, and on target.[9]

Work is being redefined at the individual level, as well. As jobs move where they can be performed most efficiently, the universe of people and perspectives to which workers are exposed is exponentially greater, due in large part to ICT.

It is more and more likely now that, instead of walking down the hall to get a question answered by a colleague, workers will send an email or instant message or both, or initiate a videoconference with a colleague nearby or halfway around the world whom they may never meet in person. With the increasingly connected nature of business today, especially in large and complex organizations,

workers also often have the same level of visibility into internal systems and data whether they share the same employer, or work for a supplier or partner.

Better work-life balance through technology

The power, storage capacity, and input/output speed of PCs plus email, instant messaging, file-sharing programs, IP telephony, videoconferencing, and wireless technologies in multiple form are combining to expand personal workspaces. Online collaborative workspaces facilitate the timely sharing and manipulation of ideas among team members, regardless of location or time zone. Workers have the capacity to inform themselves like never before because of sites such as Google or Yahoo!, leading to better decisions; equally, they can participate in informing others by contributing to blogs or wikis.

The personal and business benefits of this flexible new world are many: rather than being confined to an office eight hours a day, having secure, mobile access to work data from anywhere, anytime, better balances work and life. Because workers can be as connected and productive at home as at work, mothers and fathers have the flexibility to tend to a sick child or more easily manage home responsibilities.

From a business perspective, connected work environments lead to greater productivity and innovation, as diverse perspectives and ideas are shared and combined in ways never thought of before. Especially in larger, more complex organizations, career options are expanded as workers interact across functions and departmental boundaries, build personal "networks" of colleagues with whom they trust and cooperate, and develop new skills that will pave the way to greater job satisfaction.

In this sense, the Shell Group (Shell Oil), a global energy and petrochemical provider, has created a process and a collaborative workspace for engineers from multiple disciplines worldwide to troubleshoot problems in near real-time. Not only can pipeline engineers and corrosion engineers solve problems more quickly by working together, but their collaboration often leads to new ideas that help them prevent problems before they arise, deliver products to market faster, and improve Shell's competitive position.[10]

Another example can be found in California's Silicon Valley. Here, an increasingly culturally and linguistically diverse population has led to challenges in providing accurate and well-understood medical information and patient care. In response, local hospitals created the Health Care Interpreter Network, where bilingual or multilingual hospital employees in functions as diverse as accounting or facilities management can be called upon to provide translation services to doctors and patients. This network was initially implemented via "old-fashioned" phone calls to the translators, who would then have to walk or drive to where the patient and doctor were located. The addition of voice and videoconferencing capabilities has enabled interpreters to instantaneously provide support from wherever they are located without interrupting their regular work routines; the collaborative technologies also enable them to provide their services as many as 25 times in a given day, compared with an average of only 8 times per day previously.

Work is what we do, not where we are

Increasingly, work is something we do rather than a place we go. In a sense, the workday never ends as it flows between businesses, over networks, person-to-person and machine-to-machine, around the globe and around the clock. Workers enter and exit the stream of productivity as needed, unbound by time or geography, expanding perspectives, knowledge, and career options. Organizations enjoy unprecedented opportunity for an infinite variety of partners with whom to work and markets to enter.

Work will undoubtedly continue to evolve. New skills will be required, ones that build on the ability to manage and shape communication and interactions. "Knowledge work" will continue to grow in importance and take new and unimagined forms, as organizations continue to access the power of ICT to satisfy the human passion for learning and discovery. It is how tomorrow's workers will find fulfillment in work, instead of counting the minutes to workday's end.

ICT: Changing the way we live

With the advent of powerful PCs featuring easy-to-use browsers, the Internet has experienced rapid growth. More countries than ever, recognizing the advantages and opportunities available through such technologies as the Internet and mobile phones, are engaged in some form of advanced ICT. It is how governments worldwide are increasing productivity and revenues while citizens experience improvements in the quality of their lives.

From email to e-banking, eBay to e-health, and e-learning to e-government, citizens in developed nations are just clicks away from abundant information and services, resulting in greater knowledge and better, faster communication. Citizens can also improve the quality of their personal health and development as health agencies, schools, and employers increasingly rely on ICT to enable citizens to communicate, learn, and work from home.

But the revolution is not limited to developed countries only; popular search engines such as Google receive a billion searches each day, more than half in languages other than English; nine trillion emails zip across the globe annually.

Success and challenges in developed and emerging countries

Business models worldwide are rapidly changing to capitalize on the advantages to governments and citizens promised by ICT. More than 500 e-government initiatives have been launched around the world within the last five years, a remarkable increase since 1996, in which only 3 were launched.[11]

Governments are quickly finding that making services available online improves productivity and revenue while also improving citizen experiences.

Though rapidly developing nations such as China, India, and Saudi Arabia are building or improving infrastructures to support new initiatives, they face numerous challenges, among them: locating the investment capital required to build a stable communications infrastructure, setting regulations and guidelines needed to establish secure transactions and establish a unified platform, and providing education and public awareness to its citizens to access these technologies.

Education and professional development

For this new Internet-driven culture to succeed, governments are focusing attention on education, promoting Internet benefits, and finding ways to motivate and enable citizens to access and use e-government services.

To increase the level of computer literacy in Dubai, for instance, the government launched an eCitizen certification program aimed at increasing public awareness and e-government services usage. This program provides training on email and Internet fundamentals as well as helping the citizens of Dubai take advantage of online government services in areas such as education, health care, transportation, and housing.

There must also be sufficient numbers of IT-skilled workers on hand to build and maintain the ICT infrastructure. Malaysia has joined forces with world-class IT companies such as Electronic Data Systems, Hewlett-Packard, and Microsoft to employ and train local people. In Hong Kong, citizens are offered computer training in community centers, and advertisements promoting Internet usage are widespread throughout the economy.[12]

Around the world, enormous benefits for citizens and public and private businesses are being realized, and more are yet to be uncovered as countries continue to move further into an Internet-driven culture that delivers communication, services, and information at an almost instantaneous rate.

Improving and extending e-government

Once ICT infrastructures are established, many countries turn their focus to expanding service reach and improving network reliability and capability. Some governments are simply putting information about basic services online,

27

while others are looking ahead, planning multiphase approaches to build upon existing e-government programs.

In South Africa, for example, the government has a long-term strategy that includes the launch of its "People First" Internet gateway, providing citizens with a single point of entry to obtain government services and information. In the next phase, the portal will expand to enable users to make online transactions. High-demand government services—such as pension payouts and tax payments—will be the first to be available online, followed by additional services, the ability to conduct transactions, and more. In the final phase, advanced and complex transactions—such as applying for a passport—will be available, as well as access to small business and education resources.

To fully capitalize on the potential benefits of implementing e-government services, governments must assess current practices and reorganize processes with priority given to the needs of citizens and businesses.

Governments are already reaping the benefits of automating repetitive tasks, merging departmental functions, and streamlining processes to optimize efficiencies. Powerful search engines such as Google have helped enhance services delivery and offered citizens wider access to those services. Over the past decade, a handful of public-sector organizations around the world—schools, public-welfare agencies, health-care systems, postal and transit systems, and militaries—have improved performance by 5 to 30 percent or more.[13]

By streamlining processes, eliminating duplicate work, and improving interdepartmental coordination, governments can offer a single portal where many services can be handled at the same time. In Singapore, for example, citizens access all government services at one website, eCitizen (www.ecitizen.gov.sg). The result is a "one-stop shop" for government services—an approach that, for some governments, can reduce costs 35 to 45 percent while dramatically improving service quality and convenience.[14]

These one-stop shops are empowering citizens to complete many tasks at one time that once may have required trips to multiple physical locations and time wasted standing in line. For example, a single website (or "mega-portal") that allows people to file a change of address with the postal service may also allow them to select a moving company, notify friends and periodicals of the new address, open a bank account, and change utility services.

Mega-portals expand Web services by merging public and private sectors, and also provide a helpful boost to commerce. Because e-government websites are heavily used and publicized, they are likely to attract huge numbers of visitors; vendors linking to them could experience more traffic than Amazon and eBay combined, resulting in significant revenue for both public and private sectors.[15] An example of this kind of hybrid approach is the United

States government's official Web portal, FirstGov (www.firstgov.gov), which offers more than 100 government services and provides links to more than 150,000 websites.

Governments focused on delivering the advantages of Internet access to greater numbers of citizens are finding new methods to extend their reach. In Hong Kong, the government is building e-government kiosks in shopping malls, supermarkets, and railway stations. In Bangladesh, where most citizens may be categorized as rural poor, the government has introduced ICT Access Centers. These centers are equipped with modern PCs and Internet facilities, their purpose to guide rural people into adopting mainstream development principles. In Dubai, the government is utilizing mobile technology to provide an entirely new level of mobile communication: the m-government portal allows users to access information and conduct transactions instantly via handheld mobile devices, such as mobile telephones and personal digital assistants (PDAs).

Saudi Arabia recently completed an ambitious national ICT network to help the government guide 2.5 million touring religious pilgrims annually. One of the largest such undertakings worldwide, the privately funded network connects ministries, embassies, tour operators, and travel agents, automating what once was the laborious work of accommodating visitors. Pilgrims to the Kingdom now can arrange visits online through a one-stop Internet portal—many receive visas in four to eight hours, rather than the three to six weeks that was formerly typical. Moreover, this online approach will allow Saudi Arabia to accommodate twice the number of visitors over the next decade.[16]

In Germany, a decentralization of operational responsibilities and a redesign of local agencies and their service offerings have resulted in successful new prototypes—citizens now find waiting times cut in half and time available for counseling doubled. The result has been much higher overall user satisfaction.

Challenges ahead: More to be accomplished

The expansion of ICT has had an astounding impact on cultures and societies. In Africa, many citizens spend scarce money on mobile telephones rather than shoes and basic household items. School children around the world with Internet access at home demonstrate an advantage over their peers in doing research for homework. And village leaders in Punjab, India, with Internet access use their ability to monitor weather patterns in real time to manage pest control (www.nextbillion.net).

But many of ICT's benefits remain yet to be seen—billions worldwide are still without easy access to the Internet. The result? In the United Kingdom, setting up a new business online can be accomplished in less than 24 hours; in most emerging nations, the same task may take weeks. Or months.

As countries deploy advanced technologies, many will need to consider developing channels outside the home to enable all citizens to access the Internet. The spread of mobile telephones, broadband wireless networks, and low-cost IT technologies have played significant roles in improving lives by providing additional channels to access to information and services.

In addition, governments must also continue to play a role in providing public education and awareness of the benefits of using ICT to obtain services and information, and increase the quality of life for their citizens. For example, as health and education programs continue to use advanced technologies to bring services online, users enjoy access to better health care and educational opportunities.

Governments using e-government portals are just beginning to enjoy the benefits of increasing productivity and boosting revenue. Even greater savings and quality improvements could come from enhancing government productivity, which in 10 years is estimated to increase 5 to 20 percent.[17]

There is still much more ahead for the role of ICT; advances in wireless technologies, mobile telephones, and PDAs will bring advanced communications to mobile users and further its reach to provide communication, information, and education services online to citizens in every corner of the globe.

Evolution of the village storyteller: ICT at play

Huddled for warmth around a snapping fire, the small band of primitives listen intently to the one they call "Storyteller." The old man holds them enraptured with spellbinding tales of bold hunts, jealous animal-gods, and ancestors long dead.

Since the dawn of humanity, such entertainment has stitched together the seams of society's fabric. Even with the advent of the modern age, with its radio, television, and motion pictures, most people obtained the largest share of their popular entertainment from a comparatively small number of content sources, and often shared the experience in groups.

But over the most recent two decades highly customized and individualized entertainment choices have proliferated, targeted to increasingly stratified audiences.

The new storyteller

Mornings start in much of the developed world with family members choosing from hundreds of television channels to enjoy over bran flakes and a steaming beverage. They then hop into cars, buses, planes, or trains clutching portable music players packed with thousands of songs, movies, podcasts, or television programs. At work, PCs do more than accomplish work-related tasks—they also serve up an endless supply of Internet-distributed video, audio, and text entertainment, carefully selected and independently consumed.

Though the number and variety of choices has grown, little has changed about the nature of electronic entertainment itself. For the most part, the fundamental aspects of radio, television, movies, and games remain unchanged. What has changed—dramatically—is the way entertainment is created, delivered, and consumed.

In the United States, average households spend more than 57 hours weekly watching television, up from around 43 hours per week in 1975.[18] Surprisingly, much of the content flowing into homes is neither requested nor viewed; the average household may receive 96 channels yet regularly view as few as 15.

Today, several technologies help consumers gain better access to only the content they desire. Digital video recorders (DVRs) allow viewers to capture only the programming in which they have an interest to watch when it is convenient. Viewers can view these recorded programs on any television set in the home, on PCs, or even download them onto portable players for portable entertainment.

The ever-expanding reach of the Internet offers some intriguing new options for television program distribution. Internet protocol television (IPTV) uses IP to send video signals over high-speed network connections, offering consumers control over the content entering their homes and how they interact with it.

This distribution method is still in its infancy, but with its many advantages, it may soon follow the path that Voice over IP (VoIP) has taken: many consumers will choose IPTV over satellite, cable, and other delivery options. A strong motivation among service providers for IPTV is the thirst for revenue growth. By pairing entertainment services with telecommunications, revenue opportunities could expand significantly.

With more than 500,000 subscribers and large deployments in Taiwan, Japan, and China, Asia is leading the global IPTV market; Hong Kong currently leads the world in IPTV penetration.[19] Contributing to the region's rapid IPTV technology deployment is the deep penetration of broadband Internet access. As broadband speeds accelerate worldwide through such technologies as fiber to the home/curb/premises (FTTx) and worldwide interoperability for microwave access (WiMAX), IPTV deployment will likely spread further—and faster.

Perhaps the biggest news in television is mobile TV. In March 2006, Japan launched a commercial mobile TV service using radio frequencies reserved for digital television. However, Japan will have to catch up to the Republic of Korea, where mobile TV subscribers topped the 1 million mark in June 2006; there, average mobile TV subscribers watch about an hour of television per day on their phones.

29

How rapid is the adoption of mobile TV? The number of mobile TV subscribers will increase from 3.4 million today to 102 million by 2010,[20] according to forecasts by research firm In-Stat, driving global revenues past US$10 billion by 2011.[21]

There's no denying that—as evidenced by the rapid adoption of mobile TV—television and video hold center stage as today's storyteller, leading some to claim that radio is dead. But reports of radio's death were exaggerated. Despite emphasis on the visual, audio entertainment still plays a significant role—and new technologies are making it more relevant than ever. Radio remained essentially unchanged for decades until the 1990s, when it experienced a rapid evolution punctuated by the advent of streaming delivery over the Internet. By the end of the decade, streaming was joined by another form of delivery: satellite. These new forms of delivery in many ways parallel what cable television brought to broadcast television—wide distribution, often with fewer regulatory restrictions.

Internet radio is gaining in popularity. The US audience is now 53 million per month, up from 37 million in 2005.[22] Satellite radio is also enjoying healthy growth with the two major United States–based satellite radio companies reporting a combined subscriber base, at the end of 2005, of more than 9 million users.

Podcasting, close cousin to radio, is another exciting new development. Unlike streamed audio, podcasts are audio recordings downloaded via the Internet to a user's computer. Listeners can subscribe to individual podcasts, delivered automatically from the podcaster's host site.

But podcasting is more than just a means to deliver content; selections are often transferred to portable music players, such as the Apple iPod, for listening on the go. Users hear only the content they want, usually commercial free, whenever and wherever they choose.

Podcasting has also pioneered user-generated content. Although many podcasts are simply recordings of radio broadcasts, many are composed entirely of fresh content created exclusively for the medium, often by nonprofessional enthusiasts. This is a significant new development, as podcasting has fostered tremendous amounts of content, narrowly focused on niches neglected by other media.

Services such as iTunes, Rhapsody, and Urge all deliver a wide range of music—legally—to millions of users on home and mobile music players.

Beyond the convenience of instantly purchasing custom-selected music, this new business model has yielded additional benefits to both content consumers and providers. Musicians unable or unwilling to sign recording contracts with major music labels are afforded a direct route to fans, delivering content the market could not otherwise support. In fact, this trend has led to the emergence of numerous "boutique" sites dedicated to rare or specialty music, such as zimaudio.com, which proclaims

itself "… the biggest database of Zimbabwean music and artists past, present and future."

In all these cases—radio, podcasting, and music—IP networks play a vital and evolving role in delivering and distributing content.

Accessing the latest celluloid hits from Hollywood via the Internet is hardly news; today, it is possible to buy dozens of movie titles from iTunes and sync them to an iPod. What is new is the emerging business model that makes online distribution of movies practical, profitable, and secure; however, as of this writing there is no clear outcome for this new model.

There is another angle to the distribution of movies via IP-based networks that has gone largely unreported, yet offers a number of tantalizing opportunities. IP networks are helping to overcome many of the logistical, marketing, and economic challenges faced by movie theater owners—and changing the way consumers enjoy an evening at their local cinema.

The costs associated with producing and distributing the film prints themselves can be staggering, from between US$1,200 and US$2,000 per theater. These costs can be slashed through the digital distribution and exhibition of content over secure, broadband connections. Digital content delivered over IP networks can also feature integrated digital rights management, reducing the threat of films being pirated or used inappropriately. Moreover, content can be tracked, and adherence to licensing agreements can be documented. Of course, as every copy of digital content retains all the fidelity of the original, movies distributed via the Internet would be free of the scratches, stains, and audio drop-outs that characterize celluloid films that have taken one too many trips around the projector reel.

Finally—and most interesting of all—digital content distributed via the Internet to theaters can be customized, by the theater, in any number of ways: by geography, audience demographics, interest, gender, and more. It is easy to imagine a theater creating a Sunday afternoon matinee at the movies featuring a professional sporting event, followed by a sports-themed feature film, and concluding with a sports-themed comic performance.

Gaming

The popularity of massive multiplayer online role-playing games (MMORPGs) has grown in recent years following rapid growth in high-speed broadband Internet access. Not surprisingly, MMORPGs are particularly popular in Korea and Japan, where affordable multi-megabit-per-second access is commonplace (see Figures 3 and 4).

Every day across the globe, thousands of MMORPG participants join forces to perform tasks and meet objectives to win advancement. The only limitation is broadband Internet access. For example, a small group of teenagers may form a team tasked with defeating a similar group of

Figure 3: Asian MMORPG market: Peak concurrent users (January 2001–July 2006)

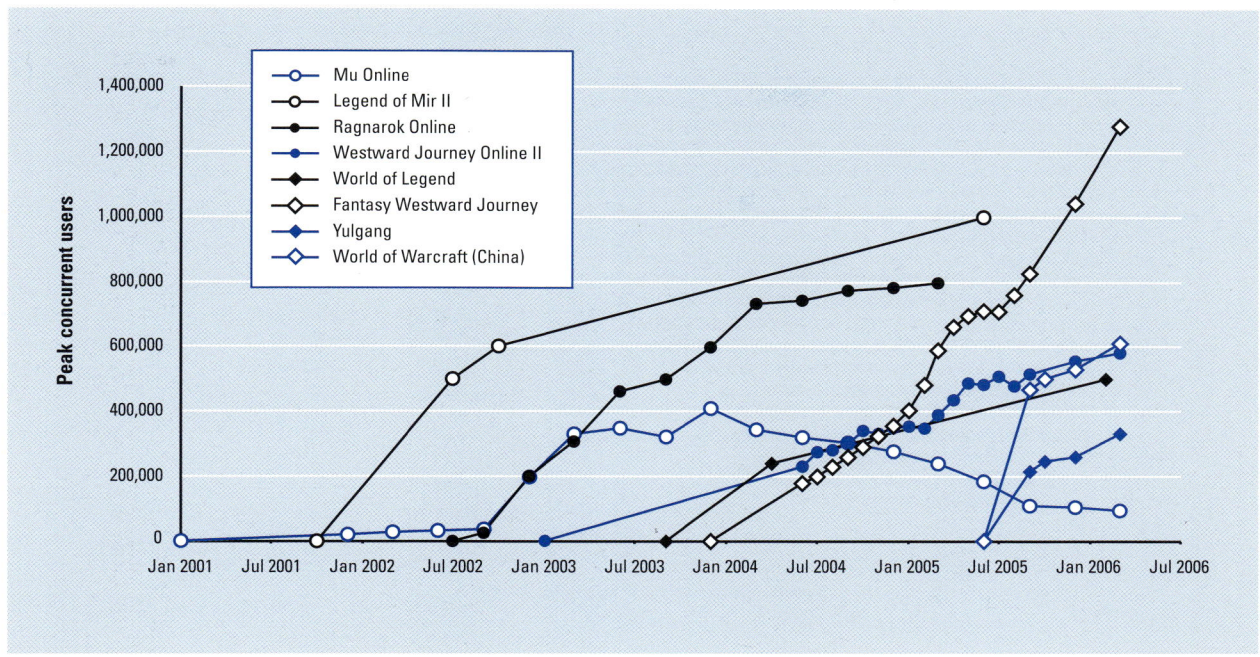

Source: Woodcock, 2006.

Figure 4: Total active worldwide MMORPG subscriptions (January 1997–July 2006)

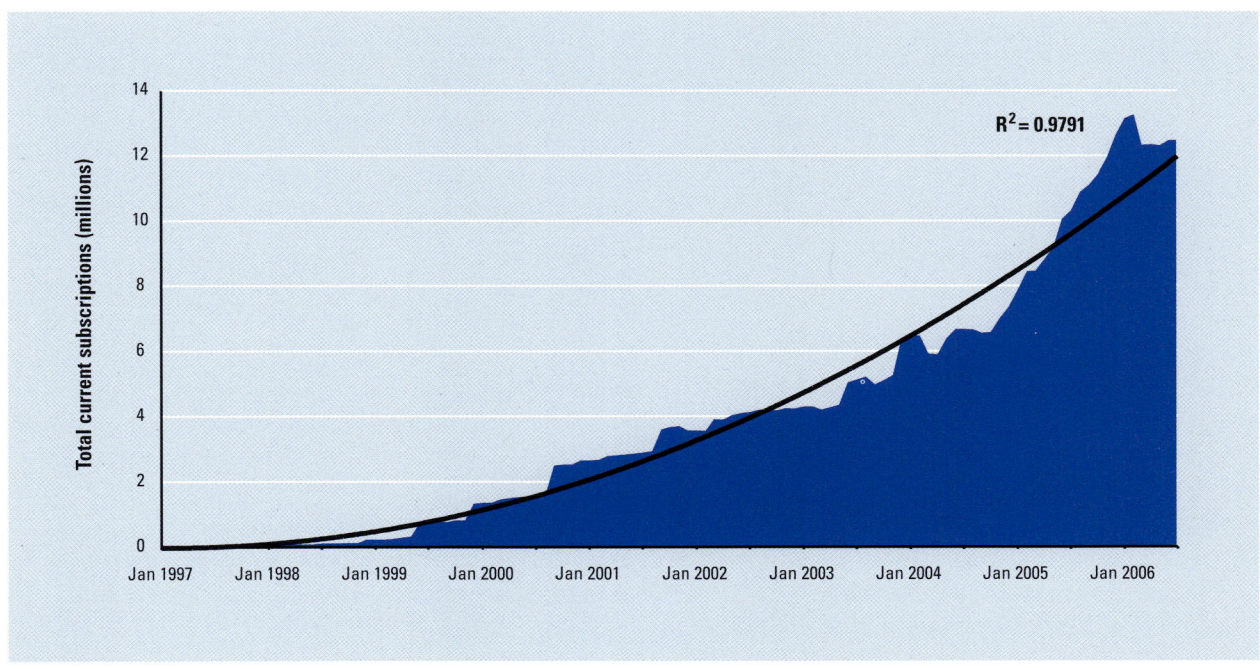

Source: Woodcock, 2006.

Figure 5: Broadbank penetration by technology, top 20 economies worldwide (January 1, 2005)

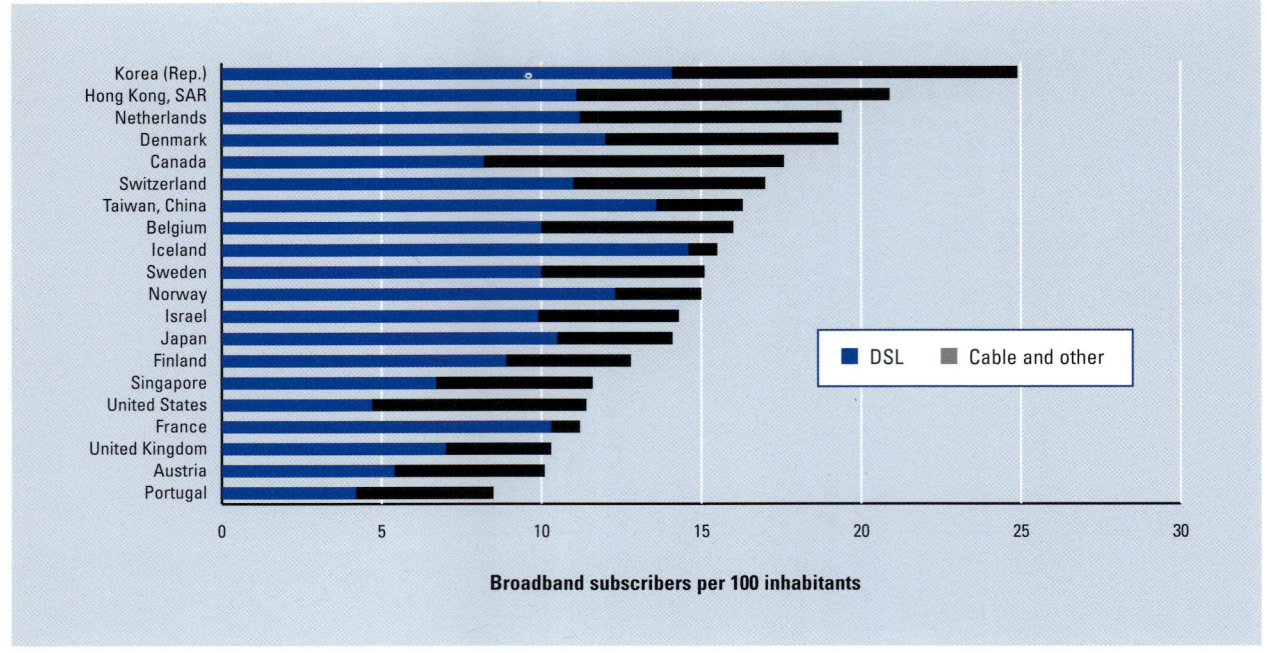

Broadband subscribers per 100 inhabitants

Source: International Telecommunication Union, adapted from national reports
Note: Data exclude mobile cellular broadband (such as 3G).

middle-aged business professionals to save all of humankind in *Halo 2* on Microsoft's Xbox Live.

Also curious is the phenomenon known as "Second Life," where players are represented by avatars (graphical images that represent people) in a virtual world. According to creators, there are more than 70,000 Second Life users, with more than 20 percent of users self-identifying as "residents" of their virtual world. They chat with other residents, form relationships, buy and sell products and, ironically, attend entertainment events in virtual venues.

IP network technology: Let us entertain you

Many of these exciting new developments in entertainment have been made possible by the deployment of new technologies, particularly high-speed Internet access in the home, where penetration levels have only recently made Internet delivery of rich media content practical and affordable (see Figure 5). Increased use of WiFi for the last mile and such high-speed technologies as WiMAX and FTTx will serve only to accelerate this trend.

One area of technology that casts a shadow over the proliferation of high-quality content available through the Internet is digital rights management (DRM). The goal of DRM technology is to protect the intellectual property rights of content creators, preventing original works from being copied and distributed without authorization or compensation. A number of mechanisms have been tried,

and failed, primarily because of the onerous restrictions placed on consumers' ability to use the content they have purchased. Apple seems to have found a workable compromise with a combination of iPod devices, iTunes Store, and FairPlay DRM technology. However, the proprietary nature of this solution precludes it from being a universal DRM solution.

We enjoy new ways to access a far greater range and format of entertainment, thanks to widespread, far-reaching IP networks that connect us physically and wirelessly. Through such technologies, we can enjoy the freedoms of modern society yet still fulfill the basic human desire for the social experience shared by our prehistoric ancestors, around a leaping fire, millennia ago.

Silhouetted by the cool glow of the giant plasma screen, the solitary figure is entranced by the parade of colorful images, his heart quickened to the staccato beat of stereo audio, his imagination captured by bold tales of love, triumph, defeat, and sorrow. He is the modern equivalent of the ancient storyteller.

Networks foster "student-centric" learning environments

For hundreds of years, the traditional schoolhouse has prevailed as a basic model for education. This model now faces disruption by an Internet-based world of high-speed networks, PCs, and collaborative educational programs that connect teachers, students, and content worldwide.

Education for the masses

Traditional education usually meant a sole authority figure presenting a static curriculum to students gathered together in one place. The curriculum typically attempted to meet the general needs of the group, with little emphasis on addressing individual students' needs, abilities, and learning styles. Over the decades, class sizes in public schools have expanded (usually the result of budget limitations), and today often yield student-to-teacher ratios as large as 30 to 1—or higher. These class sizes have further limited the ability to customize lesson plans to meet the diverse needs of individual students.

Today, however, IP networks are helping build virtual schoolhouses that can scale teaching resources and educational content beyond the physical classroom to reach many more students. Perhaps as important, the network's support of multimedia content also drives interactive student participation and exploration, self-learning, and peer education for more customized educational experiences.

The collaborative, network-based model for teaching and learning represents a fundamental shift from teacher-centric to student-centric learning environments that accommodate individual learning styles and circumstances. The reach of the Internet and its rich, multimedia content is liberating students from many of the geographic, demographic, and economic constraints of the past.

In the Netherlands, for example, educational organizations established the Kennisnet Foundation to increase Internet use in education. The foundation connects more than 11,000 schools and other educational institutions, libraries, and museums to Internet-based educational content and services.[23] By combining infrastructure and services with innovative content, Kennisnet Foundation schools attempt to maximize the potential of both students and technology.

Fast, global networks can make educational content and resources available to more people. A network that uses standard protocols and interfaces supports electronic educational curricula and learning tools that can transcend geographic borders, offering the potential of bringing educational parity to the masses. A student in Africa, for example, can access course material from the Massachusetts Institute of Technology as easily as a student on the campus—provided that network access is available to the African student.

Government- and business-sponsored initiatives are playing a critical role in making network access available to developing countries and their lower socioeconomic classes. Technology companies, federal and local governments, and nonprofits have joined forces to deliver lower cost—or even free—Internet access, for example. A number are also donating or subsidizing the cost of PCs for students.

Many of these initiatives target an ideal 1-to-1 PC-to-student ratio, which can be cost-prohibitive for some communities. The good news is that research conducted by the Organisation for Economic Co-operation and Development (OECD) in 2005 indicates that even a 5-to-1 ratio of students-to-device in middle and secondary schools offers very similar educational benefits at a much more feasible cost basis.[24]

The Jordan Education Initiative exemplifies an Internet-enabled joint project in a developing country. Math and English as a Second Language are currently offered over the network to 100 Jordanian schools. A goal is to provide assistance to neighboring countries in setting up similar educational programs. The project—a collaborative effort between the Jordanian government and Cisco Systems, facilitated by the World Economic Forum—is helping Jordan meet its objective of creating a knowledge-based, entrepreneurial economy.

The educational equality among groups and nations that can eventually result from such efforts helps to dissolve traditional socioeconomic barriers to opportunities and improves standards of living. Network-based education holds the potential of bridging the wide economic and social divide that differentiates affluent countries from the developing world.

While the traditional teacher-centric educational setup remains intact in many places—particularly in public elementary schools—introducing the network as a learning platform connects, personalizes, and deepens educational experiences (Figure 6). Networks provide communication; access to extended resources; and the use of sound, graphics, video, text, interactivity, and other digital capabilities to support the customized pairing of an individual student's learning style with the ideal medium.[25]

Examples of network-based customized learning

Project LISTEN at Carnegie Mellon University developed software that functions as an electronic "reading tutor" to children in grades two through four. The program listens to each student's reading and responds. An assessment study in 2004 showed significant gains in reading fluency and timed sight word recognition.[26]

Similarly, St. Lucie County in Florida is piloting a wireless program using small handheld computers. Students in elementary, middle, and high schools record written passages into their devices, which teachers download and grade, creating individualized and confidential instruction.

St. Joseph's Academy, an all-girls' school in Baton Rouge, Louisiana, uses laptops instead of textbooks for classes such as physics. A website containing articles, video clips, and simulations makes the material more realistic—and easier to grasp. St. Joseph's also has a videoconferencing project with a school in Mexico for students in its advanced Spanish classes. The program, a type of "audio/video pen pal," enables English and Spanish students

Figure 6: Technology can enhance learning retention

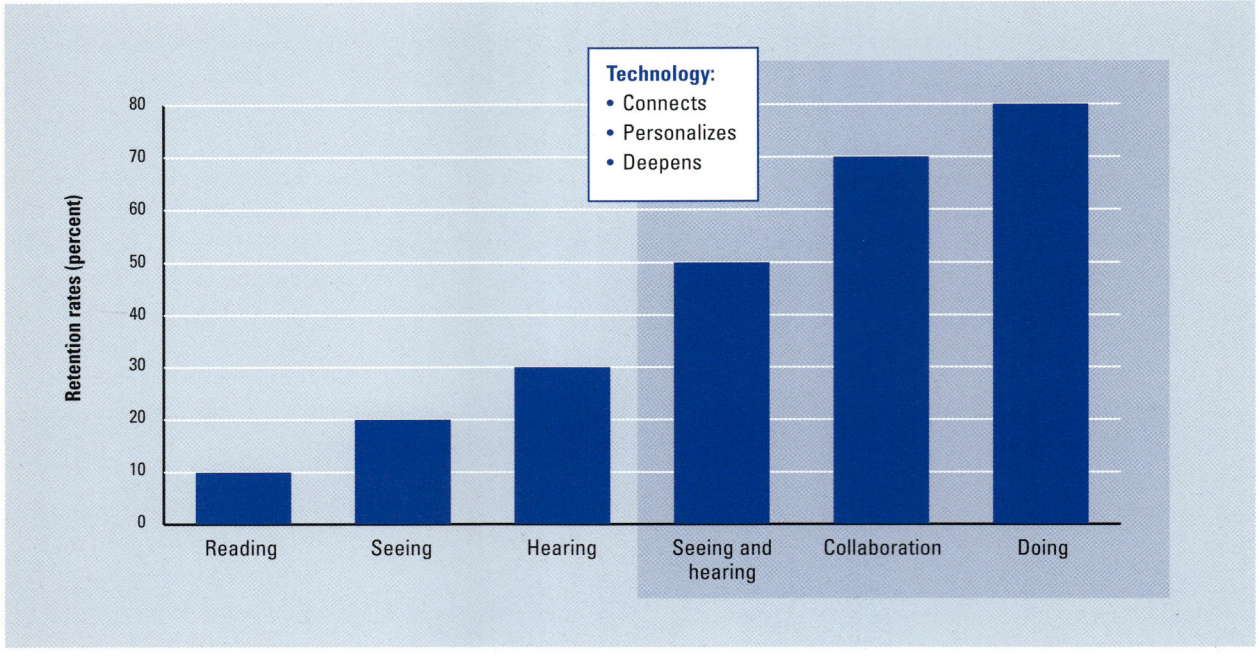

Source: Chi et al., 1989.

34

to practice their second language live by network-based videoconference with a native speaker.

There is a social networking aspect to these network-based exchanges. Not only do students improve their fluency in other languages, they gain valuable cultural perspectives about other countries.

Social networking is a big contributor to education, which is increasingly viewed as a lifelong, ongoing process rather than an activity reserved for certain scheduled hours of the day. Formal schools aren't the center of many learners' worlds; rather connections, relationships, and popular culture often make a bigger impact.[27] IP networks foster these aspects of learning, as students are highly motivated to communicate via technology: text messaging, email, talking, or videoconferencing.[28]

Close to home: Advantages of the new educational model

The emerging educational model involves a world of self-learning and peer-to-peer teaching not possible prior to the emergence of networks. The extended reach of the traditional classroom can be international, regional, or local. Consider the local network impact on:

- Home schooling: According to the US National Center for Education Statistics, the number of home-schooled children has been rising steadily for several years. Home-schooled students and their parents

can now draw upon online curricula and resources as educational tools. Traditional text might be the best fit for one student or a particular subject; simulation or interactivity might work better for others.

- Ill and at-risk students: There are students who must be outside the classroom short- or long-term due to illness or disruptive behavior. Virtual learning from a distance allows these students to continue to participate in class and access assignments and class notes. Research shows that virtual learning—attending class electronically via a network—is generally equivalent to in-person learning in terms of its educational effectiveness, successfully extending the learning experience to those who would otherwise be left out.[29]

Broward County Public School District in Fort Lauderdale, Florida offers a Hospital Homebound Program and Expulsion Abeyance program in which remote students phone in to a class with a teacher and other homebound students. Some classes, such as geometry, combine voice conferencing with Web collaboration tools—students participate by voice while concurrently viewing in their Web browser words or images the teacher draws on a tablet.

Also interesting are Webcast lectures that record educational content, allowing students who miss a key concept or phrase to listen or watch several times to better understand the material. Having access to the recorded material often improves comprehension by repetition, which has been shown to assist in learning and memory.

- Parent involvement: In many communities, parents now have Web-based access to their children's assignments, the status of project completion, grades, and other information. As a result, they remain knowledgeable, involved, and interested in students' homework and subjects.

- Supplemental tutoring: In a creative example of scaling teaching resources, the Maricopa County Community College system in Phoenix, Arizona, has used the network to supplement traditional teaching with senior citizen tutors. In this case, the small efforts of many seniors complement the large efforts of the few teachers.[30]

- Outdated textbooks: Textbooks, particularly in the sciences, can quickly become obsolete. Recently, for example, the International Astronomical Union re-categorized Pluto from planet to dwarf planet. Millions of textbooks will have to be amended, consuming time and financial resources. Even CDs must be re-mastered, copied, and distributed. However, a network-based online learning curriculum can change hundreds of references to Pluto within a matter of hours.

Discovery-based learning

In many parts of the world, the current generation of students has grown up with PCs and the Internet, resulting in a level of comfort with technology that their predecessors didn't have. As a result, formal, authority-based learning is giving way to "discovery" learning that is more action-oriented and personal.

The network empowers students to "learn to learn" through virtual experiences. They witness others performing a task online then try it themselves, rather than waiting to be formally taught. As the need to be physically present has been removed, this self-discovery approach can extend to all kinds of educational activities. For example, students can meet virtually and brainstorm with experts and peers in chat rooms and other virtual communities of interest.

The network also supports educational tools that allow for more customized and discovery-based learning styles:

- White boarding: Placing shared files in an onscreen space that can be virtually marked up over the network—mimicking a chalkboard or white board—

extends learning through interactivity and the use of more engaging materials to explain concepts.

- Portable devices: Laptops, personal digital assistants, and smartphones can be coupled with high-speed WiFi and cellular networks to help extend connectedness and learning beyond regular school hours, making learning an ongoing part of life. There are a number of free, corporate-sponsored WiFi networks in metropolitan areas; in addition, many communities, cities, and individuals around the world have set up free or low-cost, subsidized, wireless access networks.

- Simulations: Research conducted in 2004 by the British Educational Communications and Technology Agency (BECTA) suggests that simulations in the natural sciences result in increased learning and retention.[31] Students explore "what-if" scenarios and build schemas of understanding, which helps them retain the material.

- Videoconferencing: This makes virtual fieldtrips and other experiential learning possible. Bergen County Technical Schools in New Jersey, for example, have conducted virtual fieldtrips by videoconference with Washington, DC's Smithsonian Institute.

The pervasiveness of the network, extended by wireless access links and devices, are helping take education "on location." Wireless tablet PCs are in use in graduate students' field research at Wake Forest University in Winston-Salem, North Carolina, for example, which has created what it calls the "traveling classroom."

Graduate students can record survey results and other data electronically, such as biological observations, while in the field. Wake Forest University has asserted that having a mobile, network-attached device that captures information and thoughts, stores them on a server for calculation, and fosters real-time electronic exchanges with professors has made the final product from students "significantly better" than handwritten notes.

The openness of an IP-based network is making the deployment of such applications easier. Wake Forest University has said that applications are written to work on the end devices, but because of industry-standard IP networking technologies, there's nothing "network-specific" that has to be done.[32]

Learning as large as the world, designed for each student

Online content that has been enhancing the educational experience in schools through PCs and servers is finding its way onto worldwide, standards-based networks, creating student-centric learning experiences. The network-based model represents a shift from a teacher-centered environment that leaves little room for individual learning abilities and styles to deliver multimedia and interactive experiences that best matches each student's learning needs.

The networked virtual classroom has the potential to scale instruction globally and to bring educational parity to many parts of the world. As most experts and teaching resources have been geographically concentrated in developed—and often affluent—parts of the world, such a vision has never before been realized. For the network-based virtual classroom to close the digital divide, network connectivity must also become available in developing areas. The computer and telecommunications industries are joining the effort, with local and federal governments, to deliver these necessities.

Just the beginning: The impact of the network continues to grow

Recently, particularly over the past half decade, a quiet revolution has been taking place. The impact of ICT—once the exclusive provenance of developed nations, wealthy corporations, and the technologically advanced—is spreading worldwide to connect people, places, and things in ways that have never before been imagined, much less attempted.

Like a rising sun that warms the chill morning earth, these technologies have nurtured the growth of communications networks as platforms upon which we can build richer, healthier, and better-balanced life experiences. In the years to come, we can look forward with anticipation to even more practical and satisfying ways the network enhances how people work, live, play, and learn.

Notes

1 See Bryan and Joyce (2005).

2 See Perez (2002).

3 See World Internet Usage and Population Statistics (2007).

4 Friedman 2006.

5 See the Voluntary Interindustry Commerce Solutions Association (2007).

6 See Tohamy (2006).

7 See Bryan and Joyce (2005).

8 See Heffner (2005).

9 See Friedman (2006).

10 See SiteScape (2001).

11 See Bryan and Joyce (2005).

12 See Bryan and Joyce (2005).

13 See Bryan and Joyce (2005).

14 See Bryan and Joyce (2005).

15 See Bryan and Joyce (2005).

16 See Mountford (2006).

17 See Bryan and Joyce (2005).

18 See *eMarketer* (2006).

19 See Barrett (2006).

20 See du Pre Gauntt (2006).

21 See du Pre Gauntt (2006).

22 See Rose and Rosin (2006).

23 See Baldwin (2003).

24 See OECD (2005).

25 See Brown (2000).

26 See Poulsen (2004).

27 See Findlay et al. (2004).

28 See Fadel (2006).

29 See Blomeyer et al. (2004).

30 See Brown (2000).

31 See Abbott et al. (2004).

32 Information provided by J. Dominick, Assistant Vice President for Information Systems, Wake Forest University, September 2006, in a live interview.

References

Abbott, C., T. Beauchamp, B. Blakeley, M. Cox, V. Rhodes, and M. Webb. 2004. "A Review of the Research Literature Relating to ICT and Attainment." BECTA ICT Research. January. Available at www.becta.org.uk/page_documents/research/ict_attainment04.pdf.

Baldwin, H. 2003. "The New Frontier." Cisco iQ magazine September/October: 38–47.

Barrett, J. 2006. "The IPTV Conundrum in Asia." Parks Associates White Paper.

Blomeyer, R., C. Cavanagh, K. J. Gillan, M. Hess, M., and J. Kromrey. 2004. *The Effects of Distance Education on K-12 Student Outcomes: A Meta-Analysis.* Jacksonville, FL: Learning Point Associates.

Brown, J. S. 2000. "Growing Up Digital: How the Web Changes Work, Education, and the Ways People Learn." *Change* magazine March/April: 11–20.

Bryan, L. and C. Joyce. 2005. "The 21st-Century Organization." *The McKinsey Quarterly* 3. Available at www.mckinseyquarterly.com/article_page.aspx?ar=1628&L2=18&L3=30.

Chi, M., M. Bassok, M. Lewis, P. Reimann, and R. Glasser. 1989. "Self-Explanations: How to Study and Use Examples in Problem Solving." *Cognitive Science* 13: 145–82.

du Pre Gauntt, J. 2006. "Mobile TV: Big in Japan (and South Korea and China)." *eMarketer* Research Perspective, September 8.

eMarketer. 2006. "Turning On, Tuning Out." *eMarketer* Research Perspective, March 14.

Fadel, C. 2006. "Technology in Schools: What the Research Says." White paper published by Cisco Systems/Metiri Group. Available at www.cisco.com/web/strategy/docs/education/TechnologyinSchoolsReport.pdf.

Findlay, J., R. N. Fitzgerald, and R. Hobby. 2004. "Learners as Customers." Proceedings of the International Conference on Educational Technology (ICET), Singapore, September 9–10.

Friedman, T. 2006. *The World Is Flat.* New York: Farrar, Straus and Giroux.

Heffner, R., 2005. *Digital Business Architecture: IT Foundation for Business Flexibility.* Forrester Research Report, November 7. Available at www.forrester.com/Research/Document/Excerpt/0,7211,36927,00.html.

Mountford, P. 2006. Emerging Today, Transformed Tomorrow. ExecNet video presentation, November 2006, Cisco. Available at http://news-room.cisco.com/dlls/tln/execnet/.

OECD (Organisation for Economic Co-operation and Development). 2005. *Are Students Ready for a Technology-Rich World? What PISA Studies Tell Us.* Paris: OECD. Available at www.oecd.org/dataoecd/28/4/35995145.pdf.

Perez, C. 2002. *Technology Revolutions and Financial Capital: The Dynamics of Bubbles and Golden Ages.* Cheltenham, UK: Edward Elgar Publishing.

Porter, M. E. 1985. *Competitive Advantage.* Chapter 1: 11–15. New York: The Free Press.

Poulsen, R. 2004. "Tutoring Bilingual Students with an Automated Reading Tutor that Listens: Results of a Two-Month Pilot Study." Unpublished Masters Thesis, Department of Education, DePaul University, Chicago, IL.

Rose, B. and L. Rosin. 2006. *The Infinite Dial: Radio's Digital Platforms.* Arbitron/Edison Media Research study. New York and Somerville, NJ: Arbitron Inc./Edison Media Research. April.

Sharma, S. 2005. "VOIP and Broadband: Technical, Economic, and Market Issues." January 5, 2005. International Telecommunication Union.

SiteScape. 2001. "Collaboration + Workflow = e-Process." SiteScape white paper.

Tohamy, N., 2006. *The State of Global Supply Chain Management.* Forrester Research Report, April 11. Available at www.forrester.com/Research/Document/Excerpt/0,7211,39230,00.html.

Voluntary Interindustry Commerce Solutions Association. 2007. Collaborative Planning, Forecasting & Replenishment Committee webpage. Available at www.vics.com/committees/cpfr.

Woodcock, B. S. 2006. "An Analysis of MMOG Subscription Growth – Version 21.0." Available at www.mmogchart.com.

World Internet Usage and Population Statistics. 2007. January 11, Miniwatts Marketing Group. Available at www.internetworldstats.com.

CHAPTER 1.3

Opportunities and Challenges of Next-Generation Networks in Telecommunications

SCOTT BEARDSLEY, McKinsey & Company Inc., Belgium

LUIS ENRIQUEZ, McKinsey & Company Inc., Belgium

MEHMET GÜVENDI, McKinsey & Company Inc., Turkey

DUARTE BRAGA, McKinsey & Company Inc., Portugal

WIM TORFS, McKinsey & Company Inc., Belgium

SERGIO SANDOVAL, McKinsey & Company Inc., Belgium

What has been building for several years is coming to pass: major telecommunications operators are finally rolling out their next-generation networks (NGNs).

For operators, these all–Internet protocol (IP) networks will bring several benefits. First NGNs will reduce the operational costs of multiple services by an estimated 35 percent through their simpler architectures and economies of scale. Second, they will speed the development of new services, including video telephony, white boarding, and multimedia conferencing with file and application sharing. Third, they will help protect operators' businesses from cable and other service providers such as MSN, Skype, Google, and Yahoo!

But for all these advantages, building the NGNs will require huge investments in infrastructure. These will create significant financial risks for their operators. Moreover, NGNs will force changes to the economics of the telecommunications industry through changes in traditional pricing models. These changes could find previous cash cows, such as voice service, drying up.

On top of these challenges, however, is an overriding concern: are the regulatory regimes of most nations ready? Are the regulators looking back to the issues of the 20th century, or are they looking forward to the new era—one that demands a clear head to determine how to regulate this unprecedented convergence of voice, television, and mobile communications—so that all will benefit? Getting regulation wrong can put up to 45 percent of an operator's four-year EBITDA (earnings before interest, taxes, depreciation, and amortization) at risk.[1] Such economic disincentives could limit the full deployment of the NGNs and reduce their chances for success.

What, then, are the costs and benefits of the NGNs? What are the regulatory battles surrounding them? And what actions are required—by both the industry and its stakeholders—to face these critical challenges?

In the discussion that follows, we will explore these issues. But one point is already clear to us: both the policy makers and the industry players must work together. They must manage this transition to NGNs wisely—so that the investors are encouraged to invest, and so that the stakeholders can capture the full benefits of this new era.

Social benefits and new capabilities of next-generation networks

Why are NGNs being built? Because regardless of the size of the investment, the expected benefits are great. These benefits, as we noted earlier, include substantial operational cost savings and a greater ability to deliver high-end multimedia services such as Internet protocol television (IPTV) to consumers.

Although the acronym *NGN* is a simple label, it signifies different levels of service and network upgrades

Figure 1: Next-generation network investments: Three types of upgrades

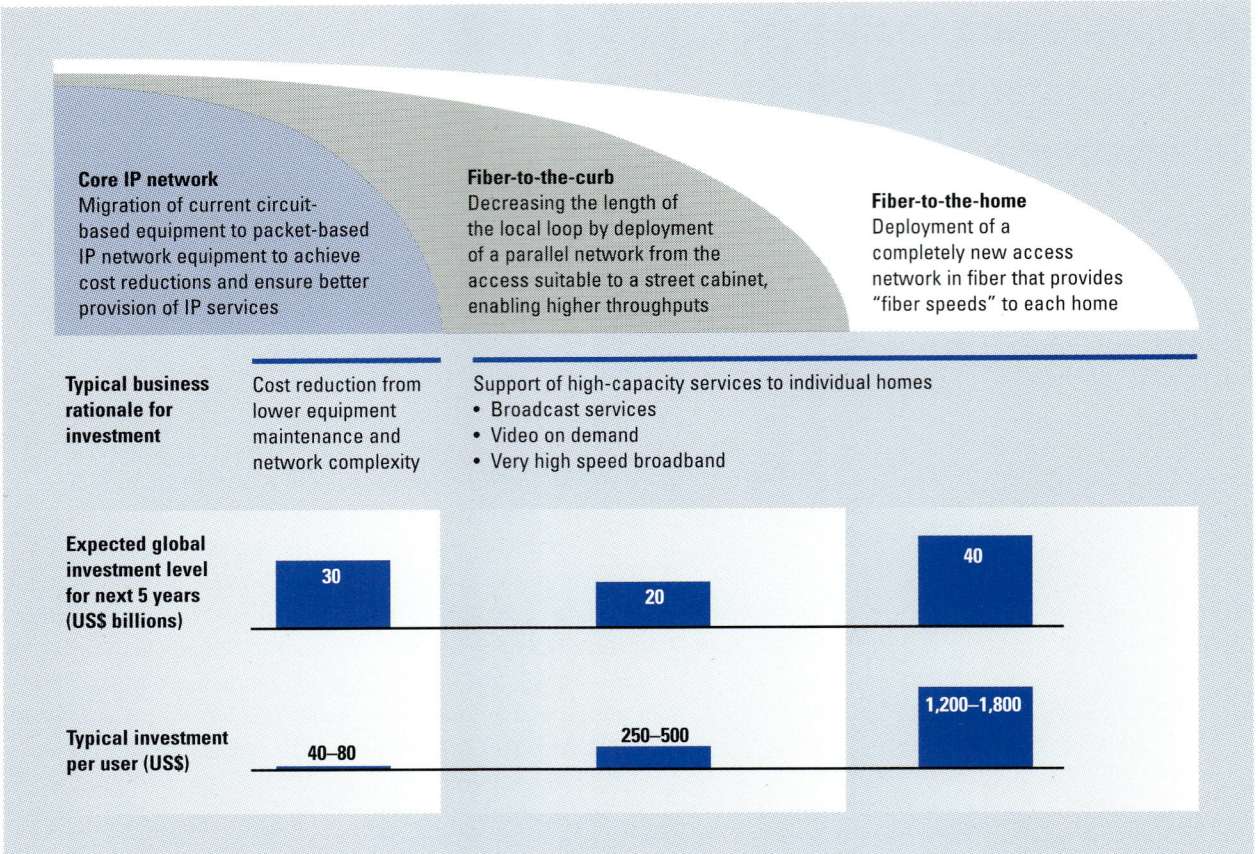

Source: IDATE, 2005a; OECD, 2005; McKinsey analysis.

to different operators (as shown in Figure 1). NGN upgrades range from the core IP network level (in which operators and manufacturers are pushing for all-IP backbone transport and low-cost distribution solutions) to the local loop (in which fiber-to-the-home [FTTH] solutions can delivery seemingly unlimited bandwidth). The different "flavors" of NGN offer different customer benefits—and imply different costs.

Core network upgrades: Simplification and lower cost

Most major telecommunications operators seeking the new cost curve promised by equipment vendors are either seriously considering or actively deploying new core networks. British Telecom (BT) became a pioneer in this area when it announced a multibillion dollar investment in its 21st Century Network "21CN," and most other operators followed suit. Once its new network is in place in 2009, BT says it expects its operating expenses to fall by 30 percent.[2]

Core IP networks eliminate the multiple networks, layers, and protocols that currently plague the average incumbent. For the industry, this is the long-searched-for "network grail" that asynchronous transfer mode (ATM) technology promised but failed to deliver. For this we can thank the switches that optimize bandwidth and support IP directly over optical fiber, thereby ensuring the enormous flexibility of the IP protocol. The benefits? Less network complexity, reduced needs for specialized engineering and maintenance teams, and improved network reliability. If any more incentive is needed, it comes from the many vendors who have announced that they will progressively discontinue traditional network equipment (along with an affordable investment of around $40 to $80 per user). Thus this transition is a near-certainty, in the near term, for all operators.

During roll-out, operators will compete aggressively on price while rapidly deploying the more "intelligent" of the IP-based services such as TV multicast or network-based personal video recording. This will increase their ability to compete with wireless and cable players.

New higher-speed services through optical fiber solutions

In terms of multiple bandwidth–hungry services to the home, many operators are betting on building optical fiber solutions in the local loop. Fiber, based on traditional copper loops, enables much faster and more reliable speed than current digital subscriber line (xDSL) technologies. The hope of operators deploying this type of solutions is that the high-quality, triple-play services that these investments will deliver will turn the competitive game in their favor.

The first type of solution is fiber-to-the-curb (FTTC) or fiber-to-the-node (FTTN). It consists of deploying fiber in only *a section* of the last mile to deliver faster speeds than other xDSL technologies. Operators building FTTC, such as Deutsche Telekom and South Bell Canada (SBC), are balancing less available bandwidth (and related services) with significant lower investment per user of about $250 to $500. FTTC offers operators considerable savings because it avoids investments in the "last section of the last mile," a distance that is always the most expensive section to connect with fiber.

The second type of fiber solution being deployed is fiber-to-the-home. This solution connects the client and the access switch through a dedicated fiber cable. This enables consumers to enjoy almost unlimited bandwidth, while operators get the lower operating costs associated with its reliable, noise-free technology. The downside is in the investment, however: a whopping $1,200 to $1,800 per client for full FTTH solutions (such as those deployed by Verizon and Nippon Telegraph and Telephone Communications Corporation, or NTT).

What is the business case, then, for optical fiber architectures (FTTx)? Against the growing threat of wireline attackers and increasingly aggressive cable and mobile players, this technology allows operators to retain their large client base. It also allows providers to up-sell the new value-added services provided by these platforms to their customers. However, this benefit is not obvious to most operators; for this reason, optical fiber deployment is still on the drawing board for the majority of them. Verizon's recent announcement of its commitment to FTTH, and the subsequent battering of its shares, is a cautionary tale. Currently, the announced worldwide commitments to FTTx solutions are still at around US$60 billion, only double the amount announced for Core IP Networks despite the 50-fold difference in investment per client.

Huge attention on next-generation networks at a global level, but largest amounts of capital expenditures continue to be day-to-day

The Economist featured a cartoon recently in which a saloon customer is sitting at a bar with bottles labeled "Broad Band," "Telephony," and "TV" before him. The bartender on the other side of the counter, however, is busily shaking a tumbler with all those tasty ingredients mixed within: it's labeled "Info Heaven."

In another *Economist* cartoon, readers see a group of masked men labeled "AT&T," "Verizon," and "other telecom operators," punching and gouging at each other's eyes. The caption reads: "LIVE: All-In Convergence Wrestling."

A third cartoon might have established the final frame of the story: government regulators, wearing the striped uniforms of referees, trying to enforce some rules that will keep everyone on the straight and narrow.

These cartoons capture the essence of today's NGN issues: great opportunity—but also great regulatory and business battles over the conditions in which industry investments will take place. Many telecommunications firms want to capture the benefits from the different types of NGN deployments, which is why they are planning to invest heavily on NGNs. In the United States, Verizon is deploying a network in parts of 18 states (about 400 communities). By 2009, they hope to provide fiber connection to 18 million homes with speeds of 100 megabits. To reach that target, Verizon said it will spend $22 billion.[3]

In Europe, Deutsche Telekom, BT, and KPN have committed $3.8 billion, $19 billion, and $1.9 billion respectively to build networks.[4] Central European operators in Slovakia, Bulgaria, Hungary, Croatia, and the Czech Republic have also announced investments in NGNs.

But investment is not the only news. There is news also in the regulatory battles over the conditions of those and other investments. Telstra, the Australian incumbent operator, for instance, has been seeking a regulatory break for its planned NGN investments. In early 2005, the company warned that state regulators were stepping beyond their authority. "Our investment in a decade-long build-out of a fiber optic network is an issue of national policy first, and accompanying legislation perhaps, and this is the difficulty I have with the regulator," said Telstra's (former) Chief Executive Officer, Ziggy Switkowski. "The regulator is there to ensure compliance with the rules; I don't really look to the regulator to help architect an industry." He added that Telstra's decision to invest in a high-capacity network, that might cost many billions of dollars and would dramatically increase broadband speeds to 4 million homes in the country, is an important decision, one "where we would expect to make above-average returns for making the investment and taking the risk. And the regulator's role, in all that, is secondary."[5]

But Helen Coonan, Australia's minister of communications, disagreed. Referring to Telstra's plan to increase earning margins to 52 percent over five years, she remarked, "With those kinds of margins, Telstra doesn't need regulatory breaks. They should get on with it, and let's see what they can do to turn the company around."[6]

Figure 2: Telecommunications industry as a global investment driver

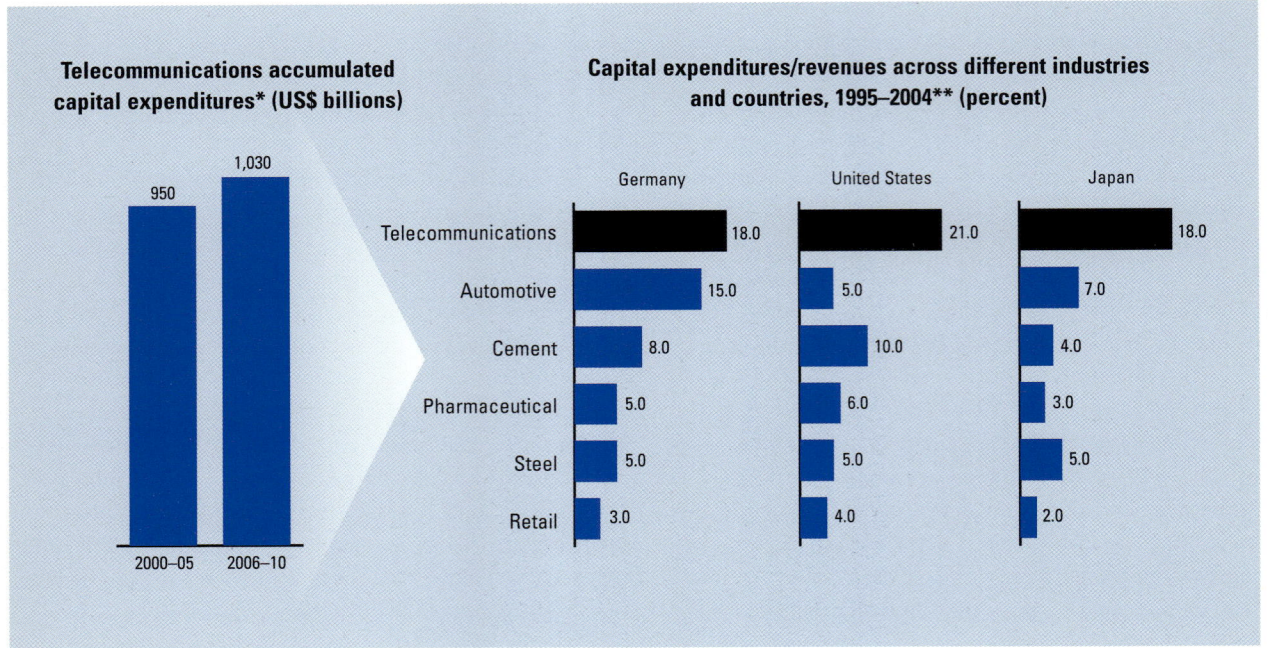

Source: IDATE, 2005a, 2005b; OECD, 2005; Bloomberg online database; McKinsey analysis.

* Includes enterprise networking equipment investments
** Weighted average for listed companies

In Germany, meanwhile, the government offered a draft of its new telecommunications law that would postpone the policing of Deutsche Telekom's $3.8 billion high-speed network by two to five years. That raised opposition from Viviane Reding, the European Union commissioner for telecommunications. It is "against the European interest and the German interest" she argued recently,[7] adding that the "EU rules do deliberately not provide for regulatory holidays, precisely in order to prevent a re-monopolization of markets."[8] In particular, she said, in network-based economies "effective competition does not prevent, but drives investment."[9]

Indeed, the NGNs have brought a new paradigm to the industry. And with it, operators and regulators alike must grope approach that will deliver the right level of investment and industry competition.

The potential impact of the next-generation network debate on capital expenditure investments

The prominence of these and other regulatory battles has obscured an important fact: overall capital expenditures investment in the telecommunications industry today is significantly higher than NGN investments alone.

On a global level, in fact, the telecommunications industry is one of the most important drivers of all industry investments. Over the next five years, their capital expenditures will reach the US$1 trillion mark (Figure 2).

When compared to the automotive, pharmaceutical, steel, cement, retail, and other investment-heavy industries, telecommunications outstrip them all in terms of capital expenditures–to-revenue ratios. This can be explained by the speed of technological change in the telecommunications industry, which compels operators to investment heavily in order to remain both competitive and innovative.

As shown in Figure 3, the main components of infrastructure investment in the telecommunications industry are mobile infrastructure, operations support systems, access networks, and switching and routing elements, which together account for 80 percent of total spending. The composition of these investments is not expected to change greatly in the next five years with the exception, perhaps, of mobile infrastructure spending.

Figure 3 also puts NGN investment in perspective within the industry as a whole. Today NGN investments account for about 11 percent of total industry investments. This is expected to grow to approximately 16 percent by 2010. NGN investment is therefore only a small fraction of all telecommunications investments and will remain so for the foreseeable future.

The industry needs to consider the relative size of these NGN investments and put them into perspective in view of the level of media attention this topic has recently received. Without such perspective, the increased interest in NGNs and the uncertainty around it may not affect

Figure 3: Infrastructure investment in the telecommunications industry, 2005–10 (US$ billions, percent)

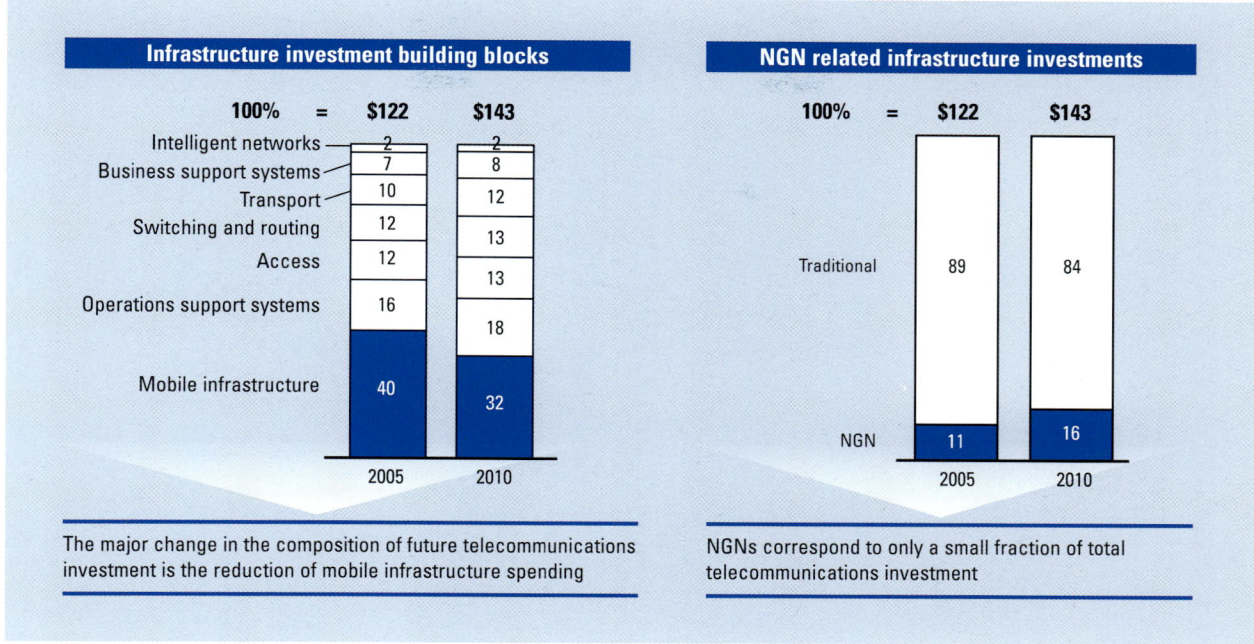

Source: Data from Gartner, 2006b; McKinsey analysis.

only the level of NGN investment but also, and more broadly, overall investments in the industry. This offers a warning to the industry and policymakers alike: there is a lot at stake, and the success, or failure, of NGNs in the next few years could have wide repercussions for the industry.

Next-generation networks: Effects on pricing and revenue models

Operators have traditionally relied on the healthy margins generated from fixed voice telephony. These have enabled the enormous investments in network capital expenditures that we find today. Moving to NGNs could drastically change the current voice-pricing model for both retail and wholesale services, potentially cannibalizing the main revenue source of most operators.

Playing a key role in this shift in technology and pricing will be the world's telecommunications regulatory authorities. Their decisions will not only shape the future returns of new investments, but will largely determine the adoption or not of NGNs.

New pricing structure for voice services

Voice telephony on the fixed network has always been characterized by high margins. These margins have been derived mostly from voice traffic: the per-minute charges that consumers pay for making a call. The monthly rental fees for telephone services, on the other hand, have

experienced low and even negative margins (even after numerous countries have repeatedly rebalanced the tariffs).

The high margins on voice calls have largely covered more than the high investment costs of rolling out a telecommunications backbone. They have also covered the losses of access to the network and the costs of Universal Service Obligations. Moreover, these margins have allowed incumbents to invest heavily in upgrades to the existing networks (as well as in such new network infrastructures as mobile networks, broadband infrastructure, television services, and so on). High returns, together with the opportunities brought by the liberalization of the telecommunications markets, are responsible for the vibrant telecommunications industry of today.

Going forward, however, NGNs will significantly change the way voice is priced to consumers and other operators. Historically, voice calls have been priced per minute. Moreover, the longer the distance between the two calling parties, the more expensive the call.

With NGNs, time-based billing will become irrelevant (in a world where costs are driven by bandwidth usage). Also, the distance part of pricing will become obsolete since IP packages do not follow a fixed route. Instead, voice calls will be made over the IP data network. They will share the "pipe" with several bandwidth-intensive services (such as digital television and broadband services), and will utilize only a small fraction of the network. But

Figure 4: Telecommunications industry fixed revenues, 2005–10 (US$ billions, percent)

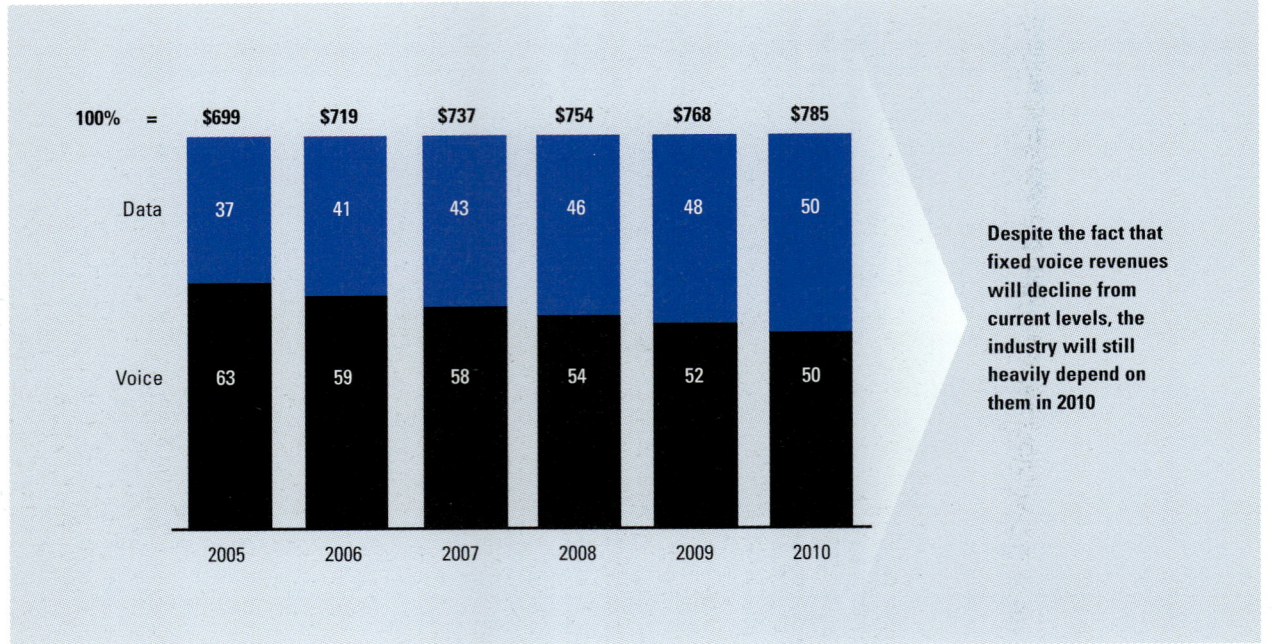

Source: IDC, 3Q-2006; Gartner, 2006a; Yankee Group, 2006a, 2006b; McKinsey analysis.

considering that, on average, more than 60 percent of an operator's revenues comes from voice, the future prof-itability of the industry will rest on the pricing for voice (Figure 4). Although projections continue to show voice as an important contributor to revenues, its rate of decline will depend heavily on the impact of new pricing models on voice revenues.

Passing on price reductions: Limited incentives for network operators

As we said earlier, if operators price their retail voice services according to a bandwidth-based, non–distance-related formula, the price for a voice call could collapse, leading to a cannibalization of the bulk of the fixed operator's revenues. This would destroy the incentive for deploying NGNs unless alternative revenue sources were identified, which is not currently the case.

Could telecommunications firms actually destroy their own best source of profits? What makes this unlikely is that incumbents control their own retail prices, and therefore would be careful not to cannibalize their voice revenues (as long as competition does not force prices down). Rather, they would most likely price their NGN voice services at a healthy premium. Most of their competitors would like-wise not be eager to instigate a downward price spiral. The danger, as we will explain below, could come from asset-light competitors—those reselling wholesale minutes.

Although incumbent operators control retail prices, they usually do not control wholesale prices (for example, voice interconnection rates, unbundling rates, and so on). Rather, in almost all liberalized countries, wholesale rates have been heavily regulated, either through telecommunications-specific regulation or through the intervention of general competition. These costs are usually set on the basis of the theoretical costs of an "efficient" operator (long-run incremental costing, or LRIC).

Offering wholesale services (priced per bandwidth, non–distance dependant) to competing operators (based on NGN costs), however, will allow asset-light voice providers to drastically undercut the incumbent's voice prices. Why? Because the actual wholesale fee paid to the incumbent still constitutes a large portion of their cost base. These price decreases would result either in large market share losses for the incumbent or an overall reduction of price levels in the market (or both).

At the same time, until a full switchover to NGN is completed, incumbent operators will have to continue to support their old copper networks. During this transition —which may take several years—the incumbents will have to cover both their NGN costs and some of their old public switched telephone network (PSTN) costs. Unless the network operators can quickly find alternative, high-margin revenue streams, this could significantly affect their profitability and potential for financing.

Figure 5: Value at stake for incumbents from NGN transition (percent)

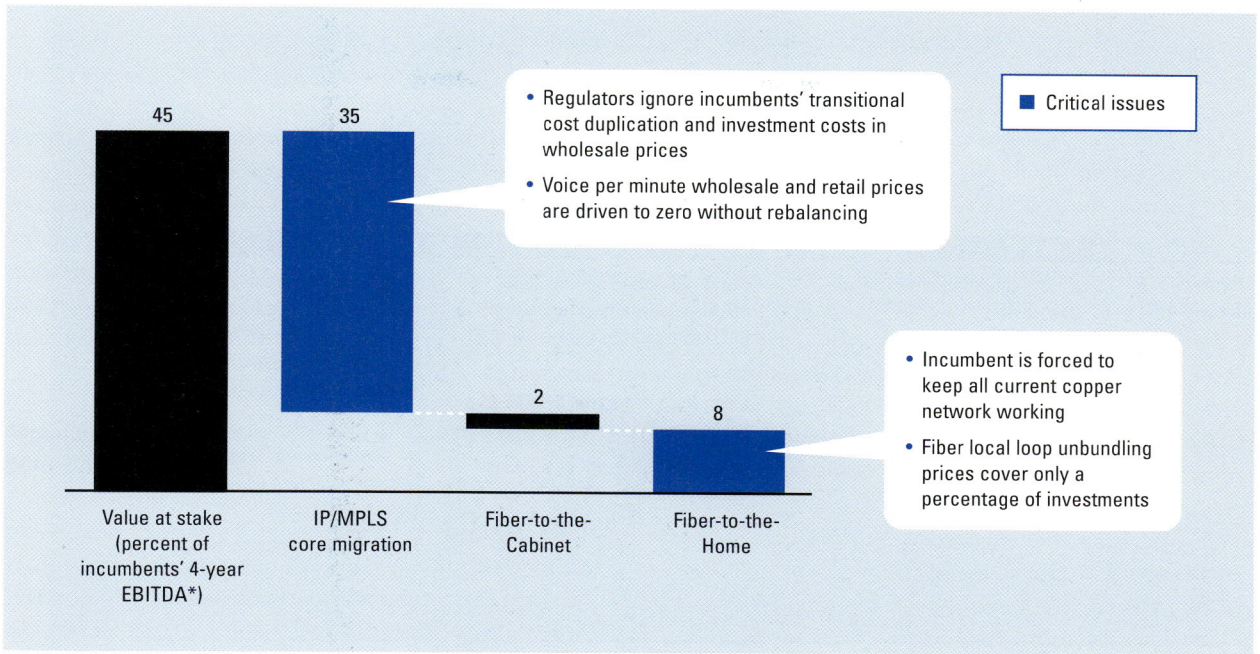

Source: McKinsey analysis.
* Constant prices; 2006 estimated EBITDA

In the absence of new revenue streams (such as increased access fees or new services), McKinsey estimates that declining retail prices, coupled with a period of transitional cost duplication, could reduce incumbents' four-year EBITDA by up to 45 percent (Figure 5).

Revenue declines will jeopardize broader industry investments

Network operators worldwide are assessing the feasibility of NGNs. As discussed above, this presents a double-edged sword. On the one hand, operators are tempted to deploy NGNs in order to offer new services and further reduce costs. On the other, the possibility of a loss in the incumbents' main revenue source (voice) creates a reluctance to invest heavily in the new technologies.

The fall in voice revenues that NGN wholesale pricing could bring, in fact, could seriously affect the future of telecommunications investment both for voice services and broader infrastructure investments. Over the next four years, the amount of investment under threat could be as high as $100 billion.[10] Without the steady revenue streams from the voice business, much of the past investment in broadband infrastructure, digital broadcasting services, and even mobile telephony might never have been made.

The effect of NGN on voice revenues, then, casts a big question mark over the future of the industry: will NGN allow the current levels of telecommunications

investment to continue? A reduction in those investments could reduce new benefits for consumers and innovation in the sector as a whole. Funding for NGNs could decline —not only on the part of telecommunications operators, but also on the part of equipment manufacturers, content providers, and related service providers as well.

Current regulatory regimes: Issues of NGN deployment

For all theses reasons, NGN regulation is one of the hottest issues discussed in industry and policy circles. And it should be—considering the size of the investments and the impact of the new pricing models on the industry. This issue will not only shape the revenue streams of the operators, but it will also affect the operating models and the timetable for the roll-out of services.

What is at the heart of the issue? Essentially, the question is whether a mere extension of the current regulatory framework (that of conventional PSTN networks) to the NGNs will allow the NGNs to thrive.

The conventional regulatory framework focuses on cost orientation, price controls, and service quality—with the ultimate objective of creating a functioning and sustainable competitive environment (often in the form of a "service-based" competition model). This framework, however, has evolved in the context of a widely deployed infrastructure—one that already exists and whose costs have been largely recovered. The challenge of NGNs is

demonstrate a willingness to tolerate temporary but meaningful sources of competitive advantage as part of the normal functioning of a market ("When in doubt, do not regulate.")

Can the industry lead a strategic dialogue with policymakers?

For industry players, the challenge is twofold: first, they must manage the strategic and tactical issues they face from NGNs. These encourage the long-term deployment of the NGN, but at a cost to their existing revenue models. Second, they must convince policymakers to rethink their approach to the regulation of the sector. This is essential to ensure returns to the infrastructure that will enable the transition to the IP-based revenue models. This will not be easy, as the entire regulatory toolkit and the policy debate has centered on how to "open up" the new networks (with some exceptions, such as the regulatory model in place in the United States).

In this context, industry players must:

- **Develop a clear economic and strategic plan around the deployment of NGNs, including the management of the transition process.** This plan would include not only a business plan that maps out the economics of deployment, but also a strategy for migrating users of existing networks to NGNs. It should also consider how to price wholesale and retail products during the transition for both traditional and NGN products.

- **Engage proactively in discussions with policymakers and other stakeholders on the long-term evolution of the industry, including the changes in market structure.** This would entail engaging in meaningful discussions around fact-based scenario analyses that could enlighten the future development of the industry. Understanding the potential impact of today's decisions in tomorrow's market structure is crucial to communicate clearly what is really at stake from the new NGN paradigm.

Conclusion

For the last 20 years and more, regulation and deregulation have transformed the telecommunications industry, making it one of the most vital markets in the world. But now the industry is at the cutting edge of technologies and opportunities that are still beyond our ability to imagine fully.

It is at this time that regulators, policymakers, and the industry must find the wisdom to think beyond what we can even see. They must set rules that will encourage the funding of this vastly expensive infrastructure—and give it time to take wing. Yet they must also find the appropriate

level of competition, so that it will remain vital and sensitive to the needs of society and consumers. That is the challenge ahead.

Notes

1 This assessment of risk based on an estimate by McKinsey.

2 See The Economist (2006b).

3 See Reading (2006a).

4 See Deutsche Telekom (2005); British Telecom (2005); KPN (2005, p. 61).

5 *Australian Financial Review* 2005a, p. 11.

6 *Australian Financial Review* 2005b, p. 3.

7 Reuters 2006.

8 Reading 2006b.

9 *Financial Times* 2006, p. 15.

10 See Gartner (2006b).

References

Allen, J. 2006. "The Death of Price per Minute for Fixed Interconnect?" Available at www.analysys.com/default_acl.asp?mode=article&iLeftArticle=2238.

Australian Financial Review. 2005a. "Switkowski Warns Against Break-up at Any Cost." March 12: 11.

———. 2005b. "Stop Trying to Alter the Rules, Telstra Told." November 18: 3.

Beardsley, S., I. Beyer von Morgenstern, L. Enriquez, and C. Kipping. 2002. "Telecommunications Sector Reform: A Prerequisite for Networked Readiness." *The Global Information Technology Report 2001–2002: Readiness for the Networked World*. New York: Oxford University Press.

British Telecom. 2005. "BT Announces Preferred Suppliers for 21st Century Network Programme." April 28. Press release. Available at www.btplc.com/News/Articles/ShowArticle.cfm?ArticleID=c15461a1-6d24-448b-8e5d-d6c967ba720c).

Deutsche Telekom. 2005. "Deutsche Telekom Launches High-Speed Broadband Network." September 1. Press release. Available at www.telekom3.de/en-p/medi/2-pr/2005/09-s/050901-ifa-trade-fair-ar.html.

The Economist. 2006a. "A Survey of Telecoms Convergence." October 12. Available at www.economist.com/surveys/displaystory.cfm?story_id=E1_SJJVPTN.

———. 2006b. "Your Television Is Ringing." October 12th, 2006 Available at www.economist.com/surveys/displaystory.cfm?story_id=E1_SJJVPTN.

Financial Times. 2006. "All Eyes on D-Tel's VDSL Fight. It's a Regulatory Tussle with Europe-wide Implications." August 21: 15.

Gartner Inc. 2006a. July 2006. *Forecast: Fixed Public Network Services, Worldwide, 2004–2010*. Report No. G00142658.

———. 2006b. August 2006. *Forecast: Global Telecommunications Market Take, July 2006*. Report No. G00142027.

IDATE. 2005a. *World Telecom Equipment Market 2005*. December.

———. 2005b. *World Telecom Services Market 2005*. December.

IDC. 3Q-2006. *Worldwide Telecom Spending 2006–2010 Forecast*. Report No. IDC1408547.

Kirkman, G., P. Cornelius, J. Sachs, and K. Schwab, eds., *The Global Information Technology Report 2001–2002: Readiness for the Networked World*. New York: Oxford University Press.

KPN (Koninklijke KPN N.V.). 2006. *Annual Report 2005*. The Hague: KPN.

OECD (Organisation for Economic Co-operation and Development). 2005. *Communications Outlook 2005*. Paris: OECD.

OFCOM (United Kingdom Office of Communications). 2006. *Regulatory Challenges Posed by Next Generation Access Networks*. November. OFCOM.

Reading, V. 2006a. "Building Tomorrow's Networks: Defining the Long Term Policy." Paper presented at the European Telecommunications Network Operators (ETNO) Association 4th Annual Conference, Brussels, November 7.

———. 2006b. "From Service Competition to Infrastructure Competition: The Policy Options Now on the Table." Paper presented at the European Competitive Telecommunications Association (ECTA) conference, Brussels, November 16.

Reuters. 2006. "EU still Wants D.Telekom to Open Broadband Network." September 28. Helsinki.

Yankee Group. March 2006a. *Global Consumer Fixed-Line Forecast*. Report, Boyd Peterson.

Yankee Group. March 2006b. *Global Business Fixed-Line Forecast, March 2006*. Report, Camille Mendler.

The Next Frontier of E-Government: Local Governments May Hold the Keys to Global Competition

BRUNO LANVIN, World Bank

ANAT LEWIN, World Bank

In 1950, one-third of the world population lived in cities. Half a century later, the proportion had increased to one-half, and it is estimated that, by 2050, six billion people (that is, two-thirds of the world population) will live in cities. Currently the urban population of developing countries is projected to double in 30 years, increasing from 2 billion in 2000 to 4 billion in 2030. In less than 10 years from now, most of the "mega-cities" emerging from that process will be located in developing countries (see Figure 1).

Such projections obviously raise questions about the ability of the cities of the future to sustain this type of growth while maintaining adequate levels of production and delivery of key public services such as water, transport, electricity, sanitation, education, and containment of crime and pollution. There is, however, another side to this equation, often overlooked. It relates to the emerging role of cities (and of subnational entities generally) to become global players—as attractors of foreign investment, competitiveness hubs, and/or platforms for the combination of local and international components of global production and supply chains.

At the same time, more and more governments around the world are seizing opportunities to move to "e-government" as a way of enhancing the effectiveness and efficiency of their national public sectors, in particular through outsourcing the production and delivery of public services to the private sector. This trend compounds another one, by which central governments have been delegating an increasing number of their traditional responsibilities to subnational entities such as states, regions, municipalities, or cities. Many phrases and philosophies have been coined and formulated to describe or justify such a process, including *new federalism* in the United States, *de-centralization* and *de-concentration* in many European countries, and even *subsidiarity* in the EU context.

We are hence witnessing the rapid convergence and combination of three trends: (1) the growth of the size and economic weight of local entities such as cities; (2) the increasing ability and will of governments to use information technologies and outsourcing to fulfill their tasks and serve their citizens better through e-government; and (3) the growing potential (and obligation) of local entities (typically cities) to act as global players, designing and implementing their own policies and strategies to attract investment and carving out their share of benefits from the emerging global economy.

51

Figure 1: Projected population size of mega-cities in 2015

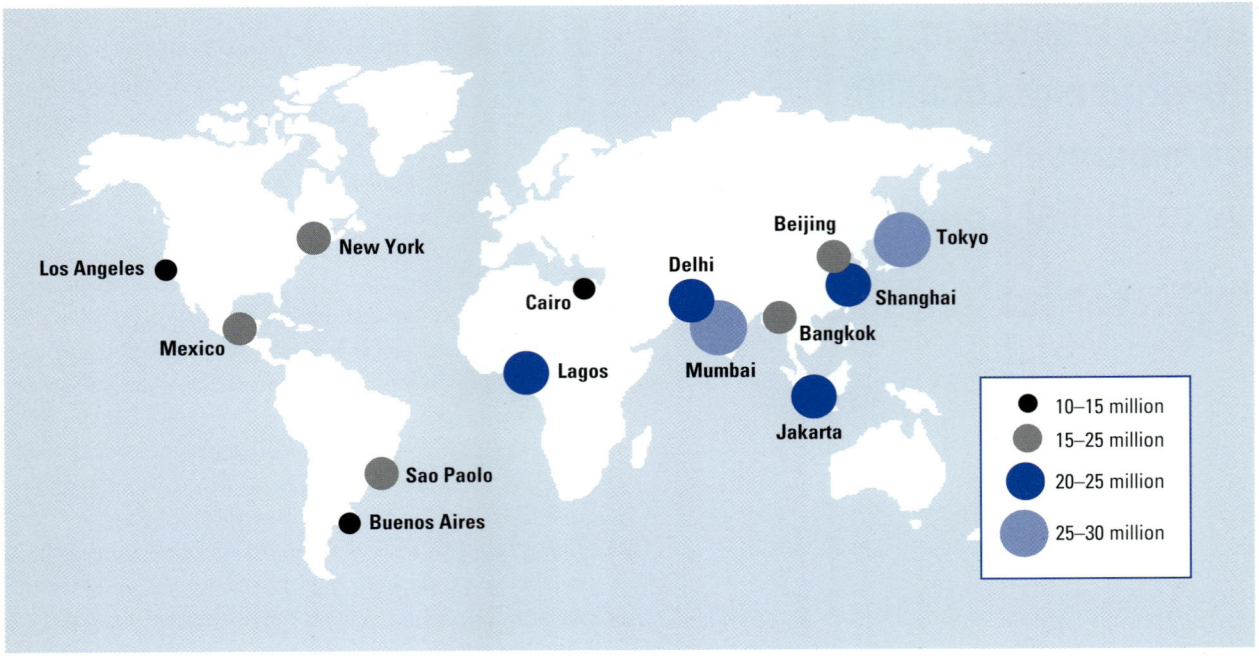

Source: UN Habitat and authors' calculations.

The following sections will consider the convergence of these three trends from the point of view of policymakers (local and central) by addressing the following issues:

- Do we see the emergence of "local global players" (LGPs), and, if so, what are their characteristics?

- What are the analytical tools available to measure the relative performance of local entities (typically cities) compared with the performance of nations when it comes to networked readiness and e-government?

- What do these entities allow (or not allow) us to identify as worldwide best practices in deploying e-government centrally and locally? What additional tools (indicators, data) are required to address this issue in an action-oriented fashion?

The emergence of local global players

Both economic and urban literatures have long ago identified cities as key players in global competition, and even as central engines in shaping and spreading globalization itself. Phrases such as *global cities*,[1] *world cities*,[2] or *networked cities*[3] have been coined in the process.

In this context, the importance of telecommunications (and e-readiness in general) is certainly not new. Already in 2001, Townsend noted that telecommunications networks had been "an essential component of urban infrastructure in the 1990s, enabling the coordination of increasingly complex, multilocation, and time-sensitive production systems as well as fractured social networks."

Nowadays, a growing number of local governments are emerging as LGPs, competing for international markets and investments. Newspapers are replete with advertisements aimed at attracting companies and talented individuals to "knowledge hubs" around the world. Regularly, international magazines publish rankings of cities worldwide, according to cost of living and quality of life. Sometimes called *e-cities*, *Internet cities*, or *Knowledge Cities*, new "e-ready" hubs seem to spring up around the world.

Successful LGPs (such as Singapore, for example, or Andhra Pradesh in India) have combined superior levels of connectivity, a capable pool of human resources, and an innovative private sector. All of these can be furthered by local government policy; however, the quality and efficiency of local efforts and governance that are key determinants of the success and the competitiveness of "local global hubs" are less often noticed or quantified. Local e-government is emerging today as a powerful tool by which such LGPs have enhanced and will continue to enhance their own competitiveness and that of their respective countries (see Box 1 for definitions of *local government* and *e-government*).

In many parts of the world, building and promoting local champions of e-readiness is perceived as a national priority by central governments. In countries as diverse as Tunisia, Morocco, Senegal, Russia, the United Arab Emirates, Mexico, Qatar, or Saudi Arabia (see Box 2), major plans are being designed and launched to build local versions of IT parks, business process off-shoring (BPO) centers, and Internet/knowledge cities in an effort to capture part of the increased foreign direct investment, employment, and economic growth that a deepening of globalization is expected to bring.

It is increasingly recognized that it is not only a national government-led policy decision to support a certain industry such as ICT over others—as in the case of a localized IT park or a municipal decision to implement a city strategy for global excellence. In other words, it is not only a top-down or supply-driven approach that is causing local performance to gain in relevance. Yet relatively little attention has been given so far to analyzing on a globally comparative basis the role of e-government services in successful LGPs.

How do ICT and e-government benefit local global players?

Apart from typical national e-government services such as registrations, customs, taxation, and elections, it is local governments that have direct contact with citizens for a multitude of services; these local governments also attend to a large number of citizens' needs. Specific e-government services are increasingly handled at the local rather than national level. This is the case, for instance, for small- and medium-sized enterprise (SME) registration, vehicle and drivers' licenses, enrollment at educational institutions and vocational programs, furthering human resources skills, or professional authorizations and licenses (for example, for shops, pharmacies, and so on). The provision of increasing local e-government services contributes to e-readiness and competitiveness at the global level.

More generally, developing information infrastructure and ICT services can also assist in creating relative strength

Box 1: A few definitions

Local government

For the purpose of this chapter, we shall rely on the encompassing definition provided by Shah, for whom "local government refers to specific institutions or entities created by national constitutions (Brazil, Denmark, France, India, Italy, Japan, Sweden), by state constitutions (Australia, the United States), by ordinary legislation of a higher level of central government (New Zealand, the United Kingdom, most countries), by provincial or state legislation (Canada, Pakistan), or by executive order (China) to deliver a range of specified services to a relatively small geographically delineated area."[1]

Local government can hence be considered here as comprising governments that are not central, national, or federal. The term includes state, provincial, regional, municipal, and city governments. For methodological and availability purposes, the data for this paper are cities-based; however, the lessons inform other local government structures as well.

E-government

E-government, according to the World Bank, refers to the use by government agencies of information technologies (such as Wide Area Networks, the Internet, and mobile computing) that have the ability to transform relations with citizens, businesses, and other arms of government.

These technologies can serve a variety of different ends: better delivery of government services to citizens, improved interactions with business and industry, citizen empowerment through access to information, or more efficient government management. The resulting benefits can be less corruption, increased transparency, greater convenience, revenue growth, and/or cost reductions.

Traditionally, the interaction between a citizen or business and a government agency has been taking place in a government office. With emerging information and communication technologies it is possible to locate service centers closer to the clients. Such centers may consist of an unattended kiosk in the government agency, a service kiosk located close to the client, or the use of a personal computer in the home or office.

Analogous to e-commerce, which allows businesses to transact with each other more efficiently (B2B) and brings customers closer to businesses (B2C), e-government aims to make the interaction between government and citizens (G2C), government and business enterprises (G2B), and inter-agency relationships (G2G) more friendly, convenient, transparent, and inexpensive.

It cannot be understated, however, that e-government is not mainly about informatization of government-related transactions, but foremost about better government.

1 Shah 2006.

Box 2: Building a Knowledge City in the Saudi desert

On Saturday, June 18, 2006, King Abdullah bin AbdulAziz of Saudi Arabia launched the Knowledge Economic City (KEC) in Medina, the second Holy Place of Islam after Mecca. This new city will be developed on 4.8 million square meters of land, while the constructed area will encompass 9 million square meters, attracting some 25 billion Saudi riyals (US$6.7 billion) worth of investments. The project is expected to add 20,000 new jobs to the region.

The city will offer a range of new entities, including a technology and knowledge–based industries (KBI) zone; an advanced IT studies institute, a campus for medical research and life sciences, an integrated medical services zone, a retail zone, a business district, and a residential area. A monorail will be tethered to a planned train station, thus tapping into the railway access to Mecca, Yanbu, and the King Abdullah Economic City (to be built soon), as well as the port city of Jeddah.

As mentioned by Amr Dabbagh, governor of SAGIA (Saudi Arabia General Investment Authority, which leads the project), "This vital project falls well within plans to upgrade the Kingdom's regions in a sustainable fashion taking into consideration each region's

competitive advantages. Given Medina's historic stance as the launch pad of Islamic culture more than 14 centuries ago, the KEC project is a renaissance of sorts."[1]

Medina's Knowledge Economy City will be the fourth such economic city planned for Saudi Arabia as the country looks to diversify its economy and provide employment. The three economic cities launched to date by SAGIA are in line with its strategy to promote investments into the country's sectors that have been identified as offering the best competitive advantages— namely, energy, transportation, and knowledge-based industries. The first economic city (at an estimated cost of 100 billion Saudi Arabian riyals, launched in December 2005 in Rabegh)—the King Abdullah Economic City near Jeddah—focused on promoting energy- and transportation-related industries. The second economic city (at a cost of 30 billion riyals, also launched in June 2006 in Hayel)—the Prince Abdul Aziz bin Mousaed Economic City—is designed around transportation and logistical services. This latest city in Medina captures the essence of SAGIA's third focus, knowledge-based industries.

1 AME Info 2006.

for the local compared with the national level. ICT firms such as broadband service providers (who, unlike telecom service providers, do not necessarily face universal access obligations) can roll out services with a focus on regions or cities. The focus on the subnational may be written into a contract won through a tender issued by a local government.

Similarly, recent technological progress in "last mile connectivity" (such as WiFi and WiMAX) tends to benefit local governments directly, since they are typically concerned with a more limited geographical area (see the definition of *local government* in Box 1). Some city governments, such as those of Seattle and Philadelphia in the United States, have been quick to seize such opportunities in their efforts to provide the entire city boundaries with access to WiMAX services; these cities can reap benefits in e-readiness beyond the local level.

Clearly, such opportunities cannot be fully seized by local governments in the absence of a combination of factors, including: (1) a good and cost-effective basic information infrastructure; (2) a strong and visible political will at the local level to provide appropriate legal, regulatory,

and competition bases for good governance and a vibrant business climate; (3) adequate levels of education and "IT-savviness" in the local population; and (4) an innovative ICT private sector.

In this respect, e-government has the potential to be both an engine and a contributor to competitiveness at the local as well as at the international level. Through outsourcing contracts (private-public partnerships, or PPPs), for example, the dynamics between local government and local ICT firms can provide new markets and job opportunities for domestic firms, including SMEs as providers of local e-government services. The experience of Riga, Latvia, (see Box 3) is an interesting example in this context, and it may explain the relatively high ranking of that city in terms of e-government (discussed in the next section).

The importance of providing sufficient attention to e-readiness and e-government components in assessing the global and local performance of cities and local governments is progressively making its way in the work of analysts and practitioners.

Many of the conceptual and analytical frameworks attached to cities' strategies identify competitiveness and good governance as key pillars for sustainable and successful approaches and policies. Under those two headings, ICT and hence networked readiness have critical roles to play.

In the World Bank's approach, for instance, "modern communications and technology services" are quoted among the components and preconditions for cities' competitiveness, and "public access to information about local government decision making and actions" is mentioned as a tool to enhance good governance (see Table 1).

A tale of many cities

Although analytical efforts have been made to describe local e-government initiatives and their good practices, remarkably little attention has been granted to measuring the e-readiness of subnational spaces, including cities. Two of the more systematic attempts to measure "urban performance" or competitiveness have been made recently—one by Kaufmann et al. at the World Bank and the other by Rutgers University in collaboration with South Korea's Sungkyunkwan University in Seoul.

While the first attempt by Kaufmann et al. focuses on governance (including indicators on state capture, informal money laundering, red tape, and trust in politicians, as well as bribery in affecting utilities, laws, and permits), it also includes indicators that are vital to determining a city's level of e-readiness, such as access to electricity, telephone

lines, mobile telephones, and Internet in schools.[4] In contrast, the Rutgers-SKKU e-Governance Performance Index 2005 aims at ranking cities in terms of e-government performance, leaving out some indicators of e-readiness such as access to ICT and the enabling environment (see Box 4).

But why would one think that the e-readiness of an individual city can be significantly different from that of the country in which it is located? After all, looking at the way in which the Networked Readiness Index (NRI) of the present *Report* is built, one can find a significant number of key indicators that are relevant to the local level—for example, regulatory environment, intensity of local competition, firm-level technology absorption, and protection of property rights.

By comparing the overall city e-governance score of the Rutgers-SKKU dataset with the overall country networked readiness score of this GITR edition, one finds that, indeed, the e-government performance of individual cities is not straightforwardly linked to the e-readiness of the respective countries in which they are located (see Figure 2a).

In Figure 2a, the overall trend demonstrates that the majority of countries exhibit a degree of local e-government performance (the Rutgers-SKKU e-Governance Performance Index on the vertical axis) in line with what one could expect from their national networked readiness score (the

Table 1: The World Bank's Strategic Vision & Actions to Support Sustainable Cities

Goals	Components and preconditions	Enabling policy and institutional framework	Instruments and supporting actions
COMPETITIVENESS			
• Growth and increased productivity of city output, broad-based employment, investment, and trade in response to market opportunities	• Efficient factor markets (land, labor, capital), well integrated between rural and urban economies • Efficient and demand-responsive markets for infrastructure • Efficient local public administration that is business-friendly • Healthy industry structures (with fluid entry and exit for firms of all sizes) that integrate informal sector firms • Investment and industrial development corresponding to the comparative advantage of the city economy • Modern communication and technology services	• Legal and regulatory frameworks that support appropriate business incentives and impose minimal transactions costs • Public-private partnerships to identify market opportunities and remove bottlenecks in developing land, infrastructure, and cultural heritage assets • Land, real estate, and transport planning that supports spatially efficient land use and adequate supply of developed land for business and residential uses • Rule of law and property rights protected	• National urban strategies and action plans • City development strategies and action plans • Housing and real estate development programs and housing finance reforms • Urban regulatory audits (land, housing, business, labor) • Infrastructure subsector investment and reform programs, including public-private infrastructure framework analysis • Macroeconomic dialogue and Structural Adjustment Lending frameworks for stabilization and economic liberalization
GOOD GOVERNANCE AND MANAGEMENT			
• Accountability, transparency, and integrity of local government • Local government institutions sensitive to the needs of poor and disadvantaged residents and to gender differences in service requirements • Cost-effective fulfillment of local government service obligations	• Broad participation of all groups in urban governance, through both formal and informal channels and institutions • Clear incentives for performance by all levels of government affecting urban development • Strong capacity to ensure the delivery of services through a variety of mechanisms • Strong public trust and trust of high levels of government in local government • Public access to information about local government decision making and actions	• Clear frameworks for intra- and intergovernmental assignment and delegation of functions, responsibilities, revenues, and expenditures • Mechanisms for objective, independent review of local government performance • Good collaboration among local government agencies and informal institutions such as community-based organizations • Development and application of management tools and best practices • Professionalization and training of local government staff • Appropriate public-private partnership frameworks implemented • Regular public consultation and oversight in budget and local government decision making processes • Incentive structures for public representatives and employees that encourage integrity and minimize corruption	• National urban strategies and action plans • Self-standing advisory services • City development strategies and action plan • Urban management projects • Support to associations of municipalities for sharing best practices, technical assistance, training, and city twinning • Identification and dissemination of best practices in urban management • City awards for managerial excellence • Municipal management and public integrity training • Institutional reviews and anticorruption surveys including local government

Source: World Bank Urban Department, Infrastructure Group, 2000 (available at www.worldbank.org/html/fpd/urban/publicat/cities_in_transition.pdf).

Box 4: Rutgers-SKKU E-Governance Performance Index 2005

The data used by this study to evaluate e-readiness at the local level are based on a survey on digital governance in large municipalities worldwide in 2005. The survey was conducted by the E-Governance Institute of Rutgers-Newark and the Global ePolicy/ eGovernment Institute of Sungkyunkwan University in Seoul, Korea, and was co-sponsored by the United Nations Division for Public Administration and Development Management and the American Society for Public Administration. In this survey, *digital governance* includes both digital government (delivery of public service) and digital democracy (citizen participation in governance). Specifically, the study analyses security, usability, and content of websites; the type of online services being offered; and citizen response and participation through websites established by city governments. The instrument for evaluating city and municipal websites consisted of five equally weighed components: (1) Security and Privacy, (2) Usability, (3) Content, (4) Services, and (5) Citizen Participation. For each of the

five components, 18–20 measures were applied, each coded on a scale of two to four points.

International Telecommunication Union (ITU) data were used to select the sample based on population size, the total number of individuals using the Internet (>160,000), and the percentage of people using the Internet. Websites were evaluated by two evaluators between August and November 2005 in their native languages.

In this research, the main city homepage is defined as the official website where information about city administration and online services are provided by the city. The city website includes websites about the city council, mayor, and executive branch of the city. If there are separate homepages for agencies, departments, or the city council, evaluators examined whether these sites were linked to the menu on the main city homepage. If the website was not linked, it was excluded from evaluation.[1]

1 More information is available in Holzer and Kim (2005).

Figure 2a: City e-government vs. overall networked readiness: World

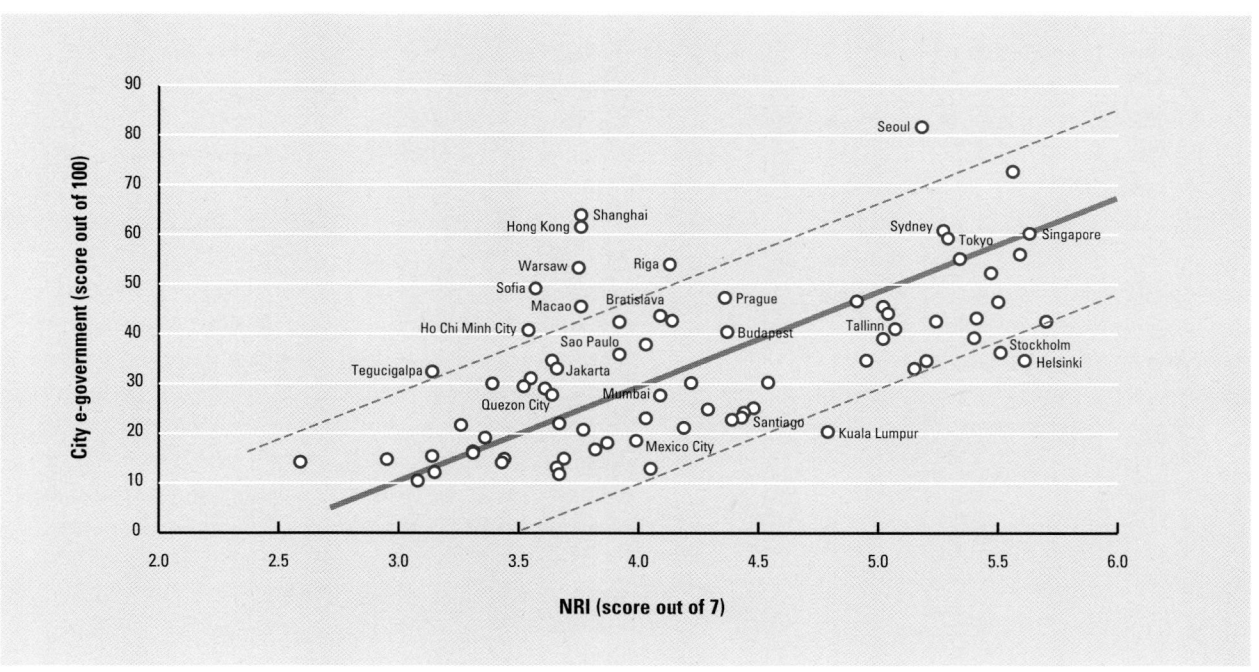

Source: NRI 2006–07; Rutgers-SKKU e-Governance Performance Index 2005; and authors' calculations.

Figure 2b: City e-government vs. overall networked readiness: Asia and the Pacific

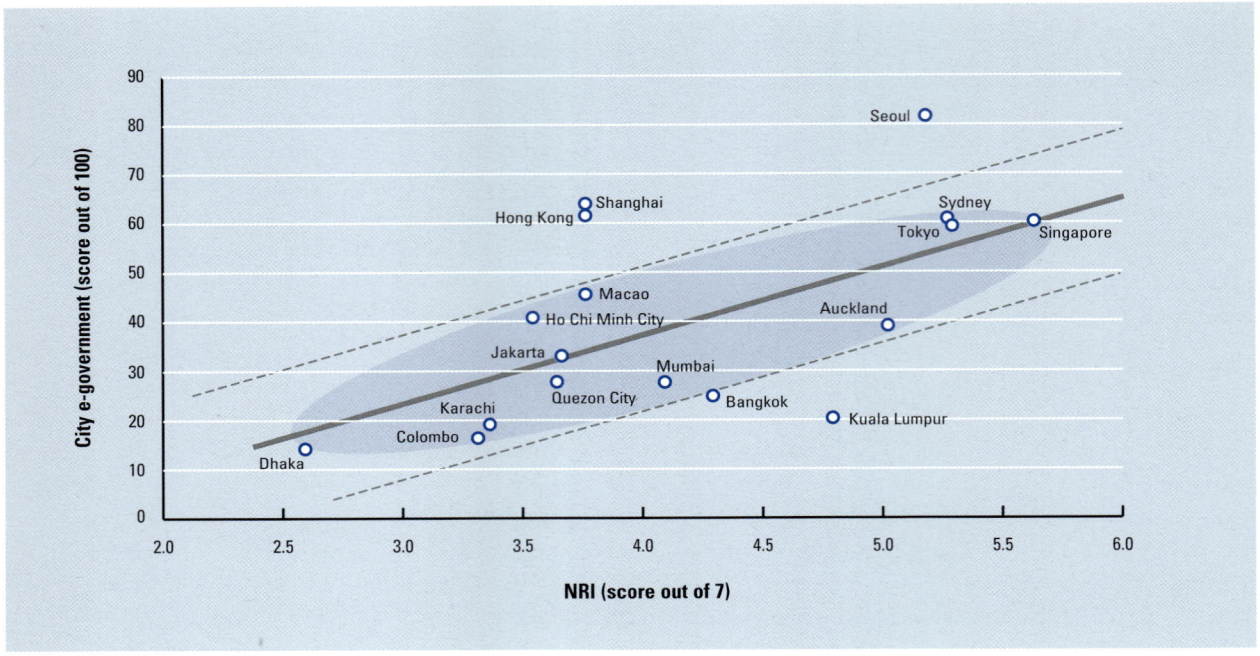

Source: NRI 2006–07; Rutgers-SKKU e-Governance Performance Index 2005; and authors' calculations.

NRI on the horizontal axis). However, several cities seem to be performing less successfully at the local level than their overall networked readiness would indicate (for example, Kuala Lumpur, Stockholm, and Helsinki). Others, on the contrary, perform better as local e-government hubs than the networked readiness of their respective countries would suggest. This is the case for Tegucigalpa (Honduras), Ho-Chi-Minh City (Vietnam), Warsaw (Poland), Macao, Hong Kong, Shanghai (China), Sofia (Bulgaria), and Riga (Latvia). In particular, three Eastern European cities in developing countries—Sofia, Warsaw, and Riga—are scoring close to or above 50 on the city axis; they are joined by three additional Eastern European cities—Tallinn, Bratislava, and Budapest—once the bar is lowered to scores or 40 or higher. Clearly there is a story to be told on city-level successes in Eastern European local e-government.

Indeed, the picture becomes more interesting and somewhat different when one considers regional subsamples of the same data. Because the overall sample—which is based on the common subset of NRI and the Rutgers-SKKU e-Governance Performance Index, making a total of 76 countries—is small, such a disaggregation cannot be pushed too far. Taking it to the level of broad regions (North America, South America and the Caribbean, Western Europe, Eastern and Central Europe, Africa, the Middle East, and Asia and the Pacific), a few interesting observations emerge.

For Asia and the Pacific, we find an ellipse that is flat (see Figure 2b), indicating a stronger correlation between overall networked readiness and municipal e-government performance. However, there are notable exceptions. Shanghai and Hong Kong as cities rank higher (Shanghai at 63.93 and Hong Kong at 61.51) than the NRI score of China as a whole (3.68) would suggest. The same is true for Seoul, the undisputed champion of the Rutgers-SKKU index with a score of 81.70; while Korea scores "only" 5.14 in this year's NRI. The opposite story seems to affect Kuala Lumpur, which—as a city—performs less well than Malaysia as a country. At roughly the same level of overall networked readiness, the cities of Quezon City (Philippines), Jakarta (Indonesia), Ho Chi Minh City (Vietnam), and Macao, Hong Kong, and Shanghai (China)[5] show stark differences in local e-government performance. To some extent, the same can be said about Tokyo and Sydney, which rank closely on both measures, but when compared with similarly nationally networked Seoul, they differ with a markedly lower local e-government score.

Moving to South America, a richer set of data offers interesting insights about the relation between city and country performances. Figure 2c shows that the dispersion of South American countries along the spectrum of networked readiness is broader than that of the corresponding countries along the axis of city e-government performance—translating visually in a rather flat ellipse covering

Figure 2c: City e-government vs. overall networked readiness: South America

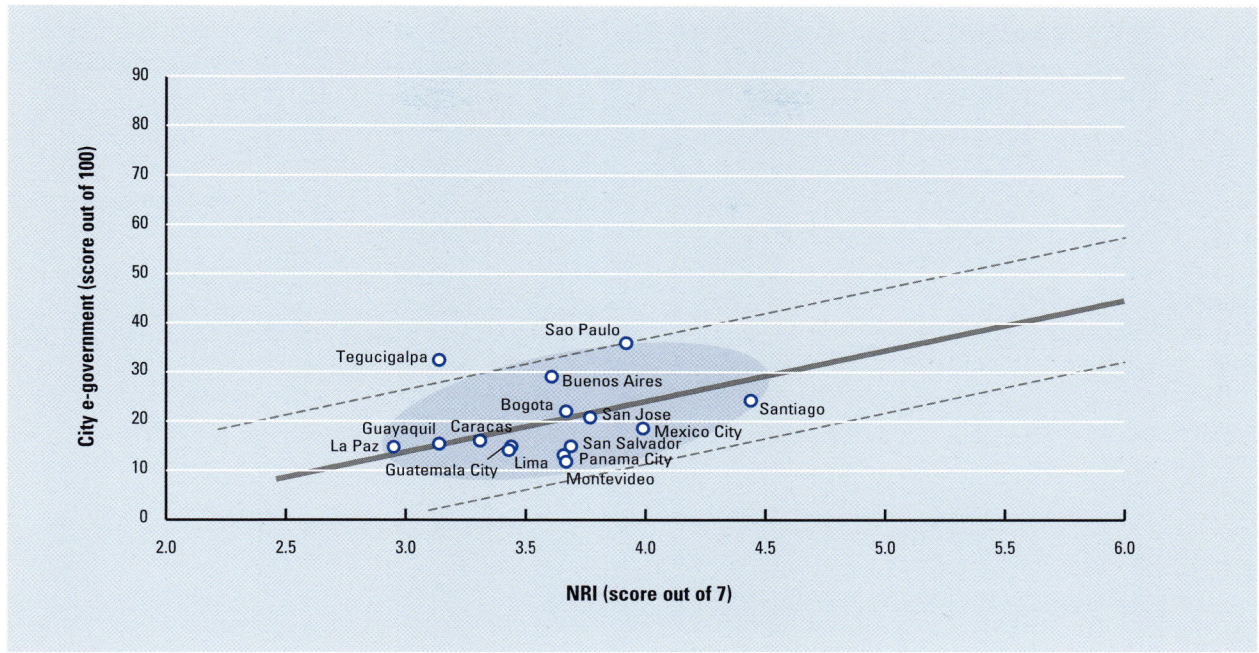

Source: NRI 2006–07; Rutgers-SKKU e-Governance Performance Index 2005; and authors' calculations.

the cloud of points. Tegucigalpa (Honduras) and Sao Paolo (Brazil) clearly outperform their respective countries, while Santiago (Chile) seems to tell the opposite story. The difference is striking between the respective city-level e-government performances of cities such as regional high-performer Sao Paolo on one hand and Mexico City on the other, although both cities operate with very similar levels of overall networked readiness.

As one could suspect, Europe offers a slightly complex picture, even if one separates Western Europe from Eastern and Central Europe (Figure 2d). A first conclusion is that the difference between "old Europe" and "new Europe is much less visible from the point of view of cities' performance than it is from that of overall networked readiness. At the national level, Estonia, the birthplace of Skype, remains the networked readiness champion among emerging European economies, but most of the other Eastern European economies considered also compare well with the laggards of Western Europe (for example, the Czech Republic, Hungary, Lithuania, Slovenia compare favorably with Cyprus, Greece, and Italy. However, on the city scale, Eastern Europe has a number of superior performers, including Bratislava (Slovakia), Prague (Czech Republic), Sofia (Bulgaria), and, above all, Warsaw (Poland) and Riga (Latvia), who are leaders in the European region as a whole. The e-government performance of those last three cities is clearly higher than their respective overall networked readiness levels would indicate.

Table 2: Eastern European cities in Rutgers-SKKU subindexes

City	Privacy and security rank	Usability rank	Content rank	Service delivery rank
Prague	n/a	7	n/a	9
Riga	n/a	5	5	n/a
Tallinn	n/a	n/a	2	n/a
Warsaw	n/a	n/a	7	5

Source: Rutgers-SKKU e-Governance Performance Index 2005.

The city-level dataset offers further insight into the leadership of Warsaw, Prague, Riga and Tallinn with regard to its subindexes on usability, content, and service delivery. Notably, no Eastern European city scored well in the privacy and security subindex; however, they performed well (often being in the top 10) in the other categories (usability, content, and service delivery).

Finally, the dataset used in this study offers only a small set of cities for three regions (two cities in North America, three in the Middle East, and four in Africa—if one includes Cairo in Africa rather than in the Middle-East)—a sample that is not sufficiently large to make significant observations. One can notice, however, that in those three regions the correlation between the NRI and the Rutgers-SKKU index is strong (see Figure 2e).

Figure 2d: City e-government vs. overall networked readiness: Eastern and Western Europe

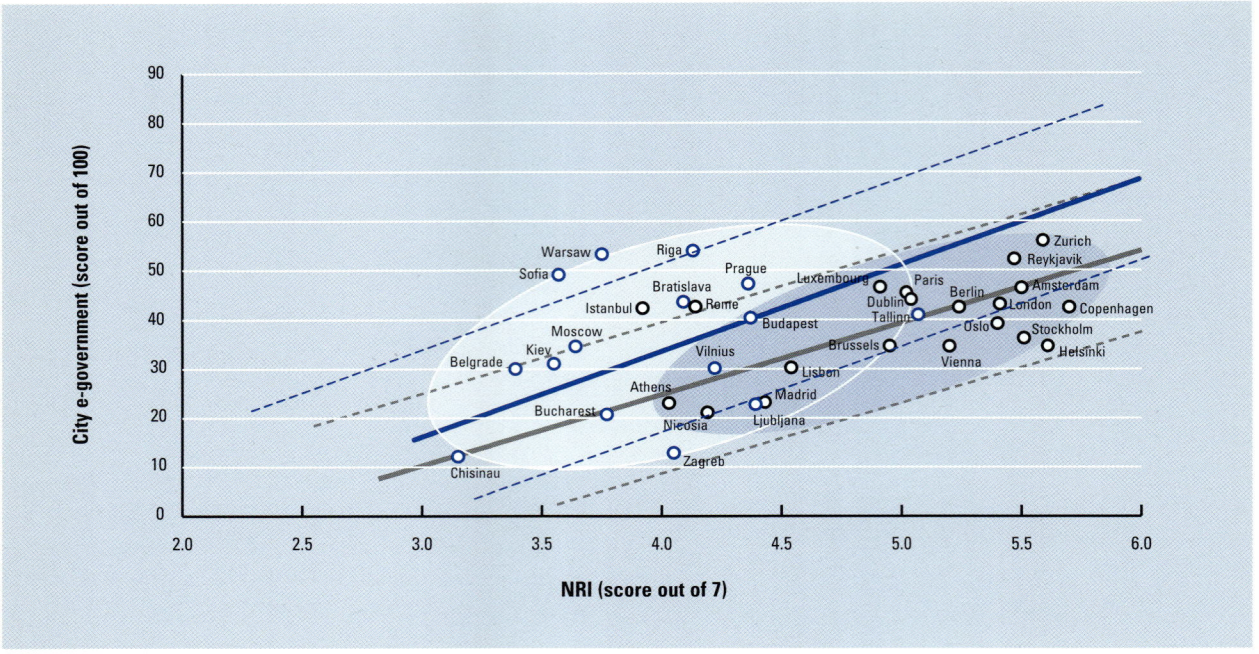

Source: NRI 2006–07, Rutgers-SKKU e-Governance Performance Index 2005; and authors' calculations.

Figure 2e: City e-government vs. overall networked readiness: Africa, Middle East, and North America

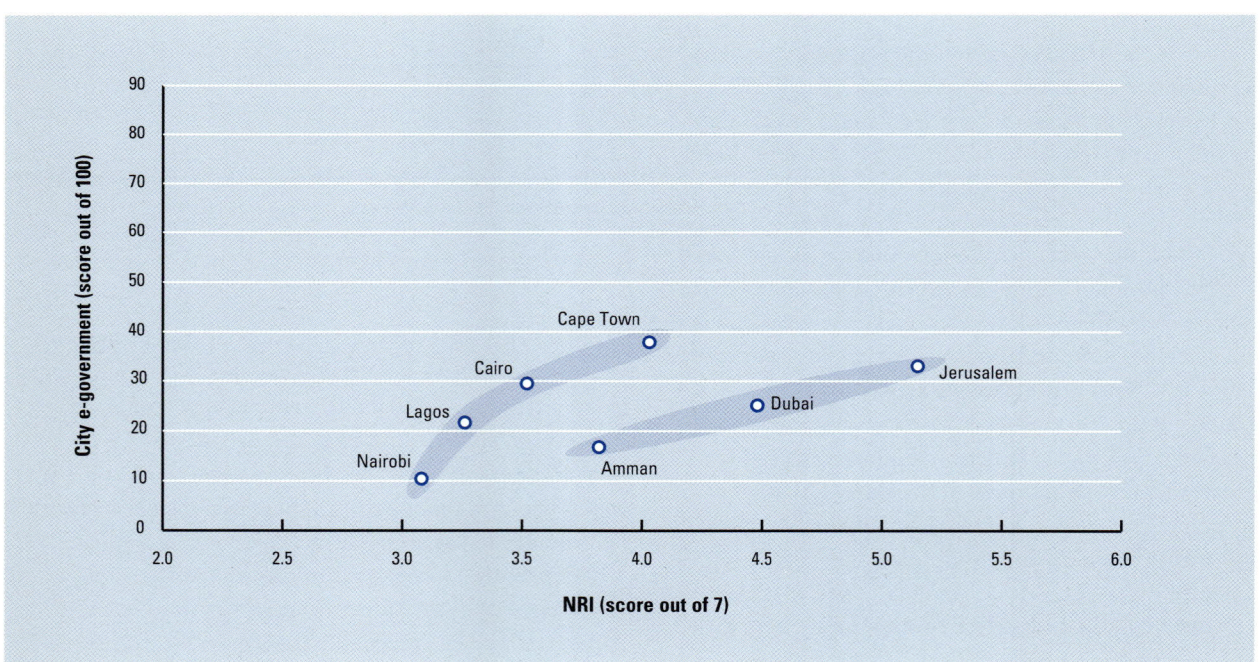

Source: NRI 2006–07; Rutgers-SKKU e-Governance Performance Index 2005; and authors' calculations.

The next frontier: Local e-readiness

Studying local government in more detail is important given the subnational characteristics of the e-readiness agenda. The analysis above has led us to three major conclusions:

1. Subnational economic spaces (cities in particular) have played a central role in shaping the current wave of globalization. The emergence of LGPs can be seen as a revenge of geography, whereby the benefits of the "death of distance" (which have allowed international operators to invest, produce, and sell across global networks of cooperation) have been combined with those of the physical proximity or congregation of local players (for example, ICT hubs in India, or more complex combinations of talents such as in London's City).

2. The dynamics of the ICT sector, and of ICT infrastructure and services in general, tend to reinforce the influence and roles of the local level in the overall process of globalization. The advent of short-range telecommunications technologies such as WiFi or WiMAX, combined with the regulatory space offered to broadband providers generally, are allowing the emergence of new business models that provide information-intensive services (including e-government) at the local level. In countries where most of the steps have been taken to establish e-government at the national level (as is the case in many Latin American countries, for instance), possibilities for taking advantage of new advances in IT seem to be even more significant at the local (and particularly municipal) level. For the next few years, and for all those reasons, the local level can truly be seen as the next frontier of e-government on a worldwide scale.

3. The various regions of the world tell different stories about the respective abilities of national economies and cities to enhance their respective levels of networked readiness, and to use e-government as a tool for competitiveness, good governance, and improvement of the quality of life of their citizens. However, they all show (even if at varying degrees) that the digital divide is less broad between cities than it is between countries.[6] This results not only from the superior agility of smaller economic spaces to seize opportunities in rapidly changing environments, but also from the fact that LGPs tend to network almost naturally with each other—the result of

common technical constraints (for example, international ports need to adopt common procedures and technical norms to accommodate certain types of vessels, or deal with multi-modal transport), or of the emergence of standard practices in the ways in which global business is being carried out across national borders. In all regions, some world cities emerge from the pack, showing higher rates of e-readiness (and e-government readiness) than their respective countries.

How should those conclusions affect the policies, strategies, and day-to-day management of decision makers in government (central and local) and in the private sector (domestic and international)? Can best practices be identified across regions to demonstrate how a city, region, or other local entity can outperform its competitors (and even its own country) in terms of e-readiness? Can such practices lend themselves to identifying a corpus of knowledge that could then be shared with other cities and subnational entities in various parts of the world? Can developing countries, in particular, benefit from such practices and knowledge to accelerate their own e-strategies by fostering the emergence on their soil of LGPs? If so, should they start building knowledge cities from scratch, using brand-new technologies and state-of-the-art approaches to urban planning and city development strategies, or should they build on the existing comparative advantages of some of their poles of competitiveness such as ports or economically specialized areas?

All of those questions are of strategic importance for central and local governments, as well as for their partners from the private sector, both domestic and international. However, to address them in a meaningful manner, one would need more than the remarkably scarce data currently available.

The search for meaningful indicators at the local level

Five years ago, analysts were already lamenting the absence of relevant data linking the world of telecommunications and information services with that of urban and local management and competitiveness. As Townsend (2001) put it, "The recent rapid growth of the Internet has avoided scrutiny from urban planners as little information is available from which to assess its impact on cities and regions. As a result, explanations of the relationship between telecommunications and urban growth are overly simplistic.... In general, the literature on global cities is long on speculation and short on specification when addressing the telecommunications issue."

It is fair to note, however, that significant international efforts have recently been launched to generate more internationally comparable data describing relationships between ICT and *national* competitiveness. This was precisely the purpose of efforts made by the World Economic Forum, initially as part of its *Global Competitiveness Report* work, and soon as the basis for the first edition of the present *Global Information Technology Report*. More recently, and as a result of increased awareness generated by the United Nations World Summit on Information Societies (WSIS), a significant number of international organizations have decided to join forces to produce better data in this domain.[7]

A similar effort is now needed to extend such efforts to the local level.

Several important questions hence need to be examined urgently, regarding (1) the possibility of using existing methodologies (generally developed at the national level) for the collection and analysis of data concerning local governments and local entities, and (2) the necessity of creating specific data sets (and collection methods) to better understand the ways in which e-readiness is being built at the local level.

In the context of the present *Report*, it is worth considering the following issues: does it make sense to try and bring the NRI methodology to local levels? For which variables can this be done (either because the data are available, or because they can be collected or approximated)? What additional data (specific to the local level) would need to be identified, collected, and integrated in a local e-readiness and e-government index? A first attempt has been made here (see the appendix to this chapter) to address those questions, but clearly much more work is needed to provide them with the answers they deserve.

A proposal for action

For all players involved in or affected by globalization, identifying best strategies requires the ability to identify where major decisions will be made and by whom. The present chapter has tried to show that (1) local governments (for example, cities) will play an increasingly important role in shaping their own global competitiveness and that of their respective countries, and that (2) in so doing, they will make an increasing use of ICT, especially as a tool for e-government. In that sense, local governments may indeed hold the keys to the next wave of global competition.

If this is the case, investors and policymakers will need to better understand and measure the performance of local governments with regard to e-readiness in general, and to e-government in particular. It is hence proposed here that a definite effort be launched as early as possible to identify and provide relevant data and indicators in that area.

Existing efforts and methodologies (such as this *Report*'s Networked Readiness Index) should be used to their full extent. As an indication of what might be needed in this regard, the appendix to this chapter offers a first census of which NRI data would need to be generated at

the local level. An e-readiness ranking of cities worldwide could be obtained as a result, which would be of value for private and public decision makers, either as investors looking for adequate and competitive locations or as public decision makers trying to identify best practices and to stimulate the development of their own cities and economies.

By acknowledging the fact that national indicators often hide striking differences between different geographical locations, and by providing new tools to measure and address such differences (and share best practices wherever they emerge), the international community would contribute not only by enhancing the value of its current work on competitiveness, investment climate and e-readiness, but also by casting new light on ways to make globalization a tool for development and poverty reduction.

Notes

1 See Marcuse and Van Kempen (1999).

2 See Sassen (1995) and Knox (2002).

3 See Townsend (2001).

4 See Kaufmann et al. (2006).

5 It must be noted here that, for the purposes of this chapter, Hong Kong, Shanghai, and Macao have been treated in the same manner: e-governance indicators (Rutgers-SKKU data) have been mapped against the country NRI rating for China. This choice was made both for consistency reasons (treating all Chinese cities in the same fashion), but also because it befits the overall purpose of this section, which is to identify cities for which local e-governance performance is above (or below) what the NRI performance of their respective national environments would suggest.

6 Obviously, this statement needs to be kept within the limits of the sample selected here—that is, the sample of global cities and the countries that host them. However, one could convincingly argue that a broader sample of cities/countries would probably reinforce rather than weaken this conclusion, be it only because countries that are lagging in terms of connectivity generally tend to display a stronger digital divide between urban centers (typically the capital city) and rural areas.

7 A result of the World Summit on Information Societies (WSIS), a "Partnership on Measuring ICT for Development" was launched in November 2005, with the goal to "accommodate and develop further the different initiatives regarding the availability and measurement of ICT indicators at the regional and international levels" and "provide an open framework for coordinating ongoing and future activities, and for developing a coherent and structured approach to advancing the development of ICT indicators globally, and in particular in developing countries." Current partners include the ITU, the OECD, UNCTAD, UNESCO Institute for Statistics, the UN Regional Commissions (UNECLAC, UNESCWA, UNESCAP, UNECA), the UN ICT Task Force, the World Bank, and EUROSTAT. See http://measuring-ict.unctad.org/QuickPlace/measuring-ict/Main.nsf/h_Toc/281E7067B40AD764C1256EE80048DACC/?OpenDocument.

References

AME Info. 2006. "King Abdullah bin AbdulAziz launches Knowledge Economic City." Press release, June 18. Available at http://www.ameinfo.com/89077.html.

A. T. Kearney and *Foreign Policy*. 2006. *The Globalization Index. Foreign Policy* 157 (November/December): 75–81.

Bairoch, P., J. Batou, and P. Chèvre, P. 1988. "La population des villes européennes de 800 `a 1850." Geneva: Librairie Droz.

Cairncross, F. 1997. *The Death of Distance*. Cambridge, MA: Harvard Business School Press.

Cities Alliance. 2006. *Guide to City Development Strategies: Improving Urban Performance*. Available at www.citiesalliance.org.

Cross, M. and R. S. Moore. 2002. *Globalization and the New City: Migrants, Minorities and Urban Transformations in Comparative Perspective*. Basingstoke, Hampshire, New York, and Houndmills: Palgrave.

Friedmann, J. 1986. "The World City Hypothesis." *Development and Change* 17: 69–83.

Glaeser, E. L. and J. Gaspar. 1998. "Information Technology and the Future of Cities." *Journal of Urban Economics* 43: 136–56.

Hall, P. 1966. *The World Cities*. London: Weidenfield and Nicolson.

———. 1997. "Modelling the Post-Industrial City." *Futures* 29 (4): 311–22.

Holzer, M. and S-T. Kim. 2005. *Digital Governance in Municipalities Worldwide: A Longitudinal Assessment of Municipal Websites Throughout the World, 2005 Rutgers-SKKU E-Governance Performance Index*. Rutgers University and Sungkyunkwan University.

Gugler, J. 2004. *World Cities beyond the West: Globalization, Development and Inequality*. New York; Cambridge, UK; Port Melbourne, Australia; and Cape Town, South Africa: Cambridge University Press.

Kaufmann, D., F. Léautier, and M. Mastruzzi. 2004. "Globalization and Urban Performance." In *Cities in a Globalizing World*, ed. F. Léautier, pp. 27–68. Washington, DC: World Bank.

Knox, P.L. 2002. "World Cities and the Organization of Global Space." In *Geographies of Global Change*, 2nd edition, ed. R.J. Johnston, P.J. Taylor, and M.J. Watts, pp. 328–38. Oxford: Blackwell.

Lanvin, B. 1995. "Why the Global Village Cannot Afford Information Slums." *The New Information Infrastructure*, ed. W. Drake. New York: Twentieth Century Fund.

———. 2006. "How Governments Shape Global Sourcing." In *Six Billion Minds: Managing Outsourcing in the Global Knowledge Economy*, ed. M. Minevich, F. J. Richter, and F. Hoque, pp. 169–87. Stamford, CT: BTM Institute/Aspatore.

Léautier, F. ed. 2004. *Cities in a Globalizing World*. Washington, DC: World Bank.

Marcuse, P. and R. Van Kempen. 1999. *Globalization Cities: A New Spatial Order?* London and Cambridge: Blackwell Publishing.

Moon, M. J. 2002. "The Evolution of E-Government among Municipalities: Rhetoric or Reality?" *Public Administration Review* 62 (4): 424–33.

Sassen, S. 1995. "On Concentration and Centrality in the Global City." *World Cities in a World System*, ed. P.L. Knox and P.J. Taylor, pp. 63–78. Cambridge, UK:Cambridge University Press.

Shah, A., ed. 2006. *Local Governance in Developing Countries. Public Sector Governance and Accountability Series*. Washington DC: World Bank.

Tat-Kei Ho, A. 2002. "Reinventing Local Governments and the E-Government Initiative." *Public Administration Review* 62 (4): 434–44.

Torres, L., V. Pina, and B. Acerete. 2006. "E-Governance Developments in European Union Cities: Reshaping Government's Relationship with Citizens." *Governance* 19 (2): 277–302.

Townsend, A. M. 2001. "The Internet and the Rise of the New Network Cities, 1969–1999." *Environment and Planning B: Planning and Design*, volume 28: pp. 39–58.

Zhou, X. 2004. "E-Government in China: A Content Analysis of National and Provincial Web Sites." *Journal of Computer-Mediated Communication* 9 (4). Available at http://jcmc.indiana.edu/vol9/issue4/zhou.html.

63

Appendix A: Availability of networked readiness data at the subnational level

NRI indicators	Question	Published data available at the subnational level	Ability to publish or collect data for subnational level (main cities or regions)	Kaufmann et al. data available	Rutgers-SKKU data available
ENVIRONMENT COMPONENT					
Market environment					
1.01 Venture capital availability, 2006	Entrepreneurs with innovative but risky projects can generally find venture capital in your city (1 = not true, 7 = true)	No	Yes	No	No
1.02 Financial market sophistication, 2006	The level of sophistication of financial markets in your city is (1 = lower than international norms, 7 = higher than international norms)	No	Yes	No	No
1.03 Technological readiness, 2006	Your city's level of technological readiness (1 = generally lags behind most other countries, 7 = is among the world leaders)	No	Yes	No	No
1.04 State of cluster development, 2006	Strong and deep clusters are widespread throughout the economy (1 = strongly disagree, 7 = strongly agree)	No	Yes	No	No
1.05 US utility patents, 2005	Number of utility patents (i.e., patents for invention) granted between January 1 and December 31, 2005, per million population	No	Yes	No	No
1.06 High-tech exports, 2004	High-technology exports as percentage of total exports, 2004	No	Yes	No	No
1.07 Burden of government regulation, 2006	Complying with administrative requirements (permits, regulations, reporting) issued by the government in your city is (1 = burdensome, 7 = not burdensome)	No	Yes	No	No
1.08 Extent and effect of taxation, 2006	The level of taxes in your city (1 = singificantly limits the incentives to work or invest, 7 = has little impact on the incentives to work or invest)	No	Yes	No	No
1.09 Time required to start a business, 2006	Number of days required to start a business, 2006	Yes	Yes	No	No
1.10 Number of administrative procedures required to start a business, 2006	Number of administrative procedures to start a business, 2006	No	Yes	No	No
1.11 Intensity of local competition, 2006	Competition in the local market is (1 = limited in most industries and price-cutting is rare, 7 = intense in most industries as market leadership changes over time)	No	Yes	No	No
1.12 Freedom of the press, 2006	In your city, can the media publish/broadcast stories of their choosing without fear of censorship or retaliation? (1 = no, 7 = yes—whatever they want)	No	Yes	No	No
Political and regulatory environment					
2.01 Effectiveness of law-making bodies, 2006	How effective is your national parliament/congress as a law-making and oversight institution? (1 = very ineffective, 7 = very effective—the best in the world)	No	Yes	No	No
2.02 Laws relating to ICT, 2006	Laws relating to the use of information and communication technologies (ICT) (electronic commerce, digital signatures, consumer protection) are (1 = nonexistent, 7 = well developed and enforced)	No	Yes	No	No
2.03 Judicial independence, 2006	Is the judiciary in your city independent from political influences of members of government, citizens, or firms? (1 = no, heavily influenced, 7 = yes, entirely independent)	No	Yes	No	No

Appendix A: Availability of networked readiness data at the subnational level *(cont'd.)*

NRI indicators	Question	Published data available at the subnational level	Ability to publish or collect data for subnational level (main cities or regions)	Kaufmann et al. data available	Rutgers-SKKU data available
ENVIRONMENT COMPONENT *(cont'd.)*					
Political and regulatory environment *(cont'd.)*					
2.04 Intellectual property protection, 2006	Intellectual property protection in your city is (1 = weak or nonexistent, 7 = equal to the world's most stringent)	No	Yes	No	No
2.05 Efficiency of legal framework, 2006	The legal framework in your city for private businesses to settle disputes and challenge the legality of government actions and/or regulations (1 = is inefficient and subject to manipulation, 7 = is efficient and follows a clear, neutral process)	No	Yes	No	No
2.06 Property rights, 2006	Property rights, including over financial assets, are (1 = poorly defined and not protected by law, 7 = clearly defined and well protected by law)	No	Yes	No	No
2.07 Quality of competition in the ISP sector, 2006	Is there sufficient competition among Internet sevice providers (ISPs) in your city to ensure high quality, infrequent interruptions, and low prices? (1 = no, 7 = yes, equal to the best in the world)	No	Yes	No	No
2.08 Number of administrative procedures to enforce a contract, 2006	Number of administrative procedures to enforce a contract, 2006	No	Yes	No	No
2.09 Time to enforce a contract, 2006	Number of days to enforce a contract, 2006	No	Yes	No	No
Infrastructure environment					
3.01 Telephone lines, 2005 year available	Main telephone lines per 100 inhabitants, 2005 or most recent	No	Yes	Yes (1998)	No
3.02 Secure Internet servers, 2005 recent year available	Secure Internet servers per 1 million inhabitants, 2005 or most	No	Yes	No	Proxy
3.03 Internet hosts, 2004	Internet hosts per 10,000 inhabitants, 2004	No	Yes	No	No
3.04 Electricity production, 2003	Per capita electricity production (kWh), 2003	No	Yes	Yes (1998)	No
3.05 Availability of scientists and engineers, 2006	Scientists and engineers in your city are (1 = nonexistent or rare, 7 = widely available)	No	Yes	No	No
3.06 Quality of scientific research institutions, 2006	Scientific research institutions in your city (e.g., university laboratories, government laboratories) are (1 = nonexistent, 7 = the best in their fields internationally)	No	Yes	No	No
3.07 Tertiary enrollment, 2004	Gross tertiary enrollment rate, 2004 or most recent year available	Yes	Yes	No	No
READINESS COMPONENT					
Individual readiness					
4.01 Quality of math and science education, 2006	Math and science education in your city's schools (1 = lag far behind most other countries, 7 = are among the best in the world)	No	Yes	No	No
4.02 Quality of the educational system, 2006	The educational system in your city (1 = does not meet the needs of a competitive economy, 7 = meets the needs of a competitive economy)	No	Yes	No	No

Appendix A: Availability of networked readiness data at the subnational level *(cont'd.)*

NRI indicators	Question	Published data available at the subnational level	Ability to publish or collect data for subnational level (main cities or regions)	Kaufmann et al. data available	Rutgers-SKKU data available
READINESS COMPONENT *(cont'd.)*					
Individual readiness *(cont'd.)*					
4.03 Quality of public schools, 2006	The public (free) schools in your city are (1 = of poor quality, 7 = equal to the best in the world)	No	Yes	No	No
4.04 Internet access in schools, 2006	Internet access in schools is (1 = very limited, 7 = extensive —most children have frequent access)	No	Yes	Yes	No
4.05 Buyer sophistication, 2006	Buyers in your city are (1 = unsophisticated and make choices based on lowest price, 7 = knowledgable and demanding and buy based on superior performance attributes)	No	Yes	No	No
4.06 Residential telephone connection charge, 2005	One-time residential telephone connection charge (US$) as a percentage of GDP per capita, 2005 or most recent year available	Yes	Yes	No	No
4.07 Residential monthly telephone subscription, 2005	Residential monthly telephone subscription to the public switched network (US$) as a percentage of monthly GDP per capita, 2005 or most recent year available	Yes	Yes	No	No
4.08 High-speed monthly broadband subscription charge, 2006	High-speed monthly broadband subscription charge (US$) as a percentage of monthly GDP per capita, 2006	Yes	Yes	No	No
4.09 Lowest cost of broadband, 2006	Lowest sampled cost (US$) per 100 kbits/s as a percentage of monthly income (GNI), 2006	No	Yes	No	No
4.10 Cost of mobile telephone call, 2005	Cost of 3-minute local call during peak hours (US$) as a percentange of monthly GDP per capita, 2005 or most recent year available	Yes	Yes	No	No
Business readiness					
5.01 Extent of staff training, 2006	The general approach of companies in your city to human resources is (1 = to invest little in training and employee development, 7 = to invest heavily to attract, train, and retain employees)	No	Yes	No	No
5.02 Local availability of specialized research and training services, 2006	In your city, specialized research and training services are (1 = not available, 7 = available from world-class local institutions)	No	Yes	No	No
5.03 Quality of management schools, 2006	Management or business schools in your city are (1 = limited or of poor quality, 7 = among the best in the world)	No	Yes	No	No
5.04 Company spending on research and development, 2006	Companies in your city (1 = do not spend money on research and development, 7 = spend heavily on research and development relative to international peers)	No	Yes	No	No
5.05 University-industry research collaboration, 2006	In its R&D activity, business collaboration with local universities is (1 = minimal or nonexistent, 7 = intensive and ongoing)	No	Yes	No	No
5.06 Business telephone connection charge, 2005	One-time business telephone connection charge (US$) as a percentage of GDP per capita, 2005 or most recent year available	Yes	Yes	No	No
5.07 Business monthly telephone subscription, 2005	Business monthly telephone subscription to the PSTN (US$) as a percentage of monthly GDP per capita, 2005 or most recent year available	Yes	Yes	No	No

Appendix A: Availability of networked readiness data at the subnational level *(cont'd.)*

NRI indicators	Question	Published data available at the subnational level	Ability to publish or collect data for subnational level (main cities or regions)	Kaufmann et al. data available	Rutgers-SKKU data available
READINESS COMPONENT *(cont'd.)*					
Business readiness *(cont'd.)*					
5.08 Local supplier quality, 2006	The quality of local suppliers in your city is (1 = poor, as they are inefficient and have little technological capactiy, 7 = very good, as they are internationally competitive and assist in new product development)	No	Yes	No	No
5.09 Computer, communications, and other services imports, 2004	Computer, communications, and other services as percentage of total commerical services imports, 2004	No	Yes	No	No
Government readiness					
6.01 Government prioritization of ICT, 2006	Information and communication technologies (ICT) (computers, Internet, etc.) are an overall priority for the government (1 = strongly agree, 7 = strongly disagree)	No	Yes	No	No
6.02 Government procurement of advanced technology products, 2006	Government purchase decisions for the procurement of advanced technology products are (1 = based solely on price, 7 = based on technological performance and innovativeness)	No	Yes	No	No
6.03 Importance of ICT to government's vision of the future, 2006	The government has a clear implementation plan for utilizing ICT to improve the city's overall competitiveness (1 = strongly disagree, 7 = strongly agree)	No	Yes	No	No
6.04 E-participation index, 2005	The e-participation index assesses the quality, relevance, usefulness, and willingness of government websites for providing online information and participartoy tools and services to the people, 2005	No	No	No	Proxy
6.05 E-government readiness index, 2005	The e-government readiness index assesses e-government readiness based on website assessment, telecommunications infrastructure, and human resources endowment	No	No	No	Proxy
USAGE COMPONENT					
Individual usage					
7.01 Mobile telephone subscribers, 2005	Mobile telephone subscribers per 100 inhabitants, 2005 or most recent year available	No	Yes	Yes	No
7.02 Personal computers, 2004	Personal computers per 100 inhabitants, 2004 or most recent year available	No	Yes	No	No
7.03 Broadband Internet subscribers	DSL Internet subscribers, per 100 inhabitants, 2005	No	Yes	No	No
7.04 Internet users, 2005	Internet users per 100 inhabitants, 2005 or most recent year available	No	Yes	No	No
7.05 Internet bandwith, 2004	International Internet bandwith (Mbps) per 10,000 inhabitants, 2004	No	Yes	No	No

Appendix A: Availability of networked readiness data at the subnational level *(cont'd.)*

NRI indicators	Question	Published data available at the subnational level	Ability to publish or collect data for subnational level (main cities or regions)	Kaufmann et al. data available	Rutgers-SKKU data available
USAGE COMPONENT *(cont'd.)*					
Business usage					
8.01 Prevalence of foreign technology licensing, 2006	In your city, licensing of foreign technology is (1 = uncommon, 7 = a common means of acquiring new technology)	No	Yes	No	No
8.02 Firm-level technology absorption, 2006	Companies in your city are (1 = not able to absorb new technology, 7 = aggressive in absorbing new technology)	No	Yes	No	No
8.03 Capacity for innovation, 2006	Companies obtain technologies (1 = exclusively from licensing or imitating foreign companies, 7 = by conducting formal research and pioneering their own new products and processes)	No	Yes	No	No
8.04 Availability of new telephone lines, 2006	New telephone lines for your businesses are (1 = scarce and difficult to obtain, 7 = widely available and highly reliable)	No	Yes	No	No
8.05 Availability of mobile telephones, 2006	Mobile or cellular telephones for your business are (1 = not available, 7 = as accessible and affordable as in the world's most technologically advanced countries)	No	Yes	No	No
8.06 Extent of business Internet use, 2006	In your city, companies use the Internet extensively for buying/selling goods and services and for interaction with customers (1 = strongly disagree, 7 = strongly agree)	No	Yes	No	No
Government usage					
9.01 Government success in ICT promotion, 2006	Government programs promoting the use of ICT are (1 = not very successful, 7 = highly successful)	No	Yes	No	No
9.02 Availability of online services, 2006	In your city, online government services such as personal tax, car registrations, passport applications, business permits, and e-procurement are (1 = not available, 7 = extensively available)	No	Yes	No	Proxy
9.03 ICT use and government efficiency, 2006	In your view, ICT use by the government has improved the efficiency of government services and has facilitated interaction with business and civil society (1 = strongly disagree, 7 = strongly agree)	No	Yes	No	No
9.04 ICT pervasiveness, 2006	The presence of ICT in government offices in your city is (1 = very rare, 7 = commonplace and pervasive)	No	Yes	No	No

Note: These data were compiled based on best knowledge of availability of cross-national, comparative, subnational datasets. The term *city* has been substituted for *country* in the variable definitions

68

Reluctant Gatekeepers: Corporate Ethics on a Filtered Internet

JOHN G. PALFREY, JR., Berkman Center for Internet & Society,

Harvard Law School

It is getting harder to be a successful technology company. In the earliest days of the Internet, the relevant markets were modest in size and close to home. A local Internet service provider, for instance, once could profit by offering a dial-up Internet access service over plain old telephone lines to people who lived close by to the corporate headquarters. Few of the big players involved were large, publicly traded entities. Revenue projections commonly looked like hockey sticks pointing toward bright blue skies. And, most important for the purposes of this chapter, states throughout the world left alone both the Internet and the companies that plied it for profit. The prevailing orthodoxy was that a state that required too much of companies doing business on the Internet would be making a dire mistake by restricting the early growth of online activity. Few states placed any kind of liability or responsibility on intermediaries. Many states even made the Internet a tax-free haven to promote its growth.

Now that we are more than 10 years into the Internet revolution, these are no longer the key facts on the ground. The Internet is big business in which large, entrenched players—and not just what were once called dot-coms—with colossal market capitalizations compete with one another over multibillion-dollar revenue streams. The relevant markets that they reach span much of the globe. But the most important fact that has changed: states have increasingly begun to force companies that provide Internet services to do more to regulate activity in the Internet space. This approach applies a new kind of pressure on nearly every corporation whose business involves information and communication technologies (ICT).

Internet filtering and surveillance put this phenomenon into stark relief. *Internet filtering* refers to the practice by which states restrict citizens from accessing or publishing certain information on the Internet. Closely related, *Internet surveillance* refers to the means by which states record, listen in on, or track down conversations that take place over the Internet. Over the past five years, the OpenNet Initiative (ONI)—a collaboration that joins researchers at the University of Toronto, the University of Cambridge, the University of Oxford, and Harvard Law School—has tracked the steady rise of Internet filtering practices from only a handful of states in 2002 to more than two dozen in 2007. The most extensive of these filtering regimes are found in states in three regions of the world: the Middle East and North Africa; Asia and the Pacific; and Central Europe and Asia. In the context of this *Report* on the promotion of networked readiness, it is worth noting that this trend cuts directly against the general guidance by Internet development experts in favor of deregulation of the ICT environment in the interest of growth.

The states that employ these filtering and surveillance regimes cannot do the work alone. This simple fact sets up the ethical quandary at the heart of this chapter. Virtually none of the two dozen or so states that filter the Internet have a network controlled entirely by the state. The most successful strategies for accomplishing state-mandated filtering and surveillance, like the Internet itself, are highly distributed in architectural terms. In almost every case, states have to rely upon private actors to carry out most of the censorship and surveillance. The means by which states call upon private actors, and for what purpose, vary from state to state. But the trend points toward greater expectations placed by states on private actors to help get the online censorship and surveillance job done.

For global technology companies, this scenario sets up a hard problem. The shareholders in large technology companies reasonably expect continued growth of market share, improved margins, and so forth. The shares in these firms are often publicly traded by investors in the state in which they are chartered. The pull of markets further from home is obvious and powerful. In many instances, the social norms and conceptions of civil liberties in the new target market are dissonant with the norms and liberties enjoyed where the senior executives and most powerful shareholders of the corporation live. An everyday act of law enforcement in an authoritarian market looks like a human rights violation to a more liberal one. Sometimes, that act may in fact contravene international human rights standards—and some shareholders, concerned about matters beyond growth and profits, are starting to ask hard questions of corporations about their involvement in such practices.

Corporations are increasingly finding themselves caught in the crosshairs as they are asked by local authorities to carry out censorship and surveillance online. This chapter describes this growing, thorny problem and some possible means to resolve it. The most promising approach is neither local law nor a new international covenant, but rather a strong, enforceable code of conduct created by the corporations themselves, in concert with nongovern-mental organizations (NGOs), academics, states, and other stakeholders.

Control of the information and communication technologies environment

We are still in the early stages of the struggle for control on the Internet. Early theorists, citing the libertarian streak that runs deep through the hacker community, suggested that the Internet would be hard to regulate.[1] "Cyberspace" might prove to be an alternate jurisdiction that the long arm of the state could not reach. Online actors, the theory went, would pay little heed to the claims to sovereignty over their actions by traditional states based in real-space.

An emerging trajectory: More state control, greater pressure on private parties

As it turns out, states have not found it so very hard to assert sovereignty where they have needed to do so. The result is the emergence of an increasingly balkanized Internet, and the theory of "unregulability" no longer has currency. Many scholars have described the present-day reality of the reassertion of state online control, despite continued hopes that the Internet community itself might self-regulate in new and compelling ways.[2]

The dynamic of online control has changed greatly over the past 10 years, and it is almost certain to change just as dramatically in the 10 years to come. The "technologies and politics of control" of the Internet, as Jonathan Zittrain has put it, remain in flux.[3] Members of the Internet Governance Forum (IGF), chartered via the process that produced two instances of the World Summit on the Information Society (WSIS), continue to wrestle with a broad set of unanswered questions related to control of the online environment. At a simple level, the jurisdictional question of who can sue whom (and where that lawsuit should be heard, and under the law of which jurisdiction the conflict should be adjudicated, for that matter), remains largely unresolved, despite a growing body of case law. A series of highly distributed problems—spam, spyware, online fraud—continues to vex law enforcement officials and public policymakers around the world. Intellectual property law continues to grow in complexity, with some degree of harmonization underway among competing regimes. Each of these problems leaves many unresolved issues of global public policy in its wake.

A key aspect of online control—and one that is empirically proven through the work of ONI—is that states have, on an individual basis, defied the cyber-libertarians by asserting control over the online acts of their own citizens in their home state. The manner in which this control is exercised varies. Sometimes the law bans citizens from performing a certain activity online, such as accessing or publishing certain material. Sometimes the state takes control into its own hands by erecting technological or other barriers within the state's confines to stop the flow of bits from one recipient to another. Increasingly, though, the state is turning to private parties to carry out the online control. Many times, those private parties are corporations chartered locally or individual citizens who live in that jurisdiction. The emphasis of this chapter is yet another instance, in which the state requires private parties—often intermediaries whose services connect one online actor to another—to participate in online censorship and surveillance as a cost of doing business in that state.[4]

Legitimate state online control

The need for states to be able to exercise some measure of online control is broadly accepted. Likewise, states ought to be able to provide rights of action—ordinarily, the right to sue someone—to their citizens to enable them to seek redress for harms done in the online environment. That presumption is not challenged in this chapter. The easiest, perhaps most universal case is the common abhorrence of child pornography. Most societies share the view that imagery of children under a certain age in a sexually compromising position is unlawful to produce, possess, or distribute. The issue in the context of child pornography is less whether the state has the right to assert control over such material, but rather the most effective means of combating the problem it represents and the problems to which it leads without undercutting rights guaranteed to citizens. The prevention of online fraud or other crimes, which often target the elderly or disadvantaged, likewise represents a common purpose for some measure of state control of bits online. Some would argue that intellectual property protection represents yet another such example, though the merits of that proposition are hotly contested.

Where the state cannot effectively carry out its mandate in these legitimate circumstances, the state reasonably turns to those best positioned to assert control of bits. Often, though not always, the state turns to Internet service providers (ISPs) of one flavor or another. The law enforcement officer, for instance, calls upon the lawyers representing ISPs to turn over information about users of the online service who are suspected of committing a common crime, such as online fraud. As criminals use the Internet in the course of wrongdoing, states need to be able to access the increasingly useful store of evidence collected online.

The strongest form of this argument is that online censorship and surveillance is a legitimate expression of the sovereign authority of states. Saudi Arabia, for instance, which implements one of the most extensive and longest-running filtering regimes, did not introduce Internet access to its citizens until the state authorities were comfortable that they could do so in a manner that would not be averse to local morals or norms. In particular, the Saudi regime has concerned itself with blocking access to online pornography, which it has done with a startlingly high degree of effectiveness over the past five years. A state has a right to protect the morality of its citizens, the argument goes, and unfettered access to and use of the Internet undercuts public morality in myriad ways. Many regimes, including those in Western states (including the United States), have justified online surveillance of various sorts on the grounds of ordinary law enforcement activities, such as the prevention of domestic criminal acts. Most recently, states have begun to justify online censorship and surveillance as a measure to counteract international terrorism

concerns, or more simply as the unalterable right of a state to ensure its national security. Whether or not states are right that they invariably have this sovereign authority is an open question—and beyond the scope of this chapter.[5]

Drawing a line: Where state online control implicates human rights standards

Some state-mandated acts of online control are not straightforward acts of local law enforcement. As the practice of online censorship and surveillance become more commonplace and more sophisticated, human rights activists and academics tracking this activity have begun to question whether some regimes of this sort violate international laws or norms. Quite often, the states that carry out online censorship and surveillance are signatories to international human rights covenants or have their own rules that preserve certain civil liberties for their citizens. The United States is home to a controversy of this sort, as the Electronic Frontier Foundation and others have filed a class action lawsuit against telecommunications giant AT&T for collaborating with the National Security Agency in a wiretapping program.[6]

The hardest puzzles are those cases where acts of local law enforcement seem to members of the international community to be violations of international norms. Consider a sovereign, jealous of its power, that disables access to opposition websites in the lead-up to an election —and then relents once the threat of losing control is abated. Or a state that routinely uses censorship and surveillance as a key element of a campaign to persecute a religious minority group. Or a state that relies upon online surveillance for the purpose of jailing political dissidents whose acts the state has committed to respect by international treaty. What about when a state is trying to protect public morals by keeping citizens from looking at garden-variety online pornography, but in so doing also block information on culturally sensitive matters, such as HIV/AIDS prevention or gay and lesbian outreach efforts?

What's at stake: Why Internet filtering and surveillance give rise to an ethical quandary

Just as states have a forceful claim to their right to exert sovereignty over their citizens, Internet censorship and surveillance prompt legitimate legal and normative concerns. The most straightforward of these concerns involve civil liberties. The online environment is increasingly a venue in which personal data are stored and across which personal communications flow. The basic rights of freedom of expression and individual privacy are threatened by the extension of state power, aided by private actors, into cyberspace. When public and private actors combine to restrict the publication of and access to online content, or to listen in on online conversations, the hackles of human rights activists are understandably raised. Some argue that the right of free association is likewise violated by certain Internet censorship and surveillance regimes that are emerging around the world. Most complaints cite the Universal Declaration of Human Rights or the International Covenant on Civil and Political Rights as grounding ideals to which many states have agreed.

Even if one agrees with the strong form of the state sovereignty argument, and sets aside the notion that Internet censorship and surveillance can represent a violation of international laws and norms, one might still contend that these regimes are unwise or unethical. Internet censorship and surveillance, the technologist might argue, violate the "end-to-end principle" of network design. The end-to-end principle stands for the proposition that the "intelligence" of the network should not be placed in the middle of the network, but rather at the end-points. The extraordinarily rapid growth of Internet throughout the world is chalked up to this simple idea. By imposing control in the middle of the network—say, at the "great firewall" that surrounds China or proxy servers in Iran or at ISPs in dozens of states around the world—rather than at the user level, the censors will stymie the further growth of the network.

Jonathan Zittrain makes a related—but at once more subtle and more compelling—argument against unwarranted intrusion into online environments by pointing to the importance of "generative" platforms in the context of ICT. Rather than hewing to the original design of the network, he argues, the decision maker should favor those technical decisions that enable acts of innovation on top of the existing layers in the ecosystem—including not just those layers in the middle of the network, but also those at the edges. The kinds of individual creativity made possible by the personal computer (PC), including self-expression in the form of the creation of user-generated content, might be thwarted by the presence of a censorship and surveillance regime. The on-again, off-again blockage of the user-generated encyclopedia, Wikipedia, makes this case clearly. The sporadic use of filtering regimes to block the use of Voice over Internet Protocol (VoIP), often to protect the monopoly in voice communications of a local incumbent, also stands for this proposition.[7]

A third argument against the use of online censorship and surveillance regimes, and the participation of foreign technology companies in their instantiation, is the impact that these actions may have on the emergence of democracies around the world. The Internet has an increasing amount to do with the shape that democracies are taking in many developing states. The Internet is a potential force for democracy by increasing means of citizen participation in the regimes in which they live. The Internet can open the information environment to voices other than those organs of the state that have traditionally had a monopoly

on the broadcast of important stories and facts, which in turn gives rise to what Fisher refers to as "semiotic democracy." The Internet can give a megaphone to activists and to dissidents who can make their case to the public, either on the record or anonymously or pseudonymously. The Internet can help make new networks, within and across cultures, and can be an important productivity tool for otherwise underfunded activists. Likewise, the Internet can function as a force for semiotic democracy—the notion that the control of cultural goods and the making of meaning are placed in the hands of many rather than few. Not least, the Internet is a force for economic development and the creation of a technologically sophisticated, empowered middle class, often in the form of local technology entrepreneurs. The Internet, in this sense, might function as a generative network in human terms, by helping to give rise to a more empowered citizenry.

New markets, new challenges

Technology, media, and telecommunications firms must decide whether to compete in markets where Internet censorship and surveillance are taking place against this contested backdrop. Internet filtering occurs in three regions of the world in particular: the Middle East and North Africa, Asia and the Pacific, and Central Europe and Asia. China continues to be the case that garners the most public attention, because of the size of its market and the extent to which the state has set in motion the world's most sophisticated filtering regime. But China is far from alone, as more than three dozen states carry out some form of Internet censorship and surveillance online.

How Internet censorship and surveillance works

To add to the complexity of the matter, the mode and extent of censorship and surveillance varies substantially from one state to another. States rely upon a combination of types of controls to accomplish filtering and surveillance. The most apparent mode is through the use of technology. In its simplest form, the state places special code on computers that lie between the individual end-user and the broader network. The job of the code is to block certain data packets from reaching their destination or simply to learn and record the contents of those requests and who made them. Sometimes it is apparent to the end-user that his/her request for a certain Web page has been blocked by the state; more often, it is not so apparent. The manner and extent to which censorship takes place online is easier to prove, while surveillance is more elusive—though, from the perspective of the state, it is not necessarily any harder to accomplish.

Online censorship (less so, surveillance) is carried out through nontechnical means as well. These controls are sometimes imposed by law: end-users are disallowed to access or to publish certain information that is deemed to undermine public order or other state interests, for instance. The laws are typically very broad, hard to understand, and even harder to follow with any degree of precision. These controls are also imposed most effectively through "soft controls," whereby cultural norms drive censorship or surveillance into the home or local community, often resulting in extensive self-censorship.

Integrated modes of online control: Combining the technical and the legal

For the purposes of this chapter, the most salient form of filtering is a combination of technical and legal control, trained on private actors with access to services that lie between an end-user and the network at large.[8] The state, unable to carry out filtering effectively on its own, requires private actors to carry out the censorship and surveillance for it. This requirement comes as a formal or informal condition of holding a license to provide Internet-related services in that state. So, for a large search engine, the mandate from the state might be to ensure that search results provided to citizens of that state do not include links to online content that is banned in that jurisdiction. Likewise, the provider of a weblog-publishing tool might be prompted to include controls that disallow an individual publisher from including certain words in the title of a blog post. An Internet service provider might be required to keep records of the online activity of all or some of its subscribers, or to monitor the group of people who seek to access certain kinds of content. The provider of a Web-based email service might be required to turn over the email messages of a user suspected of a crime, or who is simply believed to be a member of the political opposition. The owner of a cybercafé, who is required to maintain logs of who uses what computers in their big open room, might be called upon to report on the identity of a certain Web surfer who used a given PC during a given time interval, or to call a special number on the fly if the online activity of a customer sets off certain alarm bells. The reach of the state is far greater in the online space when private actors can be enlisted to cooperate closely with the state's enforcers.

Two taxonomies of private actors facing this quandary

Different ICT-related firms are called upon by states to carry out quite different online censorship and surveillance tasks. In seeking to fashion a policy response, it helps to disaggregate the firms implicated in this matter. Two taxonomies offer ways to disaggregate these firms. The first approach is to consider the type of business line of the firms, which is most useful for determining which firms might get drawn into an ethical controversy of this sort. The second, and more useful, taxonomy considers the

nature of the involvement of the firms in the online censorship and surveillance regimes. The second taxonomy points the way forward more clearly toward a solution.

Types of firms

Several types of corporations might find themselves snared in this net. The first corporations to find themselves involved in the censorship and surveillance controversy were technology hardware providers that sold the switches and routers involved in these regimes. In many parts of the world, Internet security firms sell the services and products used in the censorship and surveillance regimes. More recently, content and online service providers, whose customers are typically end-users, have been implicated. Looking ahead, as technologies and forms of digital content converge, other telecommunications service providers may well find themselves in a similar position.

Hardware providers

First, technology hardware manufacturers face scrutiny for their sales of routers, switches, and related services to the regimes that carry out online censorship and surveillance practices. According to the critique of human rights activists, companies that profit from the sale of the hardware that blocks the flow of packets online or enables states to trap and trace online communications are acting unethically. The problem, the critique goes, is akin to the Oppenheimer problem in the context of nuclear technologies. Although nuclear technologies can provide energy efficiently to those who need it, it can also power weapons of mass destruction of previously unprecedented power. The hardware manufacturers respond that the technologies sold to regimes that censor and practice surveillance are precisely the same as those technologies sold to firms and governments in states that do not carry out such regimes. This issue is not new, these firms respond. Dual-use technologies present this issue in an untold number of contexts. And the blame should be placed on those who implement the dual-use technologies in the suspect manner, not on those who produce the "neutral" technologies.

Software providers

The second class of firms implicated in this matter includes those corporations that sell the software and services that determine what gets blocked, recorded, or otherwise impeded. Internet security firms often serve states, corporations, and other institutions that seek to impede the free flow of packets for one reason or another. A library, for instance, might wish to block underage patrons from accessing pornography online. A similar software package could enable a state to configure a proxy server between a citizen and the wider Internet to block or track certain packets. Many of the states in the Middle East and North Africa that have filtering regimes in place rely upon software packages, and corresponding lists of banned sites, developed and compiled in the United States. These firms make arguments similar to those of the hardware providers: their technologies and services are dual-use in nature. The tool that can protect a child from seeing a harmful image can also keep a citizenry away from politically or culturally sensitive information online. The human rights critique, the firms argue, should be trained on the regimes that apply the services in a manner that violates laws and norms, not on the service providers who make the tools and update the lists. But, some observers suggest, the lists of banned sites include some NGOs that have no place there if the notion is just to protect children, for instance.

Online service providers

Most recently, the providers of Internet-based applications have found themselves facing hard questions about their activities in such regimes. A wide range of firms fall in this category: ISPs, email service providers, blog-hosting firms, search engines, and others. ISPs are asked to route traffic in certain ways in order to prevent citizens from accessing or publishing certain content; likewise, ISP data retention policies are a hot topic of debate in many jurisdictions, as the personal data they keep about citizens are at once sensitive and potentially useful in the context of law enforcement activities. Email service providers are routinely asked to turn over information related to subscribers. The makers of weblog software and hosting services are asked to block certain information from being published and told to take down the postings or entire blogs of subscribers. Search engines are required to limit the results that appear in response to certain queries entered by citizens. The nature of the ethical questions each of these types of firms face varies with the nature of the service they provide and the type of participation the state asks of them. In most instances, corporations respond that they have an obligation to obey local law with respect to services they offer in all jurisdictions. Corporations often perceive that they do not have the option of resisting the demands of law enforcement officials, for fear that the corporation or their local employees will face sanctions or that their license to operate will be revoked. Some corporations, recognizing the risks inherent in doing business in certain regimes, have limited the types of services that they offer in those contexts to avoid being placed in an uncomfortable role.

Online publishers

Corporations that publish information online are also caught up in this issue, though their situation is somewhat more straightforward. As a general matter, online publishers are treated as are other publishers in the states in which they operate, so the ordinary media restrictions that attach to newspapers and other traditional media also attach in the online space. Likewise, the notion of providing a single news or information service from one place in the world that is accessible from any other place, so long as it is not censored, remains a viable model. Large media companies, such as the BBC or CNN, tend to adopt this posture. Sometimes their content is filtered at the state level, but in those instances, the censorship is performed within the affected state. The ethical issue would arise only for those firms with local offices and offerings targeting a state that censors online material.

Telecommunications and other content delivery providers

On the horizon, one might imagine that additional classes of corporations could soon be drawn into this controversy. For instance, as mobile telecommunications providers continue to thrive and begin to function as digital content providers, it is only a matter of time before these intermediaries will be pressed into service by states as a requirement of their licenses to operate. Providers of VoIP services have already found that their services are sometimes blocked; filtering and surveillance, though posing new technical challenges, may follow. Firms that serve other businesses in delivering online content— including rich media, such as streaming audio and video, in additional traditional Web pages—also may be subject to such restrictions. Any large-scale intermediary that plays a role in delivering digital information to an end-user might find itself an arm of the state in the online environment—and will have to answer to the same questions as their peers in the hardware, software, and Internet services industries.

Types of involvement

Another way to categorize the firms that face increasingly difficult ethical questions in this context is to assess not the type of firm, but the type of involvement that a given firm has in the censorship or surveillance regime in question. Though the first taxonomy is simpler, this second taxonomy makes the ethical questions come into greater relief than assessing simply the type of firm involved. This second taxonomy provides a basis for the different types of ethical obligations that might apply to various firms.

Direct sales to states of software or services:

- **to filter online content**
  This category includes those firms that seek to profit from selling software or online services, including constantly updated block lists, that states use to implement their online censorship regime.

- **for surveillance**
  This category includes those firms that seek to profit

from selling software or online services, including suites of Internet security systems, that states use to implement their online surveillance regime.

Direct sales of dual-use technology used in:

- **filtering online content**
  This category includes those firms that seek to profit from selling Internet-related hardware, including related software and services, that states use to implement their online censorship regime.

- **online surveillance**
  This category includes those firms that seek to profit from selling Internet-related hardware, including related software and services, that states use to implement their online surveillance regime.

Offering a service:

- **that is subject to censorship**
  This category includes those firms that seek to profit from providing online services that result in a citizen of a state accessing information in a manner that is censored, such as through a search engine with results omitted or an ISP that refuses access to certain parts of the Internet.

- **that censors publication**
  This category includes those firms that seek to profit from providing online services that disallow a citizen of a state from publishing certain information online or that takes down previously published information at the behest of a state.

- **with personally identifiable information, subject to surveillance**
  This category includes those firms that seek to profit from providing online services that capture personally identifiable information about a citizen of a state and where that information may be monitored, searched, or turned over to state authorities upon request.

In certain contexts, the executives of a firm in any of these categories might believe that they do not face a hard ethical question. For instance, in the case of an email service provider that turns over information to a law enforcement officer about a subscriber in a manner that prevents commission of a crime, the corporation may have few qualms about its actions. By contrast, when the information sought by the state is related to a political dissident whose every action is lawful or protected by international norms, the ethical landscape is transformed. The same is true with respect to censorship: the blocking or taking down of hate speech may well be viewed differently than the blocking or taking down of the expression of certain

religious beliefs, for instance. The ethical question in any given instance may ultimately turn less on the precise role of the corporation in the digital ecosystem and more on the nature of the information or the manner in which it is requested of the corporation.

Potential responses

Reasonable people disagree as to the best means of resolving these emerging ethical concerns. One might contend that there is no ethical problem here—or, at least, that the ethical problem is nothing new. If an Internet censorship and surveillance regime is entirely legitimate from the perspective of international law and norms, the argument goes, then a private party required to participate in that regime has a fairly easy choice. If the executives of a corporation based in Europe disagree on a personal level with a censorship and surveillance regime, then they should simply exercise their business judgment and refuse to compete in those markets. Alternatively, those executives could decide to refuse to comply with the demands that they believe put their firm in a position in which their ethics are compromised and then accept the consequences—including possibly being forced to leave the market—that befall them as a consequence of their resistance. One option, then, is to do nothing, to accept the status quo, and to let the trend play itself out. In the unlikely event that online censorship and surveillance were to cease across the globe, or if states were to stop calling upon private actors to get the job done, or if corporations were to stop expanding into other markets, the problem might be resolved cleanly. But absent such changes in the facts as they stand, the stakeholders who care about these issues have a series of possible ways to move forward to resolve the conflicts.

Industry self-regulation

The most likely—and most desirable—means of resolving this problem would be for the relevant industries themselves to come up with a sustainable manner of ensuring that they operate ethically in these charged contexts. One or more groups of industry members might come up with a voluntary code of conduct that would govern the activities of individual firms in regimes that carry out online censorship and surveillance. This process would profitably include additional nonstate actors, such as NGOs and academics, as well as regulators with relevant expertise and authority. Corporations might further refuse to do business in regimes that put them in a position where they cannot comply with local laws while also honoring the voluntary code. Alternately, individual firms could come up with their own principles, much like a privacy policy on today's Internet, with statements to clarify to users, shareholders, and others how the firm will handle these situations. Last, an outside group might come up with a set of principles to which firms could be encouraged to subscribe, on the model of the Sullivan Principles and the Apartheid-era South Africa, and based upon which an institution might emerge to support the principles. As in the case of the Sullivan Principles, one or more states might ultimately take the principles and convert them into national law once they have reached a point of stability and acceptance.

The elements of such a code or set of principles might be general—a set of core commitments such as transparency, rule of law, the rights of free expression and individual privacy, and so forth—or more specific, according a taxonomy of the second sort described above. The more specific the code, the more useful, almost certainly, though the reality of getting competing businesses to agree to detailed business practices of this sort is daunting.

A critical part of such a voluntary code would be either to enact them into law or to develop an institution charged with monitoring adherence to the code and enforcing violations. This institution—perhaps not a new institution, but a pre-existing entity charged with this duty—ought to include among its participants representatives of NGOs or other stakeholders without a direct financial stake in the outcome of the proceedings. This institution might or might not have state regulators involved as partners to ensure compliance. The institution would play an essential role in ensuring that the voluntary code of conduct not only has force over time, but also that it continues to address the ethical issues as they evolve.

Law

The legal system might provide one or more ways to resolve the ethical dilemmas facing corporations in the context of states that censor or carry out surveillance online, though classic state regulation is unlikely to be the most effective means of addressing the problem over time. Individual states might require corporations chartered in their jurisdiction to refrain from certain activities when operating in other states. The analogy in the US context runs to the Foreign Corrupt Practices Act, which disallows corporations chartered in the United States from bribing foreign officials and other business dealings that would violate United States law if carried out in the home market. A "hands-tying" regulation of this sort might be combined with other approaches that might attack particular parts of the problem, but would be unlikely to resolve the conflict outright. Such approaches might include funding for pro-democracy activities in the online context, banning the sale of certain technologies, banning the location of servers in certain places, or applying pressure in the context of trade negotiations on those states that are placing the corporations in a difficult ethical position.

The reasons not to rely upon traditional legal mechanisms in this context are that such mechanisms will likely be blunt instruments and will almost certainly take so long to put in place that the contours of the problem will have changed beyond recognition by the time of enactment. Changes to the statute or treaty may be equally hard-won. Laws fashioned in this fast-moving environment will function as a hopelessly trailing indicator. Law should be seen as a component of a solution, but not the primary approach.

International governance

Problems in cyberspace have rarely been solved by coordinated international action, though there is no inherent reason to believe that international cooperation or governance could not play a meaningful role in resolving these ethical dilemmas. The United Nations has not been involved in extensive regulation of the online space, perhaps with the exception of the role of the International Telecommunication Union (ITU) in related telecommunications contexts. The Internet Governance Forum, ably chaired by Nitin Desai and under the secretariat of Markus Kummer, has the authority to conduct an international dialogue on issues related to the information society. An international treaty process, though cumbersome, could emerge as the way ahead. Some activists have considered litigation under existing human rights agreements.

Other modes of pressure

Human rights activists, academics, and shareholder advocates have played an important role to date in the public discourse related to this issue. The US Congress has held hearings on this matter in order to draw attention to the actions of large technology firms. The New York City Comptroller has recently filed shareholder actions with certain technology firms to prompt action on these topics. Human rights organizations and investor groups around the world have hosted forums to shine a spotlight on corporate involvement in filtering and surveillance regimes. Although the involvement of NGOs and other outsiders in the process of addressing these ethical issues is not a solution in itself, it is clear that these stakeholders play an important role in any next steps.

Conclusion

The most promising approach to addressing the ethical dilemma facing multinational corporations doing business in states that carry out online censorship and surveillance is for the information technology community to work together to develop a voluntary code of conduct, and possibly to enact that code into law over time. That code must be coupled with the establishment of a reliable mechanism for monitoring and compliance assurance, whether through traditional state-based enforcement or an institution created for this purpose. This approach could, at once, be responsive to the nuanced issues involved, flexible over time as the technologies and politics shift, and sustainable over the long term. Such a process ought to include the NGO community at the table in a supportive, nonadversarial, mode. State regulators might also be drawn into the process in constructive ways. A process to establish such a code is well underway, with Google, Microsoft, Vodafone, and Yahoo! working with two dozen investor, human rights, and academic groups, such as the Center for Democracy and Technology, Business for Social Responsibility, and Amnesty International.[9] The affected industry need not—and ought not—go it alone.

Though the environment is too complex and unstable for the standard modes of law-making to work in the near term, states do have a role to play in helping to resolve this tension. A patchwork of competing state laws that restrict corporations chartered in one locale in how they do business in this regard in other locales could be counterproductive. The challenges inherent in framing the Global Online Freedom Act of 2006 and 2007, in the US context, point to some of the many the hazards of this approach.

The proper role of the state in the context of addressing this problem is twofold. First, those states that are more concerned with what their corporations are doing elsewhere than they are with what these corporations are doing at home should support these corporations as they seek to act responsibly in a complex global environment. That support might come in the form of state involvement and encouragement for participation in the voluntary code as the industry works with the NGO and academic communities to derive a set of ethical guidelines. Support might also mean using leverage in trade negotiations—by raising this issue in bilateral negotiations with key states, for example—to lessen the extent that corporations are placed in this position in the first place. Where constructive, states might consider rule-making that ties the hands of their corporations to provide support for their refusal to operate outside of the bounds of these ethical constraints. And states might enact laws that codify the principles that the industry comes up with through the collaborative process that is underway. But states alone are unlikely to be able to lead constructively and quickly enough to address this problem.

On a fundamental level, the states that are increasing Internet filtering and surveillance themselves are best positioned to resolve this tension. In some instances, the primary driver for change might be a careful review of the human rights obligations, whether these obligations come through treaty or otherwise, that place limits on state sovereignty to act in this manner. Human rights activists may prompt this review through litigation if states do not undertake it themselves. In other instances, the driver might be economic: there is little argument that the development of a competitive environment for businesses using ICT is a positive factor in economic growth, particularly of developing economies. In either event, states that place restrictions on Internet usage and seek to leverage network usage for purposes of surveillance outside the bounds of human rights guarantees do so at some political and economic peril. And multinational corporations have every incentive to work hard toward an industry-led, collaborative approach to resolving the tension in the meantime.

Notes

1 See Barlow (1996).

2 The trajectory of this struggle for control has been well documented. See, for example, Zittrain (2003) and Goldsmith and Wu (2006, pp. 65–86).

3 See http://cyber.law.harvard.edu/is02/ (last accessed December 26, 2006).

4 It has not yet been determined conclusively whether states would force foreign corporations to leave the jurisdiction for disobeying these edicts.

5 Note that Dutta and Jain (2006) consider, on p. 14, that "the number of Internet users in 2003 exceeds the number of personal computers on a global level, as compared to 1999, when the situation was reversed." To the extent that this phenomenon is due in part to shared Internet connections, such as cybercafés, (no doubt in addition to mobile devices, among other factors), these points of presence become increasingly important to the story of censorship and surveillance. It is worth noting that this chapter does not seek to address all forms of online control carried out by private parties at the behest of states: for instance, much online control is carried out by local firms, such as ISPs or cybercafés, that provide online services in their home markets.

6 See www.eff.org/legal/cases/att/ (last accessed December 26, 2006) for details about this lawsuit.

7 Consider the relevant arguments set out in "The Infrastructure Challenge in Telecommunications: A Role for Regulation," Chapter 1.2 of *The Global Information Technology Report 2006–2007.*

8 See Reidenberg (2004); see also Palfrey and Rogoyski (2006).

9 See www.socialfunds.com/news/release.cgi/7272.html for a full list of all groups involved as of January, 2007.

References

Barlow, J. P. 1996. *The Declaration of Independence of Cyberspace.* Available at http://homes.eff.org/~barlow/Declaration-Final.html (last accessed December 26, 2006).

Beardsley, S., L. Enriquez, M. Guvendi, M. Lucas, and A. Marschner. 2006. "The Infrastructure Challenge in Telecommunications: A Role for Regulation." *The Global Information and Technology Report 2005–2006: Leveraging ICT for Development.* Hampshire: Palgrave Macmillan. 25–37.

Dutta, S. and A. Jain. 2006. "Networked Readiness and the Benchmarking of ICT Competitiveness." *The Global Information and Technology Report 2005–2006: Leveraging ICT for Development.* Hampshire: Palgrave Macmillan. 3–24.

Fisher, W. W. 2004. *Promises to Keep: Technology, Law, and the Future of Entertainment.* Palo Alto: Stanford University Press. Chapters 1 and 6.

Goldsmith, J. and T. Wu. 2006. *Who Controls the Internet: Illusions of a Borderless World.* New York: Oxford University Press.

OpenNet Initiative. 2002–2007. Available at www.opennet.net.

Palfrey, J. and R. Rogoyski.. 2006. "The Move to the Middle: The Enduring Threat of 'Harmful' Speech to Network Neutrality." *Washington University Journal of Law and Policy* (21): 31–65.

Reidenberg, J. R. 2004. "States and Internet Enforcement." *University of Ottawa Law & Technology Journal* 1 (213): 213–30.

Zittrain, J. 2003. "Internet Points of Control." *Boston College Law Review* 44 (2): 653–88.

Part 2
Access to ICT: Selected Case Studies

Estonia: A Sustainable Success in Networked Readiness?

SOUMITRA DUTTA, INSEAD

I would not consider it an exaggeration to say that "e" has put Estonia back on the world map. Living in a small country with limited resources, the pressure to make public administration as efficient as possible forced our Government to look for opportunities to take advantage of modern technology and turn Estonia into eEstonia.

— Meelis Atonen, Minister of Economic Affairs and Communications[1]

An astonishing amount of innovation has emerged from Estonia, a country of 1.4 million inhabitants that has been a formal state for less than 20 years. E-leadership has proven to be instrumental in helping Estonia through the painful transition from centralized state planning to the model of modern governance it is today. The country has pioneered and developed unique solutions and systems that have become an integral part of the life of most Estonians. Estonia's clear vision and leadership in information and communication technologies (ICT) have led to results that often surpass those achieved by the older democracies of Western Europe. This is especially remarkable when one notes that the nation was ruled by foreigner powers— including Denmark, Germany, Sweden, and Russia—for centuries. The merger of e-leadership and political vision has been one of the critical factors in its economic growth, the spreading of democracy and its resulting accession to the European Union (EU).

Nevertheless, all is not perfect in the land of e-cabinets, Skype (the peer-to-peer Internet telephony network), mobile payments, and electronic ID cards. Despite many major and very visible successes (see Box 1), Estonia still faces a number of challenges and must overcome certain weaknesses. The question that remains open today is whether the country can leverage its knowledge and best practices and turn this advance into a truly sustainable model.

The support of Alexandre Brodbeck and Cyril Gueche, MBA students at INSEAD in 2006, for supporting the research and writing of this chapter is gratefully acknowledged.

Box 1: E-facts about Estonia

- Fifty-four percent of the population (aged 6–74 years) are Internet users (*TNS EMOR*, Spring 2005).

- Thirty-four percent of the households have a computer at home, and 82 percent of home computers are connected to the Internet (e-Track Survey, *TNS EMOR*, Spring 2005).

- All Estonian schools are connected to the Internet.

- There are over 700 Public Internet Access Points (PIAPs) in Estonia, 51 per 100,000 people (one of the highest numbers in Europe).

- Income tax declarations can be made electronically via Internet. In 2005, 76 percent of Estonian taxpayers declared their income tax via the Internet.

- Government expenditures can be followed on the Internet in real-time.

- Cabinet meetings have been changed to paperless sessions using a Web-based document system.

- Seventy-two percent of Estonian Internet users conduct their everyday banking via Internet (e-Track Survey, *TNS EMOR*, Spring 2005).

- Ninety-three percent of the population are mobile telephone subscribers (National Communication Board, Spring 2005).

- Estonia is completely covered by digital mobile telephone networks.

Source: Ministry of Foreign Affairs, Government of Estonia. Available at www.vm.ee/estonia/kat_175/pea_175/1163.html.

Estonia's unique ICT advances in e-services

The entire notion of the internet and the information society has attained an unusual attraction at all levels in Estonia. One would for instance not find many nations where both the political and economic elite and more ordinary citizens share the same vision to the extent that seems to be the case in Estonia on developing the most advanced information society.[2]

E-government

The government of Estonia has been a leader in the country's use of ICT. Not only has the government created the appropriate regulatory and policy environment to support the widespread adoption of ICT, it has also been a pioneer in using ICT for its own processes. For example, in August 2000, the government of Estonia, in a world pioneering advance, changed its cabinet meetings to paperless sessions using a Web-based document system. The Web-based system automated the preparation process and the proceedings of the cabinet meetings and enhanced greater communication and information sharing across different branches of the government.

In a sight rarely seen in other countries, Estonian ministers peruse draft bills and regulations, make comments and suggestions, and vote entirely online at computer terminals. The system, coupled with the use of digital signatures, eliminates the need for sending volumes of papers across government offices. Ministers are also free to participate in cabinet meetings remotely. The Web-based automation saves the government of Estonia approximately €192,000 per year in paper and copying costs.

Estonia was also one of the first countries in the world to introduce electronic voting. E-voting has been in development in Estonia since 2002 and was used for the first time in local council elections in October 2005. The system allows citizens to sign their ballots electronically via the Internet. Electronic voting does not eliminate traditional voting, but it provides a convenient alternative option for voting.

Electronic voting takes place only on advance polling days (sixth to fourth day before election day), and government-issued ID cards are used for voter identification. Today, nearly 90 percent of voters have an ID card. The success of Estonia in effectively developing and implementing e-voting is remarkable when one notes the challenges that even technology leaders such as the United States have in implementing automated voting machines.

Estonia has also started using ICT for enhancing the participation of its citizens in formulating government policy. In the summer of 2001, the government created Täna Otsustan Mina ("I Decide Today"), a Web page where ministries upload all their draft bills and amendments. People are encouraged to go to the site and review and comment on the legislative process and suggest amendments to proposed legislation. Approximately 5 percent of all suggested amendments are incorporated into the final enactment.

A large range of government services are offered online to Estonian citizens. E-Citizen is a nationwide project aimed at enhancing online interactions between Estonian citizens and the public sector. An Internet portal called the Estonian State Web Centre was created in 1998. It contains links to all governmental institutions' websites and everybody has access to almost all the official documents. In comparison with other EU members, Estonia stands out for the ease of applying for various environment-related permits on the Internet, of registering a new business, of filing a statement with police, and of conducting various health-related and social security matters.

E-society

"Internet—100 metres" can be read on a traffic sign that points to a potholed village road on the Estonian island of Hiiumaa. Since 1997, public Internet access points have been set up all over Estonia. They are located in local libraries, post offices, and even rural general stores. "Free Internet access should be available within cycling distance," as is stated in the Countrywide Equal Access programme.

In 1997, the Tiger Leap programme started reshaping the Estonian educational system so as to fulfil the needs and requirements of an information society. A whole generation is growing up with the Tiger Leap programme. Information technology is actually being implemented in the educational system, since all Estonian schools are now connected to the Internet. This unprecedented access to knowledge is promoting openness and tolerance to such an extent that it is changing the entire future of Estonia. It is impossible to predict that future, but we do know that these well- informed children will one day be adults.[3]

Critical enablers for the creation of an effective e-society are access to the Internet and the necessary skills to use the technology. The government has visibly stimulated increased access to PCs and the Internet over the years. For example, in 2001, the government—in a public-private partnership—announced the "Look @ the World" project, aimed at increasing Internet users in Estonia to over 90 percent within three years. As part of this partnership, private companies committed to invest in the project a sum equal to the government's annual information technology (IT) budget. In another initiative in April 2002, the Look @ World Foundation launched a project to provide basic computer and Internet skills to adult citizens. By the end of the project, on March 31, 2004, 102,697 people (10 percent of the adult population of Estonia) had completed the training.

Since January 2002 Estonia has also been a pioneer in the introduction and use of electronic identification cards. In addition to many advanced security features, the card has a machine-readable code and a microchip containing the visual data on the card as well as two security certificates (long number series) to verify the individual and supply digital signatures. The card currently doubles as a passport and visa valid in several partner countries.

As early as February 2000, the Estonian parliament approved a proposal to guarantee Internet access to each of its citizens just like any other constitutional right. People all over the country can access the Internet free of charge from over 700 public Internet access points (PIAPs), or 51 PIAPs per 100,000 people. Most PIAPs are located in libraries and other municipal buildings across the country. Thanks to the Tiger Leap program, Estonian school children are above-average users of the Internet.

Seventy-two percent of Estonian Internet users conduct their everyday banking via the Internet.[4] Internet banking has become a common channel through which people perform transfers, pay for services, pay taxes, communicate with the Tax Board, and so on. Access to various information and bank services through mobile telephones using Wireless Application Protocol (WAP) are popular; for example, as of June 2005, the financial institution Hansapank alone has 36,000 WAP clients.

The research presented in this year's *Global Information and Technology Report* reveals the good standing of Estonia with an overall Networked Readiness Index rank of 20 (see Chapter 1.1). For the fifth consecutive year, Estonia is also first among Central and Eastern European countries on the NRI ranking. Estonia's performance is driven by its high performance in government and citizen e-readiness and the strength of its regulatory framework for ICT. The Estonian Penal Code has included articles on security-related computer and data crimes since 1997. Data-related criminal legislation is constantly updated and revised according to international standards and the progress of technology.

Key aspects of Estonia's information policy are highlighted in Box 2. Additional e-services and mobile services (m-services) are listed for purposes of brevity in Boxes 3 and 4.

Box 2: Key aspects of Estonian ICT policy

- E-services: developing e-services for citizens, the business sector, and public administration.

- E-democracy: creating and analyzing ICT solutions that can help the development of e-democracy.

- Public-sector: increasing the effectiveness of the public sector.

- E-education: increasing the computer literacy of the population.

- E-security: improving and developing the Information Technology Security Policy.

- International reputation: sustaining the international reputation of Estonia as a nation with rapid and effective ICT development.

- E-involvement: increasing the opportunities for society to use ICT and digital solutions.

Box 3: Additional e-services for citizens

Income tax declaration

Thanks to the use of electronic ID cards as a safe way to access private data, citizens can declare their taxes online. They can also have constant access to their files and submit requests for value-added tax refunds.

Social security

All the necessary forms and regulations about the social security system are accessible on the Estonian social fund website. A dedicated portal gives practical information about the social rights and obligations of the people living in Estonia, in addition to tips about dealing with Estonian state institutions.

Family subsidies

The online Parental Benefit service allows people to submit applications for the family support program. The entire system is connected to different state databases and eliminates the need for citizens to submit data already known by the state.

Medical insurance

Thanks to the use of electronic ID cards, citizens can use e-services available through the national portal to check the validity of their health insurance, their address (and if necessary they can correct it), the name of their family physician, and the payment of sickness benefits.

Library services

The complete catalogue of books available in Tallinn libraries is available online. It is also possible to reserve the books online.

Education

It is possible to enroll online in higher education institutions. The processing, decision-making, and informing students of their acceptance (or otherwise) is made in a single environment on the Internet.

E-voting

Voters can use the Internet for voting in local elections. Voters are authenticated through their electronic ID cards and their selections are confirmed with their digital signatures.

Key success factors: What is so special about Estonia?

Estonia has clearly become a frontrunner in developing ICT. It is useful to understand what about its background and its direction make this aspect of Estonian progress so spectacular.

Historical background

Estonia has a long history of occupation by foreign powers. Shortly after the fall of the Iron Curtain, Estonia was among the first countries from the former Soviet bloc to regain its independence, in 1991. In the early 1990s the Estonian government decided to scrap all of its existing outdated Soviet hardware and replace it with a new infrastructure. Since this symbolic date Estonia has worked hard to build democratic institutions, independence, and recognition on the world stage. Rewards for these efforts were quick, culminating in Estonia's accession to NATO and to the European Union in 2004.

Among other things, the Soviet legacy included certain important fundamental building blocks. Indeed, the Soviet regime fostered high levels of literacy in the country (over 99.8 percent of the population), and established the expertise of the Baltic countries in scientific research. Within the Soviet bloc, Estonia was assigned the task of conducting advanced electronic and software research (which was applied mainly in military weapon development and manufacturing). Moreover, the Institute of Cybernetics was also created in Estonia, along with one of the most important centers of Artificial Intelligence research.

This technological heritage is a key factor in explaining the Estonians' receptiveness to advanced technology (for example, 54 percent of Estonians use their mobile telephones for purposes other than communication, such as for m-payment). Having been educated in a high-tech environment, Estonian entrepreneurs later grew out of an energetic, youthful society that embraced technology as the fastest way to catch up with the West.

Box 4: Mobile services available

Mobile payment

By sending an SMS, including the amount of the transaction, a Tallinn customer can pay for products and services in more than 500 locations in the Estonian capital.

SMS Gateway

The software utility SMS Gateway promotes entrepreneurship and mobility by letting everybody make their own SMS-service, through a user-friendly Web-interface, and earn money.

Telematic services

Telematics enable one to control various devices via a mobile telephone developed by the Tallinn-based Oskando company. Examples of applications include gate controller (open and close parking gates via a telephone call or an SMS message), home controller (control electrical heating, internal and external heating, security systems, saunas, and so on, via an SMS message) or car controller (control security and auxiliary devices such as central locking control, heating device control, and car blocking).

Mobile positioning

The Regio company developed a software that increased the accuracy of mobile positioning between two and eight times, to a precision of between 100 and 500 meters. The service allows employees of emergency call centers to see a computer map showing the location of the individual who is making the emergency call. Professionals also use the service to locate and manage their fleets, thus optimizing routes and collecting useful information to feed an activity-based costing (ABC) accounting system.

M-parking

Among the many mobile solutions pioneered in Estonia, the mobile parking is arguably the most famous and well known outside the country. Mobile parking is a fairly recent innovative joint initiative launched by the Estonian Mobile Telephone (EMT) and the parking surveillance authorities. In order to use the system, after parking a vehicle, the driver sends an SMS to the parking center; the driver sends another when leaving the parking senter. The parking fee is subsequently added to telephone bill or deducted automatically from client's mobile bank account.

M-voting at events

SMS-poll participants can use their mobile telephones to answer surveys or give immediate feedback to conference presenters.

M-teacher

The m-teacher service provides teachers with an interface through which they can send text messages to parents when important information needs to be forwarded (class events such as excursions, theater trips, and so on, as well as individual messages about child's progress—about good grades, bad grades, skipping school, and so on). The goal of m-teacher is to simplify and foster the dialogue between school and home, teachers and parents.

M-neighborhood watch

Taxi and bus drivers, security companies, and other active people can receive SMS notifications on issues (missing persons, stolen cars) that require watchful eyes. Messages are sent by the police control center.

M-library

A library sends out notifications about waiting lists to the readers' mobile telephones; if a person wants to borrow a book, movie, or audiotape that is currently not available, he or she can register and receive an SMS when it becomes available.

M-ticket

M-ticket is a solution developed by Mobi Solutions Ltd, Estonia, enabling public transportation clients to buy tickets via their mobile telephones. The customers receive a text message that serves as the mobile ticket. The solution includes a simple and effective method for checking the validity of the mobile tickets.

85

The role of the government

The key message we can teach is that e-government is not only making governance more effective and transparent, but it gives the possibility to develop a real partnership between the government and people.

— Mart Laar,
Prime Minister (1992–2002) of Estonia[5]

Estonian leaders played a major role in promoting and implementing significant advances in ICT. Their strong leadership ensured the long-term commitment of resources and expertise and the cooperation of disparate factions. Mart Laar, Estonia's prime minister from 1992 to 2002, was able to articulate a unifying theme that propelled the e-government initiative through all the necessary steps. "His courageous program as Estonia's prime minister created the 'Baltic Tiger,' a free and prosperous nation that is a model for the world to emulate," notes Ed Crane, the president of CATO institute, which recently awarded Mart Laar the Milton Friedman Prize for Advancing Liberty.[6]

Estonia could not have made the progress in e-leadership that it did had it not been for Mart Laar. Mart Laar made a number of risky decisions, from a political standpoint—such as introducing mandatory electronic ID cards and online income tax payments—setting the scene from an early stage in his mandate for e-governance, the key to Estonia's future. For example, by declaring Internet access a human right, the Estonian government created strong national support for the implementation of their Tiger Leap program. Several commentators have identified Laar and members of his government as the cause of Estonia's current e-leadership. Indeed Laar surrounded himself with advisers who shared his vision and brought strong technological skills, such as Ivar Tallo (now head of the E-Governance academy in Tallinn) and Linaar Viik (now a partner responsible for strategic development at Mobi Solutions, and a member of the board of the Information Society and Innovation of the Government of Estonia), to name but a few. For the information society unit at Estonia's ministry of economic affairs and communications, support from the political leadership of the country has been crucial. The small size of the country enabled such visionary leaders to mobilize the whole territory around a technological message and target.

Public-private partnerships

To implement its strategy, the Estonian government was required to call upon the private sector. According to Mart Laar, "Most disadvantageous [from the government's perspective] was that we did not have large enough resources to build up e-government, and this was very good because then we had to build public-private partnerships."[7] The creation of such strong partnerships, the liberalization of the telecommunications industry, and the early and widespread adoption of Internet inside the government created a virtuous circle in which the government acted as a driving force for the ICT industry.

The public and private initiative largely enabled the spread of new e-services. The government's push for low Internet connection costs and widespread access was picked up by benevolent individuals. "Mr. Haamer, one of Estonia's unofficial geeks, is largely responsible for a level of WiFi connectivity—even in remote areas—that puts the biggest cities in America to shame. For the last three years, he and a handful of volunteer evangelists with the WiFi.ee organization have successfully lobbied Estonian cafes, hotels, hospitals, city parks, local governments and even major gas stations to start offering Net access, helping to design and set up the networks."[8] Today WiFi reaches hundreds of public locations (city squares, hotels, pubs, coffee-shops, gas stations, and so on), and more than two-thirds of the locations are free of charge.

The main scientific mobile applications and technology development centers have been developed in cooperation with the two largest universities and telecommunications companies in Estonia, which provide facilities for experimental development and testing new applications. The favorable attitude of the Estonian public toward new technologies also makes it a good place for introducing new services. The development centers unite the graduate students of ICT and provide opportunities for international cooperation.

Estonian banks themselves are interested in the wide dissemination of ICT knowledge and skills amongst citizens, and are participating actively in these government projects, donating significant financial aid, and even distributing free hardware. Wider ICT awareness directly benefits banks by significantly reducing transaction costs.

The ICT sector

Considerable investments in ICT networks by foreign, mostly Nordic, investors have been made to modernize the ICT infrastructure in Estonia. Strength in telecommunications technologies has made Estonia an attractive partner for world leaders such as Ericsson and Nokia. There are about a dozen key mobile value-added service developers and providers in the Estonian market.[9] The new and advanced infrastructure provides a good test-bed for new products and services.

The Estonian ICT sector is small but dynamic and oriented toward innovative high-tech applications. The sector generated approximately 9.2 percent of the Estonian GDP in 2004.[10] There are more than 400 ICT companies operating in Estonia, employing about 9,000 people. Estonian firms have developed comparative competitive advantages within banking software systems, encryption solutions, and mobile solutions.

Estonia's strength is not in low costs of labor or mass production, and Estonia cannot compete with large countries in top technology development. Rather Estonia is strong in sectors that require above-average professional development and more advanced requirements such as high-quality, advanced features, and quick delivery.[11]

The role of large Estonian companies

The key role of Estonian banks in public-private partnerships has already been mentioned. Large institutions such as Hansapank have invested in cutting-edge electronic and mobile platforms, the success of which have depended on achieving high levels of adoption among their customers and, more generally, the population's e-readiness. Estonian bank's online clients total nearly 1 million—an astonishing proportion when one considers the total population of 1.4 million. Estonian banks have been pioneers in rolling out many mobile services. For example, Hansapank and Uhispank have offered the opportunity for their customers to pay for purchases (at motels, stores, pharmacies, and so on) with their mobile telephones since 2002.

On the business side, Estonian Mobile Telephone (EMT)—Estonia's largest mobile operator—has acted as an incubator for ICT startups, providing funding and technology support for little or no immediate financial return. For example, EMT's revenues from m-parking fees, a service co-founded by EMT, Regio, and Mobi Solutions, amounted to barely more than a few hundred thousand euros. Although revenues from mobile services remain very small, this is not considered a problem because the increasing availability of mobile services increases the customers' willingness to use services and create a favorable environment for innovation. Not only does EMT act as a socially responsible company but it has also effectively anticipated the explosion of the demand for mobile content and the convergence of devices (mobile telephones, computers, PDAs, and so on)—both domains in which investment needs must be satisfied now, focusing on long-term growth rather than short term profitability. Although EMT's dominant position in the local market (EMT's largest competitor represents only 50 percent of EMT's market share) could have led it to maintain the competitive status quo, instead it has actively promoted innovation and competitiveness by supporting promising local ventures and entering into technology-exchange agreements with competitors.

An inclusive information society

It is no longer possible to speak of technology as something, which creates an insurmountable chasm between the work opportunities of different people. Quite the opposite—the computer has now become an everyday tool which connects society.[12]

The concept of the information society has always been high on the political agenda in Estonia. The challenge of developing a truly inclusive information society very often hinges upon the ability to make the various Internet and mobile services available in all walks of life and especially in rural and remote regions. It is usually not a problem to roll out WiFi and mobile solutions in capitals and major cities of a nation. However, the challenges come when attempting to roll out WiFi in remote villages. In Estonia the initiative Village Road has, in a very operational manner, addressed this challenge of reaching out. Village Road brings together multiple stakeholders—government institutions, commercial entities, citizens, companies, and financial institutions. In order to make it work across the nation, a national collaboration network has been formed. This network is led by a national Village Road council and operated by working groups across the country. Information society tools such as an interactive Web page have been put in place to help coordinate efforts and exchange lessons learned. The Village Road project is not only about technical solutions and information society. It is also about building up social capital on a national level and should be counted as a success factor. At a more specific level the Village Road has implemented traditional measures such as linking farmers and poorer families up with Internet connections and providing them with cheap PCs.[13]

Regional integration

From the 1990s onward, relations between urban regions of Helsinki and Tallinn have enormously intensified. The integration of Estonia in the European Union has also facilitated this networking.

The Baltic Pallet is an example of a joint regional project for cross-border cooperation among the cities of Stockholm, St. Petersburg, Tallinn; the Häme, Helsinki, Leningrad, and Riga regions; the Åland Islands; and southwest Finland. The Baltic Pallet is creating a region in the frontline of the information society and aims both to provide sustainable development of the urban areas and to develop corridors connecting the urban systems. This project focuses on tourism, environmental issues, infrastructure, and information technology, among other things. It allows information about leading projects, in the ICT sector for example, to be exchanged between the project members, allows experiences to be shared, and encourages replication of successful implementations to be facilitated.

Can Estonia continue to be an e-leader?

Estonia has succeeded in branding itself as an e-leader over the last decade. It has been repeatedly quoted in the leading international media as an exemplary case of a knowledge-based society. In fact, Estonia has become synonymous with mobility and the Internet. This is remarkable given that before Skype, there was not a single internationally renowned company that had emerged from Estonia's ICT sector. Many attribute this successful e-branding of the country to the synergy across multiple mobile initiatives that have been adopted by the population.

Investment in R&D

However, maintaining this e-leadership will not be easy. Estonia's business environment is dominated by small- and medium-sized businesses that are often suppliers to international companies and do not invest sufficiently in basic R&D. The success of everyday mobile applications in Estonia masks the low level of R&D investment (0.8 percent of GDP 2005) in the country. [14] Furthermore, about three-quarters of this R&D funding comes from public sources (as compared with about two-thirds of R&D from private sources in other European nations). The EU Commission concludes in its 2005 *Innovation Trend Chart* report on Estonia that the major obstacles for R&D progress in Estonia are:

- insufficient awareness of the need for innovation policy among politicians,

- different understanding of innovation policy among various ministries, and

- lack of resources. [15]

Estonia will face challenges in continuing its development into an advanced information society if it does not invest more in R&D. Businesses and universities need to partner closely to create commercially successful innovations.

Clusters of innovation

Many relish the attention that Skype has brought to Estonia. But Estonia cannot build a long-lasting technology industry on a single hit or even a few hits. Hotbeds of technology entrepreneurship such as Silicon Valley are home to clusters of innovative companies that interact closely with each other and share knowledge and skills. Skype is a closed company, with proprietary software and owners who were so secretive about their plans that for a time local journalists did not know where its offices were.

The company's two founders are not even Estonian. Niklas Zennstrom is a Swede, and Janus Friis is a Dane. Skype's legal headquarters are in Luxembourg; its sales and marketing office is in London. Although Estonian developers wrote Skype's basic code, only a fraction of the Skype bonanza went into Estonian pockets.

Part of the problem for Estonia's entrepreneurs is the nation's inexperience in capital markets. It regained its independence only in 1991, after the collapse of the Soviet Union. Estonia's entrepreneurs do not yet have the Rolodexes. Then, too, there is its small size. Estonia's entire software development industry employs roughly 2,500 people, less than the research and development staff at a major global technology company. Landler points out that "some people contend that Estonia's success is a function of hard work and happy circumstance rather than raw talent." [16]

The government of Estonia has not adopted a cluster development policy, preferring rather to support horizontal measures addressing a wide range of enterprises with the same (or similar problems). However, a triple-helix cooperation model—across science, industry, and government—has been developed via a Competence Center program that brings together key enterprises and academic research centers in specific technologies and sectors. Many experts question whether Estonia can continue on its path of technology leadership without a focused approach to developing clusters of innovation in key sectors such as ICT and biotechnology. Many other economies—such as India and the United Arab Emirates—are successfully utilizing a cluster-based approach to stimulating technology leadership and entrepreneurship.

The infrastructure

Estonia is ranked a very creditable 20th in this year's Networked Readiness Index. This is in no small measure due to the modern ICT infrastructure that Estonia put in place as a replacement for the antiquated telecommunications infrastructure that it inherited from the Soviet years. However, Estonia remains a "relatively poor" country, with a GDP per capita about a third of the EU average. There are also significant improvements in infrastructure and access that need to be made.

According to the *IT in Public Administration of Estonia, Yearbook 2004*, 30 percent of households were equipped with a PC (as compared with an average of 47 percent of households for the EU15) and 22 percent have a PC with an Internet connection. Despite these relatively low figures, 54 percent of the adult population use a PC and 49 percent use the Internet (higher than the EU15 average of 41 percent). Estonian citizens have found ways to use the Internet despite the absence of connected computers in their homes. Whether it be through office computers, PIAPs, Internet cafés, or some other vehicle, this is a tribute to the quality of the services provided online (online banking on the private side and e-voting, online tax payments, and so on, on the public side) and the desire of the end users to take full advantage of the availability of these services. A clear illustration of this dynamism is the level of participation in online tax filing, which exceeds 90 percent of the population.

A possible weakness on the infrastructure side is the risk of obsolescence. A victim of its own success perhaps, Estonia, after a rapid phase of technical evolution and implementation, must now cope with aging equipment and outdated systems. If the technology cycles are the same for everyone, why should things be different for Estonia? Technological change will have a disproportionately large impact on Estonia as its economy is less developed and its companies are able to invest less in ICT than other European nations.

A Dearth of IT graduates

In Estonia, 500 engineers graduate every year. In India, the number is closer to 500,000. There is a serious lack of engineers in Estonia, and a handful of companies are hiring almost anyone that comes out of university with a computer science degree. Estonian universities do not have the capacity to produce more of the much-needed qualified engineers. In addition, salaries remain low in Estonia, and nearby Westernized markets such as Finland represent a huge draw for young and qualified Estonians. The government is reluctant to promote studying abroad for fear of accentuating the brain drain, which is still under control at the moment.

As a unique public-private partnership, the Estonian government, the two largest Estonian universities (Tallinn Technology University and Tartu University), and the Estonian ICT industry proposed the creation of the Estonian Information Technology College, a private institution established and financed by the Estonian Information Technology Foundation. The main aim of this college is to prepare highly qualified ICT specialists and to support the ICT-related development in Estonia. However, the results of any efforts made today will be felt only five or ten years from now. Estonia needs to find creative ways to attract the talent it needs today to continue growing in the ICT sector. Much of this talent will have to come from other countries—from European nations such as Russia and possibly other Asian countries such as China, India, and the Philippines. Attracting such talent is not easy, however, as demonstrated by the failure of the "green card" residency scheme (for attracting skilled ICT workers) in Germany.[17] Estonia will need to win in this global war for talent to ensure its future eleadership.

In-house innovations may never leave the "house"

Enormous innovations have occurred in the private sector, and in banking in particular. Most of these applications are conceived in-house by the firms' development teams. Hansapank, Estonia's largest consumer bank, does not view itself as a software developer and is focused on its core business of delivering banking services. In addition, the online and mobile offerings are seen as a source of competitive advantage. It is therefore unlikely that most of the cutting-edge mobile and online applications developed for and by Estonian banks will be "sold" internationally or adapted to other businesses.

A continuing challenge for Estonia is how to export the products of its private enterprises. Indeed, even with 100 percent penetration, local companies can reach only 1.4 million consumers domestically. Hence, the need to scale geographically is present. The e-government wave has fostered innovation and growth of the private sector through private-public partnerships. Although the nascent state would not have been able to achieve its objectives without the help of private firms, government orders have stimulated the sector and contributed to the offer of innovative solutions. Indeed, many Estonian innovations involve public markets or services. Despite Estonia's membership in the European Union, public markets outside Estonia will be difficult for Estonian companies to enter. To be recognized as a global e-leader, Estonia will need to package and sell its innovative ICT services abroad.

Conclusions

Estonia has in a short time caught up with advanced countries in terms of information and communication technology (ICT) infrastructure and in the use of ICT in society. Attitudes favouring ICT, innovative thinking and progressive ICT entrepreneurship, have developed a strong technological infrastructure in Estonia. These factors combined with strong economic growth as well as macroeconomic stability form a favourable basis for further development.

— Ministry of Foreign Affairs, Government of Estonia, 2005[18]

When Estonia declared its independence in August 1991, few would have predicted that in the short span of 15 years Estonia would rank as one of the leading countries in the networked economy. The power of visionary political leadership combined with strong public-private partnerships has been demonstrated amply in the rapid transformation of Estonia over the last decade. Today Estonia has become a role model for others—and has triggered a surge of interest in the potential of ICT for development within other nations. Estonia cannot rest on its laurels, however. It faces significant challenges in sustaining its position of e-leadership in the networked economy of the future.

Notes

1 Estonian IT Policy: Towards a More Service-Centred and Citizen-Friendly State. Principles of the Estonian Information Policy 2004–2006, May 2004.

2 Rikman and Andersen 2006.

3 Viik 2003.

4 2005 TNS EMOR Survey, Spring 2005.

5 Quoted in Poluck (2005).

6 See Cato Institute (2006).

7 Poluck 2005.

8 Borland 2005.

9 See Rikman and Andersen (2006).

10 See "Estonian Info Society Development Plan 2010" Ministry of Economic Affairs and Communications 2006.

11 See Vaho Klaamann, "Estonian ICT sector 2004"; www.itl.ee.

12 Viik 2003.

13 See Rikman and Andersen (2006, pp. 18–19).

14 See Rikman and Andersen (2006, pp. 18–19).

15 See European Commission (2005).

16 Landler 2005.

17 See Vijapurkar (2003).

18 Available at http://www.vm.ee/estonia/kat_175/pea_175/1163.html.

References

Borland, J. 2005. "Perspective: Estonia Sets Shining Wi-Fi Example." *CNET News*, November 1. Available at http://news.com.com/ Estonia+sets+shining+Wi-Fi+example/2010-7351_3-5924673.html.

Cato Institute. 2006. "Former Estonian Prime Minister Mart Laar Wins Friedman Prize for Liberty." News Release, April 20. Available at www.cato.org/new/pressrelease.php?id=22.

European Commission. 2005. *European Trend Chart on Innovation. Annual Innovation Policy Trends and Appraisal Report: Estonia, 2004–2005.* Innovation/SMEs Programme, European Commission. Available at http://trendchart.cordis.lu/reports/documents/Country_Report_Estoni a_2005.pdf .

Government of Estonia. Foreign Office Website. Available at http://www.vm.ee/estonia/kat_175/.

IT in Public Administration of Estonia. Available at http://www.riso.ee/ en/pub/2004it/.

Landler, M. 2005. "Hot Technology for Chilly Streets in Estonia." *New York Times Online*, December 13.

Poluck, M. 2005. "Good Things Come in Small Packages." *The Guardian*, November 23. Available at http://society.guardian.co.uk/e-public/ story/0,,1648381,00.html.

Rikman. E. and J. B. Andersen. 2006. *Europe Innova, mWatch Tallinn: A Survey on Mobile Readiness in the City of Tallinn and their Regional Context.* Prepared by the mClusters Consortium for BDA Estonia. Available at www.livinglabs-europe.com/documents/tallinn/mWatch Tallinn 2006.pdf.

Viik, L. 2003. "The Internet Connects People not Computers." *Estonian Ministry of Foreign Affairs: Modern Estonia.* September 9. Available at www.vm.ee/estonia/kat_175/pea_175/2079.html.

Vijapurkar, M. 2003. "Lukewarm Response to Germany's Green Card Scheme." *The Hindu* (newspaper), September 28. Available at www.hinduonnet.com/thehindu/2003/09/28/stories/ 2003092801650900.htm.

Access to Communications Services in Sub-Saharan Africa

MARKUS HAACKER, International Monetary Fund

Over the last decades, the global economy has embarked on a dramatic transformation characterized by shifts in the location of economic activities, increased fragmentation of production processes, and the emergence of some new types of trade, most notably in services.

From a company's perspective, the decision of where to locate production is primarily driven by the costs of production. Much of the ongoing re-location of economic activities on the global scale is driven by differences in labor costs across countries (reinforced by the opening of the economies of the former Soviet bloc, China, and India, which—as Freeman (2005) puts it—have doubled the size of the global labor force). To the extent that agglomeration economies matter for an industry, countries may be able to sustain or reinforce a competitive advantage that initially is driven mainly by labor costs.

At the same time, the ongoing transformations are enabled by technological progress, in particular the rapid advances in information and communication technologies (ICT). These have resulted in the emergence of new categories of trade in services (for example, call centers, programming, and back-office operations). Regarding commodity trade, the role of ICT as an engine of the ongoing global transformations is at least threefold:

1. Improved communications facilitate trade in traditional commodities (such as textiles) and improve the availability of market information.

2. By making possible an increased fragmentation of production processes, ICT has have also increased the potential for trade in components.

3. Much of the value of ICT equipment is embodied in electronic components, which travel lightly.

However, the benefits of globalization have not accrued to all developing economies in a similar fashion. Specifically, most of the "success stories" have been written in Asia, and few in Africa. Therefore, Asia's share in global GDP has more than doubled between 1950 and 2001 (from 18 percent to 38 percent), whereas the share of Africa in world GDP has declined from 3.8 percent to 3.2 percent (Maddison 2003).[1]

As a consequence, some observers conclude that Africa has—at least so far—"missed the boat," and that Asia now has the double advantage of a vast pool of cheap labor coupled with an increasingly sophisticated know-how of production processes and other agglomeration economies.[2]

The views expressed herein are those of the author and should not be attributed to the IMF, its Executive Board, or its management.

Figure 1: Imports of telecommunications equipment

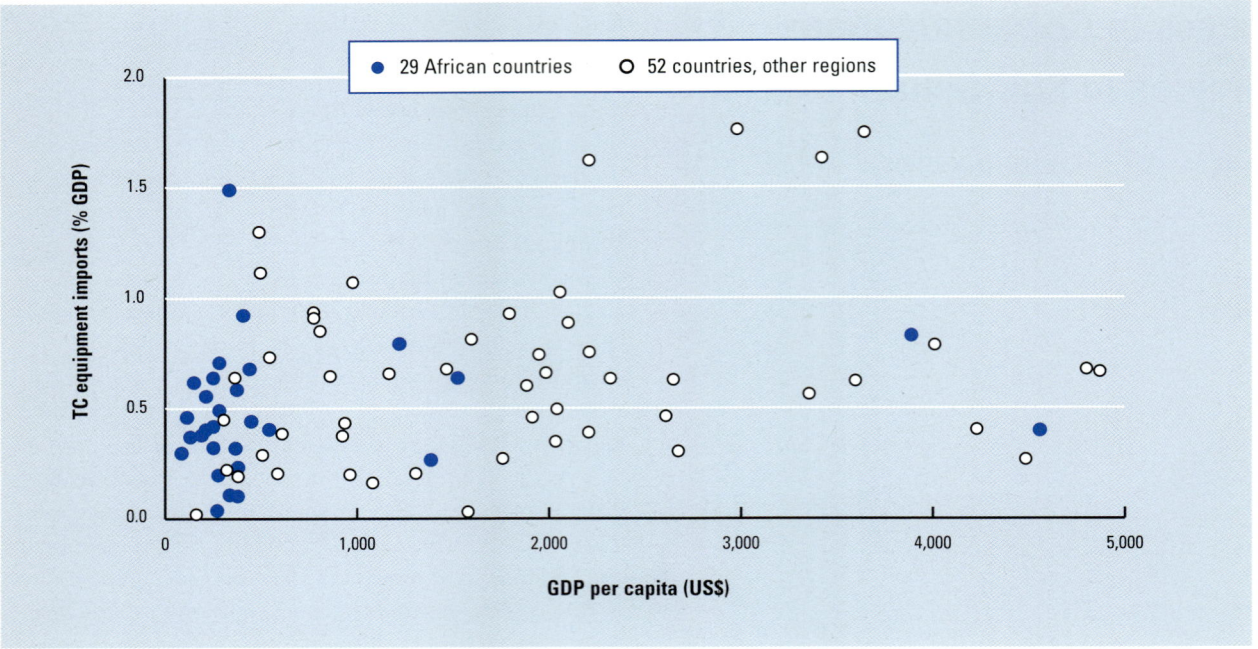

Source: Author's calculations, based on United Nations, 2006; IMF, 2006.

On the other hand, progress in communication technologies means that sub-Saharan Africa's access to the global economy—in terms of the ease of doing business between Africa and other regions—has immensely improved. Indeed, some of our findings suggest that advances in communication technologies are having a disproportionately positive impact in sub-Saharan Africa.

Although our focus is on a specific aspect of globalization, we arrive at a differentiated assessment of the forces underlying ongoing transformations in the global economy. Clearly, most economies in Africa have not been as successful as several "star" performers in Asia in taking advantage of structural shifts in the global economy. At the same time, we observe that advances in communication technologies are leveling the field to the advantage of African economies. While "agglomeration forces" may give the most successful developing economies to date an edge and some long-lasting advantages, we find that there are powerful trends that are working in the opposite direction, improving access to the global economy for some of those developing economies that were not among the first movers in terms of taking advantage of the opportunities of an increasingly globalized world economy.

Trends in equipment spending

Some important insights can also be gleaned from an analysis of the pattern of spending on ICT-related equipment. As these data are not available for many low-income countries, we instead use data on ICT-related imports (a good measure of spending on IT equipment, because most low-income countries do not produce such equipment).[3] Figure 1 shows telecommunications import data for 29 African countries as well as 52 countries from other regions with similar income levels.

We find that for most countries, imports of communications equipment lies within a band between 0.2 and 1.0 percent of GDP. Most African countries are located toward the left of Figure 1, with income levels up to US$1,000. Although imports of communications equipment in low-income countries are lower on a per-capita basis than in higher-income countries, there is essentially no difference in the tele*density* of economic activities across countries with different income levels (as measured by ICT imports or spending as a percentage of GDP). This result is reinforced if one considers that the five countries with the highest level of imports relative to GDP are small island economies, a factor that likely drives up their costs of communication. Also, we see that the teledensity of economic activity in African economies is similar to that of other economies at similar or somewhat higher income levels. This finding suggests that African

Figure 2: Imports of IT equipment

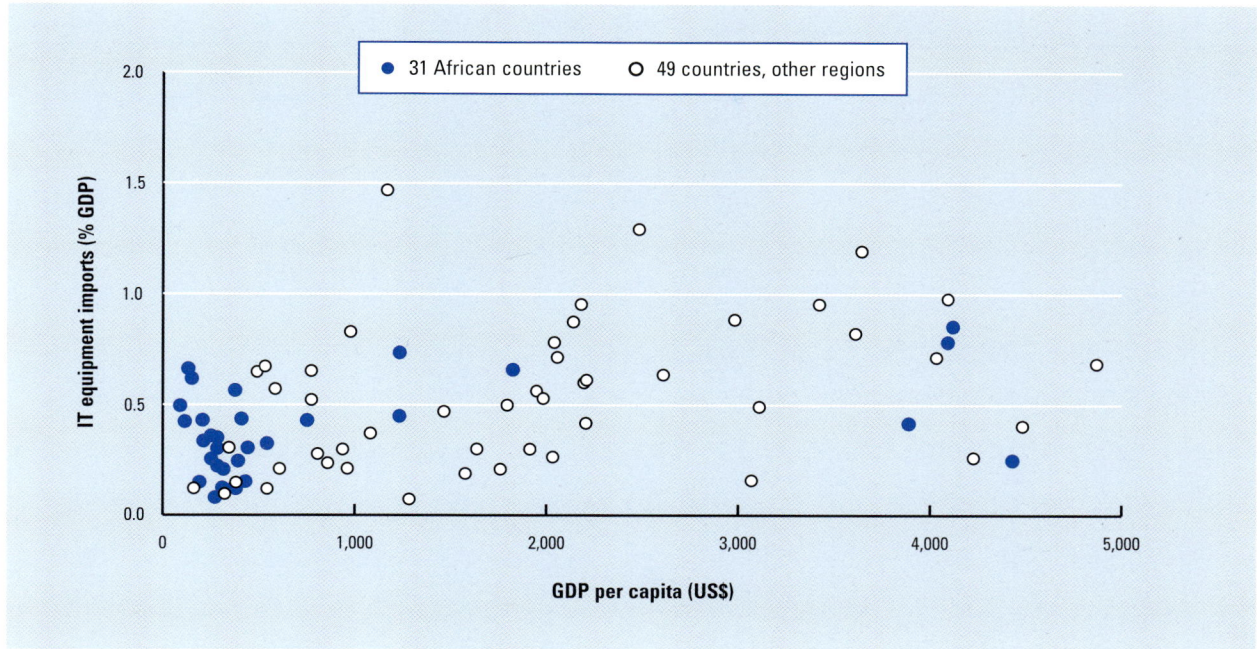

Source: Author's calculations, based on United Nations, 2006; IMF, 2006.

economies benefit from advances in communication technologies in a fashion similar to that of low- or middle-income countries from other regions.

Because—for many applications—IT equipment complements the use of communications equipment and services, we also report the corresponding data for IT equipment. We find that the IT density of economic activity increases with income level. However, as before, we do not find a difference in IT density between African economies and those from other regions.

Thus, our spending data suggest that African economies benefit from advances in ICT in a fashion similar to low- and middle-income countries, as far as the tele- and IT-density of economic activities is concerned.

To complement our analysis of spending patterns across countries, we have conducted an empirical analysis, including some structural and policy variables. We find that the composition of GDP is relevant for the teledensity of economic activities. ICT-related spending, as a percentage of GDP, is significantly lower in oil-exporting countries than elsewhere. We also find that policy matters. ICT-related imports are negatively related to overall tariff rates. The latter also means that some of the positive association between the IT-density and income per capita (Figure 2) may reflect the fact that tariff rates tend to be lower in countries with higher income per capita.

Networked readiness in sub-Saharan Africa

A different approach to measuring access to communication services is the Networked Readiness Index (NRI) developed by the World Economic Forum and INSEAD (described in Chapter 1.1). The Index includes ratings for 23 African countries out of a total of 122 countries (see Table 1). Overall, African countries appear between rank 47 (South Africa) and 122 (Chad), with an average rank of 99.5. Beyond this, some useful lessons can be learned from analyzing the pillars of the Index. First, the 23 African countries tend to rank much lower in the infrastructure environment pillar than they do for the political and regulatory environment (and, to a lesser extent, the market environment). Second, the individual usage pillar tends to have lower ranks than the business and government usage pillars. Our interpretation of these differences is that actual usage of communication services is lagging behind indicators for, among other things, the market environment (availability of trained people, ease of doing business) the regulatory environment (property rights, legal framework), and technology absorption. Both the infrastructure environment pillar (including the number of telephone lines or Internet hosts) and the individual usage pillar, on which the scores for sub-Saharan Africa are relatively low, are the most intensive in terms of measures of actual usage of communication equipment and services (see Appendix A, Chapter 1.1).

93

Table 1: Networked readiness: 23 African countries

Country	Market environment	Political and regulatory environment	Infrastructure environment	Individual readiness	Business readiness	Government readiness	Individual usage	Business usage	Government usage
Angola	120	116	122	120	108	118	108	116	109
Benin	90	99	111	113	114	98	106	112	71
Botswana	66	45	93	56	61	83	64	94	81
Burkina Faso	93	75	104	117	99	96	116	90	52
Burundi	117	117	116	115	116	122	117	121	107
Cameroon	115	119	112	102	92	115	98	109	105
Chad	122	122	117	122	121	121	120	122	115
Ethiopia	98	112	119	118	122	111	122	118	92
Kenya	100	66	90	108	79	97	105	81	91
Lesotho	96	109	107	111	120	110	100	115	114
Madagascar	94	86	99	104	106	90	118	102	78
Malawi	105	71	114	112	96	108	119	111	106
Mali	101	70	98	119	112	87	114	87	45
Mauritius	51	46	61	49	54	41	49	67	70
Mauritania	83	94	118	109	103	51	95	57	21
Mozambique	118	113	121	116	107	89	110	101	112
Namibia	64	38	115	91	72	106	87	78	102
Nigeria	59	62	95	103	87	75	101	82	67
South Africa	34	25	70	70	31	40	60	43	59
Tanzania	85	49	100	107	98	77	112	85	75
Uganda	107	63	88	121	113	82	113	76	56
Zambia	113	85	113	101	115	116	109	86	120
Zimbabwe	119	111	103	110	81	119	94	119	121
Average	**93**	**82**	**104**	**104**	**96**	**94**	**102**	**94**	**86**

Source: NRI 2006–07.

Access to communication services in sub-Saharan Africa: Some indicators

Communication services are unique in terms of data availability because utilizing them usually requires some form of subscription, which can be easily collected. Thus, for both main telephone lines and mobile telephone services, data on the number of subscribers are available for essentially all low-income countries. Another variable we will look at is the number of Internet users (this variable includes not only subscribers but also an allowance for Internet access through Internet cafés and the like). Figure 3 provides the latest available data (2004) on the usage of these key communication technologies by region.

We see—as expected—that access to communication services in sub-Saharan Africa is very limited compared with other world regions. Access to main telephone line services (1.8 per 100 inhabitants) is only about 9 percent of the global average. For mobile telephone services, the level of access in sub-Saharan Africa is much higher (7.2 per 100 inhabitants, or 25 percent of the global average), and there are 1.9 Internet users per 100 inhabitants (14 percent of the global average).

The role of mobile communication services warrants further attention. From Figure 3, it is very clear that mobile telephone services in sub-Saharan Africa have helped to increase access to communication services unlike in any other world region. Since 1991, when mobile telephone services were available in Mauritius and South Africa only, the number of mobile subscribers in the region has grown at an average annual rate of 91 percent; it has grown at a rate of 50 percent between 1999 and 2004. Over the same period, main telephone line subscriptions in sub-Saharan Africa grew at an annual rate of 6 percent and 4 percent only. However, growth of mobile telephone services was so strong that total telephone subscription grew at an annual rate of 21 percent (1991–2004) or 31 percent (1999–2004), making sub-Saharan Africa the fastest-growing market for communication services over these periods.

Relative to the global average, sub-Saharan Africa has therefore been able to very substantially narrow the gap in access to communication services, as total subscriptions (per 100 inhabitants) almost doubled—from 10 percent of the global average in 1991 to 19 percent of the global average in 2004.

These impressive gains, however, certainly understate the progress made in sub-Saharan Africa relative to the rest of the world. In the technologically most advanced countries (which can be proxied by Western Europe and North America in Figure 3), mobile telephone services have primarily increased the depth of access to communication services (because a mobile telephone most commonly complements or replaces a main telephone line subscription).

Figure 3: Access to communication services by region, 2004

3a: Subscriptions to main telephone lines

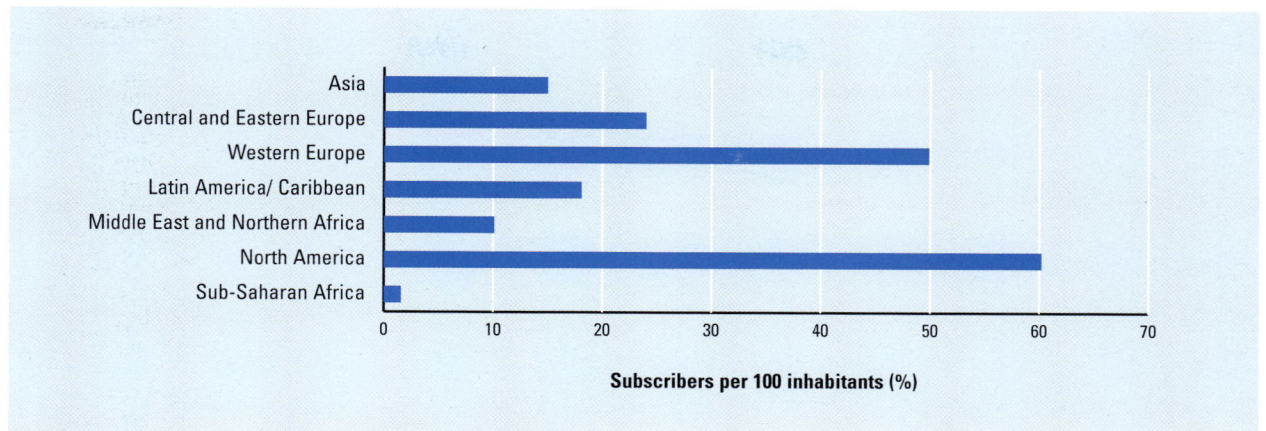

3b: Subscriptions to mobile telephone services

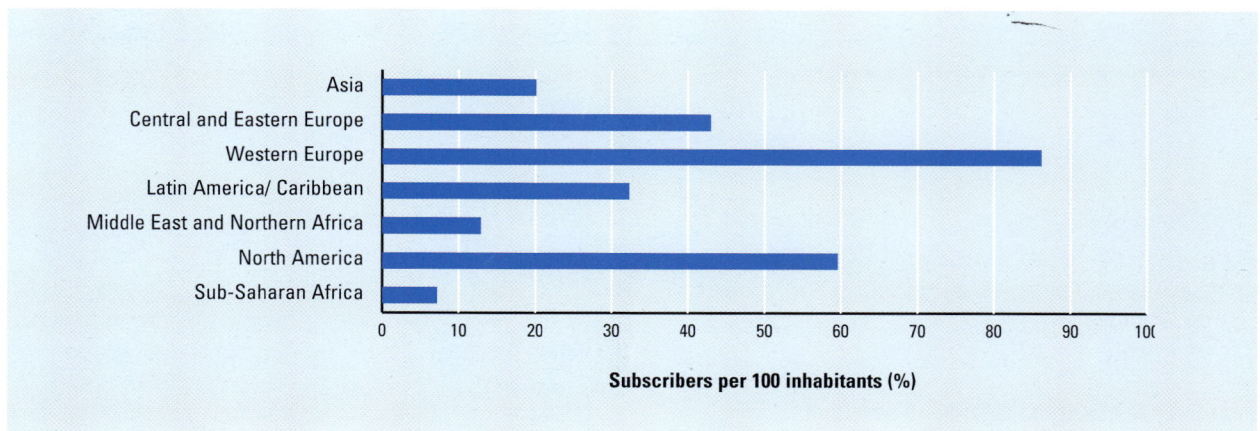

3c: Internet users

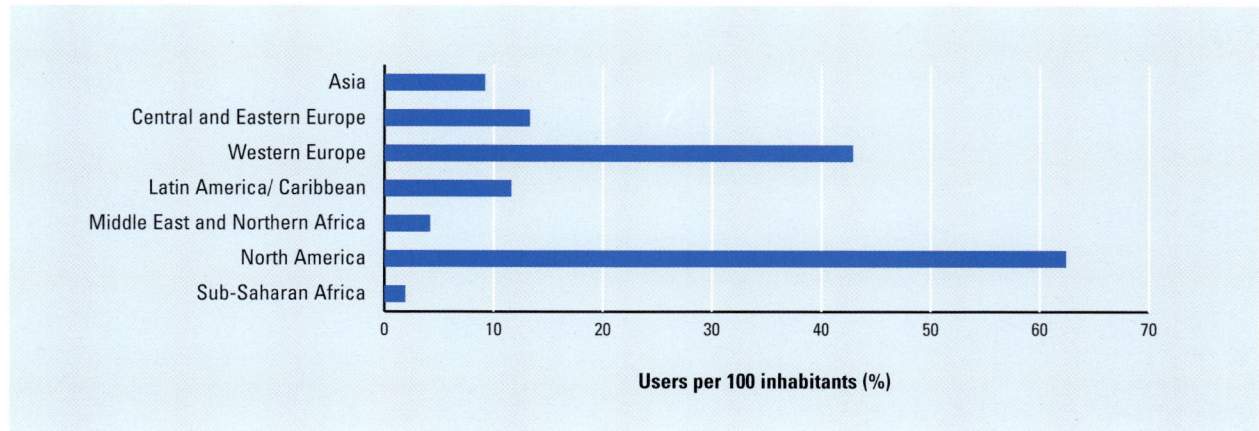

Source: Author's calculations, based on ITU, 2006.

Figure 4. Main telephone line and mobile telephone services

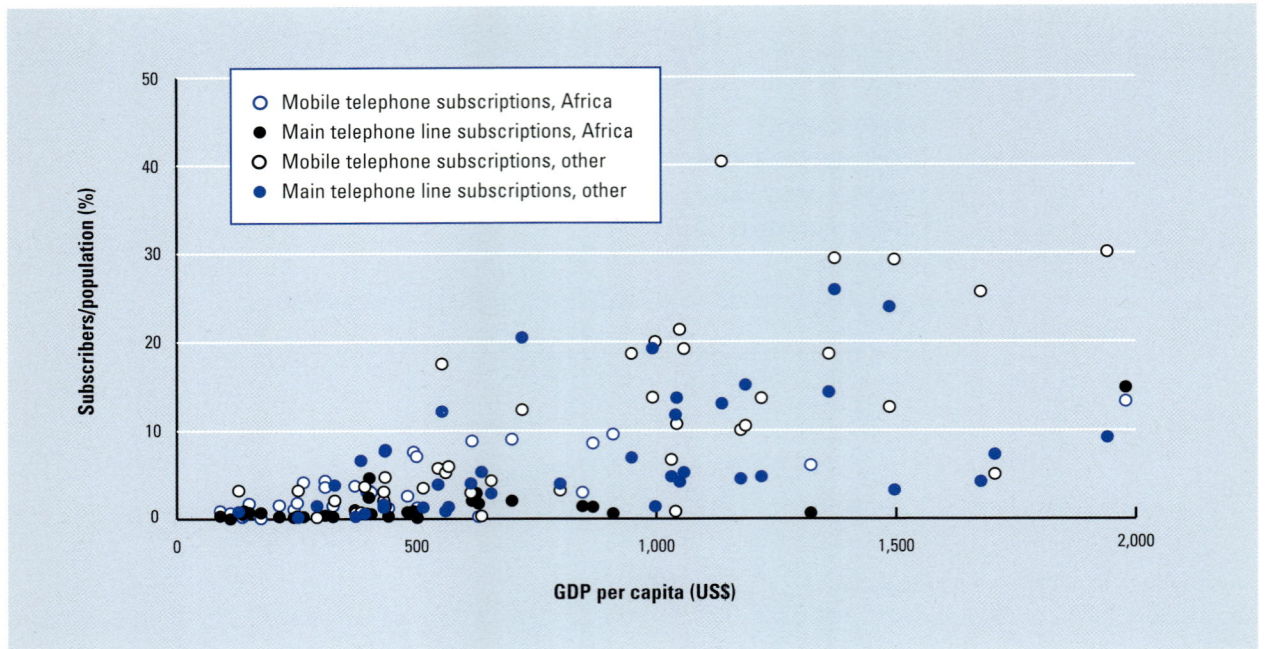

Source: Author's calculations, based on IMF, 2006; ITU, 2006.

In sub-Saharan Africa, however, the majority of mobile telephone subscriptions were purchased by individuals or businesses who previously did not subscribe to main telephone line services.

A closer look at access to telephone services

Figure 4 shows the number of subscribers to telephone services for countries with a GDP per capita of US$2,000 or less, including 33 countries from Africa and 39 from other regions.[4] The most important lesson from this kind of analysis is that access to telephone services is clearly correlated with GDP per capita.

In order to—literally—get a better picture of the role of mobile telephone technologies, Figure 5 illustrates the share of mobile telephone subscribers among total telephone subscriptions. Regarding the average share of mobile subscribers, our findings are consistent with our overview of access to communication services (which shows that mobile telephone technologies have substantially increased access to telephone services): mobile telephone services account for about two-thirds of total subscribers for the countries shown. Although low-income countries tend to have a higher share of mobile telephone subscribers, the other notable feature of the data is the high dispersion of this share, ranging from about 10 percent to close to 100 percent. We discuss possible determinants of the share of mobile telephone services below.

Teledensity in sub-Saharan Africa

A different way of addressing the role of communication services in economic activity is by looking directly at the teledensity of economic activity. We use this term to relate indicators measuring access to communication services to indicators of the scale of economic activity. Through this device, we hope to gain a better understanding of the importance of communication services for a country's economy. Specifically, we look at the number of subscriptions or users relative to GDP (measured as units of services per US$ million of GDP).

We find that the teledensity of economic activity in sub-Saharan Africa holds up well relative to other regions. Overall, for each US$ million of value-added created in sub-Saharan Africa, there are 118 phones—a number that holds up to teledensity in Asia (120), is similar in magnitude to teledensity in Latin America and the Caribbean or in the Middle East and North Africa, and is over 60 percent higher than the global average of 72.

Remarkably, we find that the teledensity of economic activity in the United States or Western Europe is much lower than in these four regions, while it is much higher in Eastern Europe, possibly reflecting the ongoing adjustment and transformation there.

In line with our previous discussion of main telephone lines versus mobile telephone services, we also find that the high density in sub-Saharan Africa is predominantly

Figure 5: Mobile telephone subscriptions (percent of total)

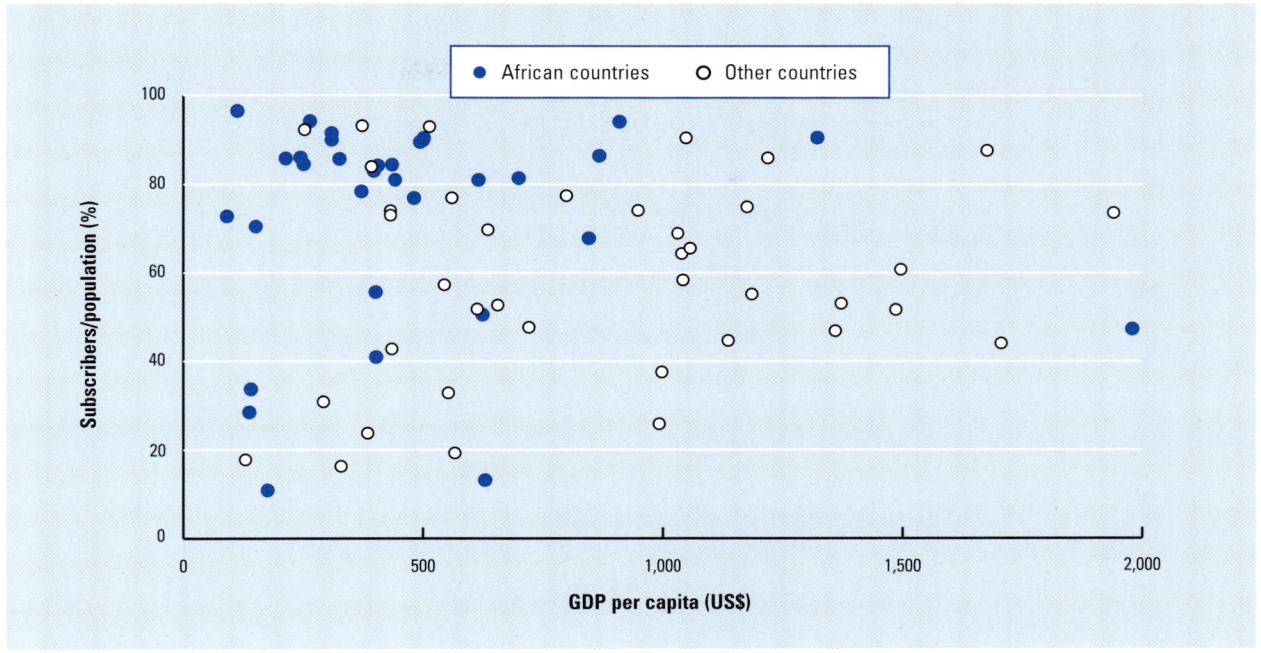

Source: Author's calculations, based on IMF, 2006; ITU, 2006.

the result of mobile telephone services, accounting for about 80 percent of total telephone subscriptions, a ratio higher than in any other region. Figure 6 also shows the density of Internet users, with similar results—for sub-Saharan Africa, each US$ million in value-added is associated with 25 Internet users, a level that is higher than the global average.

A note is in order here on our measure of usage of communication services, which is based on telephone subscription or the number of Internet users. Our measures reflect the width of usage, but, owing to data limitations, we cannot address the intensity of usage (for example, hours or data transfer per user). Thus, in terms of the absolute level of usage, our measures likely overstate the level of usage in low-income countries.

To understand the consequences of applying the concept of teledensity to the intensity of usage rather than to its width, consider the following example. Recall that the density of access to communication services in sub-Saharan Africa, measured as subscribers per US$ million of value added, is 64 percent higher than the global average. If the intensity of usage in sub-Saharan Africa was only one-half or one-quarter of the global average, then the number of hours of telephone communications per US$ million of value added would be 20 percent or 60 percent lower in sub-Saharan Africa, relative to the global average.

Which measure of teledensity is most appropriate depends on the context. If the most important aspect of communication is accessibility and the exchange of essential market or order information, then our measures—based on the number of subscriptions or users—is the appropriate one, and we find that sub-Saharan Africa has disproportionally benefited from advances in communication technologies. For activities that require continuous and high-density data flows, the measures based on the intensity of usage are appropriate. While our reasoning here is somewhat speculative, in light of data limitations, we believe that sub-Saharan Africa may be lagging behind the global average in intensity of usage, which could explain why certain communication-intensive economic activities (for example, call centers) are usually not located in Africa. Our preliminary conclusion therefore is that sub-Saharan Africa is among the primary beneficiaries of advances in communication technologies as far as width of access is concerned (measured by the number of users), but it is less clear that the region is a primary beneficiary in terms of the intensity of utilization of communication services.

What drives investment in mobile communication services in sub-Saharan Africa?

Primarily, the decision of a provider to enter a market is driven by expectations of profitability. Conversely, the number of providers a market can sustain is determined by

Figure 6: Teledensity of economic activity by region, 2004

6a: Density of main telephone line subscriptions

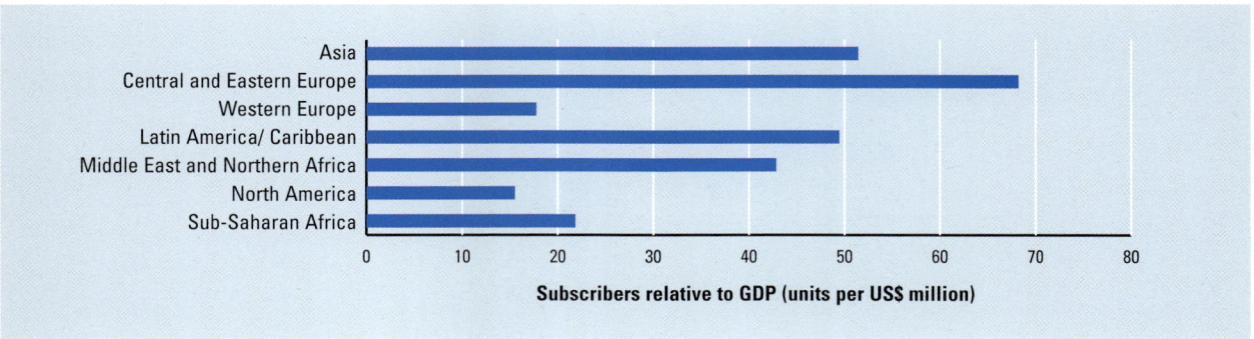

6b: Density of mobile telephone subscriptions

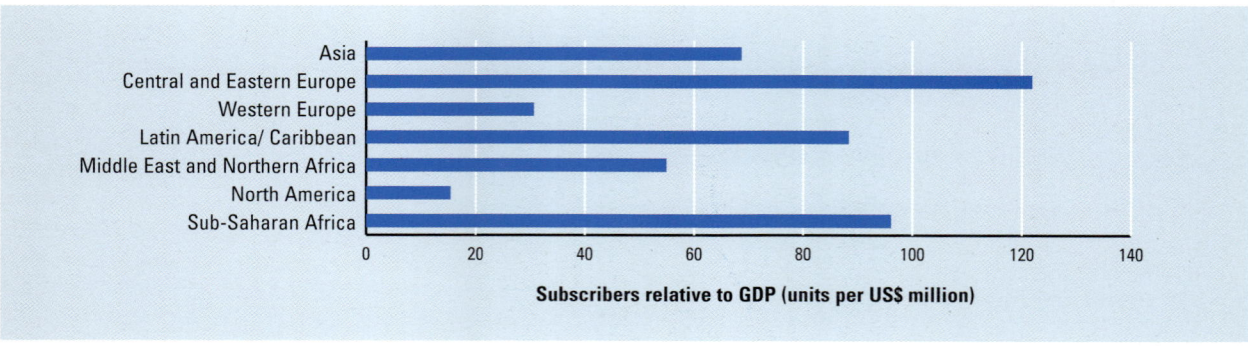

6c: Density of total telephone subscriptions

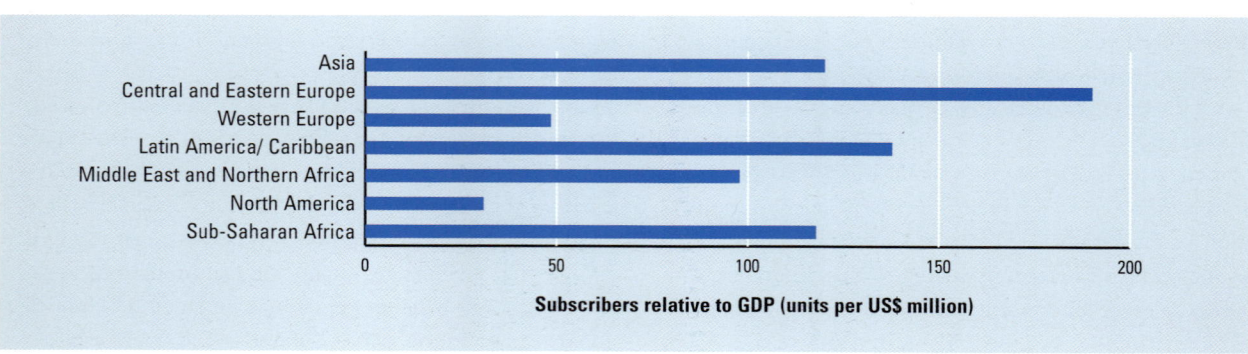

6d: Density of Internet users

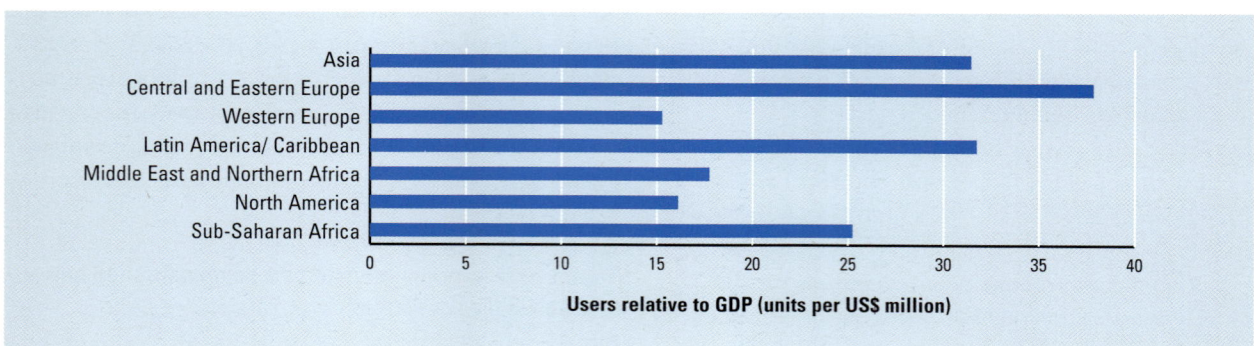

Source: Author's calculations, based on IMF, 2006; ITU, 2006.

Figure 7: Mobile telephone providers and GDP

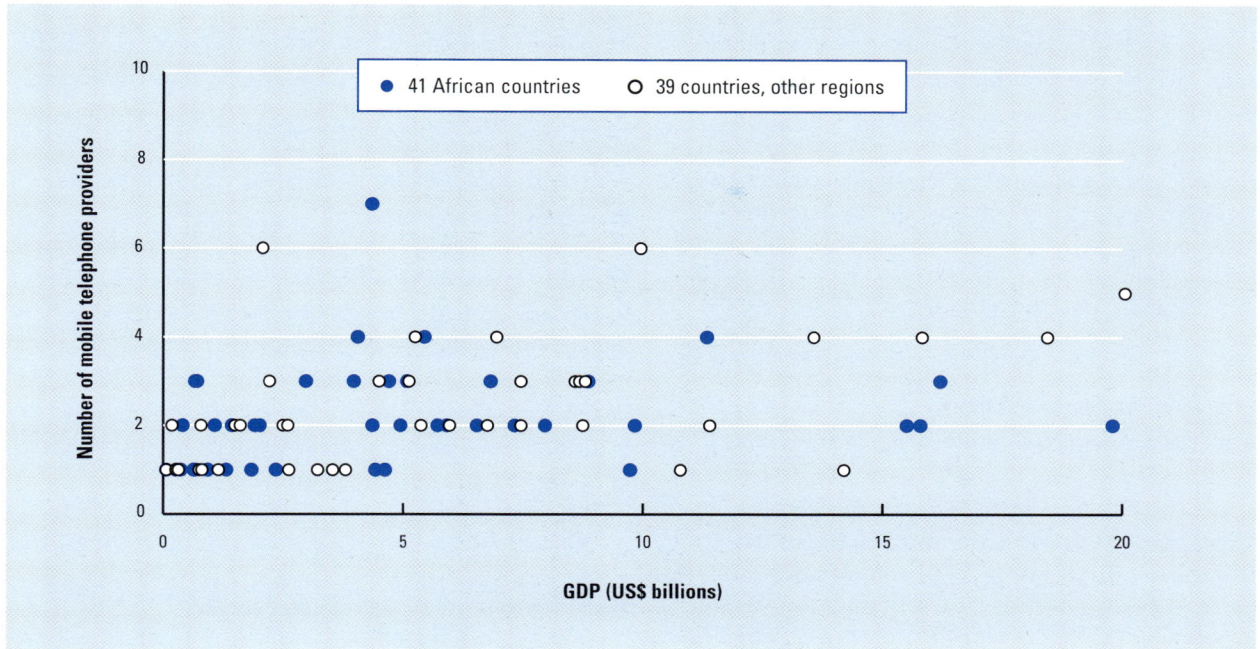

Source: Author's calculations, based on IMF, 2006; ITU, 2006.

the size of the market. One aspect that is relevant for many low-income countries is that a large proportion of the population may not be able to afford a subscription to telephone services.

This line of reasoning was succinctly summarized by Andrew Mhembu, the Deputy CEO of Vodacom, when his company won a license to operate a GSM network in Mozambique in 2002:

> As is the case with most African countries, the vast majority of the population is very poor. However, there is a wealthy segment of the population prepared to pay for cellular services in US dollars and large enough in size for our investment to be very worthwhile.[5]

While these considerations apply to any kind of investment, the market for mobile telephone services is peculiar because market entry requires a license, which gives a direct role to government. At the same time, as the number of competitors is limited, the market structure and market regulation also affect profitability and outcomes in terms of access to communication services.

How much do these considerations matter? To address this question, we have created a database covering providers of mobile telephone services.[6] Figure 7 relates the number of providers to GDP. We see what appears to be a positive (though not very clear) correlation between the level of GDP and the number of providers. This is

confirmed by an empirical analysis, which suggests that a difference in GDP of US$50 billion is associated with one additional provider. At the same time, GDP per capita (and some other proxies for market size) do not explain the variations in the number of providers.

This finding, however, does not explain much of the variation between countries with low levels of GDP. We therefore also look at whether there is a monopoly or "competition" (crudely defined as two or more providers) in the market. Although GDP matters, it turns out that the most important predictor of competition is the World Bank's index of regulatory quality.

Thus, we find that many countries in sub-Saharan Africa are at a disadvantage in terms of attracting providers of mobile telephone services, because the size of the respective market (measured by the level of GDP) is relatively small. However, there is a role for policy—countries that rank highly on the World Bank index for regulatory quality are more likely to feature more than one provider.

More on market structure

The markets for main telephone line and mobile telephone services do not differ only in terms of the number of providers. Haacker (2005) investigates the main shareholders of providers of main telephone line and mobile telephone services in 21 African countries.[7] In 20 of these countries, there was only one national provider for main telephone line services; in 15 of these countries, the dominant operator was controlled and frequently wholly owned by the government. The number of providers of main telephone line services averaged 2.3 in these countries; of a total of 48 national providers, 32 were controlled by multinational companies, and only 12 (including 4 monopolies) were controlled by the respective governments. This finding suggests that mobile telephone technologies are not only associated with an increase in competition in the telecommunications market, they also act as vehicles of technology transfer, with multinational providers bringing in their expertise and technological know-how.

The pricing of telephone services

A key indicator (and determinant) of access to mobile telephone services is the price of these services in relation to income (which we proxy by gross national income [GNI] per capita). As an indicator for the costs of access to communication, we use the price baskets for residential fixed lines and mobile telephone services from the World Bank (2006).

Figure 8 shows the respective price baskets in US dollar terms and as a percentage of GNI. Regarding the price baskets in US dollar terms, two features stand out. First, for low-income countries, mobile telephone services appear to be more expensive than residential fixed lines. Second, the variations across countries by income in the price of telephone services are less pronounced for mobile telephone services than for fixed lines. When the prices for telephone services are related to GNI, the message that emerges is very similar for each type of telephone service. Although the costs of each basket is generally lower than 1 percent of GNI per capita for high-income countries, the costs range from 1 percent to of GNI to 10 percent of GNI for middle-income countries, and from 5 percent of GNI to around 100 percent of GNI for low-income countries. This means that access to telephone services of either form is essentially universal for high-income countries, but unaffordable for a wide range of the population in low-income countries.

Prepaid mobile telephone accounts

One constraint to expanding access to communication services in low-income countries is the low level of development of the financial sector in these countries (which increases the costs of monetary transfers) and their weak contract enforcement (which makes it difficult to collect unpaid bills). Mobile telephone technologies facilitate the use of prepaid service arrangements along several lines. First, they are practicable in settings where contract enforcement is weak, as providers are not at risk for having

to collect arrears. Second (this point primarily applies to prepaid cards), the transactions between the provider and the user are facilitated by the use of mobile telephone technology, involving a wireless financial transfer when the user credits his or her mobile telephone account with the value of the purchased card. Thus, there is no financial intermediation required other than the distribution of the coded cards through retailers, small merchants, and—frequently—street vendors.

Figure 9 suggests that prepaid subscriptions may indeed be a factor behind the disproportionate spread of mobile telephone services in low-income countries. Prepaid services are the dominant mode of access to mobile telephone services in low- and low-middle-income countries, accounting for a share of almost 90 percent in low-income countries and about three-quarters in middle-income countries, but only about half in high-income countries.

Market density and geographical access

Above, we discussed the prices of main telephone line and mobile telephone services, and concluded that they are unaffordable for a large share of the population. To some extent, sharing a telephone (and its costs), or using public telephones, provides a way around this, although these arrangements do not offer the same degree of accessibility.

A second form of exclusion from communication services works through *aggregate* demand for such services on the regional or local level. Telephone providers may not extend service to areas where the market is too thin to recover the costs of installing or operating the network. This could be the case because the income level (total, not average) of the population of an area is too low, either because average incomes are low or because the population density is low.

The geographical coverage of mobile (and fixed-line) telephone services is limited in low- and low-middle-income countries. A survey of 184 companies operating in these countries shows that individual companies cover, on average, 27.2 percent of the area of the respective country (the median is 10.0). The overall rate of geographical coverage, however, is higher, because there is more than one provider in most low- and low-middle-income countries (although the areas covered by the respective networks tend to overlap strongly).[8] Another important distinction is between geographical coverage and access measured as a percentage of the population. As the areas covered appear to include the areas with the highest population density, the share of the population of a country within reach of a mobile telephone network is higher than the geographical coverage rate.

Figure 8: Costs of access to telephone services

8a: Price basket for mobile telephone services

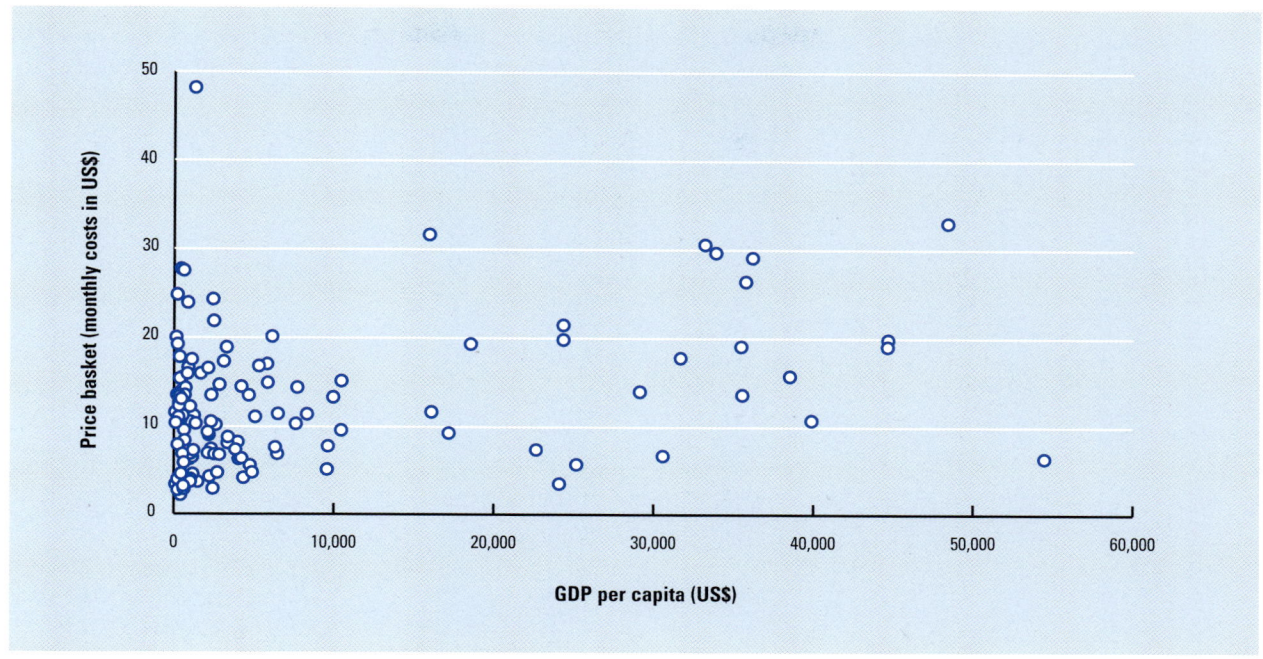

Source: Author's calculations, based on IMF, 2006; ITU, 2006; and World Bank, 2006.

8b: Price basket for residential fixed telephone lines

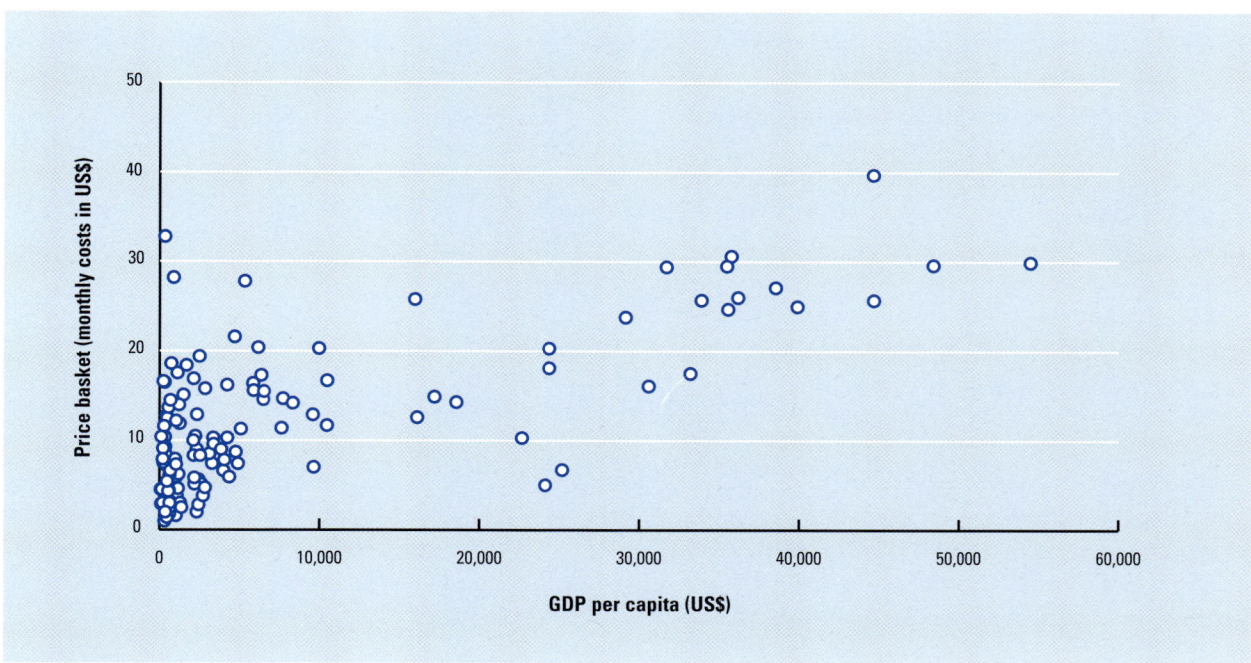

Source: Author's calculations, based on IMF, 2006; ITU, 2006; and World Bank, 2006.

Figure 8: Costs of access to telephone services *(cont'd.)*

8c: Price basket for mobile telephone services (annualized, % GNI, log scale)

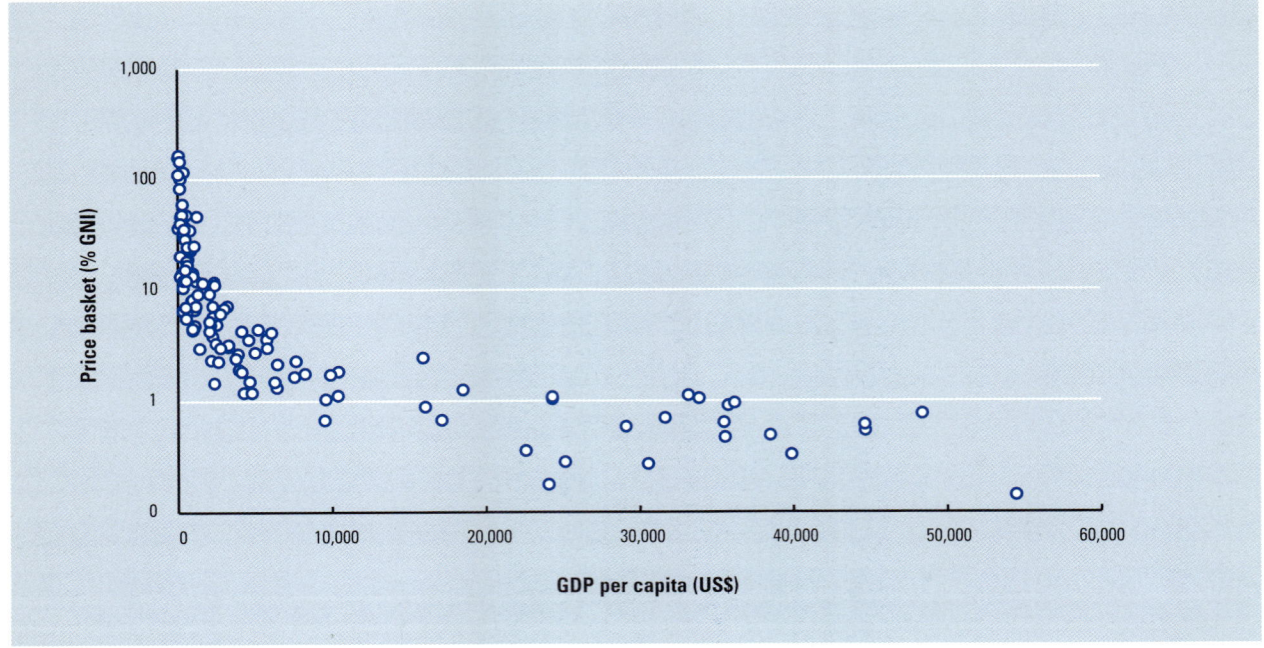

Source: Author's calculations, based on IMF, 2006; ITU, 2006; and World Bank, 2006.

8d: Price basket for residential fixed telephone lines (annualized, % GNI, log scale)

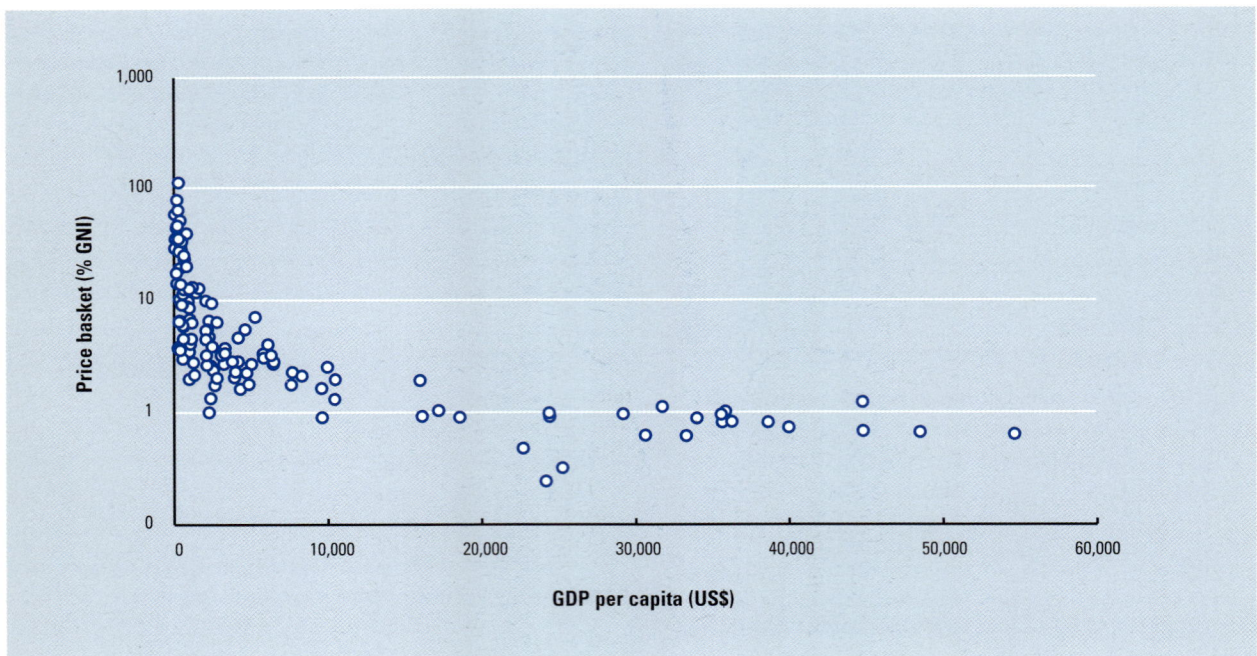

Source: Author's calculations, based on IMF, 2006; ITU, 2006; and World Bank, 2006.

Figure 9: The role of prepaid subscriptions

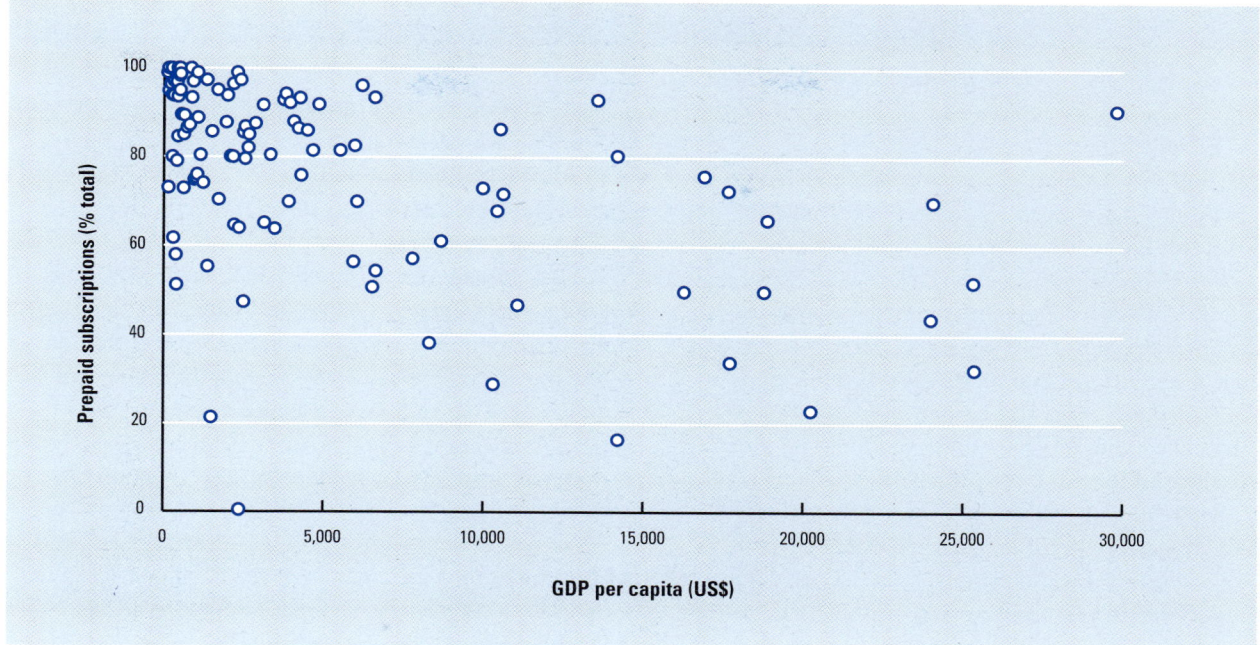

Source: Author's calculations, based on IMF. 2006; ITU, 2006.

What determines the costs of access?

We already observed, based on Figure 8, that the costs of access to communication services on average differ little across countries from different income categories, even though the difference between individual countries can be substantial. We are now going to take a closer look at the determinants of prices of telephone services, using the prices for bundles of mobile and main telephone line services compiled by the World Bank (introduced above), examining in particular the role of the market structure and various measures of economic activity.

For mobile telephone services, the most important determinant of the price of services is the number of providers (and its square), suggesting that there are substantial savings to customers as the market for mobile telephone services becomes more competitive. The estimated coefficients imply that the 2nd, 3rd, and 4th entrants to the market mean incremental annual savings to the subscribers of about US$25, US$20, and US$15, respectively (or accumulated savings of US$25, US$45, and US$60), which are substantial compared with a median annual cost of US$133 for the sample. The potential savings from increased competition then taper off with about six competitors in the market (at an annual savings of US$ 80 compared with a monopoly). Whereas the coefficient of GDP per capita is small and insignificant across various specifications, we also find that some proxies for the

overall scale of economic activity (GDP, population size) return negative coefficients, which could imply that there are some economies of scale at work.[9]

Regarding the prices of fixed-line services, the picture is somewhat less clear. GDP per capita explains little; there is some negative association between the level of GDP and the price of telephone services. However, the price of fixed-line services appears to depend negatively on the number of providers of mobile telephone services—moving from a monopoly to a highly competitive environment in mobile telephone services is associated with a fall in the bill for main telephone line services of US$40. This may imply that a more competitive market from mobile telephone services puts pressure on the price for main telephone line services too, but it could also be the result of some common factors that have a bearing on both markets (for example, regulation).

Determinants of access

We have touched on various aspects of access to communication services in sub-Saharan Africa above. Now, we will pull together these different strands to provide a more comprehensive empirical analysis of determinants of access. Using the respective coverage rates of mobile and main telephone line services as dependent variables, we start out with a broad specification that includes scale variables such as GDP, GDP per capita, population size, the prices for the respective services, and the number of providers (for mobile services). As indicators for the market and institutional environment, we use the World Bank's governance indexes; to capture some effects related to the geographical density of economic activity, we add the rate of urbanization and interact it with each of the scale variables.

We find that the level of economic development is the most important determinant of access, but that the structure of the market for communication services also plays an important role in explaining cross-country differences. Unsurprisingly, the coverage rate of mobile telephone services increases with GDP per capita. However, through different specifications, we find that it is not the levels of GDP or GDP per capita that matter most, but these variables multiplied with the rate of urbanization. This probably indicates that in low- and low-middle-income countries, it is generally not cost effective for providers to cover rural areas; the interaction terms may represent the effect of "effective demand" for mobile telephone services, depending on the level of economic activity and the share of the population that can be reached at low cost.

The other determinant of access to mobile telephone services we can identify is the extent of competition and the regulatory environment. We find that each additional competitor is associated with a 1.8 percentage point increase in access to mobile telephone services, and countries ranked by the World Bank as featuring an effective regulatory environment also tend to have higher coverage rates. The own-price has a negative effect on coverage of mobile telephone services, while the coefficient of the price of main telephone line services is positive. However, these price effects are not significant at the 10 percent level.

In a similar fashion, we analyze the determinants of the coverage rates of main telephone line services. Successively eliminating insignificant variables yields a simple representation—the product of GDP per capita and urbanization explains 60 percent of the variation in the coverage rate of main telephone line services. Adding the prices of the different types of services to the regression, the coefficients show the expected sign (the own-price is negative and the price of mobile telephone services negative, although only the own-price is significant).

Some conclusions

What did we learn about the role of modern communication technologies in sub-Saharan Africa? What are the consequences for the ease of doing business in, with, and from Africa? And what are the implications for Africa's role in the world economy?

First, we find that Africa has benefited from recent advances in communication technologies, allowing many countries and the region as a whole to diminish the gap in access to communication services between Africa and other world regions. Indeed, if one considers the teledensity of economic activity, no such gap is apparent, largely owing to the impact of mobile telephone technologies in Africa.

Second, we find that the number of providers primarily depends on the size of the market, which puts countries with small populations or low GDP per capita at a disadvantage. However, policy can play a role in creating a competitive environment.

Third, the number of providers does have an impact on the costs of services. Our results suggest that moving from a monopolistic setting to a highly competitive environment can save about US$80 per year to a fairly regular user. The number of mobile telephone providers also appears to affect the prices of main telephone line services; however, it is not clear whether this reflects competitive or regulatory effects.

Fourth, we find—unsurprisingly—that higher levels of GDP per capita are associated with higher coverage rates of both mobile and main telephone line services. The structure of the market for communication services also plays an important role in explaining cross-country differences—the World Bank index for regulatory quality and the number of mobile telephone providers always have a positive and highly significant effect on the coverage of mobile telephone services.

Fifth, we find evidence of interactions between the markets for mobile and main telephone line services. A more competitive market for mobile telephone services also drives down prices for main telephone line services, and—through price effects—also affects the coverage rates of main telephone line services.

Sixth, our findings regarding the role of urbanization point toward some structural impediments to expanding access to communication services in sub-Saharan Africa. Where the level of GDP per capita is low, as in rural areas, the density of potential demand may not be sufficient to profitably expand access to telephone services. As a consequence, lower-income and/or isolated segments of the population may not benefit from the innovations in communication services.

Although it is beyond the scope of this chapter, it is important to note that the benefits of improved access to communication services occur both in the formal and in the informal sectors of the economy. For the formal sector, benefits include the ability to improve the management of production and procurement processes and to expand marketing and trading activities. For the informal sector, mobile telephone services extend some of the benefits of a stationary office to independent contractors, helping them to attract additional business.[10]

Africa's role in the global economy

We come back to the broader issues we discussed at the outset—the implications of advances in communication technologies for the role of sub-Saharan Africa in the global economy. Modern communication technologies, especially mobile telephone services and the Internet, have enormously improved the ease of doing business in and with Africa.

Thus, expanding on the line of reasoning offered by Collier (2006), sub-Saharan Africa may be lagging behind Asian competitors in terms of agglomeration economies, and may have no significant advantage in terms of labor costs. Nevertheless, we find that Africa has caught up impressively over the last two decades in terms of access to communication services, which—in turn—improves the capabilities of African businesses to participate in global production processes. We also note that many African countries are much closer to the European market than their Asian competitors, and that "all of the specific misfortunes that impeded coastal Africa from entering global markets are now over" (Collier 2006).

Although Collier fears that Africa may have to wait several decades to initiate an economic transformation similar to the one Asia is experiencing, we arrive at a more differentiated conclusion. First, it is worth pointing out that the argument that Asia has gained a permanent advantage in terms of agglomeration economies is too broad—many agglomeration economies occur on the industrial or regional level,[11] and achieving such economies in sub-Saharan Africa is conceivable. At the same time, our findings suggest that the playing field in one important area has shifted to the benefit of Africa, as sub-Saharan Africa has made major strides in utilizing modern communication technologies, considerably improving the ease of doing business in Africa. Thus there are opportunities now for benefiting from the ongoing transformations in the global economy, and Collier's "several decades" may turn out shorter than expected.

At the same time, there are roles for policy to create conditions for businesses to take advantage of these opportunities. First, we find that access to communication technologies depends on the quality of policies, by creating a competitive environment in the telecommunications sector. Second, important elements of the costs of trading and exporting are the ease and the costs of transportation—not only in terms of distance and the physical infrastructure but also in terms of the effectiveness of transportation services. This is an area where many coastal African countries are at a disadvantage vis-à-vis their Asian competitors, and government policies can help addressing these bottlenecks, helping to fulfill the new economic opportunities that have developed in recent years.

Notes

1 It is important to bear in mind that these averages mask very substantial differences across countries within the respective regions. For example, sub-Saharan Africa includes fast-growing economies such as Botswana and Mauritius, while Asia includes economies like Bangladesh or Nepal that grew very little over this period.

2 See Collier (2006) and Venables (2006).

3 To ensure that our import data adequately reflect spending for the countries under consideration, we eliminate all countries that do export significant quantities of ICT-related equipment from the sample. For the remainder, we net out exports of such equipment. For more details, see Haacker (2006).

4 The figure excludes five African countries with higher GDP levels for which data are available, but "zooming in" on countries with lower levels of GDP per capita allows us to get a clearer picture of trends among the vast majority of African countries.

5 Quoted from CellularOnline, http://www.cellular.co.za, in a story dated June 10, 2002 (accessed on November 11, 2006).

6 The primary data sources on providers were the websites of the GSM Association (www.gsmworld.com) and of the CDMA Development Group (www.cdg.org), cross-checked for comprehensiveness against various informal data sources.

7 Data were kindly provided by BMI-TechKnowledge Group (South Africa), based on their *Communication Technologies Handbook*.

8 Based on the sample of 361 providers operating in the 112 countries classified as low- and low-middle-income countries, we find that there is more than one provider in 88 of the 112 countries.

9 These empirical findings are documented in more detail in Haacker (2006).

10 This is a point made to the author by Keith Jefferis of Botswana, personal communication, August, 2006.

11 See, for example, Porter (1990).

References

CellularOnline. 2002. "Vodacom Gets Mozambique Contract," June 10. Available at http://www.cellular.co.za.

Collier, P. 2006. "Africa: Geography and Growth." Paper presented at the New Economic Geography: Effects and Policy Implications, a symposium sponsored by the Federal Reserve Bank of Kansas City, Jackson Hole, Wyoming, August 24–26.

Freeman, R. B. 2005. "What Really Ails Europe (and America): The Doubling of the Global Workforce." *The Globalist*, June 3.

Haacker, M. 2005. "The ICT Sector and the Global Economy: Counting the Gains." *The Global Information Technology Report 2004–2005*. New York: Palgrave Macmillan.

———. 2006. "Information and Communication Technologies in Low- and Low-Middle-Income Countries: An Economic Assessment." Unpublished manuscript. Forthcoming as IMF Working Paper, Washington, DC: IMF.

IMF (International Monetary Fund). 2006. *World Economic Outlook Database*, April 2006. Washington, DC: IMF.

Maddison, A. 2003. *The World Economy: Historical Statistics*. Paris: OECD.

Porter, M. 1990. *The Competitive Advantage of Nations*. New York: Free Press.

United Nations Statistical Division. 2006. Commodity Trade Data Base, a.k.a. COMTRADE, (data downloaded in July/August 2006).

Venables, A. J. 2006. "Shifts in Economic Geography and their Causes." Paper presented at the New Economic Geography: Effects and Policy Implications, a symposium sponsored by the Federal Reserve Bank of Kansas City, Jackson Hole, Wyoming, August 24–26.

World Bank. 2006. *World Development Indicators 2006*. Washington, DC: World Bank.

CHAPTER 2.3

Information and Communication Technologies Policy in Japan: Meeting the Challenges Ahead

HIDEO SHIMIZU, Ministry of Internal Affairs and
Communications, Japan

KUNIKO OGAWA, Ministry of Internal Affairs and
Communications, Japan

KOICHI FUJINUMA, Ministry of Internal Affairs and
Communications, Japan

The world has witnessed remarkable growth and diffusion in information and communication technologies (ICT) system usage in this decade. The further development of the ICT industry will become a major factor for economic growth. This chapter will provide some economic background related to current Japanese ICT policies. It also sets out changes in the regulatory environment and looks at the current status of ICT infrastructure in Japan. The positive outcomes of the steps that have been taken are found in the high penetration rate of the broadband, including fiber optic cable services, mobile telephones with a high Internet access rate at 87 percent (incidentally, over 60 percent of these mobile telephones are third-generation telephones). Mobile telephones in Japan can be used also for watching television, using electronic money, and purchasing electronic tickets. The final section of this chapter will highlight challenges in the ICT field for the future, such as convergence of communications and broadcasting, security and privacy, and relating government policies.

History of Japan's ICT policy up to 2005

Japan has made sequential national ICT policies since around 2000, with the aim of fostering an advanced ICT network society. The Information Technology Basic Law was enacted in November of 2000. In 1999, Japanese Internet penetration rate was just 13.4 percent, lagging behind that of Northern Europe and North American countries. Also, in April 2001 the number of broadband subscribers in Japan was 737,000, behind the United States and Korea. At the time, introducing ICT into corporate management in Japan meant simply installing information technology devices and systems. The law was established because of a sense of urgency on the part of the government about Japan's slow start in the ICT revolution.

In January 2001, based on the Law, the Advanced Information and Telecommunications Society Promotion Headquarters was established within the Japanese cabinet. Beginning in January 2001, the headquarters crafted several national "e-Japan Strategies" that would put Japan among the leading ICT nations. Specific targets were set: "Promote the upgrading of Japan's Internet network to meet the highest global standards, with super high-speed access (30–100 Mbps) possible within 5 years; and make this access available at affordable rates for all citizens."[1] Under the principle of private-sector-driven development, a number of government policies were implemented. Among these were policies to enforce fair competition in the information and communications field, policies to support the development of e-commerce systems, policies that implement e-government, and policies for human resources development.

In particular, infrastructural improvements for the expansion of broadband progressed so rapidly that by 2003

Figure 1: The framework of Japanese ICT policy

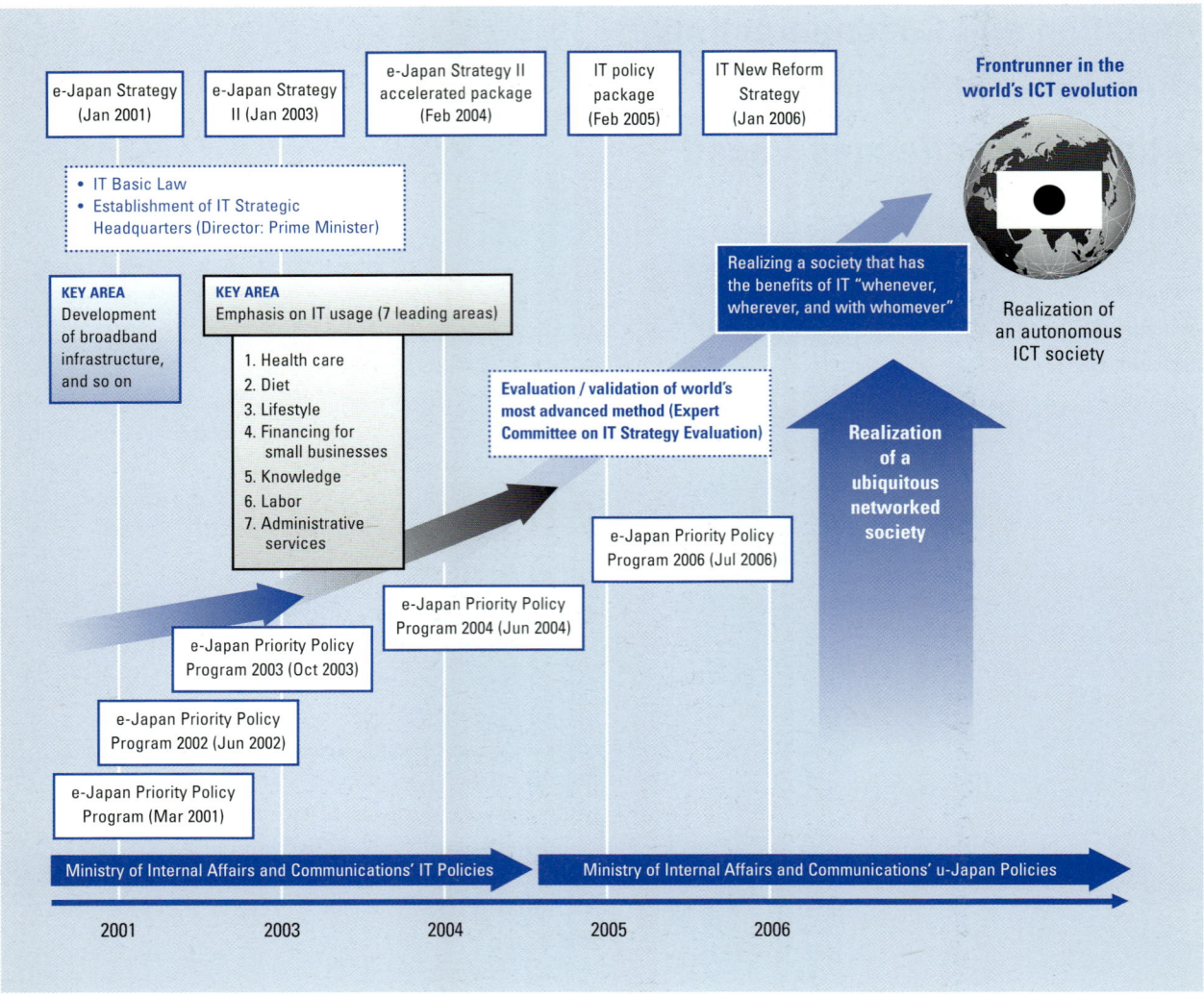

Source: MIC, Japan, 2005.

the targets of the e-Japan Strategies had been achieved. The ICT environment in Japan made it possible for 35 million households to have a constant connection to a digital subscriber line (DSL), for 23 million households to have cable TV, and for 17.7 million households to have fiber-to-the-home (FTTH) connections. As infrastructural improvements progressed, interest turned toward the promotion of usage. In 2003, the headquarters established a new strategy called "e-Japan Strategy II," settling on seven areas—health care, diet, lifestyle, financing for small businesses, knowledge, labor, and administrative services—where ICT could be used in groundbreaking ways. Again, as infrastructural improvements progressed, attention turned toward the issue of ICT usage. E-Japan Strategy II aims to continue to maintain the Japanese position as the frontrunner in terms of ICT technology beyond 2006 (see Figure 1).

Toward the ubiquitous networked society

In December 2004, just before the target year for completion of the e-Japan Strategy, the Ministry of Internal Affairs and Communications (MIC) drafted its u-Japan Policy. This policy seeks to create a ubiquitous networked society by 2010 that will permit ICT access "at any time, anywhere, with anything and by anyone."[2]

The major goal of the u-Japan Policy is to "lead the way as the world's most advanced ICT nation in 2010," to have the world's best ICT infrastructure, and to contribute to the world by providing an original and creative Japanese social model that balances infrastructure with

ICT usage. The following are the three basic elements of the u-Japan Policy:

- development of ubiquitous networks that can be used seamlessly for both wireless and fixed networks,

- advanced usage of ICT to assist in resolving social issues, and

- improvement of the environment for ICT usage in a safe and secure manner.

In January 2006 the Advanced Information and Telecommunications Society Promotion Headquarters created the Information Technology New Reform Strategy, which included follow-up on the capability of ICT to facilitate structural reforms (that is, its ability to help resolve issues facing Japanese society), ICT infrastructure improvement (that is, developing infrastructure that moves Japan toward the goal of being a ubiquitous networked society), and dissemination to the world (that is, Japan's ICT contribution to the world).

The new strategy seeks to make all areas of Japan accessible to broadband so that Japan's communications infrastructure is completely broadband-capable by 2010. The strategy also seeks to make 2011 the "First Year of Completely Digital Networks" for all communications and transmissions, with the changeover to digital terrestrial television broadcasts complete by July of 2011. All of this is to be accomplished through the principle of private-sector-driven development and a variety of government policies that seek to promote these goals.

ICT and macroeconomics: Three effects on economic growth

Japan has placed priority on ICT-related policies because ICT has an impact on economic growth in a number of ways:

- through the growth of the ICT industry,

- through a deepening of ICT capital stock, and

- by fostering productivity increases for all industries, as a result of ICT spreading across industries and corporations.

The following section provides some economic background, drawn from the Japanese information and telecommunications white paper of 2006, for current Japanese ICT policies.[3]

The first effect: The direct impact of ICT industry growth

The ICT industry accounts for an increased percentage of GDP. The per-industry real GDP of the ICT industry in 2004 was 61.9 trillion yen (US$560 billion), or 11.7 percent of Japan's real GDP. Also, the percentage change of per-industry real GDP for the ICT industry was 9.21 percent. This means that the ICT industry contributed 40 percent to the change in Japan's real GDP.[4] The effect that the ICT industry's performance has had on Japan's economic growth is significant, no matter what other industry it is compared with (Figure 2).

The second effect: ICT capital stock and economic growth

Real investments in ICT in 2004 totaled 16.4 trillion yen (US$150 billion)—21.5 percent of the total capital investments made by the private sector (see Figure 3). ICT capital stock totaled 36.9 trillion yen (US$335 billion), comprising 3.0 percent of private capital stock.[5] A look at investments in ICT capital stock over the years shows that, although they temporarily dipped in the mid 1990s, they have since rebounded and continue to grow.

Figure 4 shows the contribution of ICT capital to the economic growth rate. ICT capital stock contributed 0.21 percent to economic growth from 1990 to 1995 (economic growth was 1.51 percent); from 1995 to 2000 ICT capital stock contributed 0.54 percent (economic growth was 0.97 percent); and from 2000 to 2004 they contributed 0.21 percent (economic growth was 1.15 percent). Thus ICT capital stock deepened overall economic growth by 13.9 percent, 55.7 percent, and 18.3 percent respectively over these years. Considering that ICT capital stock comprises 2–3 percent of private capital stock, its impact on economic growth is significant.

The third effect: The spread of ICT and economic growth

The third effect is an increase in productivity resulting from the spread of ICT in industries and businesses. This increase is due to an improvement in total factor productivity (TFP), which is a variable factor and not explained in terms of changes in the input of factors of production such as capital and labor. The increase in TFP for the general ICT industry in Japan between 2000 and 2004 was 3.7 percent; it was 2.9 percent for electrical machines. Clearly TFP was high for fields related to ICT, but in other industries it was not necessarily so high, and the overall industry total stood at just 0.09 percent. It is now believed that growth in Japan's TFP is being primarily supported by the ICT industry, which is a leading factor in technological innovations.

History of changes in the regulatory environment

Over 20 years have passed since the privatization of Nippon Telegraph and Telephone Public Corporation (NTT) in 1985. Since that time a number of businesses have ventured into the communications market; coupled with an easing of regulations, this has resulted in a considerable growth of the communications market.[6] As of March 2006, there were

Figure 2: Contribution of the ICT industry to changes in real GDP, 1996–2004

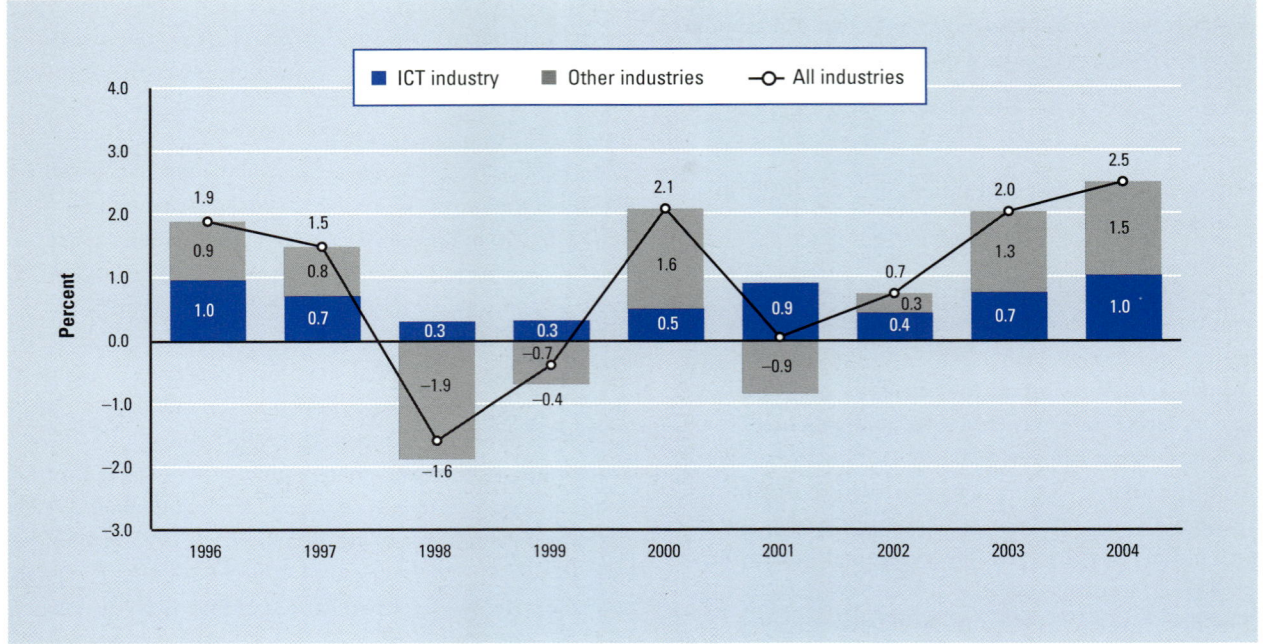

Source: Do Research Institute, 2006.

Figure 3: Evolution of real ICT investment, 1980–2004

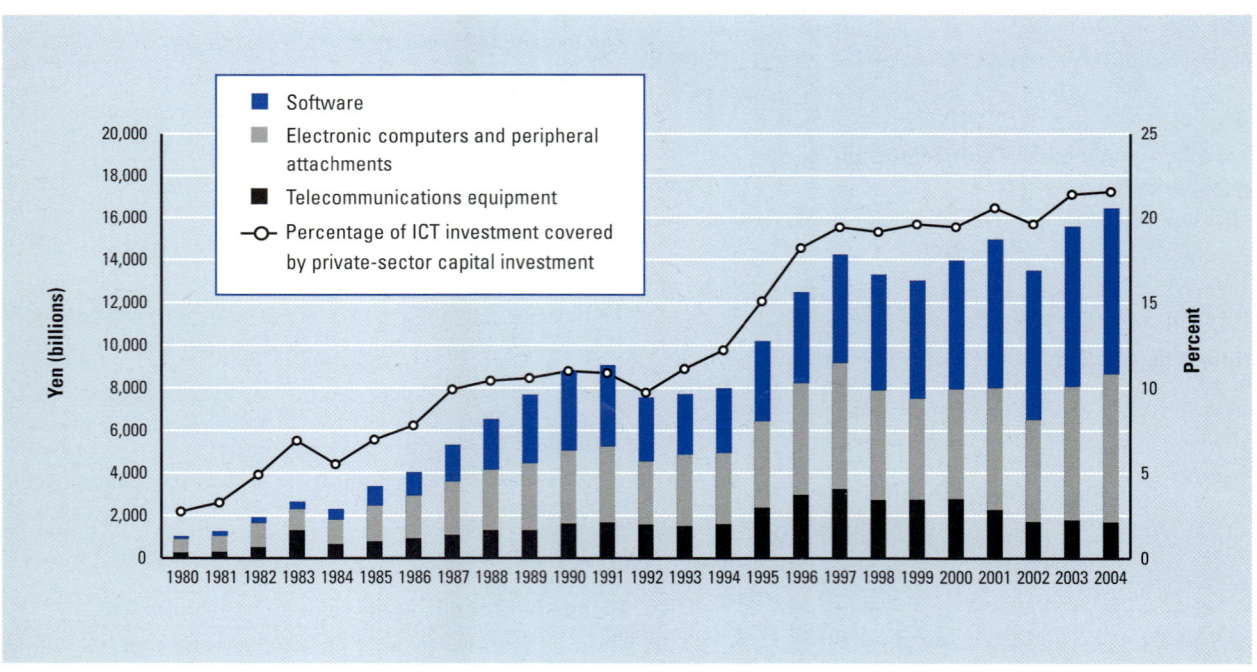

Source: Do Research Institute, 2006.

Figure 4: Contribution of ICT capital to the economic growth rate

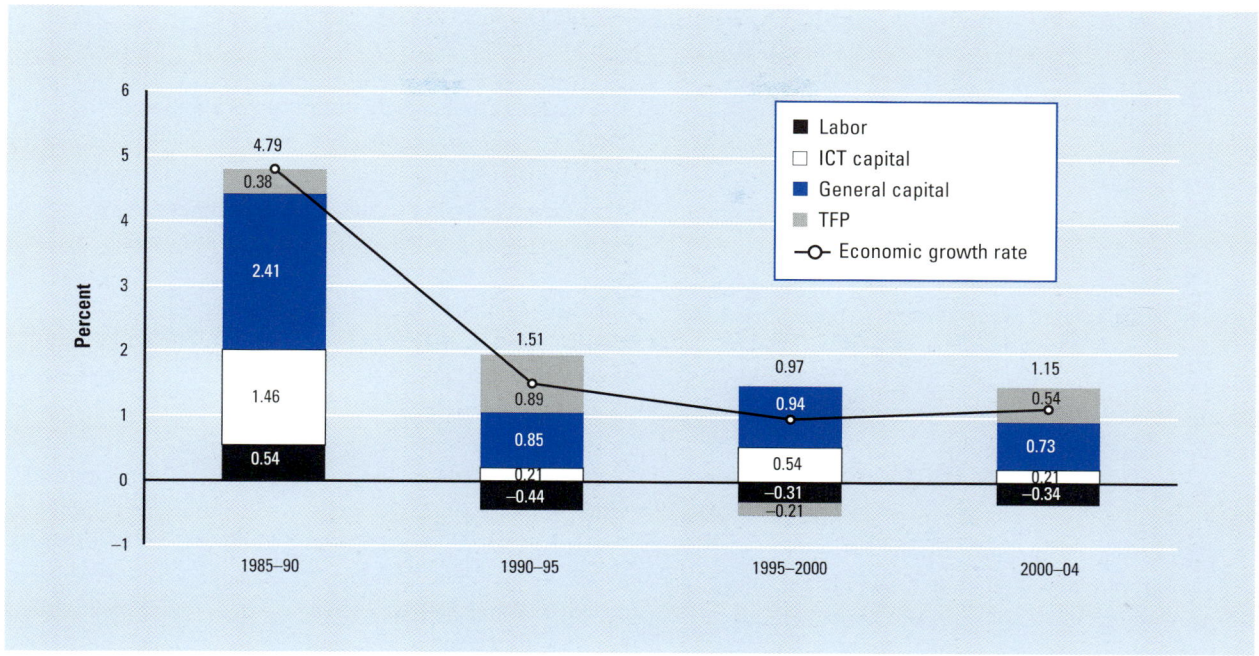

Source: Do Research Institute, 2006.

314 registered and 13,412 notified telecommunications providers offering a variety of communications services.

The regulatory framework for telecommunications businesses in Japan is continually revised to meet the needs of the telecommunications services market and to ensure that users enjoy a broad range of services through the promotion of fair competition in the facilities and services fields. These changes in the regulatory framework can be divided into three time periods, as related below.

Period one: From monopoly to competition (1985–)
In April of 1985, the Telecommunications Business Law went into effect. Based on this law, the government introduced competition into all areas of the telecommunications market; also NTT was privatized. When the law went into effect, a type 1 telecommunications business (that is, the setting up of physical facilities containing telecommunications lines and providing telecommunications services) required a massive initial investment in order to construct a network. Economies of scale, as well as other factors—such as the externality of the network—made it difficult to enter into the telecommunications market. The design of the network was based on the principle that telecommunications was an essential industry to the life and economy of the nation and that the network would play a

major role in serving the public interest, by ensuring, for instance, that people received important announcements during emergencies and crises. For these reasons, type 1 telecommunications carriers were required to obtain a license.

Period two: Promoting further competition (1997–)
In order to facilitate the entry of new carriers into the telecommunications market, revisions made to the Telecommunications Business Law in 1997 eased a requirement for obtaining permission to operate for type 1 telecommunications carriers. In 1998, restrictions on foreign investment were abolished in principle. In 1999, in order to further increase the international competitiveness of Japan's ICT industry and to create a more dynamic, competitive environment, NTT was restructured under holding corporations into three separate companies: a regional company for eastern Japan, a regional company for western Japan, and a long-distance network company.

Following the restructuring of NTT, the basic rules governing connections were revised in 2000. A long-run incremental costs (LRIC) system was implemented for telephone networks' cost calculation method, and the connection rules for DSL service were improved by creating conditions and procedures for unbundling subscriber lines and so on, and by establishing shared facilities to allow competitors to connect (a process called "colocation"). In 2001, in consideration of the degree to which competition had progressed, the Telecommunications Business Dispute Settlement Commission was established to handle disputes and other matters between telecommunications companies, and to recommend necessary improvements and changes in the rules to the MIC. In 2002, the universal service fund system was upgraded to eliminate geographical digital divide.

Period three: From ex-ante regulations to ex-post regulations (2004–)

Because of changes in the market structure, revisions were made to the Telecommunications Business Law in 2003. These revisions aimed to continue to adapt the law to changes in the market environment and network structure, as well as to encourage new companies to enter the market and to promote fair competition among them. The original business classifications were abolished and market entry regulations were greatly eased, shifting the regulations from ex-ante to ex-post. In principle, regulations governing fees and notification of contract conditions were abolished and steps were taken to ease the authorization of the relative contracts. Along with relaxing these regulations, rules to protect consumers were introduced in 2003 together with a system for evaluating competition in the telecommunications.

Evaluation of competition in telecommunications

In 2003, an analysis and evaluation of competition in telecommunications was conducted by the government. The analysis was done in each major service such as Internet access, mobile telecommunications, Internet protocol (IP) telephones, and fixed-line telephones. The following are some of the results of the analysis.

For the broadband market, attention was paid to changes in the competitive environment accompanying the shift toward the convergence of communications and broadcasting with fiber optic lines (FTTH). In the asymmetric digital subscriber line (ADSL) market, NTT East and West's share as of December 2005 was 39.1 percent; the market was judged to be functioning effectively under the rules governing connections, with vigorous competition in terms of cost and service. In the FTTH market, NTT East and West had a 60.7 percent share as of December 2005; the number of contracts was growing and video services were being increasingly realized. The mobile telephone market was switching from the 2nd generation (2G) to the 3rd generation (3G), and competition remained as vigorous as ever. NTT Docomo's share was 54.1 percent as of December 2005. It becomes difficult to predict the future trend of ICT with the innovation and changes in ICT field such as the convergence of communications and broadcasting, the convergence of fixed and mobile telephones, the further spread of broadband, IP telephones, and fiber optics.

The current status of ICT infrastructure in Japan

The following section describes the remarkable development of ICT infrastructure fostered by the above-explained Japanese policies; it also describes the new trend of convergence in broadcasting and communications.

The development of ICT infrastructure

The International Telecommunication Union (ITU)'s July 2006 edition of *The World Information Society Report* listed a Digital Opportunity Index (DOI) of 11 factors (such as population coverage for mobile telephones, Internet connection fees, Internet access from mobile telephones, and ratio of broadband coverage) for 180 economic regions. Korea had the highest DOI score, followed by Japan. Also, in a comparison of broadband fees, Japan was the cheapest at US$0.07 per 100 kilobits; it was tied with Korea as the fastest with a transmission speed of 51.2 Mbps.

Increase in the number of Internet users

In 2005, the number of Internet users in Japan stood at 85.29 million people (66.8 percent of the population). A majority of people—estimated at 48.62 million (57.0 percent of total Internet users)—used both personal computers and mobile devices (such as mobile telephones) to access the Internet. Of these, 19.21 million people (22.5 percent) used only mobile devices, and 15.85 million (18.6 percent) used only personal computers.

The rate of penetration for broadband has been increasing (see Figure 5). By the end of 2005 the number of contracts for broadband lines had reached 23.30 million (a 19.1 percent increase over the previous year). Looking at the breakdown of this figure, we see that contracts for DSL were the most common, at about 14.52 million in 2005 (a 6.2 percent increase over the previous year). Next came FTTH, with approximately 546,000 contracts (an 88.4 percent increase over the previous year). This was followed by approximately 331,000 contracts for cable Internet (an 11.8 percent increase) and approximately 20,000 contracts for fixed wireless access (FWA) (a 34.8 percent decrease).

The government's goal, based on the u-Japan Policy, is to make broadband available in every region in Japan by 2010 through measures promoting the upgrading of the broadband network. Interest in switching to FTTH Internet connection lines is increasing, and a comparison of the net increase in the number of contracts during the fourth quarter shows that the number of contracts rose for FTTH during the period, surpassing that for DSL.

Figure 5: Evolution in the number of broadband contracts

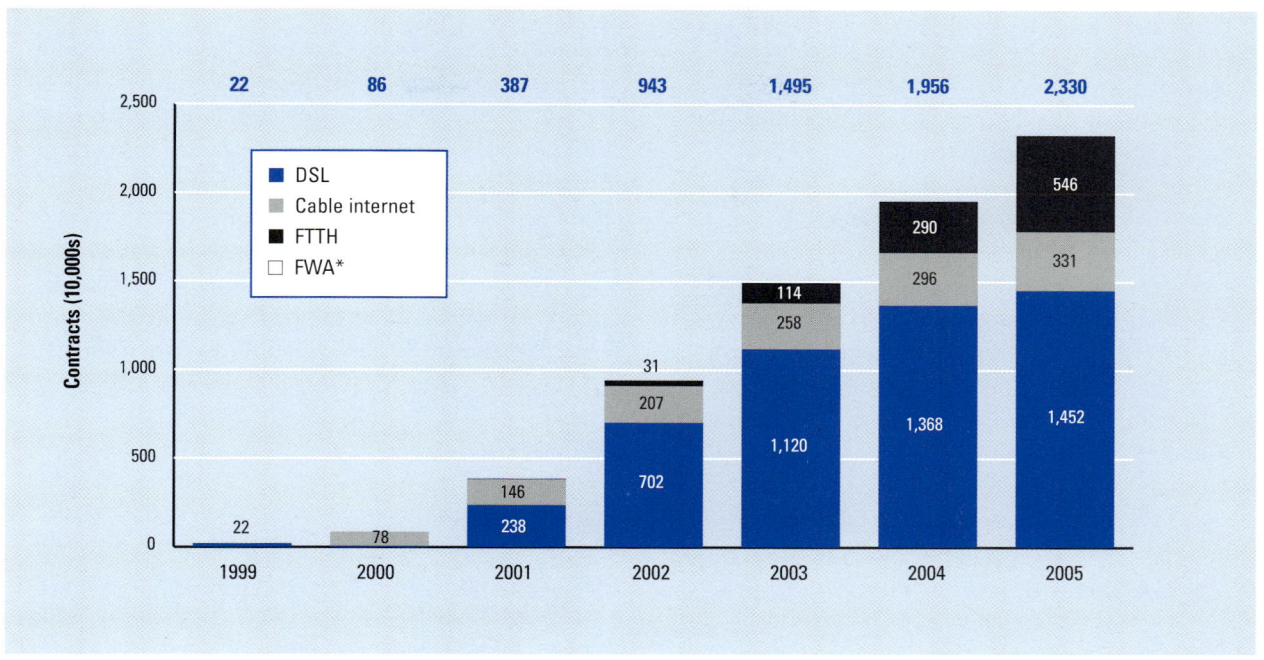

Source: Communications Usage Trend Survey, MIC, Japan.
* FWA contracts have grown from 0 contracts in 1999 to 20,000 contracts in 2005; these relatively small values are not visible in this figure.

By the end of 2005, 11.46 million people (such as DSL subscribers and fiber optic subscribers) had used IP telephones. In addition, there are peer-to-peer telephones that are cheaper and utilize IP technology, but they do not have an assigned IP phone number because they do not satisfy a quality standard. In keeping with the spread of the Internet, NTT has stopped investing in circuit switches and is instead promoting a switchover of the network to IP. Major companies in Japan have indicated a policy of switching communications networks from public switched telephone network (PSTN) to IP by 2010; it is expected that telecommunications services, including telephones, will be offered via an all-IP infrastructure.

Penetration and sophistication of mobile telephones
Mobile telephones are no longer just for talking; they have become people's most accessible mobile information terminals. One example is their Internet capabilities. By the end of 2005, the number of people accessing the Internet from mobile telephones or other mobile information terminals was 69.23 million (81.2 percent, a 7.9 percent increase over the previous year). This was the first time Internet access from mobile devices had exceeded Internet access from personal computers (the relevant figure for personal computers was 66.01 million).[7]

In addition, mobile telephone terminals offer games, music for downloading and playing, "one-segment broadcasts" (terrestrial television broadcasts aimed at mobile terminals), and video telephone. Furthermore, there are a number of mobile terminals with functions that enable the user to connect to a variety of information read from barcodes, and even to handle electronic money and tickets via smartcard or radio frequency identification (RFID). In 2004, the market size for mobile commerce (m-commerce) was 971 billion yen (US$8.8 billion), encompassing 17.2 percent of the business-to-consumer e-commerce market. People are able to do a variety of tasks whenever and wherever they want with their mobile telephones.

Currently, telecommunications carriers provide subsidies to encourage the sale of mobile telephone units; thus, users are often able to purchase relatively cheap units that offer a high degree of functionality. This structure is credited with having a great effect on helping highly functional terminals to broadly penetrate the competitive portable device market. However, to produce more competitively priced products, discussion on replacing this funding scheme with a new price plan is underway.

Dramatic increase in Internet traffic
In recent years, IP traffic in Japan has shown a tendency to double each year. The continuing increase in the number

of broadband subscribers and Internet users is the engine for this trend. It is believed that the increase in IP traffic is to the result of factors such as the strong momentum Internet use is gaining in society and the economy, the sudden rise in peer-to-peer file sharing, and the ability of individuals—not just content providers—to transmit heavy content such as music and movies.

Convergence of broadcasting and communications: The development of digital broadcasts

Terrestrial digital television broadcasts in Japan began in December 2003. The reception area for these broadcasts has expanded rapidly—by June of 2006, this area covered 32.2 million households of potential viewers, or 68 percent of the total number of households in Japan. By the end of 2006 all prefectural capitals in Japan began terrestrial digital television broadcasts; by July 2011, all broadcasts will be done digitally. Television is one of the most accessible means for citizens to obtain information, and it is hoped that switching television to digital will lead to a communications infrastructure in households. Also, in April 2006 digital one-segment broadcasts (terrestrial television broadcasts aimed at mobile terminals) began, enabling image and data broadcasts to be received on mobile devices as clearly as they appear on home televisions.

Movement toward convergence of telecommunications and broadcasting

In line with the increase in digital broadcasts and the dramatic increase in carrying capacity—the result of the spread of broadband throughout the communications network—telecommunications carriers are using their optical fiber networks as a single channel for both broadcast services (for television shows and so on) and for communications services. The movement, described as a "convergence of telecommunications and broadcasting," is based on the three phenomena of (1) full-scale online delivery of movies and music; (2) sharing of terminals, networks, and so on; and (3) the simultaneous entry of corporations into the communications and broadcast fields.

The acceleration of this convergence, as well as the situation of the telecommunications and broadcasting industries, will lead to an increase in information provided and to improved production of content. These increases are expected to contribute to strengthening the standing of Japan by, for example, disseminating Japanese culture in the world at large.

With features such as video on demand (VOD) distribution services that allow people to view videos from their personal computers (that is, Internet broadcasts); and with the common usage of the same transmission channel for both communications and broadcast services, business tie-ups continue to transcend industry boundaries.

Terrestrial digital television broadcasts have been shown in Japan since 2003. One of the goals of digitizing terrestrial broadcasting is to provide bidirectional services by connecting broadcasting with the Internet. Such bidirectional services enable the viewers and listeners to have access to a wide variety of services. Television is the medium that most people use to obtain day-to-day information and can be found in practically every household. Therefore, digitizing television will provide a basis for ICT in households. Among cable television and telecommunications corporations in recent years, an increasing number are offering a "triple play package" to consumers: Internet access, video delivery, and IP telephone service.

Challenges for the future

The world has witnessed changes to society as a result of ICT advancement. Public policies to reflect these social changes and to further ensure that the benefits of ICT trickle down are needed. Japan is restructuring its systems and policies in the ICT field, as well as promoting various measures. The following are some challenges that Japan faces in this area.

Promoting discussion on the frameworks of communications and broadcasting

The MIC held a Panel on Frameworks of Communications and Broadcasting from January to June of 2006. A comprehensive range of proposals for Japan to reach its goal of creating the world's leading communications and broadcasting infrastructure by its target of 2011 were suggested. Proposals included making use of its strengths to become the leader in the field of broadband, mobile, and television communications; preparing legal systems for the convergence of communications and broadcasting; revising its communications-related regulations; easing its broadcast regulations; and reforming Nippon Hoso Kyokai (NHK), Japan's public broadcaster. In June 2006, an agreement between government and ruling parties was reached on the framework of communications and broadcasting. Based on this agreement, a Program for Restructuring the Communications and Broadcasting Sectors was announced in September 2006. This program addressed the four sectors relating to NHK reform, broadcasting, convergence, and communications. In addition, at the end of August 2006, a study group on the comprehensive legal structure for communications and broadcasting was established. For approximately a year and a half, this study group will examine, from a technical standpoint, how to concretely orient discussion on the legal system with regard to combining and linking communications and broadcasting.

Negative issues brought about by ICT advances

As the ubiquitous network continues to develop, and as ICT continues to permeate citizens' lives, it is important to deal with concerning issues such as privacy and security, a safe and secure ICT usage, and eliminating the digital divide.

For example, ensuring security in ICT usage is becoming a major social issue. Individuals and companies may become the victims of crimes such as ID theft, skimming, or billing fraud in their daily lives and daily operations (when using an ATM, using a credit card, shopping on the Internet, and so on). These security concerns must be addressed effectively and promptly.

There is also a need to provide measures against spam email. The MIC initiated a revision to the anti-spam law (effective from November 2005) whereby anyone who uses the information of an email sender in a deceptive way is liable for prosecution. The ministry is also promoting a Spam Purging Project (begun in February 2005) that operates through the joint cooperation of the government and the private sector. The project encourages telecommunications companies to cut off and otherwise deny line access to spammers. International cooperation is also being sought; an agreement was reached between Japan and other Asian countries in April 2005, with France in May 2006, with the United Kingdom in September 2006, and with Canada in October 2006.

Furthermore, in recent years there has been a succession of incidents involving illegal or harmful information being posted on websites and electronic bulletin boards. In response, the Provider Liability Limitation Law (in effect since May 2002) and its related guidelines establish standards of restriction on providers' responsibility that enable providers to remove illegal information. Also, by creating guidelines (rather than required regulations), the government supports the voluntary efforts of ISPs to deal with this issue.

New competition promotion program 2010

The MIC held a meeting on a framework for competition rules to address the transition to IP-based networks and released a report in September 2006. This report was written in light of the changes to the market environment resulting from the development of IP, such as the spread of broadband, the switching from PSTN to an IP network, and the diversification of business models.

The measures for implementing these upgrades by 2010, with a view toward promoting greater competition and ensuring the interests of users, were released in September 2006; the MIC then set about implementing them. When competition policies are developed, the fundamentals of fair competition will be ensured, providing an appropriate balance between promoting competition to build facilities for creating networks on the one hand, and opening up the networks of dominant telecommunications companies that have bottleneck facilities on the other hand. The strategy also promotes fair competition, thus allowing for the spread of a vertical integration business model that cuts longitudinally through each network layer.

Promotion of the u-Japan Policy

The MIC is promoting the u-Japan Policy as a means of contributing to the entire government's initiatives on behalf of the IT New Reform Strategy, which aims to achieve a ubiquitous network society in Japan by 2010 as previously related in this chapter. The ministry has summarized the necessary policy packages for each year in its ICT Policy Outline; it promotes a variety of policies related to topics such as "eliminating the geographical digital divide," "developing a seamless access environment for both fixed and wireless networks," "promoting advance social system restructuring through ICT," "ensuring the safety and security of citizens through ICT," and "ensuring network reliability and security."[8]

In September 2006, the u-Japan Promotion Program 2006 was assembled as a mid-term policy to be implemented until 2010, reflecting the changes in the communications and broadcasting situation that have taken place since the creation of the original u-Japan Policy. This mid-term policy includes (1) the promotion of the convergence of communications and broadcasting, (2) strengthening the growth rate, the competition levels, and the national standing through ICT, and (3) achieving a safe and secure ubiquitous network society through ICT.[9]

Japan faces issues that stem from changes in its social and economic environment, such as a declining population —the result of a lowering birthrate and aging population. However, we believe that ICT will generate economic vitality and will encourage social and economic development through factors such as the increasing accumulation and integration of knowledge and technology, transforming the existing social and economic system, and accelerating the pace of innovation via the realization of a ubiquitous network society where ICT affects all aspects of socioeconomic activity.

Notes

1 e-Japan Strategies 2001.

2 u-Japan Policy 2005, Chapter 1.

3 See the 2006 Information and Communications White Paper, Chapter 1.

4 Of real GDP growth for 2004 (which was 2.5 percent), 1.0 percent was the result of the ICT industry.

5 The definition of *ICT capital stock* here is "electronic devices and software for use on computers which are capable of connecting networks." This definition includes electronic computers and peripheral attachments, fixed-line communications devices, wireless communication devices, and software (not including self-developed software).

6 See the supplement to the MIC report *Framework for Competition Rules to Address the Transition to IPBbased Networks*, 2006. In Japanese.

7 Communications Usage Trend Survey (2006) conducted by the MIC. Available at www.johotsusintokei.soumu.go.jp/english/.

8 See u-Japan Policy (2005, Chapter 1).

9 MIC u-Japan Promotion Program 2006.

References

Bleha, T. 2005. "Down to the Wire." *Foreign Affairs*. 84 (3).

Do Research Institute. 2006. *Survey on Economic Analysis of ICT, Japan* (in Japanese). Tokyo: Do Research Institute.

ITU (International Telecommunication Union). 2003. *World Telecommunication Indicators 2004*. Geneva: ITU.

———. 2006. *The World Information Society Report*, July edition. Geneva: ITU

MIC (Ministry of Internal Affairs and Communications), Japan. 2001–2005. Communications Usage Trend Survey. Available at www.johotsusin-tokei.soumu.go.jp/english/.

———. 2002–2006. *White Paper Information and Communications in Japan*. Tokyo: MIC.

———. 2005. u-Japan Policy. MIC. Available at www.soumu.go.jp/menu_02/ict/u-japan_en/index2.html.

———. 2006. *Framework for Competition Rules to Address the Transition to IP-Based Networks*. Tokyo: MIC.

OECD. 2004. OECD Broadband Statistics, December 2004. Retrieved July 28, 2005.

Rowen, H. et al. 2006. *Making IT: The Rise of Asia in Hi Tech*. Stanford University.

CHAPTER 2.4

Made in China: Information Technologies and the Internet

SACHA WUNSCH-VINCENT, OECD

GRAHAM VICKERY, OECD

With GDP growth rates averaging close to 10 percent over the past two decades, the size of the Chinese economy, when measured at market prices, now exceeds that of a number of major European economies; in four years time it may be exceeded by only three Organisation for Economic Co-operation and Development (OECD) member countries, namely the United States, Japan, and Germany.[1]

Chinese ICT-related trade and investment

China is forecasted by the OECD to become the world's largest exporter of goods and services by the beginning of the next decade. This paper sheds light on the current position of China in the global production, demand, and use of information and communication technologies (ICT).

China: Number one ICT exporter and growing

In 2004, China became the biggest global exporter of ICT goods, surpassing Japan and the European Union in 2003 and taking the lead from the United States (see Figure 1). In 2005 exports from China have continued to grow at an astonishing pace while other ICT export leaders have seen more modest growth (for example, the United States and Korea); Japan has even seen a fall in ICT exports. China's growing ICT trade in 2005 was bolstered by its elimination of tariffs on ICT products following its obligations under the Information Technology Agreement of the World Trade Organisation (WTO).

China's share of total world trade (exports plus imports) in ICT goods has grown rapidly. Worth less than US$35 billion in 1996, China's ICT goods trade reached almost US$418 billion in 2005, growing at more than 28 percent a year since 1996 (compound annual growth rate, CAGR). Chinese ICT exports reached US$235 billion in 2005—compared with US$155 billion for the United States (data for the European Union are not yet available for 2005, but they also could surpass ICT exports of the United States). The strong Chinese ICT export growth is reflected in a trade surplus in ICT goods of US$52 billion in 2004. Year-on-year export growth saw a relative but temporary slump from 2000 to 2001 (to 18 percent year-on-year growth, down from 44 percent in the previous year) when the sales in the ICT industry fell sharply. In 2002, however, the Chinese ICT export growth rate exceeded the high growth rates of 1999–2000, then grew by 46 percent from 2003 to 2004; more recently the growth rate has been slower for all ICT categories, with ICT exports growing by 30 percent from 2004 to 2005.

117

The views expressed in this paper are those of the authors, and shall be attributed neither to the Organisation for Economic Co-operation and Development (OECD) nor its member countries. See also the *OECD Information Technology Outlook 2006.*

Figure 1. Imports and exports of ICT goods, 1996–2005

1a: Imports

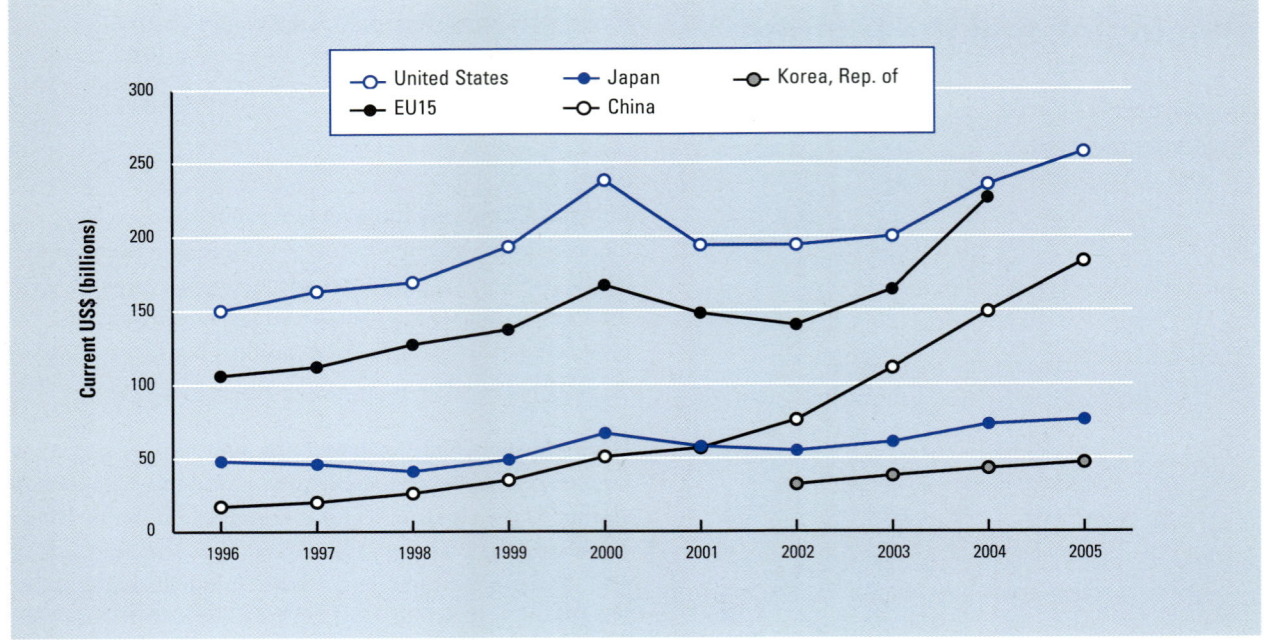

1b: Exports

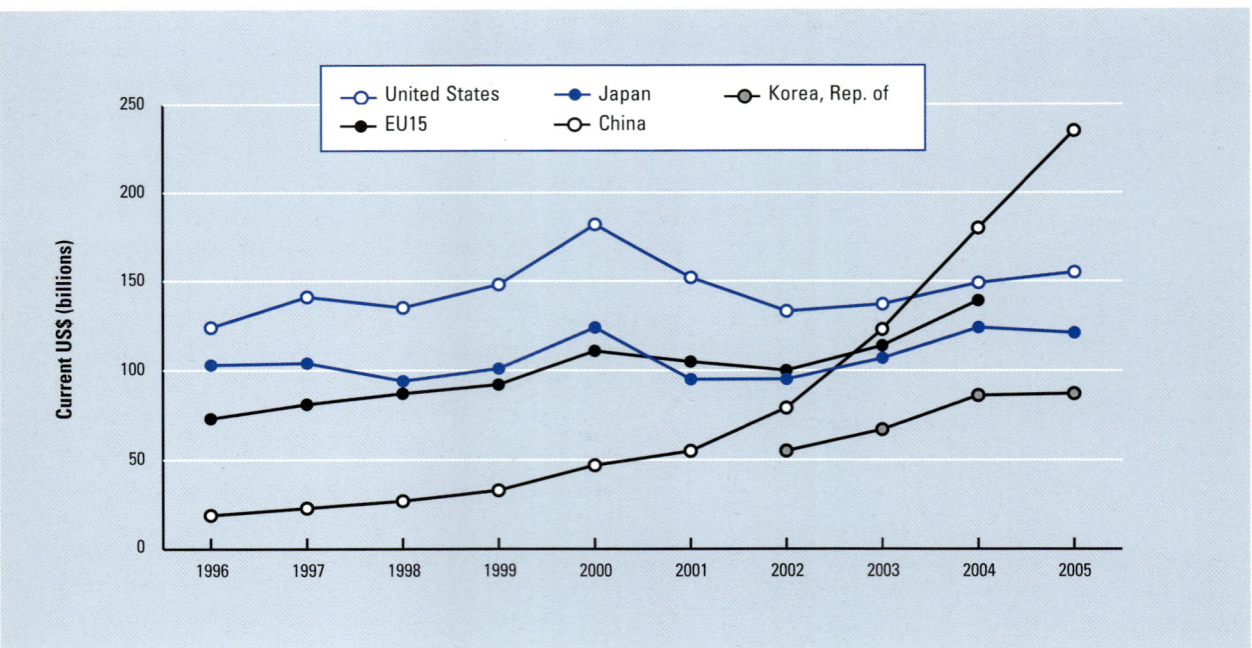

Source: OECD International Trade by Commodity Statistics (ITCS) database.
Note: Data for the European Union exclude intra-EU trade.

While China's imports of ICT goods rose rapidly between 2001 and 2005, they have remained below imports of ICT goods into the United States (US$257 billion) and most likely also the EU15 (the 2005 figures for EU15 are currently unavailable) but above those of Japan (US$76 billion) and Korea (US$47 billion). But although growth in Chinese imports of all goods was nearly halved, to 11.5 percent, from 2004 to 2005, its imports of computers, telecommunications equipment, and electrical machinery continued to rise significantly. OECD trade data for 2006 for ICT goods are not yet available, but the Chinese Monthly Customs Statistics database for the first two months of 2006 show a continued strong upward trend for ICT imports and exports.

The increase in ICT exports can mostly be traced to the transfer to China of foreign companies' labor-intensive and often low-value-added assembly and production activities of televisions, computers, handsets, and DVD players. This phenomenon became increasingly evident after the bursting of the IT bubble in 2000–01, as shown by the rapid recovery in ICT exports. The China-based affiliates of global ICT firms or third-party electronic manufacturing services (EMS) companies, which manufacture for leading global ICT firms, import intermediate products and produce mainly finished ICT goods that are exported (pure processing and assembly-related trade). For this process, high-value-added components such as central processors and memory chips are generally imported.

Data on the share of Chinese ICT-related processing trade or the share of foreign firms in total Chinese ICT exports (corresponding to the OECD definition of ICT goods) are not available. However, aggregate Chinese export figures for January to December 2005 show that 55 percent of total exports connect to processing and assembly-related trade; of total Chinese exports and imports from January to December 2005, around 58 percent are driven by foreign-invested enterprises (with 38 percent wholly foreign owned enterprises).[2] For ICT-related trade (high-technology exports or exports of electronics products, according to the Chinese definition) these shares are even more important, as processing the trade of foreign firms represents about 90 percent of Chinese ICT-related exports.[3] Hence, among the top 100 exporting foreign-invested enterprises, there are many ICT-related companies from Chinese Taipei,[4] and the United States (see Table 3 below). None of the top 10 high-technology firms in China (classified by 2005 revenues) is domestic, with Motorola, Dell, Hewlett-Packard, and Nokia leading.[5]

Despite the importance of foreign firms' ICT-related processing trade in China, there is mounting evidence that ICT-related foreign affiliates are evolving from simple assembly and manufacturing to more complex original design and production as well as fulfilling more important roles in global innovation networks. They also increasingly cater to the rapidly growing Chinese market. Furthermore, as a sign of developing domestic capacity, production and exports of Chinese ICT-producing firms are rapidly increasing in importance. In 2005 high-technology exports of national firms increased to about 12 percent of the total high-technology exports, a figure that is still low but rising.

China largely imports electronic components and exports computer and related equipment

China continues to import electronic components (68 percent of imports in 2005) while exporting computer and related equipment (45 percent of total exports in 2005) (Figure 2). In addition to satisfying local market demand, electronic components are used to assemble computer and related equipment and audio, video, and telecommunications equipment (more than 80 percent of total imports in 2005) as well as consumer electronics (for example, MP3 players).

In 2005 China was increasingly dependent on imports of the high-value components needed for the manufacture of ICT. It has significant trade deficits in electronic components: a US$68 billion deficit in integrated circuits (HS 8542)[6]—growing from US$50.5 billion in 2004—and a US$7 billion deficit in semiconductors and components (HS 8541)—growing from US$6.5 billion in 2004 and smaller deficits in products such as audio and video parts (Table 1 shows the top five Chinese ICT import and export items). Foreign semiconductor suppliers have been the main beneficiaries of rising Chinese demand for integrated circuits from abroad. China also has a trade deficit in large and medium-sized computers.

In return, China is increasingly the main export platform for final ICT products. Items for which China has the biggest trade surplus in ICT goods are computer and related equipment, with a US$58 billion trade surplus in data processing machines (HS 8471, including laptops but excluding personal computer [PC] accessories)—rising strongly from US$45 billion, along with a very considerable increase in laptop exports since 2002 and substantial surpluses in video cameras and recorders, television receivers, and telephones.

While computer and related equipment continues to be the major export item, exports of audio and video and telecommunications equipment have also soared. The most striking aspect is the strong growth in exports of telecommunications equipment, which were negligible in 1996 but grew to US$36 billion in 2004 (15 percent of total ICT exports), with the second biggest yearly growth rates from 1996 to 2005 (30 percent), a trade surplus in 2000 and exports growing by 76 percent from 2003 to 2004 and by 42 percent from 2004 to 2005. Chinese imports of

Figure 2: China's trade balance by ICT goods categories, 1996–2005

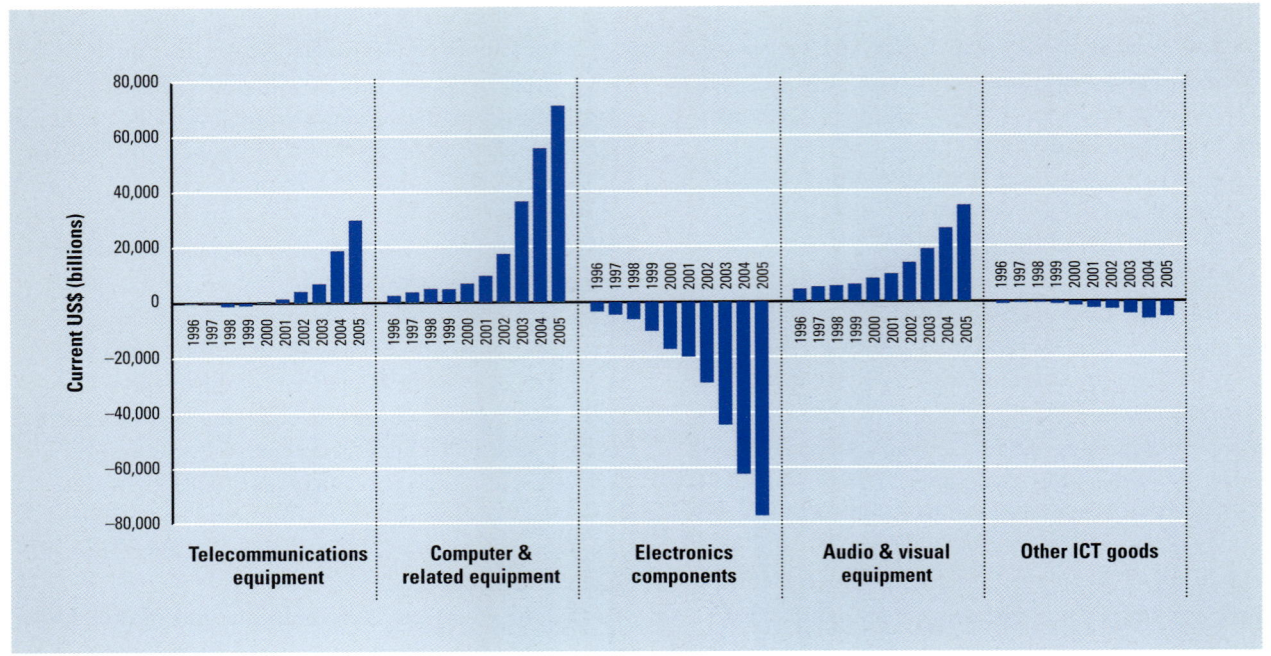

Source: OECD ITCS database.

Table 1: Top five Chinese ICT import and export items by 4-digit HS codes, 2005

HS code	Main imports	US$ (billions)	HS code	Main exports	US$ (billions)
8542	Integrated circuits	82.2	8471	Automatic data process machines, magnetic reader, etc., computer hardware	76.3
8471	Automatic data process machines, magnetic reader, etc., computer hardware	18.0	8525	Transmission apparatus for radio telephony/telegraphy/broadcasting, television	31.0
8529	Parts for television, radio and radar apparatus	16.6	8473	Parts, etc., for typewriters & other office machines, computer accessories	28.6
8473	Parts, etc., for typewriters & other office machines, computer accessories	16.5	8529	Parts for television, radio and radar apparatus	16.6
8541	Semiconductor devices	11.2	8542	Integrated circuits	14.6

Source: OECD ITCS database.

telecommunications equipment from Indonesia, Korea, Malaysia, and other Asian regions have been growing at very substantial rates and falling from countries such as the United States, suggesting new Asian intra-regional trade in telecommunications equipment.

Electronic components form the second biggest Chinese export item despite rapid growth in imports. However component imports are growing at a somewhat slower rate than growth of exports of computer and related equipment. This suggests the slow decoupling of component imports and equipment exports and may be a sign that production in China is moving up the value ladder (for example, developing its own semiconductor production). But the cause might also be different price movements in, for example, semiconductors vs. laptops.

Figure 3: China's trade balance in ICT goods, 2005

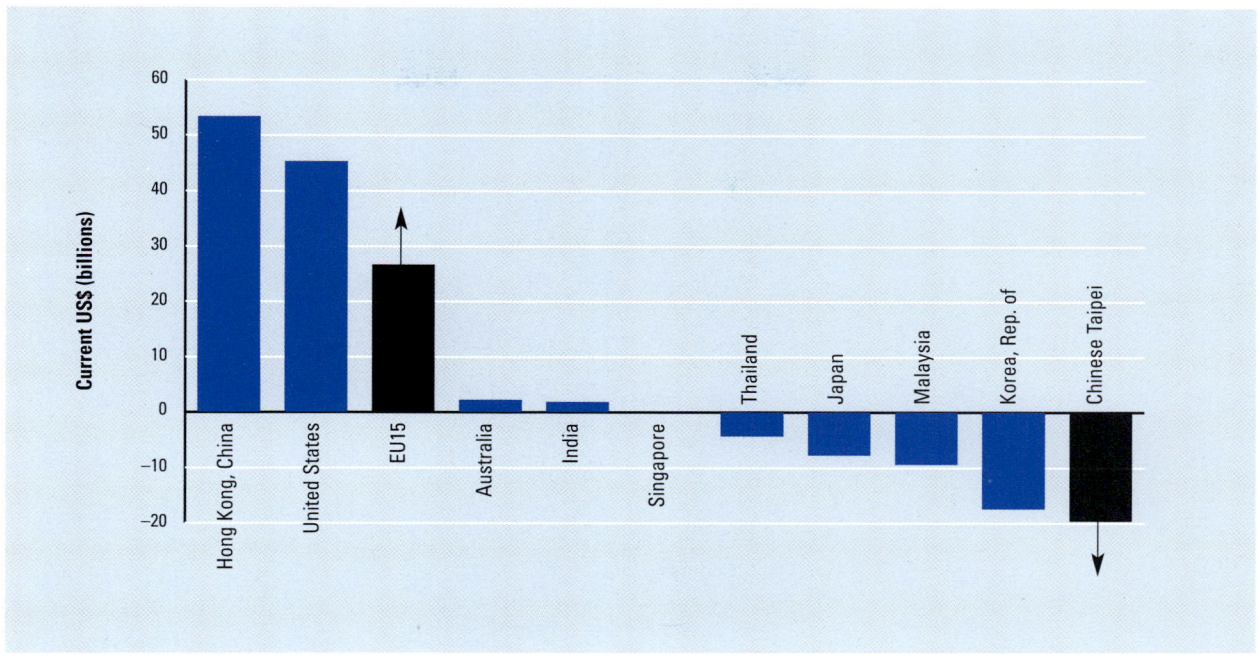

Source: OECD ITCS database. See endnote 6.
Note: In this graph the reporting country is China. Values for the EU15 and Chinese Taipei, designated with black bars, are for 2004.

Trade surpluses with the United States and the EU but deficits with Asian economies and increased Asian intra-industry trade

As shown in Figure 3, in 2005 China had large surpluses with Hong Kong, China[7] (US$53 billion, up significantly from US$37 billion) and the United States (US$45 billion, up significantly from US$34 billion); it is expected to have a large trade surplus with the EU15 again in 2005 (the 2004 trade deficit was US$27 billion and the arrows in the figures indicate the potential but unverified increase for 2005).[8] The large surplus with Hong Kong, China, underlines the significant share of Chinese ICT exports that are transhipped before being delivered to their final destination, making Hong Kong, China, one of the largest ICT exporters. Chinese exports to Hong Kong, China, were up from US$11 billion in 2000 to US$52 billion in 2005, that is, just less than a quarter of all Chinese ICT goods exports.

The United States–China trade relationship is characterized by a significant US trade deficit in computer and related equipment and a very small US surplus in electronic components. The United States is very active in supplying China with high-value components; integrated circuits and semiconductors account for the second largest items among US exports of manufactured products to China.

In contrast, China's Asian trade deficits have increased with Chinese Taipei (–US$20 billion for 2004, with data

not yet available in 2005), Korea (–US$17 billion, an increase of the US$11 billion in 2004), Malaysia (–US$9 billion), and Japan (–US$8 billion). China also has a significant trade deficit in electronic components with Japan, Korea, and Malaysia.. Chinese ICT-related intermediate goods imports, especially electronic components, increasingly come from Asia, particularly from Chinese Taipei but also from Malaysia, for example.

In 2004 the main destinations for Chinese ICT exports were the United States (24 percent of total ICT exports), Hong Kong, China (23 percent), the EU15 (20 percent), and Japan (10 percent), with Hong Kong, China, losing its place as the main export destination (Figure 4). Again the regional shift in ICT trade is apparent in the increase in exports to Asia (Chinese Taipei, Korea, Singapore, Malaysia, Thailand, and so on). The major sources of China's ICT imports are Japan (18 percent), Chinese Taipei (16 percent), Korea (13 percent) and Malaysia (8 percent). The most notable shift is the falling share of imports from the EU15 and the United States. For 2005 roughly the same distribution of export destination shares is expected, potentially with a larger share of Chinese exports going to the EU15 in 2005.

ICT-related manufacturing in Asia has grown over the past two decades, led first by Japan in the 1980s, then by Korea and Chinese Taipei in the 1990s, and now by China as other Asian countries find it increasingly difficult to

Figure 4: China's ICT goods export and import destinations (percent of total imports and exports)

4a: China imports, 2004

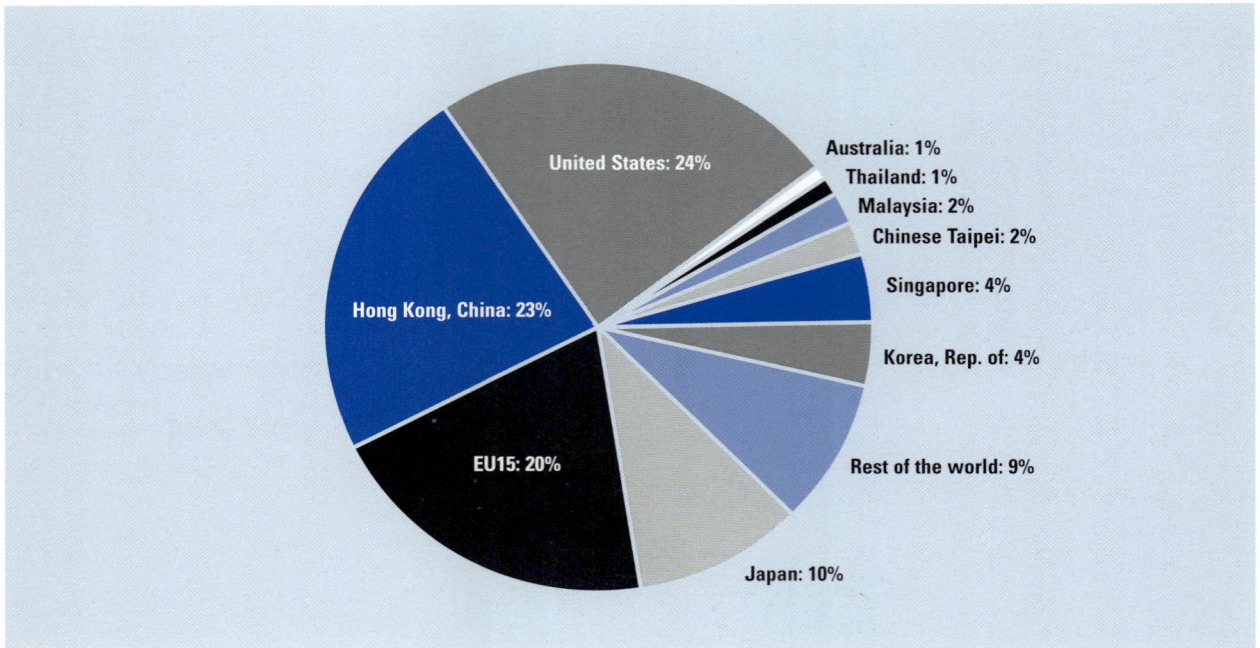

4b: China exports, 2004

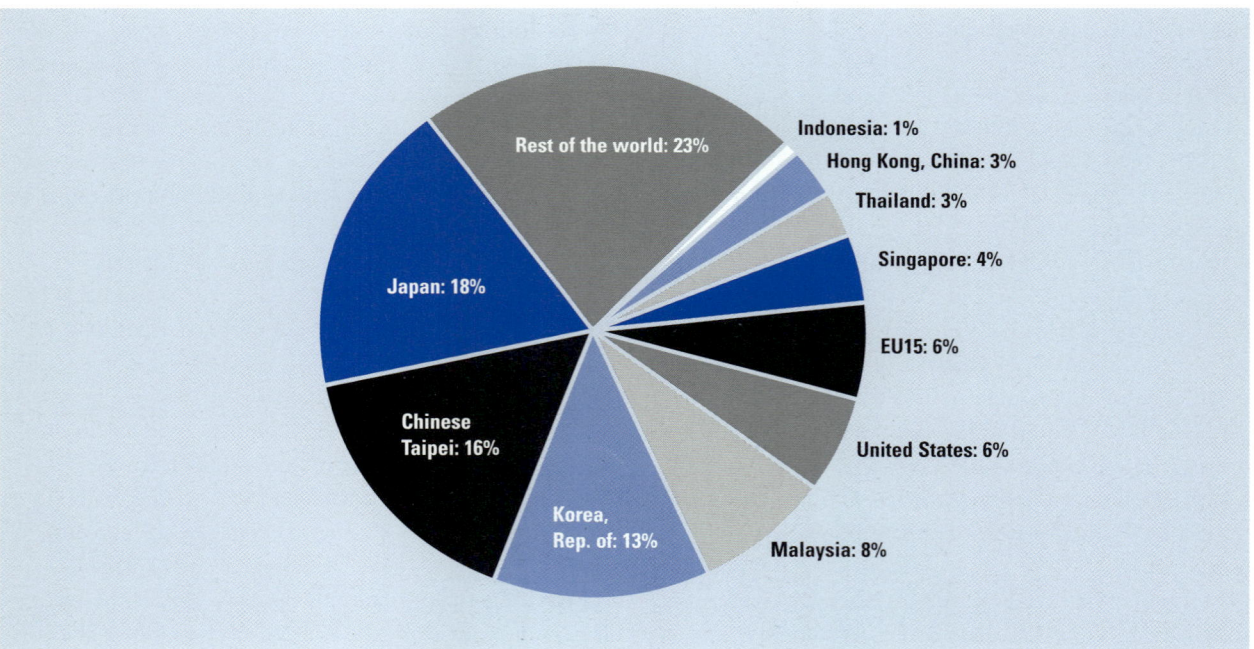

Source: OECD ITCS database.

compete with China, in particular for low-cost assembly. Chinese Taipei, for instance, was until recently the uncontested global center for electronic products and PC/laptop design and assembly. Hon Hai Precision (Foxconn), Flextronics, and Cellon from Chinese Taipei are among the world's largest contract or original design manufacturers, and produce products for the leading global ICT firms.

Recently, however, ICT firms of Chinese Taipei and Hong Kong, China, have migrated manufacturing to mainland China to reduce costs, thereby increasing China's ICT exports and making it the world's main production center for IT equipment.[9] Although Chinese Taipei remains an important production center for higher-value-added ICT goods (for example, semiconductors, flat panel displays), more technically complex activities such as design and testing tasks have also been partly shifted to China in recent years.[10]

China's role as Japan's low-cost base for export production has also substantially grown, with more complex stages of production moving to China,[11] and Asia itself becoming an integrated ICT production platform.

The growth of inward and outward ICT-related FDI

China's inward investment flow for 2005 is officially recorded as US$72 billion in foreign direct investment (FDI)[12]—compared with an estimated US$165 billion in the United Kingdom, US$6.6 billion in India, and US$3 billion in Japan. Much of the FDI is driven by ICT-related investments, which have been crucial to the growth in Chinese exports of ICT; these investments (in particular in the area of telecommunications and Internet services) have been greatly facilitated by China's accession to the WTO in 2004. In 2005, there were almost 3,000 instances of new FDI inflows with a contractual value of about US$21 billion in the telecommunications equipment, computers, and other electronic equipment sector.[13] This openness to foreign ICT-related investment is in contrast to the situation in Japan or Korea when these OECD countries were at a comparable level of ICT industry development in the 1980s and 1990s, respectively. Furthermore, China is considered not only a production site but also an opportunity for market and revenue growth that can compensate for revenue declines in more traditional markets such as Japan and Germany.

ICT-related firms from Chinese Taipei and the United States are among the top 100 firms investing in China in 2003, and Hon Hai Precision (Foxconn) is also one of the OECD's top 50 ICT companies.[14]

Available data point to an increase in the production of communications equipment, computers, and other electronic equipment by firms with direct investment in China. In 2004 there were 3,384 ICT goods firms from abroad. These accounted for 21 percent of total assets in this sector, 30 percent of total revenues, 20 percent of

profits, and 16 percent of employees. All figures, including the number of employees of ICT firms with direct investment (2.3 million in 2004), increased from 2003 to 2004; this increase occurs also when the figures are shown as a share of total FDI in all sectors. Still, mainly because of issues relating to intellectual property rights, some global ICT firms hesitate to move high-end ICT goods production to China.

Although detailed figures are hard to obtain, most Chinese sources show that FDI in services is increasing, with FDI in the telecommunications and ICT services sector (including computer services and software) growing fast from a low base.[15] However, both in general and in the ICT industry, manufacturing plays a relatively large role while that of the services sector is relatively small. Although investment in the ICT services sector is growing fast, the vast majority of investment and employment of US majority-owned foreign affiliates (MOFAs) is in ICT goods manufacturing.

Research and development–related FDI in China

Hard numbers on research and development (R&D)-related FDI in China by ICT firms are not readily available.[16] However, several sources confirm the presence in China of an increasing number of R&D units, mainly from Europe, the United States, and Chinese Taipei, and generally in the computer, communications, electronics, chemical, and automobile industries.[17] Out of 466 foreign R&D centers in China listed by the Global R&D Management Centre at Tsinghua University, 52 percent were established by companies in the ICT industry. Because of the large presence of multinationals, China has a high level of foreign ownership of domestic inventions (share of foreign-owned patents) compared with large OECD countries, a trend that is particularly strong for ICT-related patents.[18]

At present, market-driven or adaptive R&D to support local production and sales appears to predominate for foreign as well as domestic companies,[19] with R&D intensity relatively lower for foreign-owned firms.[20] However, there is a shift toward research initiatives that tap the local pool of skills and knowledge among university-educated scientists and engineers.

Outward Chinese FDI: The "go-out" strategy

The Chinese government has recently begun to encourage Chinese companies to invest overseas to gain technology, brands, and distribution channels. The "go-out" strategy encourages this through relaxed controls on overseas investment by releasing sector restrictions, abolishing the foreign exchange self-sufficiency requirements, and streamlining approval procedures.[21]

China's outward investment flow for 2005 is officially recorded as US$7 billion, including reinvested earnings, with new equity investments at about US$4.1 billion.[22] This is about 10 percent of inward investment into China and is still small as a percentage of global outward direct investment. Yet there is a shift to acquisitions of manufacturers with expertise in high-technology and market share abroad. In 2005 China's overseas direct investment was US$1.2 billion in the manufacturing sector (including ICT-related firms), accounting for 29 percent of total equity investment, and US$1.1 billion in the ICT services sector (information transmission, computer services, and software), accounting for 26.3 percent of total equity investment. The most visible signs of this trend in the past two years have been the merger of the Chinese electronics conglomerate TCL's television business with the television business of France-based Thomson in 2004, Lenovo's acquisition of IBM's PC business in 2005, and Alibaba's stake in Yahoo! China in return for Yahoo!'s purchase of approximately 46 percent of outstanding Alibaba stock in 2005.

Chinese ICT firms are now also operating research units abroad, with some companies—such as ZTE, Huawei (with, for instance, one center in Bangalore, India, with more than 500 employees)—with firmly established global R&D networks.

The ICT supply side in China

According to Chinese figures, value added of what Chinese data defines as the "information industry" (including electronics and IT goods) reached some US$118 billion in 2004 (an increase of around 30 percent from 2003) and constituted 7.5 percent of GDP. The electronics and information industry accounted for US$71 billion and the communications industry for US$47 billion.[23]

Value added of the Chinese ICT industry and the contribution of IT to economic growth

Although value added for communications services in the Chinese economy cannot be broken out, the value for post and telecommunications increased very rapidly from US$13.6 billion in 1997 to US$40.1 billion in 2003, and represented around 8 percent of value added in the services industry, up from 5 percent in 1997.[24] The services industry accounted for about 32 percent of total value added in 2004, compared with about 70 percent in most OECD countries.

Rapid increase in the production of ICT goods

The production of computers, integrated circuits, and mobile telephones—mostly by foreign firms and for export—has increased dramatically since 1995 (see Table 2), exceeding the production levels of most OECD countries. China's mobile telephone output is expected to

Table 2: Chinese production of integrated circuits, microcomputers, and mobile telephones

Year	Integrated circuits (billion units)	Microcomputers (million units)	Mobile telephones (million units)
2000	5.9	6.7	52.5
2001	6.4	8.8	80.3
2002	9.6	14.6	121.5
2003	14.8	32.1	182.3
2004	21.1	45.1	233.5
2005	n/a	80.8	303.7

Source: NBS *China Statistical Yearbook* 1996–2005; data releases from the Ministry of Information Industry (figures rounded).

attain 340 million units in 2006, of which 250 million to be exported mainly by foreign-invested companies.

As opposed to PC and mobile telephone production and as reflected by the import data, the gap between China's domestic demand for and production of integrated circuits continues to increase despite the influx of foreign producers. At present, China's domestic companies remain weak in chip designs based on their own intellectual property rights. Yet China has become a major producer of telecommunications equipment (including mobile and optical telecommunications equipment), with increasing exports in this product category.

Rise of Chinese ICT firms: Becoming global players?

Chinese ICT firms are now mainly privately held, as opposed to state-controlled. They are often listed on the Hong Kong Stock Exchange and are internationally active. Still, by 2006 most Chinese ICT firms remained small in terms of revenue and employment when compared with the most important global ICT firms.[25] Large technological and management gaps between Chinese and foreign firms, weak innovative capabilities, too much reliance on foreign technology, and trade barriers are responsible for this situation. A major challenge is moving from essentially "hosting" low-cost third-party manufacturers from foreign countries to developing Chinese providers of higher-value-added products and recognized global brands. Whereas firms in OECD economies have made a move toward high-value-added ICT services, the ICT service and software sector in China is still largely underdeveloped. Comparisons with India as a services outsourcing location are still far-fetched. Access to capital is also a problem for Chinese ICT firms.

Still, the Chinese ICT industry landscape is changing rapidly. Examples of Chinese ICT leaders include semiconductor companies (for example, Semiconductor Manufacturing International), telecommunications equipment manufacturers (for example, Huawei, ZTE Corporation), and Internet portals (for example, Baidu, Alibaba/eBay partnership).

Table 3: Top 10 Chinese electronic product providers, 2005

Rank	ICT firm	Principal activity
1	Haier Group	Consumer electronics, electronics, telecommunications equipment (handsets), and IT equipment (computers)
2	BOE Technology Group Co., Ltd	Electronics and audiovisual products (monitors, televisions)
3	TCL	Electronics, telecommunications equipment (handsets), and audiovisual products (televisions)
4	Lenovo (formerly Legend Holdings)	IT equipment (PCs and notebooks)
5	Shanghai Video & Audio Electronics	Audiovisual products, IT/telecommunications equipment
6	Huawei Technologies Co., Ltd	Telecommunications equipment
7	Midea Holding Co., Ltd	Consumer electronics
8	Panda Electronics Group Company	Consumer electronics and telecommunications equipment
9	Hisense Group	Consumer electronics, audiovisual products (televisions), IT/telecommunications equipment
10	ZTE Corporation	Telecommunications equipment

Source: MII, 2006b.

Chinese national champions in IT and electronics

Table 3 presents the top 10 Chinese providers of electronic products ranked by 2005 revenues and shows that TCL, Lenovo (formerly Legend Holdings), Huawei, and ZTE are China's leaders in ICT goods.

While annual revenues from ZTE, TCL, and Huawei generally increased from 2000 to 2005, Lenovo's revenues decreased between 2003 and 2004 and picked up again in 2005. As a point of comparison, Lenovo's revenues were US$2.9 billion in 2004, while Hewlett-Packard had revenues of US$80 billion and Dell of US$41 billion. After the takeover of IBM's loss-making PC unit, Lenovo more than quadrupled its annual revenues (US$13.3 billion in 2005); it now has about 21,000 employees and sales representatives worldwide, and a first-quarter 2006 world market share for PC shipments of about 3.5 percent (compared with 32.3 percent for Dell, 19.9 percent for Hewlett-Packard). However, following these acquisitions both Lenovo and TCL experienced revenue losses and restructuring. In the third quarter of 2006 Lenovo has lost market share in the United States and its net profit fell 16 percent from a year earlier, showing how difficult it is to perform in a very competitive globalized ICT market and a reminder of the fact that the ICT goods sector (as opposed to ICT services) operates with thin margins.

The 2005 revenues for Huawei Technologies (US$5.9 billion) and ZTE (US$2.7 billion) contrast with 2005 revenues for telecommunications equipment manufacturers such as Cisco (US$24.8 billion) or Alcatel (US$15.1 billion). Huawei's and ZTE's overseas sales are significant and growing rapidly, with a large share going now to OECD economies and competing seriously with Cisco, Alcatel, and others.

Chinese software industry revenues and exports

The Chinese software industry is small compared with its ICT hardware industry. It is also less developed than software industries in other countries (such as India). Problems associated with the Chinese software industry are a weak industrial base, a lack of international competitiveness, and thus few exports (mainly to Japan and Korea). Weak protection of intellectual property rights can also be a bottleneck to the development of the software industry in China.

According to official figures, growth in the software industry was higher, by 16.7 percent, than growth in the ICT industry as a whole. Yet foreign software producers—notably Microsoft, IBM, Oracle, and Sybase—are said to have accounted for 65 percent of China's software market in 2004, and this trend has been sustained.[26]

The top five domestic software firms with sizeable and mostly growing revenues from 2001 to 2004 are Shengyang Neusoft (revenues of US$204 million in 2004), UFSoft (revenues of US$119 million in 2004), Kingdee International Software Group (revenues of US$56 million in 2004), China National Computer Software and Technology Service Corporation, and Langchao Universal Software. ZTE and Huawei are also very active in software production. The revenues of the top five are almost negligible when compared with those of leading international software producers and small when compared with revenues of Indian software producers Tata Consulting Systems and Infosys (both saw about US$2 billion in revenues in 2004). Although Chinese software outsourcing services are still relatively small and mostly focused on Japan, this market is expected to grow rapidly.

Growing importance of Chinese Internet companies

Owing to lower technological entry barriers for Internet than for ICT goods manufacturing or ICT services, a significant number of Chinese Internet companies now compete very effectively against foreign Internet firms in China. Online auctions and person-to-person commerce (through sites such as eBay, Taobao, and Alibaba) are very popular.

China's leading Internet companies are generally increasing their revenues, some very rapidly (see Figure 5). But these revenues are still small compared to those of OECD-area Internet firms. Whereas Baidu, the largest revenue generator in 2005, had about US$320 million in revenues, the figure for Google is US$6.1 billion and for Yahoo! US$5.3 billion (that is, roughly 20 times bigger). In addition to the lesser Internet penetration and development, the small revenues are the result of the mainly national focus of Chinese Internet companies. In terms of market share in China, however, they compete vigorously with OECD-area Internet firms; for example, Baidu has a larger market share than Google in cities such as Beijing.

Chinese ICT demand and use

For global ICT firms, sales in China are a main revenue driver, and most plan to accelerate their operational expansion in China. PC and Internet penetration are growing fast and Chinese Internet users increasingly engage in creating web pages, instant communications (messaging), online computer games, and e-commerce.

China is becoming one of the most important ICT markets

Figure 6 shows the rapid rise of Chinese spending on computer hardware, software, and services (excluding communications) and the structural differences in spending on these components in China and OECD countries as exemplified by the United States. China is already the world's largest mobile telephone market and the second largest PC market. Demand for these products and for video, broadcasting, and other communications products is forecast to increase in the run-up to the 2008 Olympic Games in China.

Including communications services, China was the sixth largest ICT market in 2005 at US$118 billion (after the United States, Japan, Germany, the United Kingdom, and France), with an annual growth of 22 percent a year since 2000. It is one of the 10 most rapidly growing economies (which are rapidly catching up with major OECD markets); total spending in 2005 was still only about one-tenth of spending in the United States, but about two and a half times the ICT spending of India (US$46 billion).

In general, ICT spending as a percentage of GDP is lower in China (about 4.5 percent of GDP in 2005) than in leading OECD economies (about 9 percent of GDP in 2005), but it is catching up rapidly as Chinese firms increase their IT capital stock—especially in sectors outside manufacturing—and as household consumption increases. In 2005, spending on computer hardware was relatively high in China (it was the third largest market at US$50 billion, compared with US$145 billion for the United States but only US$10 billion for India) and represented about 70 percent of total IT spending (excluding communications). Conversely, at slightly more than 25 percent of total IT spending, spending on software (it was the seventh largest market at US$7.9 billion, compared with US$126 billion for the United States but US$1.9 billion for India) and services (the tenth largest market at US$10 billion, compared with US$287 billion for the United States but US$5.2 billion for India) was much lower than the OECD average but generally higher than India's.

Rising communications expenditures and growing Internet and broadband penetration

The share of communications expenditures in per capita annual consumption has increased significantly faster in China than other expenditure categories over the last 15 years, from almost no expenditure in 1990 to 6.3 percent in 2004.[27]

At the end of 2005, China had 37.5 million broadband subscribers (see Figure 7) (and 53 million broadband users), fewer than the 49 million broadband subscribers in the United States but more than the 22.5 million in Japan. China is expected soon to overtake the United States in number of broadband subscribers. However, only roughly 3 percent of the total Chinese population were broadband subscribers at the end of 2005, compared with 14 percent of the population in the OECD area.

In mid 2006, China had 123 million Internet users (up from 94 million in 2004). The number of broadband users stood at 77 million and exceeded narrowband/dial-up users (47.5 million). By comparison, India only had roughly 35 million Internet users in 2005, even though its population of 1.1 billion is close to China's. Nonetheless, only 8 percent of the Chinese population are Internet users and only about 4 percent are broadband users, whereas in 2004, over half of all OECD countries had Internet participation rates of over 50 percent.[28] Like PC penetration, Internet access varies greatly by region: main urban areas have much higher penetration than rural areas—for example, only 3.3 percent and 2.8 percent of the population have access to the Internet in Tibet or Guizhou. Closing this digital gap has become a government priority.

In China the Internet is most popular with well-educated young men, mostly students. The age profile of Chinese Internet users has not changed much over time. The 18–24 year age group represents 38.9 percent of

Figure 5: Chinese Internet firm growth

5a. Revenues of Chinese Internet firms, 2003–05

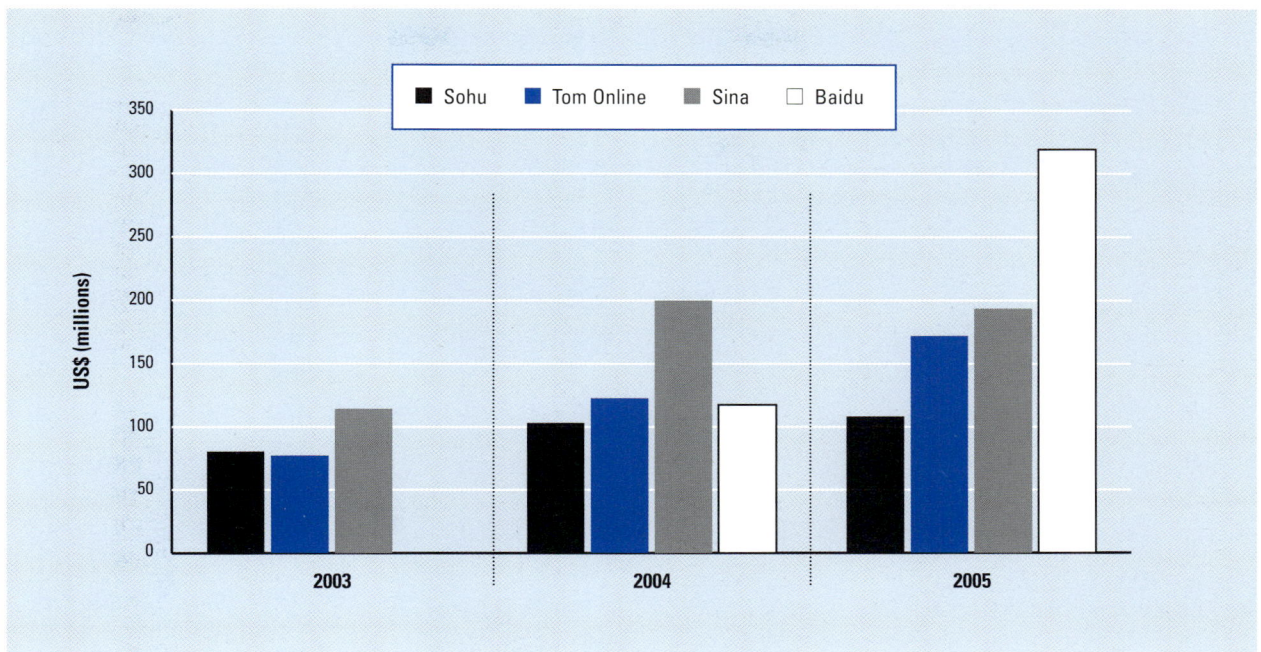

Source: Annual reports.

5b. Market share of search engines in Beijing, 2005

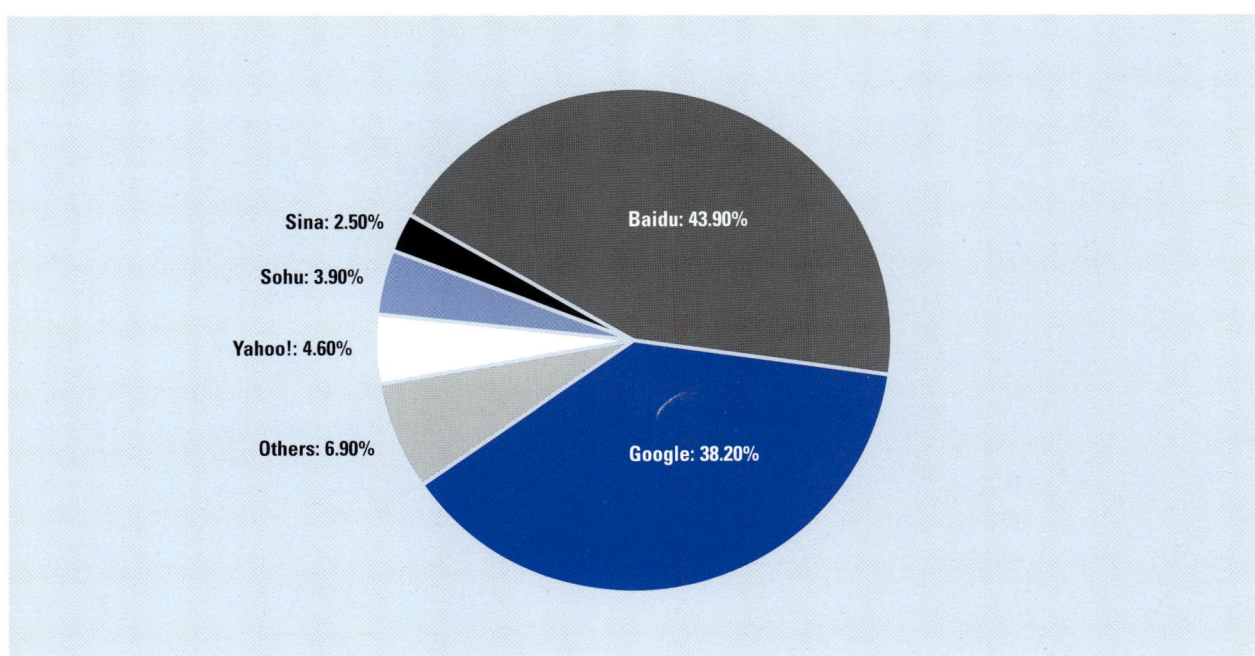

Source: China Internet Network Information Center (CNNIC)..

Figure 6: Chinese ICT market significance

6a: Chinese ICT spending, 2001–05

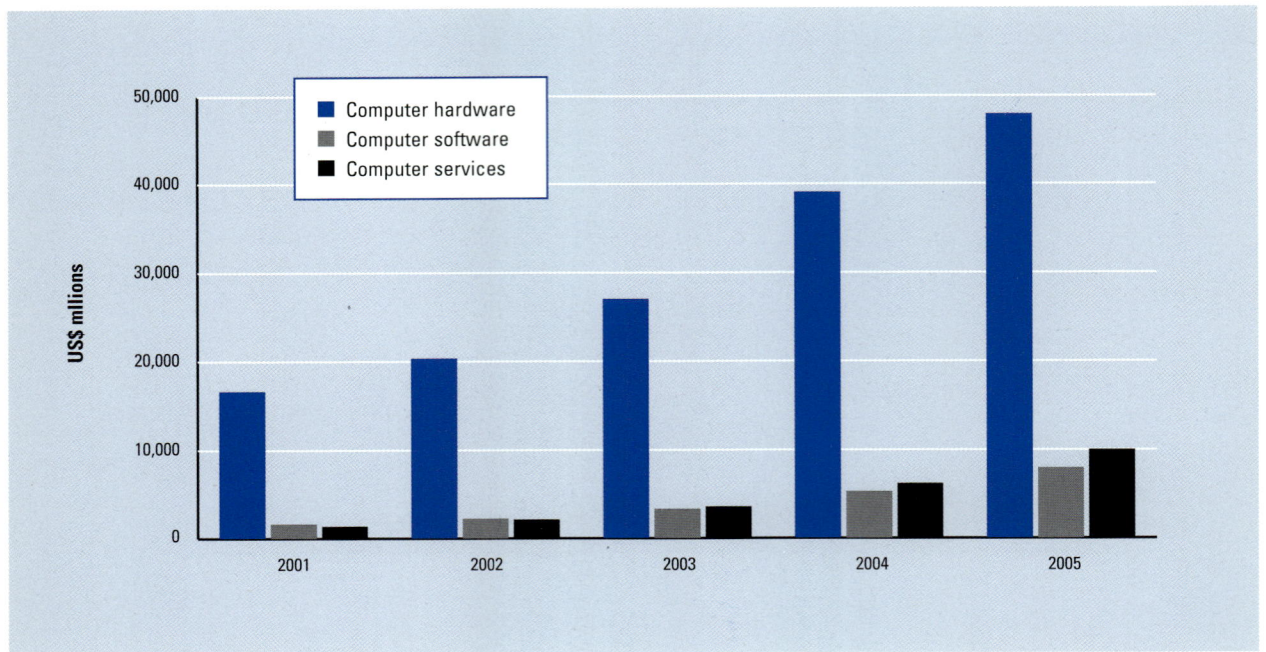

6b: Distribution of Chinese and US ICT spending, 2001–05

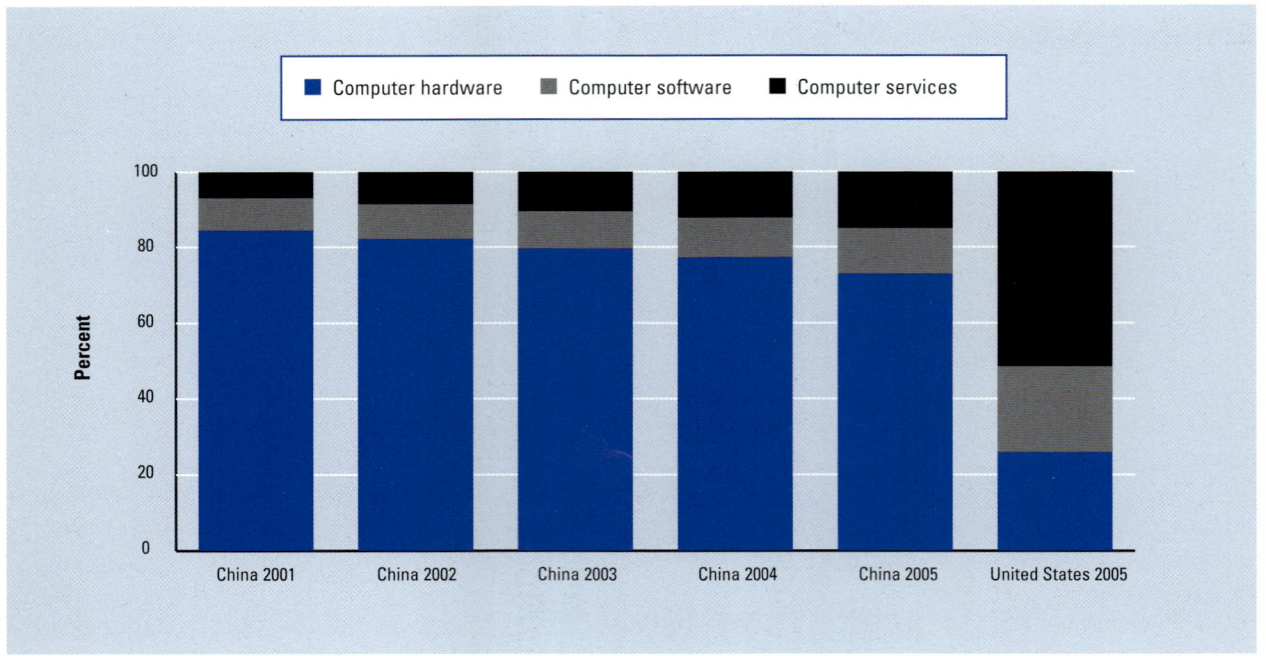

Source: OECD based on World Information Technology and Services Alliance (WITSA) data.

Figure 7: Internet and broadband subscribers in China and the OECD area, 1999–2005

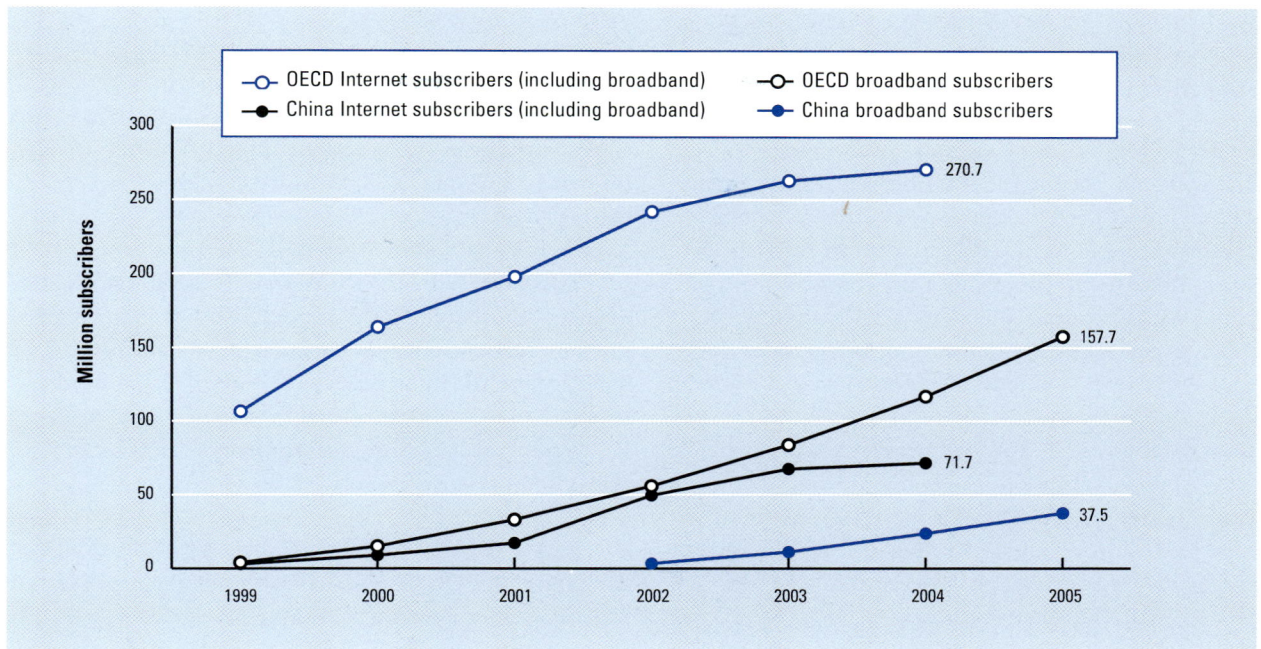

Source: OECD, 2006a; Hu, 2006.

Internet users in mid 2006 (a very rapid increase, up from 29 percent at the end of 2005), compared with roughly 10 percent of those over 40 years and only 0.8 percent of those over 60 years of age. Those under 35 years account for over 80 percent of Internet users. Information and entertainment were the two main goals of Internet use in most recent years.

One paradox is that, while low IT costs brought about by China's competitive supply has helped OECD-based firms upgrade, reorganize, and boost productivity, the actual uptake of IT within Chinese firms is lagging behind. More than half of Chinese firms surveyed in all sectors except mining are using the Internet and, depending on the sector, up to 50 percent of surveyed firms have a local area network (LAN).[29] Far fewer firms report having a website (the manufacturing sector leads, with 28 percent of surveyed firms) or conducting e-commerce (the manufacturing sector leads again, with 18 percent of surveyed firms receiving orders online and 15 percent placing orders online). Notions such as supply-chain management, resource planning, or knowledge management software that are standard currency in dynamic OECD firms are still rather undeveloped in China.

Current and future Chinese ICT supply-side policies

In sum, the Chinese ICT industry now faces the challenge of a successful transition from being low-cost manufacturers to becoming global providers of higher-value-added products and services. However, challenging the technological leadership of already-globalized ICT firms may prove difficult. Chinese firms are developing strategies to maintain a competitive advantage in the design and production of higher-value-added components and ICT end products, to avoid losing these segments of the globalized ICT value chain.

China has set itself few challenges. The future Chinese ICT supply-side policies will have the following new priorities:[30]

- accelerating structural change in the domestic information industry and moving production toward higher-value-added ICT manufacturing, software (for example, embedded software, middleware, information security products), ICT services (value-added network services, third-generation mobile communications), and information services;

- creating national ICT champions to compete on global markets and abroad;

- improving domestic innovative capabilities—for
example, through increased investment in R&D and
more domestically owned patents—in order to
decrease reliance on foreign intellectual property
rights and the associated large royalty fees paid;

- continuing government support, government procurement and subsidies in semiconductors/integrated
circuits, photoelectric displays, advanced computing
and grid computing, the development of Internet
Protocol version 6 (IPv6) and next-generation
networks, 3G wireless and a digital TV network,
domestic software development (including open
source software), and sensor networking;

- fostering Chinese ICT-related standards in 3G
wireless communications (time division-synchronous
code division multiple access), encryption for wireless
local area network equipment (wireless authentication
and privacy infrastructure), video, digital video disc
(DVD), and RFID; and

- building on the opportunities arising from increased
convergence and digital content (for example, digital
TV, animation, advertisement, online games).

Beyond these new Chinese ICT supply-side polices,
the Chinese government increasingly recognizes the
importance of the diffusion and use of ICT to spur the
industrialization and development of other non-ICT
-producing sectors (for example, agriculture but also a
modern Chinese services sector),[31] This move is character-
istic of the shift in Chinese policy attention toward
increasing ICT diffusion and use.

Conclusion

China has established itself as a leading producer and user
of ICT. On the supply side, China has become one of the
most important locations for ICT production and has
been the leading exporter of ICT goods since 2004.
Compared with economies such as Japan, Korea, and
Chinese Taipei, which have also built a significant ICT
industry in the last 20 years, China has evolved differently
by hosting foreign ICT firms or third-party contract manufacturers to conduct their final ICT product assembly in
China. In recent years, however, the rapidly developing
Chinese ICT market has triggered further investment by
foreign ICT firms to satisfy the increasing domestic
demand.

On the demand and use side, China constitutes a very
significant market and potential source of future revenues,
often compensating for declines in OECD markets.
Steadily rising disposable incomes in China have led to

increasing PC and Internet penetration and rapidly developing e-commerce activities, albeit from low levels. The
sheer scale of the Chinese ICT market and its potential to
serve as a self-supporting base for industrial development
are a key difference from other countries that have
climbed the ICT value ladder. In spite of their limited size
and technological know-how, the production and export
capacities of Chinese ICT firms are developing rapidly,
especially as providers of telecommunications equipment,
semiconductors, and Internet-related services.

The Chinese ICT industry now faces the challenge of
making a successful transition from being an industry of
low-cost manufacturers to becoming global providers of
higher-value-added products and services. However, challenging the technological leadership of already-globalized
ICT firms and competing against them in global—and
even Chinese—markets may not prove easy for Chinese
firms.

Finally, producing ICT is not helping China to reap
its full benefits. On the one hand, OECD countries are
benefiting from low-cost ICT assembly in China, which is
adding to lower global ICT prices and thus to increased
ICT use and associated productivity gains across industries.
The widespread integration of ICT offers OECD countries
a significant competitive advantage that will persist if governments pay consistent attention to removing roadblocks
to ICT production and use (especially in areas such as
ICT services, convergence, and digital content). On the
other hand, ICT uptake in Chinese firms and their efficient integration in value chains and complementary
innovations (such as organizational restructuring and
investment in skills) are lagging. To benefit fully from ICT,
its integration in the Chinese economy and society should
be high on the Chinese policy agenda.

Notes

1 See OECD (2005a).

2 See MOFCOM (2005a). This figure includes joint ventures or wholly owned foreign companies.

3 This percentage is based on industrial micro data from the Chinese National Bureau of Statistics.

4 In compliance with the policies of the authors' institution, this chapter exceptionally employs OECD's terminology for Taiwan, China.

5 See MII (2006a)

6 The HS codes are codes of the Harmonized Commodity Description and Coding System (World Customs Organization).

7 In compliance with the policies of the authors' institution, this chapter exceptionally employs OECD's terminology for Hong Kong SAR.

8 There can be significant discrepancies in bilateral trade data. In this paper when reporting exports and imports for China, we use data as reported by China. This excludes Hong Kong, China.

9 See Reed Electronics Research (2006).

10 See CAPS (2006).

11 See JETRO (2005).

12 See OECD (2006b) and MOFCOM (2006a).

13 See OECD (2006a).

14 See also MOFCOM (2004).

15 See NBS *China Statistical Yearbook* (2005), MOFCOM (2006a), and OECD (2006b).

16 The OECD's Science, Technology and Industry Directorate is undertaking a study of the Chinese innovation system, to be published in 2007.

17 See MOFCOM (2006b).

18 See OECD (2005c).

19 See MOFCOM (2005b).

20 See Motohashi (2006).

21 See OECD (2006b).

22 See MOFCOM (2006c).

23 See MII (2005).

24 See NBS *China Statistical Yearbook* (2005).

25 See OECD (2006a).

26 Cf. the US Commercial Service, Computers and Telecommunications, Export Brief under www.buyusa.gov/china/en/computers.html.

27 See NBS *China Statistical Yearbook* (1999, 2005). See also OECD (2005b) for data qualifications regarding Chinese ICT statistics.

28 See OECD (2005c).

29 Information on uptake of ICT by Chinese firms is according to statistics from the Chinese National Bureau of Statistics made available to the OECD.

30 The 11th Five-Year National Plan for 2006–2011 and the State's Medium- and Long-term Development Plan of Science and Technology lay the ground for future Chinese ICT policy.

31 See NBS (2006).

References

CAPS (Center for Strategic Supply Research). 2006. *Technology and Organisational Factors in the Notebook Industry Supply Chain*. W. Foster, Z. Cheng, K.L. Kraemer, and J. Dedrick. Tempe, AZ: CAPS. Available at http://pcic.merage.uci.edu/pubs/2006/foster2006.pdf.

CNNIC (China Internet Network Information Center). 1999–2006. *Statistical Survey Report on the Internet Development in China* (latest report from July 2006). Beijing: CNNIC.

Hu, Q. 2006. "The Internet in China." Presentation of the China Association for S&T Internet Society of China at the OECD Workshop on the "The Future of the Internet," March 8, 2006, Paris. Available at www.oecd.org/dataoecd/43/58/36274280.pdf.

JETRO (Japan External Trade Organization). 2005. "Japanese Exports of IT Products to US Decline against Backdrop of Chinese Production." Special report, M. Yoshida, Japanese Economy Division. Tokyo: JETRO. Available at www.jetro.go.jp/en/market/trend/special/pdf/jem0410-1e.pdf.

MII (Ministry of Information Industry). 2005. "Electronic Information Sector." Available at www.mii.gov.cn/mii/hyzw/tongji/2005051302.htm.

———. 2006a. *Report: Revenues of High-Tech Manufacturing Industry in China*. 18 April. Beijing: MII.

———. 2006b. *Report: Top Ten Chinese Electronic Product Providers in 2005*. Beijing: MII. MOFCOM (The Ministry of Commerce of the People's Republic of China). 2004. "100 FIEs with Largest Exports for 2003." Data release, September 27. Beijing: MOFCOM.

———. 2005a. "Import & Export Statistics by FIEs." Data release, October 21. Beijing: MOFCOM.

———. 2005b. "2005–2007 Investigation and Research Report on Industry Investment Trends by Transnational Corporation in China." Data release, March 29. Beijing: MOFCOM.

———. 2006a. "Foreign Direct Investment Tops US$60b." Data release, January 16. Beijing: MOFCOM.

———. 2006b. "Multinationals Hasten R&D Establishment." Data release, February 14. Beijing: MOFCOM.

———. 2006c. "Statistics of China's Overseas Direct Investment in 2005." Data release, February 16. Beijing: MOFCOM.

Motohashi, K. 2006. "R&D of Multinationals in China: Structure, Motivations and Regional Difference." Research Institute of Economy, Trade and Industry (RIETI) Discussion Paper No. 2006-E-005. Tokyo: RIETI.

NBS (National Bureau of Statistics of China). 1996–2005. *China Statistical Yearbook*. Beijing: NBS.

———. 2006. "The Evaluation and Research of Informatization Level in China." Informal presentation to the OECD, May 15, 2006, Paris.

OECD (Organisation for Economic Co-operation and Development). 2005a. *Economic Survey of China*. Paris: OECD.

———. 2005b. "Status and Overview of Official ICT Indicators for China." Science, Technology and Industry Working Paper No. 2005/4. Paris: OECD.

———. 2005c. *Science, Technology, Industry Scoreboard*. Paris: Directorate for Science, Technology and Industry, OECD. Paris: OECD.

———. 2006a. *Information Technology Outlook 2006*. Paris: Directorate for Science, Technology and Industry, OECD. Available at www.oecd.org/sti/ito.

———. 2006b. *Investment Policy Review of China 2006: Open Policies toward Cross-Border Mergers and Acquisitions*. Paris: OECD.

Reed Electronics Research. 2006. *Yearbook of World Electronics Data 2006, vol. 2, America, Japan & Asia Pacific*. Oxon: Reed Electronics Research.

131

Part 3
Country/Economy Profiles

How to Read the Country/Economy Profiles

THIERRY GEIGER, World Economic Forum

The following pages present the profiles of the 122 economies covered by *The Global Information Technology Report 2006–2007*. They provide a quick picture of the level of ICT development of an economy by grouping information under the following sections:

1 Macroeconomic and ICT indicators, which comprise data for population, gross domestic product (GDP) per capita, Internet users per 100 inhabitants, and Internet bandwidth per capita.[1]

2 Overall Networked Readiness Index (NRI) ranking for 2006–2007, which gives immediate insight into overall ICT readiness; one can compare this overall ranking with that of the NRI 2004–2005 and NRI 2005–2006 if the economy was included in the NRI for those years. Also shown is the economy's ranking on the World Economic Forum's Global Competitiveness Index 2006–2007.[2]

3 Three component subindexes, each consisting of a list of variables. Detailed rankings for the economy presented can be found for each of the variables listed. These variables are taken into consideration for the current NRI study.

This information, which identifies key areas of relative over- and underperformance, provides a rapid understanding of an economy's ICT readiness. For example, the rankings of the variables of venture capital availability and the state of cluster development in the environment component subindex enable the reader to identify key parameters contributing to a country's performance.

The inferences that can be derived from the ranking of a given economy can be augmented by a closer inspection of the relative performance of other economies. The numbers next to the variables refer to the numbering of the Data Tables presented at the end of this *Report*.

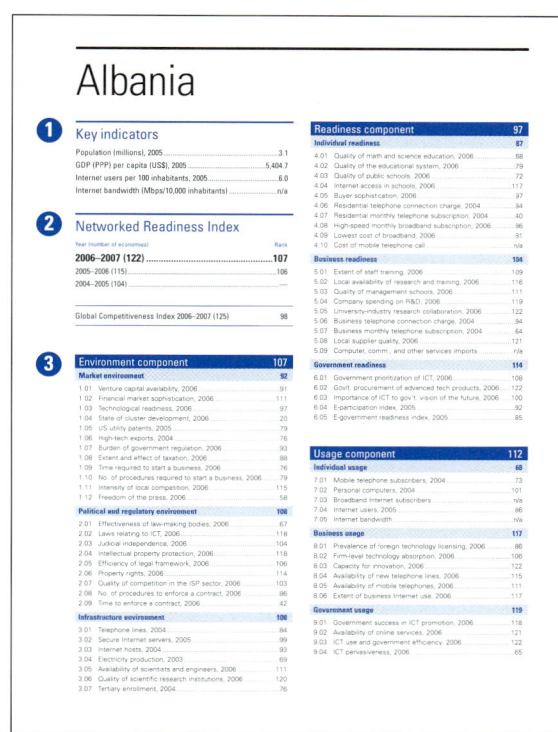

Notes

1 Sources for population include UNFPA's *State of World Population 2005*, the UN Department of Economic and Social Affairs' *Population Division Database* (June 2006), and national sources. GDP figures were obtained from the IMF *World Economic Outlook Online Database* (April and September 2006 editions). Data on Internet users and bandwidth are from the International Telecommunication Union's *World Telecommunication Indicators 2006* and from national sources.

2 See World Economic Forum, *The Global Competitiveness Report 2006–2007*. Hamsphire: Palgrave Macmillan.

List of Countries/Economies

Albania

Key indicators

Population (millions), 2005..3.1
GDP (PPP) per capita (US$), 2005........................5,404.7
Internet users per 100 inhabitants, 2005....................6.0
Internet bandwidth (Mbps/10,000 inhabitants) ....n/a

Networked Readiness Index

Year (number of economies)	Rank
2006–2007 (122) **107**	
2005–2006 (115)..106	
2004–2005 (104) ....—	

Global Competitiveness Index 2006–2007 (125)	98

Environment component | 107

Market environment | 92

1.01	Venture capital availability, 2006...91
1.02	Financial market sophistication, 2006................................111
1.03	Technological readiness, 2006...97
1.04	State of cluster development, 2006..20
1.05	US utility patents, 2005 79
1.06	High-tech exports, 2004 76
1.07	Burden of government regulation, 2006 93
1.08	Extent and effect of taxation, 2006......................................88
1.09	Time required to start a business, 2006...............................76
1.10	No. of procedures required to start a business, 2006.........79
1.11	Intensity of local competition, 2006 115
1.12	Freedom of the press, 2006...58

Political and regulatory environment | 108

2.01	Effectiveness of law-making bodies, 2006..........................67
2.02	Laws relating to ICT, 2006..118
2.03	Judicial independence, 2006 104
2.04	Intellectual property protection, 2006................................118
2.05	Efficiency of legal framework, 2006....................................106
2.06	Property rights, 2006...114
2.07	Quality of competition in the ISP sector, 2006..................103
2.08	No. of procedures to enforce a contract, 2006 86
2.09	Time to enforce a contract, 2006 42

Infrastructure environment | 108

3.01	Telephone lines, 2004..84
3.02	Secure Internet servers, 2005..99
3.03	Internet hosts, 2004 93
3.04	Electricity production, 2003 69
3.05	Availability of scientists and engineers, 2006...................111
3.06	Quality of scientific research institutions, 2006 120
3.07	Tertiary enrollment, 2004..76

Readiness component | 97

Individual readiness | 87

4.01	Quality of math and science education, 2006 68
4.02	Quality of the educational system, 2006.............................79
4.03	Quality of public schools, 2006 72
4.04	Internet access in schools, 2006..117
4.05	Buyer sophistication, 2006 97
4.06	Residential telephone connection charge, 2004 84
4.07	Residential monthly telephone subscription, 2004 40
4.08	High-speed monthly broadband subscription, 2006...........86
4.09	Lowest cost of broadband, 2006...81
4.10	Cost of mobile telephone call ....n/a

Business readiness | 104

5.01	Extent of staff training, 2006..109
5.02	Local availability of research and training, 2006 116
5.03	Quality of management schools, 2006...............................111
5.04	Company spending on R&D, 2006 119
5.05	University-industry research collaboration, 2006...............122
5.06	Business telephone connection charge, 2004 94
5.07	Business monthly telephone subscription, 2004 64
5.08	Local supplier quality, 2006 121
5.09	Computer, comm., and other services imports ....n/a

Government readiness | 114

6.01	Government prioritization of ICT, 2006..............................108
6.02	Gov't. procurement of advanced tech products, 2006......122
6.03	Importance of ICT to gov't. vision of the future, 2006......100
6.04	E-participation index, 2005 92
6.05	E-government readiness index, 2005 85

Usage component | 112

Individual usage | 68

7.01	Mobile telephone subscribers, 2004 73
7.02	Personal computers, 2004...101
7.03	Broadband Internet subscribers...n/a
7.04	Internet users, 2005 86
7.05	Internet bandwidth...n/a

Business usage | 117

8.01	Prevalence of foreign technology licensing, 2006..............86
8.02	Firm-level technology absorption, 2006.............................106
8.03	Capacity for innovation, 2006 122
8.04	Availability of new telephone lines, 2006 115
8.05	Availability of mobile telephones, 2006.............................111
8.06	Extent of business Internet use, 2006 117

Government usage | 119

9.01	Government success in ICT promotion, 2006...................118
9.02	Availability of online services, 2006...................................121
9.03	ICT use and government efficiency, 2006..........................122
9.04	ICT pervasiveness, 2006 65

Algeria

Key indicators

Population (millions), 2005..32.9

GDP (PPP) per capita (US$), 2005......................................7,189.3

Internet users per 100 inhabitants, 2005.................................5.8

Internet bandwidth (Mbps/10,000 inhabitants) ....n/a

Networked Readiness Index

Year (number of economies)	Rank
2006–2007 (122)	**80**
2005–2006 (115)..	87
2004–2005 (104)..	80

Global Competitiveness Index 2006–2007 (125)	76

Environment component — 92

Market environment — 110

1.01 Venture capital availability, 2006...111
1.02 Financial market sophistication, 2006...................................120
1.03 Technological readiness, 2006..102
1.04 State of cluster development, 2006 110
1.05 US utility patents, 2005 79
1.06 High-tech exports, 2004 95
1.07 Burden of government regulation, 2006 61
1.08 Extent and effect of taxation, 2006.......................................32
1.09 Time required to start a business, 2006...............................38
1.10 No. of procedures required to start a business, 2006.......106
1.11 Intensity of local competition, 2006 95
1.12 Freedom of the press, 2006...89

Political and regulatory environment — 81

2.01 Effectiveness of law-making bodies, 2006...........................69
2.02 Laws relating to ICT, 2006...93
2.03 Judicial independence, 2006 62
2.04 Intellectual property protection, 2006...................................72
2.05 Efficiency of legal framework, 2006......................................52
2.06 Property rights, 2006..66
2.07 Quality of competition in the ISP sector, 2006..................107
2.08 No. of procedures to enforce a contract, 2006 107
2.09 Time to enforce a contract, 2006 46

Infrastructure environment — 79

3.01 Telephone lines, 2005...87
3.02 Secure Internet servers, 2005...114
3.03 Internet hosts, 2004 113
3.04 Electricity production, 2003 77
3.05 Availability of scientists and engineers, 2006......................21
3.06 Quality of scientific research institutions, 2006 84
3.07 Tertiary enrollment, 2004...75

Readiness component — 73

Individual readiness — 74

4.01 Quality of math and science education, 2006 73
4.02 Quality of the educational system, 2006...............................91
4.03 Quality of public schools, 2006 69
4.04 Internet access in schools, 2006...92
4.05 Buyer sophistication, 2006 89
4.06 Residential telephone connection charge, 2005 60
4.07 Residential monthly telephone subscription, 2005 35
4.08 High-speed monthly broadband subscription, 2006...........90
4.09 Lowest cost of broadband, 2006...92
4.10 Cost of mobile telephone call, 2005.....................................38

Business readiness — 68

5.01 Extent of staff training, 2006..97
5.02 Local availability of research and training, 2006................100
5.03 Quality of management schools, 2006................................90
5.04 Company spending on R&D, 2006 90
5.05 University-industry research collaboration, 2006...............101
5.06 Business telephone connection charge, 2005 55
5.07 Business monthly telephone subscription, 2005 25
5.08 Local supplier quality, 2006 97
5.09 Computer, comm., and other services imports ....n/a

Government readiness — 72

6.01 Government prioritization of ICT, 2006................................34
6.02 Gov't. procurement of advanced tech products, 2006........35
6.03 Importance of ICT to gov't. vision of the future, 2006........39
6.04 E-participation index, 2005 92
6.05 E-government readiness index, 2005....................................98

Usage component — 78

Individual usage — 79

7.01 Mobile telephone subscribers, 2005 68
7.02 Personal computers, 2004..104
7.03 Broadband Internet subscribers, 2005 63
7.04 Internet users, 2005 87
7.05 Internet bandwidth...n/a

Business usage — 88

8.01 Prevalence of foreign technology licensing, 2006...............87
8.02 Firm-level technology absorption, 2006...............................68
8.03 Capacity for innovation, 2006 117
8.04 Availability of new telephone lines, 2006 82
8.05 Availability of mobile telephones, 2006 39
8.06 Extent of business Internet use, 2006 121

Government usage — 62

9.01 Government success in ICT promotion, 2006......................36
9.02 Availability of online services, 2006....................................100
9.03 ICT use and government efficiency, 2006............................72
9.04 ICT pervasiveness, 2006 27

Angola

Key indicators

Population (millions), 2005..15.9
GDP (PPP) per capita (US$), 2005......................................2,813.5
Internet users per 100 inhabitants, 2005.....................................1.1
Internet bandwidth (Mbps/10,000 inhabitants) ....n/a

Networked Readiness Index

Year (number of economies)	Rank
2006–2007 (122) **120**	
2005–2006 (115) ....—	
2004–2005 (104)...101	

Global Competitiveness Index 2006–2007 (125)	125

Environment component	121
Market environment	**120**

1.01   Venture capital availability, 2006...103
1.02   Financial market sophistication, 2006...............................116
1.03   Technological readiness, 2006...117
1.04   State of cluster development, 2006....................................120
1.05   US utility patents, 2005 79
1.06   High-tech exports..n/a
1.07   Burden of government regulation, 2006 111
1.08   Extent and effect of taxation, 2006.....................................35
1.09   Time required to start a business, 2006...........................113
1.10   No. of procedures required to start a business, 2006.........95
1.11   Intensity of local competition, 2006 121
1.12   Freedom of the press, 2006..108

Political and regulatory environment	**116**

2.01   Effectiveness of law-making bodies, 2006...........................88
2.02   Laws relating to ICT, 2006...117
2.03   Judicial independence, 2006 87
2.04   Intellectual property protection, 2006................................101
2.05   Efficiency of legal framework, 2006...................................101
2.06   Property rights, 2006 106
2.07   Quality of competition in the ISP sector, 2006..................109
2.08   No. of procedures to enforce a contract, 2006 104
2.09   Time to enforce a contract, 2006 110

Infrastructure environment	**122**

3.01   Telephone lines, 2005...114
3.02   Secure Internet servers, 2005 100
3.03   Internet hosts, 2004 112
3.04   Electricity production, 2003 101
3.05   Availability of scientists and engineers, 2006...................122
3.06   Quality of scientific research institutions, 2006 117
3.07   Tertiary enrollment, 2003..119

Readiness component	119
Individual readiness	**120**

4.01   Quality of math and science education, 2006...................117
4.02   Quality of the educational system, 2006...........................118
4.03   Quality of public schools, 2006 116
4.04   Internet access in schools, 2006......................................102
4.05   Buyer sophistication, 2006 114
4.06   Residential telephone connection charge.........................n/a
4.07   Residential monthly telephone subscription......................n/a
4.08   High-speed monthly broadband subscription charge ....n/a
4.09   Lowest cost of broadband, 2006.......................................104
4.10   Cost of mobile telephone call ....n/a

Business readiness	**108**

5.01   Extent of staff training, 2006..106
5.02   Local availability of research and training, 2006 120
5.03   Quality of management schools, 2006..............................122
5.04   Company spending on R&D, 2006 108
5.05   University-industry research collaboration, 2006...............105
5.06   Business telephone connection charge.............................n/a
5.07   Business monthly telephone subscription.........................n/a
5.08   Local supplier quality, 2006 122
5.09   Computer, comm., and other services imports, 2004...........2

Government readiness	**118**

6.01   Government prioritization of ICT, 2006..............................116
6.02   Gov't. procurement of advanced tech products, 2006......112
6.03   Importance of ICT to gov't. vision of the future, 2006......103
6.04   E-participation index, 2005 92
6.05   E-government readiness index, 2005................................113

Usage component	115
Individual usage	**108**

7.01   Mobile telephone subscribers, 2005 110
7.02   Personal computers, 2004...117
7.03   Broadband Internet subscribers.......................................n/a
7.04   Internet users, 2005 109
7.05   Internet bandwidth..n/a

Business usage	**116**

8.01   Prevalence of foreign technology licensing, 2006...............90
8.02   Firm-level technology absorption, 2006............................120
8.03   Capacity for innovation, 2006 119
8.04   Availability of new telephone lines, 2006 113
8.05   Availability of mobile telephones, 2006............................112
8.06   Extent of business Internet use, 2006 116

Government usage	**109**

9.01   Government success in ICT promotion, 2006......................91
9.02   Availability of online services, 2006..................................110
9.03   ICT use and government efficiency, 2006..........................110
9.04   ICT pervasiveness, 2006 110

Argentina

Key indicators

Population (millions), 2005..38.7

GDP (PPP) per capita (US$), 2005......................................14,108.5

Internet users per 100 inhabitants, 2005....................................17.8

Internet bandwidth (Mbps/10,000 inhabitants), 2004...............3.2

Networked Readiness Index

Year (number of economies)	Rank
2006–2007 (122) **63**	
2005–2006 (115)...71	
2004–2005 (104)...76	

Global Competitiveness Index 2006–2007 (125)	69

Environment component	77
Market environment	**91**

1.01	Venture capital availability, 2006..55	
1.02	Financial market sophistication, 2006......................................64	
1.03	Technological readiness, 2006...60	
1.04	State of cluster development, 2006..41	
1.05	US utility patents, 2005...46	
1.06	High-tech exports, 2004..53	
1.07	Burden of government regulation, 2006 102	
1.08	Extent and effect of taxation, 2006.....................................116	
1.09	Time required to start a business, 2006...............................59	
1.10	No. of procedures required to start a business, 2006.......108	
1.11	Intensity of local competition, 2006 100	
1.12	Freedom of the press, 2006..105	

Political and regulatory environment	**101**

2.01	Effectiveness of law-making bodies, 2006.........................116
2.02	Laws relating to ICT, 2006..77
2.03	Judicial independence, 2006 112
2.04	Intellectual property protection, 2006...................................86
2.05	Efficiency of legal framework, 2006...................................107
2.06	Property rights, 2006...119
2.07	Quality of competition in the ISP sector, 2006....................46
2.08	No. of procedures to enforce a contract, 2006 66
2.09	Time to enforce a contract, 2006 75

Infrastructure environment	**42**

3.01	Telephone lines, 2005...56
3.02	Secure Internet servers, 2005..57
3.03	Internet hosts, 2004..34
3.04	Electricity production, 2003...60
3.05	Availability of scientists and engineers, 2006......................65
3.06	Quality of scientific research institutions, 2006 72
3.07	Tertiary enrollment, 2003..19

Readiness component	60
Individual readiness	**65**

4.01	Quality of math and science education, 2006......................88
4.02	Quality of the educational system, 2006...............................98
4.03	Quality of public schools, 2006 85
4.04	Internet access in schools, 2006..67
4.05	Buyer sophistication, 2006 57
4.06	Residential telephone connection charge, 2005 55
4.07	Residential monthly telephone subscription, 2005 47
4.08	High-speed monthly broadband subscription, 2006...........37
4.09	Lowest cost of broadband, 2006..43
4.10	Cost of mobile telephone call, 2005....................................54

Business readiness	**55**

5.01	Extent of staff training, 2006..72
5.02	Local availability of research and training, 2006.................43
5.03	Quality of management schools, 2006..................................29
5.04	Company spending on R&D, 2006..83
5.05	University-industry research collaboration, 2006.................81
5.06	Business telephone connection charge, 2005 50
5.07	Business monthly telephone subscription, 2005 59
5.08	Local supplier quality, 2006 74
5.09	Computer, comm., and other services imports, 2004.........51

Government readiness	**67**

6.01	Government prioritization of ICT, 2006..............................113
6.02	Gov't. procurement of advanced tech products, 2006........68
6.03	Importance of ICT to gov't. vision of the future, 2006......108
6.04	E-participation index, 2005 36
6.05	E-government readiness index, 2005....................................34

Usage component	68
Individual usage	**55**

7.01	Mobile telephone subscribers, 2005 56
7.02	Personal computers, 2004...58
7.03	Broadband Internet subscribers, 2005 44
7.04	Internet users, 2005 57
7.05	Internet bandwidth, 2004 38

Business usage	**69**

8.01	Prevalence of foreign technology licensing, 2006...............61
8.02	Firm-level technology absorption, 2006...............................97
8.03	Capacity for innovation, 2006 79
8.04	Availability of new telephone lines, 2006 63
8.05	Availability of mobile telephones, 2006...............................65
8.06	Extent of business Internet use, 2006 55

Government usage	**80**

9.01	Government success in ICT promotion, 2006...................109
9.02	Availability of online services, 2006.....................................53
9.03	ICT use and government efficiency, 2006...........................58
9.04	ICT pervasiveness, 2006 89

Armenia

Key indicators

Population (millions), 2005...3.0
GDP (PPP) per capita (US$), 2005.........................4,269.6
Internet users per 100 inhabitants, 2005......................5.3
Internet bandwidth (Mbps/10,000 inhabitants), 2004.................0.1

Networked Readiness Index

Year (number of economies)	Rank
2006–2007 (122) **96**	
2005–2006 (115)..86	
2004–2005 (104) ....—	

Global Competitiveness Index 2006–2007 (125)	82

Environment component	86
Market environment	**99**

1.01 Venture capital availability, 2006..97
1.02 Financial market sophistication, 2006...................................98
1.02 Technological readiness, 2006...86
1.04 State of cluster development, 2006....................................102
1.05 US utility patents, 2005 55
1.06 High-tech exports, 2004 73
1.07 Burden of government regulation, 2006 76
1.08 Extent and effect of taxation, 2006.....................................66
1.09 Time required to start a business, 2006...............................38
1.10 No. of procedures required to start a business, 2006..........49
1.11 Intensity of local competition, 2006 107
1.12 Freedom of the press, 2006...106

Political and regulatory environment	**72**

2.01 Effectiveness of law-making bodies, 2006...........................86
2.02 Laws relating to ICT, 2006..85
2.03 Judicial independence, 2006 108
2.04 Intellectual property protection, 2006.................................92
2.05 Efficiency of legal framework, 2006 98
2.06 Property rights, 2006 71
2.07 Quality of competition in the ISP sector, 2006..................106
2.08 No. of procedures to enforce a contract, 2006 30
2.09 Time to enforce a contract, 2006 9

Infrastructure environment	**68**

3.01 Telephone lines, 2004...61
3.02 Secure Internet servers, 2005...84
3.03 Internet hosts, 2004 87
3.04 Electricity production, 2003 66
3.05 Availability of scientists and engineers, 2006......................55
3.06 Quality of scientific research institutions, 2006 69
3.07 Tertiary enrollment, 2004..68

Readiness component	96
Individual readiness	**98**

4.01 Quality of math and science education, 2006 71
4.02 Quality of the educational system, 2006............................87
4.03 Quality of public schools, 2006 82
4.04 Internet access in schools, 2006......................................100
4.05 Buyer sophistication, 2006 92
4.06 Residential telephone connection charge, 2005 72
4.07 Residential monthly telephone subscription, 2005 65
4.08 High-speed monthly broadband subscription, 2006..........101
4.09 Lowest cost of broadband, 2006......................................102
4.10 Cost of mobile telephone call, 2004...................................93

Business readiness	**97**

5.01 Extent of staff training, 2006..102
5.02 Local availability of research and training, 2006................104
5.03 Quality of management schools, 2006...............................108
5.04 Company spending on R&D, 2006.......................................99
5.05 University-industry research collaboration, 2006.................86
5.06 Business telephone connection charge, 2005 64
5.07 Business monthly telephone subscription, 2005 90
5.08 Local supplier quality, 2006 109
5.09 Computer, comm., and other services imports, 2004........86

Government readiness	**94**

6.01 Government prioritization of ICT, 2006...............................68
6.02 Gov't. procurement of advanced tech products, 2006........79
6.03 Importance of ICT to gov't. vision of the future, 2006........87
6.04 E-participation index, 2005 78
6.05 E-government readiness index, 2005.................................87

Usage component	104
Individual usage	**97**

7.01 Mobile telephone subscribers, 2005 102
7.02 Personal computers, 2004...64
7.03 Broadband Internet subscribers, 2004 85
7.04 Internet users, 2005 90
7.05 Internet bandwidth, 2004 76

Business usage	**100**

8.01 Prevalence of foreign technology licensing, 2006.............100
8.02 Firm-level technology absorption, 2006..............................44
8.03 Capacity for innovation, 2006 70
8.04 Availability of new telephone lines, 2006 107
8.05 Availability of mobile telephones, 2006 120
8.06 Extent of business Internet use, 2006 90

Government usage	**101**

9.01 Government success in ICT promotion, 2006....................93
9.02 Availability of online services, 2006..................................116
9.03 ICT use and government efficiency, 2006.........................103
9.04 ICT pervasiveness, 2006 78

Australia

Key indicators

Population (millions), 2005..20.2
GDP (PPP) per capita (US$), 2005.....................................30,897.2
Internet users per 100 inhabitants, 2005...................................70.4
Internet bandwidth (Mbps/10,000 inhabitants), 2004...............11.1

Networked Readiness Index

Year (number of economies)	Rank
2006–2007 (122) **15**	
2005–2006 (115)...15	
2004–2005 (104)...11	

Global Competitiveness Index 2006–2007 (125)	19

Environment component | 8

Market environment | **19**

1.01 Venture capital availability, 2006...15
1.02 Financial market sophistication, 2006....................................10
1.03 Technological readiness, 2006..17
1.04 State of cluster development, 2006...53
1.05 US utility patents, 2005 21
1.06 High-tech exports, 2004 41
1.07 Burden of government regulation, 2006 53
1.08 Extent and effect of taxation, 2006.......................................78
1.09 Time required to start a business, 2006..................................1
1.10 No. of procedures required to start a business, 2006...........1
1.11 Intensity of local competition, 2006 12
1.12 Freedom of the press, 2006..19

Political and regulatory environment | **7**

2.01 Effectiveness of law-making bodies, 2006............................4
2.02 Laws relating to ICT, 2006...15
2.03 Judicial independence, 2006 6
2.04 Intellectual property protection, 2006...................................10
2.05 Efficiency of legal framework, 2006.......................................11
2.06 Property rights, 2006...10
2.07 Quality of competition in the ISP sector, 2006....................27
2.08 No. of procedures to enforce a contract, 2006 9
2.09 Time to enforce a contract, 2006 6

Infrastructure environment | **5**

3.01 Telephone lines, 2005...9
3.02 Secure Internet servers, 2005 4
3.03 Internet hosts, 2004 6
3.04 Electricity production, 2003 10
3.05 Availability of scientists and engineers, 2006......................35
3.06 Quality of scientific research institutions, 2006 16
3.07 Tertiary enrollment, 2004...13

Readiness component | 17

Individual readiness | **13**

4.01 Quality of math and science education, 2006 29
4.02 Quality of the educational system, 2006...............................12
4.03 Quality of public schools, 2006 22
4.04 Internet access in schools, 2006...17
4.05 Buyer sophistication, 2006 13
4.06 Residential telephone connection charge, 2005 33
4.07 Residential monthly telephone subscription, 2005 14
4.08 High-speed monthly broadband subscription, 2006 18
4.09 Lowest cost of broadband, 2006...27
4.10 Cost of mobile telephone call, 2005......................................37

Business readiness | **22**

5.01 Extent of staff training, 2006..20
5.02 Local availability of research and training, 2006 16
5.03 Quality of management schools, 2006...............................17
5.04 Company spending on R&D, 2006 28
5.05 University-industry research collaboration, 2006................25
5.06 Business telephone connection charge, 2005 28
5.07 Business monthly telephone subscription, 2005 31
5.08 Local supplier quality, 2006 18
5.09 Computer, comm., and other services imports, 2004.........62

Government readiness | **13**

6.01 Government prioritization of ICT, 2006...............................55
6.02 Gov't. procurement of advanced tech products, 2006........30
6.03 Importance of ICT to gov't. vision of the future, 2006........41
6.04 E-participation index, 2005 9
6.05 E-government readiness index, 2005 6

Usage component | 21

Individual usage | **20**

7.01 Mobile telephone subscribers, 2005 28
7.02 Personal computers, 2004..6
7.03 Broadband Internet subscribers, 2005 29
7.04 Internet users, 2005 5
7.05 Internet bandwidth, 2004 27

Business usage | **24**

8.01 Prevalence of foreign technology licensing, 2006................8
8.02 Firm-level technology absorption, 2006..............................20
8.03 Capacity for innovation, 2006 35
8.04 Availability of new telephone lines, 2006 37
8.05 Availability of mobile telephones, 2006...............................57
8.06 Extent of business Internet use, 2006 16

Government usage | **26**

9.01 Government success in ICT promotion, 2006.....................45
9.02 Availability of online services, 2006......................................15
9.03 ICT use and government efficiency, 2006............................34
9.04 ICT pervasiveness, 2006 21

143

Austria

Key indicators

Population (millions), 2005..8.2
GDP (PPP) per capita (US$), 2005.......................................33,431.6
Internet users per 100 inhabitants, 2005...................................48.9
Internet bandwidth (Mbps/10,000 inhabitants), 2004...............66.6

Networked Readiness Index

Year (number of economies)	Rank
2006–2007 (122) **17**	
2005–2006 (115)..18	
2004–2005 (104)..19	

Global Competitiveness Index 2006–2007 (125)	17

Environment component	20
Market environment	**22**

1.01 Venture capital availability, 2006..25
1.02 Financial market sophistication, 2006...................................23
1.03 Technological readiness, 2006...25
1.04 State of cluster development, 2006 24
1.05 US utility patents, 2005 16
1.06 High-tech exports, 2004 28
1.07 Burden of government regulation, 2006 27
1.08 Extent and effect of taxation, 2006.......................................34
1.09 Time required to start a business, 2006...............................50
1.10 No. of procedures required to start a business, 2006.........49
1.11 Intensity of local competition, 2006 18
1.12 Freedom of the press, 2006..8

Political and regulatory environment	**13**

2.01 Effectiveness of law-making bodies, 2006...........................25
2.02 Laws relating to ICT, 2006...7
2.03 Judicial independence, 2006 15
2.04 Intellectual property protection, 2006...................................18
2.05 Efficiency of legal framework, 2006 9
2.06 Property rights, 2006..7
2.07 Quality of competition in the ISP sector, 2006....................11
2.08 No. of procedures to enforce a contract, 2006 27
2.09 Time to enforce a contract, 2006 36

Infrastructure environment	**17**

3.01 Telephone lines, 2005..24
3.02 Secure Internet servers, 2005 19
3.03 Internet hosts, 2004...8
3.04 Electricity production, 2003...22
3.05 Availability of scientists and engineers, 2006......................29
3.06 Quality of scientific research institutions, 2006 23
3.07 Tertiary enrollment, 2004...33

Readiness component	16
Individual readiness	**11**

4.01 Quality of math and science education, 2006 27
4.02 Quality of the educational system, 2006.............................13
4.03 Quality of public schools, 2006 12
4.04 Internet access in schools, 2006...9
4.05 Buyer sophistication, 2006 14
4.06 Residential telephone connection charge, 2005 28
4.07 Residential monthly telephone subscription, 2005 24
4.08 High-speed monthly broadband subscription, 2006 22
4.09 Lowest cost of broadband, 2006..25
4.10 Cost of mobile telephone call, 2005.....................................22

Business readiness	**9**

5.01 Extent of staff training, 2006...5
5.02 Local availability of research and training, 2006 14
5.03 Quality of management schools, 2006..................................31
5.04 Company spending on R&D, 2006 18
5.05 University-industry research collaboration, 2006.................20
5.06 Business telephone connection charge, 2005 35
5.07 Business monthly telephone subscription, 2005 22
5.08 Local supplier quality, 2006 4
5.09 Computer, comm., and other services imports, 2004...........6

Government readiness	**22**

6.01 Government prioritization of ICT, 2006................................40
6.02 Gov't. procurement of advanced tech products, 2006........22
6.03 Importance of ICT to gov't. vision of the future, 2006........31
6.04 E-participation index, 2005 24
6.05 E-government readiness index, 2005....................................16

Usage component	15
Individual usage	**18**

7.01 Mobile telephone subscribers, 2005 18
7.02 Personal computers, 2004...13
7.03 Broadband Internet subscribers, 2005 20
7.04 Internet users, 2005 21
7.05 Internet bandwidth, 2004 11

Business usage	**15**

8.01 Prevalence of foreign technology licensing, 2006..............54
8.02 Firm-level technology absorption, 2006...............................17
8.03 Capacity for innovation, 2006 10
8.04 Availability of new telephone lines, 2006 12
8.05 Availability of mobile telephones, 2006...............................10
8.06 Extent of business Internet use, 2006 17

Government usage	**14**

9.01 Government success in ICT promotion, 2006.....................24
9.02 Availability of online services, 2006.....................................12
9.03 ICT use and government efficiency, 2006............................26
9.04 ICT pervasiveness, 2006 8

Azerbaijan

Key indicators

Population (millions), 2005..8.4
GDP (PPP) per capita (US$), 2005.....................................4,600.5
Internet users per 100 inhabitants, 2005.....................................8.1
Internet bandwidth (Mbps/10,000 inhabitants) ....n/a

Networked Readiness Index

Year (number of economies)	Rank
2006–2007 (122)	**71**
2005–2006 (115)...	73
2004–2005 (104)	—

Global Competitiveness Index 2006–2007 (125)	64

Environment component	79
Market environment	**89**
1.01 Venture capital availability, 2006..63
1.02 Financial market sophistication, 2006...................................70
1.03 Technological readiness, 2006...73
1.04 State of cluster development, 2006.......................................60
1.05 US utility patents, 2005...79
1.06 High-tech exports, 2004..84
1.07 Burden of government regulation, 2006 35
1.08 Extent and effect of taxation, 2006......................................41
1.09 Time required to start a business, 2006...............................93
1.10 No. of procedures required to start a business, 2006.......108
1.11 Intensity of local competition, 2006 105
1.12 Freedom of the press, 2006...82
Political and regulatory environment	**64**
2.01 Effectiveness of law-making bodies, 2006.........................72
2.02 Laws relating to ICT, 2006..58
2.03 Judicial independence, 2006 92
2.04 Intellectual property protection, 2006...................................88
2.05 Efficiency of legal framework, 2006.....................................88
2.06 Property rights, 2006 96
2.07 Quality of competition in the ISP sector, 2006...................59
2.08 No. of procedures to enforce a contract, 2006 41
2.09 Time to enforce a contract, 2006 20
Infrastructure environment	**78**
3.01 Telephone lines, 2005..75
3.02 Secure Internet servers, 2005 94
3.03 Internet hosts, 2004 108
3.04 Electricity production, 2003 57
3.05 Availability of scientists and engineers, 2006.....................66
3.06 Quality of scientific research institutions, 2006 52
3.07 Tertiary enrollment, 2004..86

Readiness component	74
Individual readiness	**77**
4.01 Quality of math and science education, 2006.....................69
4.02 Quality of the educational system, 2006.............................85
4.03 Quality of public schools, 2006 92
4.04 Internet access in schools, 2006..85
4.05 Buyer sophistication, 2006 56
4.06 Residential telephone connection charge, 2005 89
4.07 Residential monthly telephone subscription, 2005 22
4.08 High-speed monthly broadband subscription, 2006...........82
4.09 Lowest cost of broadband, 2006..90
4.10 Cost of mobile telephone call ....n/a
Business readiness	**63**
5.01 Extent of staff training, 2006..93
5.02 Local availability of research and training, 2006 68
5.03 Quality of management schools, 2006..............................105
5.04 Company spending on R&D, 2006 54
5.05 University-industry research collaboration, 2006................72
5.06 Business telephone connection charge, 2005 90
5.07 Business monthly telephone subscription, 2005 78
5.08 Local supplier quality, 2006 77
5.09 Computer, comm., and other services imports, 2004...........1
Government readiness	**74**
6.01 Government prioritization of ICT, 2006...............................33
6.02 Gov't. procurement of advanced tech products, 2006........41
6.03 Importance of ICT to gov't. vision of the future, 2006........60
6.04 E-participation index, 2005 101
6.05 E-government readiness index, 2005....................................84

145

Usage component	62
Individual usage	**85**
7.01 Mobile telephone subscribers, 2005 84
7.02 Personal computers, 2004...92
7.03 Broadband Internet subscribers, 2005 87
7.04 Internet users, 2005 78
7.05 Internet bandwidth...n/a
Business usage	**61**
8.01 Prevalence of foreign technology licensing, 2006..............82
8.02 Firm-level technology absorption, 2006..............................62
8.03 Capacity for innovation, 2006 41
8.04 Availability of new telephone lines, 2006 59
8.05 Availability of mobile telephones, 2006 61
8.06 Extent of business Internet use, 2006 57
Government usage	**48**
9.01 Government success in ICT promotion, 2006.....................39
9.02 Availability of online services, 2006.....................................72
9.03 ICT use and government efficiency, 2006...........................50
9.04 ICT pervasiveness, 2006 46

Bahrain

Key indicators

Population (millions), 2005..0.7
GDP (PPP) per capita (US$), 2005.......................21,565.4
Internet users per 100 inhabitants, 2005.....................21.3
Internet bandwidth (Mbps/10,000 inhabitants) ....n/a

Networked Readiness Index

Year (number of economies)	Rank
2006–2007 (122) **50**	
2005–2006 (115)..49	
2004–2005 (104)..33	

Global Competitiveness Index 2006–2007 (125)   49

Environment component   65
Market environment   60

1.01 Venture capital availability, 2006..56
1.02 Financial market sophistication, 2006...................................30
1.03 Technological readiness, 2006..39
1.04 State of cluster development, 2006.......................................90
1.05 US utility patents, 2005 79
1.06 High-tech exports, 2004 82
1.07 Burden of government regulation, 2006 31
1.08 Extent and effect of taxation, 2006.......................................1
1.09 Time required to start a business ....n/a
1.10 No. of procedures required to start a business ....n/a
1.11 Intensity of local competition, 2006 54
1.12 Freedom of the press, 2006..109

Political and regulatory environment   82

2.01 Effectiveness of law-making bodies, 2006..........................90
2.02 Laws relating to ICT, 2006..51
2.03 Judicial independence, 2006 73
2.04 Intellectual property protection, 2006..................................46
2.05 Efficiency of legal framework, 2006.....................................69
2.06 Property rights, 2006..47
2.07 Quality of competition in the ISP sector, 2006..................115
2.08 No. of procedures to enforce a contract ....n/a
2.09 Time to enforce a contract..n/a

Infrastructure environment   54

3.01 Telephone lines, 2005..46
3.02 Secure Internet servers, 2005 34
3.03 Internet hosts, 2004 61
3.04 Electricity production, 2003 11
3.05 Availability of scientists and engineers, 2006.....................96
3.06 Quality of scientific research institutions, 2006 115
3.07 Tertiary enrollment, 2004...56

Readiness component   59
Individual readiness   50

4.01 Quality of math and science education, 2006 86
4.02 Quality of the educational system, 2006.............................78
4.03 Quality of public schools, 2006 68
4.04 Internet access in schools, 2006...43
4.05 Buyer sophistication, 2006 40
4.06 Residential telephone connection charge, 2005 21
4.07 Residential monthly telephone subscription, 2005 3
4.08 High-speed monthly broadband subscription, 2006 41
4.09 Lowest cost of broadband, 2006...58
4.10 Cost of mobile telephone call, 2005....................................13

Business readiness   73

5.01 Extent of staff training, 2006...59
5.02 Local availability of research and training, 2006.................96
5.03 Quality of management schools, 2006.................................77
5.04 Company spending on R&D, 2006.....................................114
5.05 University-industry research collaboration, 2006...............119
5.06 Business telephone connection charge, 2005 19
5.07 Business monthly telephone subscription, 2005 6
5.08 Local supplier quality, 2006 57
5.09 Computer, comm., and other services imports, 2004.........96

Government readiness   56

6.01 Government prioritization of ICT, 2006................................37
6.02 Gov't. procurement of advanced tech products, 2006........59
6.03 Importance of ICT to gov't. vision of the future, 2006........54
6.04 E-participation index, 2005 82
6.05 E-government readiness index, 2005 52

Usage component   39
Individual usage   32

7.01 Mobile telephone subscribers, 2005 12
7.02 Personal computers, 2004..40
7.03 Broadband Internet subscribers..n/a
7.04 Internet users, 2005 47
7.05 Internet bandwidth..n/a

Business usage   58

8.01 Prevalence of foreign technology licensing, 2006...............27
8.02 Firm-level technology absorption, 2006...............................52
8.03 Capacity for innovation, 2006 116
8.04 Availability of new telephone lines, 2006 35
8.05 Availability of mobile telephones, 2006...............................31
8.06 Extent of business Internet use, 2006 82

Government usage   50

9.01 Government success in ICT promotion, 2006.....................33
9.02 Availability of online services, 2006.....................................62
9.03 ICT use and government efficiency, 2006...........................64
9.04 ICT pervasiveness, 2006 56

146

Bangladesh

Key indicators

Population (millions), 2005..141.8
GDP (PPP) per capita (US$), 2005.........................2,011.0
Internet users per 100 inhabitants, 2005....................0.3
Internet bandwidth (Mbps/10,000 inhabitants), 2004................0.0

Networked Readiness Index

Year (number of economies)	Rank
2006–2007 (122)	**118**
2005–2006 (115)	110
2004–2005 (104)	100

Global Competitiveness Index 2006–2007 (125)	99

Environment component — 114

Market environment — 87

1.01 Venture capital availability, 2006 109
1.02 Financial market sophistication, 2006................97
1.03 Technological readiness, 2006...........................104
1.04 State of cluster development, 2006....................62
1.05 US utility patents, 2005 79
1.06 High-tech exports, 2004 93
1.07 Burden of government regulation, 2006 104
1.08 Extent and effect of taxation, 2006....................47
1.09 Time required to start a business, 2006 71
1.10 No. of procedures required to start a business, 2006........36
1.11 Intensity of local competition, 2006 66
1.12 Freedom of the press, 2006...............................69

Political and regulatory environment — 121

2.01 Effectiveness of law-making bodies, 2006.........95
2.02 Laws relating to ICT, 2006................................109
2.03 Judicial independence, 2006 99
2.04 Intellectual property protection, 2006...............116
2.05 Efficiency of legal framework, 2006...................99
2.06 Property rights, 2006 85
2.07 Quality of competition in the ISP sector, 2006....64
2.08 No. of procedures to enforce a contract, 2006 109
2.09 Time to enforce a contract, 2006 117

Infrastructure environment — 106

3.01 Telephone lines, 2005.......................................111
3.02 Secure Internet servers, 2005 119
3.03 Internet hosts, 2004 122
3.04 Electricity production, 2003 100
3.05 Availability of scientists and engineers, 2006......74
3.06 Quality of scientific research institutions, 2006 89
3.07 Tertiary enrollment, 2003...................................96

Readiness component — 118

Individual readiness — 114

4.01 Quality of math and science education, 2006 110
4.02 Quality of the educational system, 2006............101
4.03 Quality of public schools, 2006 110
4.04 Internet access in schools, 2006.......................111
4.05 Buyer sophistication, 2006 85
4.06 Residential telephone connection charge, 2005 112
4.07 Residential monthly telephone subscription, 2004 99
4.08 High-speed monthly broadband subscription, 2006....103
4.09 Lowest cost of broadband, 2006........................105
4.10 Cost of mobile telephone call, 2005...................86

Business readiness — 118

5.01 Extent of staff training, 2006............................113
5.02 Local availability of research and training, 2006 118
5.03 Quality of management schools, 2006................98
5.04 Company spending on R&D, 2006 111
5.05 University-industry research collaboration, 2006....112
5.06 Business telephone connection charge, 2005 109
5.07 Business monthly telephone subscription, 2004 91
5.08 Local supplier quality, 2006 94
5.09 Computer, comm., and other services imports, 2004....95

Government readiness — 112

6.01 Government prioritization of ICT, 2006...............72
6.02 Gov't. procurement of advanced tech products, 2006....103
6.03 Importance of ICT to gov't. vision of the future, 2006....89
6.04 E-participation index, 2005 109
6.05 E-government readiness index, 2005.................114

Usage component — 117

Individual usage — 115

7.01 Mobile telephone subscribers, 2005 111
7.02 Personal computers, 2004.................................100
7.03 Broadband Internet subscribers, 2004 97
7.04 Internet users, 2005 121
7.05 Internet bandwidth, 2004 88

Business usage — 108

8.01 Prevalence of foreign technology licensing, 2006....89
8.02 Firm-level technology absorption, 2006.............90
8.03 Capacity for innovation, 2006 115
8.04 Availability of new telephone lines, 2006 121
8.05 Availability of mobile telephones, 2006 74
8.06 Extent of business Internet use, 2006 96

Government usage — 117

9.01 Government success in ICT promotion, 2006....107
9.02 Availability of online services, 2006...................119
9.03 ICT use and government efficiency, 2006...........99
9.04 ICT pervasiveness, 2006 120

147

Barbados

Key indicators

Population (millions), 2005..0.3
GDP (PPP) per capita (US$), 2005.......................................17,610.2
Internet users per 100 inhabitants, 2005....................................59.5
Internet bandwidth (Mbps/10,000 inhabitants) ....n/a

Networked Readiness Index

Year (number of economies)	Rank
2006–2007 (122) **40**	
2005–2006 (115) ....—	
2004–2005 (104) ....—	

Global Competitiveness Index 2006–2007 (125)	31

Environment component	31
Market environment	**56**

1.01   Venture capital availability, 2006 52
1.02   Financial market sophistication, 2006...................................63
1.03   Technological readiness, 2006...34
1.04   State of cluster development, 2006 95
1.05   US utility patents, 2005 79
1.06   High-tech exports..n/a
1.07   Burden of government regulation, 2006 32
1.08   Extent and effect of taxation, 2006 43
1.09   Time required to start a business ....n/a
1.10   No. of procedures required to start a business ....n/a
1.11   Intensity of local competition, 2006 82
1.12   Freedom of the press, 2006..50

Political and regulatory environment	**30**

2.01   Effectiveness of law-making bodies, 2006...........................15
2.02   Laws relating to ICT, 2006...50
2.03   Judicial independence, 2006 16
2.04   Intellectual property protection, 2006...............................36
2.05   Efficiency of legal framework, 2006....................................20
2.06   Property rights, 2006 36
2.07   Quality of competition in the ISP sector, 2006....................66
2.08   No. of procedures to enforce a contract ....n/a
2.09   Time to enforce a contract..n/a

Infrastructure environment	**27**

3.01   Telephone lines, 2005...17
3.02   Secure Internet servers, 2005...21
3.03   Internet hosts, 2004 79
3.04   Electricity production ....n/a
3.05   Availability of scientists and engineers, 2006....................59
3.06   Quality of scientific research institutions, 2006 50
3.07   Tertiary enrollment, 2001 51

Readiness component	47
Individual readiness	**29**

4.01   Quality of math and science education, 2006.....................19
4.02   Quality of the educational system, 2006............................16
4.03   Quality of public schools, 2006 8
4.04   Internet access in schools, 2006..45
4.05   Buyer sophistication, 2006 37
4.06   Residential telephone connection charge, 2005 32
4.07   Residential monthly telephone subscription, 2005 60
4.08   High-speed monthly broadband subscription, 2006 52
4.09   Lowest cost of broadband, 2006...48
4.10   Cost of mobile telephone call, 2005....................................26

Business readiness	**59**

5.01   Extent of staff training, 2006..49
5.02   Local availability of research and training, 2006 77
5.03   Quality of management schools, 2006.................................37
5.04   Company spending on R&D, 2006.......................................58
5.05   University-industry research collaboration, 2006.................74
5.06   Business telephone connection charge, 2005 27
5.07   Business monthly telephone subscription, 2005 79
5.08   Local supplier quality, 2006 64
5.09   Computer, comm., and other services imports, 2004.........83

Government readiness	**62**

6.01   Government prioritization of ICT, 2006...............................48
6.02   Gov't. procurement of advanced tech products, 2006........43
6.03   Importance of ICT to gov't. vision of the future, 2006........59
6.04   E-participation index, 2005 82
6.05   E-government readiness index, 2005 60

Usage component	41
Individual usage	**27**

7.01   Mobile telephone subscribers, 2005 42
7.02   Personal computers, 2004...48
7.03   Broadband Internet subscribers, 2005 23
7.04   Internet users, 2005 11
7.05   Internet bandwidth...n/a

Business usage	**66**

8.01   Prevalence of foreign technology licensing, 2006...............70
8.02   Firm-level technology absorption, 2006...............................58
8.03   Capacity for innovation, 2006 85
8.04   Availability of new telephone lines, 2006 61
8.05   Availability of mobile telephones, 2006 45
8.06   Extent of business Internet use, 2006 60

Government usage	**83**

9.01   Government success in ICT promotion, 2006.....................43
9.02   Availability of online services, 2006.....................................95
9.03   ICT use and government efficiency, 2006............................93
9.04   ICT pervasiveness, 2006 74

148

Belgium

Key indicators

Population (millions), 2005...10.4
GDP (PPP) per capita (US$), 2005....................................31,243.9
Internet users per 100 inhabitants, 2005................................45.7
Internet bandwidth (Mbps/10,000 inhabitants), 2004.............112.5

Networked Readiness Index

Year (number of economies)	Rank
2006–2007 (122)	**24**
2005–2006 (115)	25
2004–2005 (104)	26

Global Competitiveness Index 2006–2007 (125)	20

Environment component	24
Market environment	**23**

1.01	Venture capital availability, 2006	26
1.02	Financial market sophistication, 2006	21
1.03	Technological readiness, 2006	20
1.04	State of cluster development, 2006	19
1.05	US utility patents, 2005	18
1.06	High-tech exports, 2004	30
1.07	Burden of government regulation, 2006	75
1.08	Extent and effect of taxation, 2006	121
1.09	Time required to start a business, 2006	46
1.10	No. of procedures required to start a business, 2006	7
1.11	Intensity of local competition, 2006	8
1.12	Freedom of the press, 2006	14

Political and regulatory environment	**24**

2.01	Effectiveness of law-making bodies, 2006	47
2.02	Laws relating to ICT, 2006	30
2.03	Judicial independence, 2006	32
2.04	Intellectual property protection, 2006	19
2.05	Efficiency of legal framework, 2006	35
2.06	Property rights, 2006	19
2.07	Quality of competition in the ISP sector, 2006	25
2.08	No. of procedures to enforce a contract, 2006	41
2.09	Time to enforce a contract, 2006	31

Infrastructure environment	**24**

3.01	Telephone lines, 2004	22
3.02	Secure Internet servers, 2005	25
3.03	Internet hosts, 2004	35
3.04	Electricity production, 2003	21
3.05	Availability of scientists and engineers, 2006	13
3.06	Quality of scientific research institutions, 2006	9
3.07	Tertiary enrollment, 2004	21

Readiness component	15
Individual readiness	**4**

4.01	Quality of math and science education, 2006	3
4.02	Quality of the educational system, 2006	8
4.03	Quality of public schools, 2006	4
4.04	Internet access in schools, 2006	23
4.05	Buyer sophistication, 2006	7
4.06	Residential telephone connection charge, 2005	14
4.07	Residential monthly telephone subscription, 2005	30
4.08	High-speed monthly broadband subscription, 2006	14
4.09	Lowest cost of broadband, 2006	18
4.10	Cost of mobile telephone call, 2005	24

Business readiness	**8**

5.01	Extent of staff training, 2006	13
5.02	Local availability of research and training, 2006	9
5.03	Quality of management schools, 2006	7
5.04	Company spending on R&D, 2006	17
5.05	University-industry research collaboration, 2006	11
5.06	Business telephone connection charge, 2005	13
5.07	Business monthly telephone subscription, 2005	20
5.08	Local supplier quality, 2006	5
5.09	Computer, comm., and other services imports, 2004	21

Government readiness	**27**

6.01	Government prioritization of ICT, 2006	50
6.02	Gov't. procurement of advanced tech products, 2006	75
6.03	Importance of ICT to gov't. vision of the future, 2006	52
6.04	E-participation index, 2005	17
6.05	E-government readiness index, 2005	18

Usage component	23
Individual usage	**16**

7.01	Mobile telephone subscribers, 2005	30
7.02	Personal computers, 2004	26
7.03	Broadband Internet subscribers, 2005	12
7.04	Internet users, 2005	26
7.05	Internet bandwidth, 2004	5

Business usage	**21**

8.01	Prevalence of foreign technology licensing, 2006	26
8.02	Firm-level technology absorption, 2006	35
8.03	Capacity for innovation, 2006	14
8.04	Availability of new telephone lines, 2006	17
8.05	Availability of mobile telephones, 2006	24
8.06	Extent of business Internet use, 2006	32

Government usage	**44**

9.01	Government success in ICT promotion, 2006	66
9.02	Availability of online services, 2006	34
9.03	ICT use and government efficiency, 2006	53
9.04	ICT pervasiveness, 2006	66

149

Benin

Key indicators

Population (millions), 2005...8.4
GDP (PPP) per capita (US$), 2005.........................1,176.0
Internet users per 100 inhabitants, 2005.....................5.7
Internet bandwidth (Mbps/10,000 inhabitants), 2004...............0.1

Networked Readiness Index

Year (number of economies)	Rank
2006–2007 (122) **109**	
2005–2006 (115)..108	
2004–2005 (104) ....—	

Global Competitiveness Index 2006–2007 (125)	105

Environment component — 104

Market environment — 90

1.01	Venture capital availability, 2006	105
1.02	Financial market sophistication, 2006	102
1.03	Technological readiness, 2006	108
1.04	State of cluster development, 2006	121
1.05	US utility patents, 2005	79
1.06	High-tech exports	n/a
1.07	Burden of government regulation, 2006	105
1.08	Extent and effect of taxation, 2006	117
1.09	Time required to start a business, 2006	56
1.10	No. of procedures required to start a business, 2006	27
1.11	Intensity of local competition, 2006	99
1.12	Freedom of the press, 2006	57

Political and regulatory environment — 99

2.01	Effectiveness of law-making bodies, 2006	48
2.02	Laws relating to ICT, 2006	96
2.03	Judicial independence, 2006	61
2.04	Intellectual property protection, 2006	82
2.05	Efficiency of legal framework, 2006	72
2.06	Property rights, 2006	95
2.07	Quality of competition in the ISP sector, 2006	98
2.08	No. of procedures to enforce a contract, 2006	107
2.09	Time to enforce a contract, 2006	99

Infrastructure environment — 111

3.01	Telephone lines, 2005	105
3.02	Secure Internet servers, 2005	113
3.03	Internet hosts, 2004	99
3.04	Electricity production, 2003	105
3.05	Availability of scientists and engineers, 2006	64
3.06	Quality of scientific research institutions, 2006	108
3.07	Tertiary enrollment, 2001	105

Readiness component — 114

Individual readiness — 113

4.01	Quality of math and science education, 2006	55
4.02	Quality of the educational system, 2006	92
4.03	Quality of public schools, 2006	65
4.04	Internet access in schools, 2006	113
4.05	Buyer sophistication, 2006	110
4.06	Residential telephone connection charge, 2005	109
4.07	Residential monthly telephone subscription, 2004	106
4.08	High-speed monthly broadband subscription, 2006	99
4.09	Lowest cost of broadband, 2006	98
4.10	Cost of mobile telephone call, 2004	99

Business readiness — 114

5.01	Extent of staff training, 2006	107
5.02	Local availability of research and training, 2006	98
5.03	Quality of management schools, 2006	63
5.04	Company spending on R&D, 2006	110
5.05	University-industry research collaboration, 2006	110
5.06	Business telephone connection charge, 2005	105
5.07	Business monthly telephone subscription, 2004	100
5.08	Local supplier quality, 2006	85
5.09	Computer, comm., and other services imports	n/a

Government readiness — 98

6.01	Government prioritization of ICT, 2006	54
6.02	Gov't. procurement of advanced tech products, 2006	42
6.03	Importance of ICT to gov't. vision of the future, 2006	78
6.04	E-participation index, 2005	101
6.05	E-government readiness index, 2005	111

Usage component — 102

Individual usage — 106

7.01	Mobile telephone subscribers, 2005	104
7.02	Personal computers, 2004	112
7.03	Broadband Internet subscribers, 2005	93
7.04	Internet users, 2005	88
7.05	Internet bandwidth, 2004	77

Business usage — 112

8.01	Prevalence of foreign technology licensing, 2006	110
8.02	Firm-level technology absorption, 2006	86
8.03	Capacity for innovation, 2006	71
8.04	Availability of new telephone lines, 2006	120
8.05	Availability of mobile telephones, 2006	116
8.06	Extent of business Internet use, 2006	103

Government usage — 71

9.01	Government success in ICT promotion, 2006	44
9.02	Availability of online services, 2006	101
9.03	ICT use and government efficiency, 2006	84
9.04	ICT pervasiveness, 2006	45

Bolivia

Key indicators

Population (millions), 2005...9.2

GDP (PPP) per capita (US$), 2005....................................2,724.4

Internet users per 100 inhabitants, 2005....................................5.2

Internet bandwidth (Mbps/10,000 inhabitants), 2004.................0.4

Networked Readiness Index

Year (number of economies)	Rank
2006–2007 (122)	**104**
2005–2006 (115)..	109
2004–2005 (104)..	99

Global Competitiveness Index 2006–2007 (125)	97

Environment component	111
Market environment	**106**

1.01	Venture capital availability, 2006	101
1.02	Financial market sophistication, 2006	92
1.03	Technological readiness, 2006	96
1.04	State of cluster development, 2006	98
1.05	US utility patents, 2005	79
1.06	High-tech exports, 2004	61
1.07	Burden of government regulation, 2006	99
1.08	Extent and effect of taxation, 2006	54
1.09	Time required to start a business, 2006	91
1.10	No. of procedures required to start a business, 2006	108
1.11	Intensity of local competition, 2006	92
1.12	Freedom of the press, 2006	49

Political and regulatory environment		**115**
2.01	Effectiveness of law-making bodies, 2006	115
2.02	Laws relating to ICT, 2006	107
2.03	Judicial independence, 2006	98
2.04	Intellectual property protection, 2006	119
2.05	Efficiency of legal framework, 2006	113
2.06	Property rights, 2006	113
2.07	Quality of competition in the ISP sector, 2006	77
2.08	No. of procedures to enforce a contract, 2006	104
2.09	Time to enforce a contract, 2006	83

Infrastructure environment		**91**
3.01	Telephone lines, 2005	89
3.02	Secure Internet servers, 2005	78
3.03	Internet hosts, 2004	76
3.04	Electricity production, 2003	95
3.05	Availability of scientists and engineers, 2006	112
3.06	Quality of scientific research institutions, 2006	116
3.07	Tertiary enrollment, 2004	43

Readiness component	99
Individual readiness	**96**

4.01	Quality of math and science education, 2006	119
4.02	Quality of the educational system, 2006	120
4.03	Quality of public schools, 2006	114
4.04	Internet access in schools, 2006	104
4.05	Buyer sophistication, 2006	115
4.06	Residential telephone connection charge, 2005	79
4.07	Residential monthly telephone subscription, 2005	92
4.08	High-speed monthly broadband subscription, 2006	87
4.09	Lowest cost of broadband, 2006	93
4.10	Cost of mobile telephone call, 2005	83

Business readiness		**100**
5.01	Extent of staff training, 2006	114
5.02	Local availability of research and training, 2006	102
5.03	Quality of management schools, 2006	113
5.04	Company spending on R&D, 2006	113
5.05	University-industry research collaboration, 2006	109
5.06	Business telephone connection charge, 2005	70
5.07	Business monthly telephone subscription, 2005	89
5.08	Local supplier quality, 2006	113
5.09	Computer, comm., and other services imports, 2004	74

Government readiness		**105**
6.01	Government prioritization of ICT, 2006	102
6.02	Gov't. procurement of advanced tech products, 2006	120
6.03	Importance of ICT to gov't. vision of the future, 2006	101
6.04	E-participation index, 2005	71
6.05	E-government readiness index, 2005	74

Usage component	101
Individual usage	**91**

7.01	Mobile telephone subscribers, 2005	85
7.02	Personal computers, 2004	89
7.03	Broadband Internet subscribers, 2005	78
7.04	Internet users, 2005	93
7.05	Internet bandwidth, 2004	61

Business usage		**107**
8.01	Prevalence of foreign technology licensing, 2006	121
8.02	Firm-level technology absorption, 2006	122
8.03	Capacity for innovation, 2006	102
8.04	Availability of new telephone lines, 2006	81
8.05	Availability of mobile telephones, 2006	75
8.06	Extent of business Internet use, 2006	104

Government usage		**86**
9.01	Government success in ICT promotion, 2006	114
9.02	Availability of online services, 2006	54
9.03	ICT use and government efficiency, 2006	59
9.04	ICT pervasiveness, 2006	97

Bosnia and Herzegovina

Key indicators

Population (millions), 2005..3.9
GDP (PPP) per capita (US$), 2005.........................6,035.2
Internet users per 100 inhabitants, 2005....................20.6
Internet bandwidth (Mbps/10,000 inhabitants), 2004.................0.2

Networked Readiness Index

Year (number of economies)	Rank
2006–2007 (122) **89**	
2005–2006 (115)..97	
2004–2005 (104)..89	

Global Competitiveness Index 2006–2007 (125)	89

Environment component	102
Market environment	**103**

1.01	Venture capital availability, 2006 83
1.02	Financial market sophistication, 2006....................94
1.03	Technological readiness, 2006..............................112
1.04	State of cluster development, 2006 106
1.05	US utility patents, 2005 79
1.06	High-tech exports...n/a
1.07	Burden of government regulation, 2006 115
1.08	Extent and effect of taxation, 2006 112
1.09	Time required to start a business, 2006................94
1.10	No. of procedures required to start a business, 2006.........89
1.11	Intensity of local competition, 2006 94
1.12	Freedom of the press, 2006...................................74

Political and regulatory environment	**107**
2.01	Effectiveness of law-making bodies, 2006.........103
2.02	Laws relating to ICT, 2006...................................111
2.03	Judicial independence, 2006 82
2.04	Intellectual property protection, 2006...................109
2.05	Efficiency of legal framework, 2006 93
2.06	Property rights, 2006 118
2.07	Quality of competition in the ISP sector, 2006................104
2.08	No. of procedures to enforce a contract, 2006.................76
2.09	Time to enforce a contract, 2006 85

Infrastructure environment	**80**
3.01	Telephone lines, 2005..52
3.02	Secure Internet servers, 2005...............................74
3.03	Internet hosts, 2004 64
3.04	Electricity production, 2003 53
3.05	Availability of scientists and engineers, 2006....................93
3.06	Quality of scientific research institutions, 2006 104
3.07	Tertiary enrollment ....n/a

Readiness component	84
Individual readiness	**68**

4.01	Quality of math and science education, 2006 45
4.02	Quality of the educational system, 2006.............................62
4.03	Quality of public schools, 2006 59
4.04	Internet access in schools, 2006..........................71
4.05	Buyer sophistication, 2006 101
4.06	Residential telephone connection charge, 2005 87
4.07	Residential monthly telephone subscription, 2004 46
4.08	High-speed monthly broadband subscription, 2006............51
4.09	Lowest cost of broadband, 2006.........................37
4.10	Cost of mobile telephone call, 2004.....................69

Business readiness	**94**
5.01	Extent of staff training, 2006................................94
5.02	Local availability of research and training, 2006.................71
5.03	Quality of management schools, 2006..................80
5.04	Company spending on R&D, 2006........................85
5.05	University-industry research collaboration, 2006.................92
5.06	Business telephone connection charge, 2005 76
5.07	Business monthly telephone subscription, 2004 87
5.08	Local supplier quality, 2006 84
5.09	Computer, comm., and other services imports, 2004.........88

Government readiness	**104**
6.01	Government prioritization of ICT, 2006.................73
6.02	Gov't. procurement of advanced tech products, 2006......109
6.03	Importance of ICT to gov't. vision of the future, 2006......115
6.04	E-participation index, 2005 101
6.05	E-government readiness index, 2005....................73

Usage component	88
Individual usage	**65**

7.01	Mobile telephone subscribers, 2005 71
7.02	Personal computers, 2004.....................................51
7.03	Broadband Internet subscribers, 2005 65
7.04	Internet users, 2005 50
7.05	Internet bandwidth, 2004 68

Business usage	**89**
8.01	Prevalence of foreign technology licensing, 2006...............92
8.02	Firm-level technology absorption, 2006.............113
8.03	Capacity for innovation, 2006 94
8.04	Availability of new telephone lines, 2006 72
8.05	Availability of mobile telephones, 2006................98
8.06	Extent of business Internet use, 2006 62

Government usage	**97**
9.01	Government success in ICT promotion, 2006....................92
9.02	Availability of online services, 2006......................86
9.03	ICT use and government efficiency, 2006..........104
9.04	ICT pervasiveness, 2006 92

Botswana

Key indicators

Population (millions), 2005..1.8
GDP (PPP) per capita (US$), 2005..11,409.7
Internet users per 100 inhabitants, 2005..................................3.4
Internet bandwidth (Mbps/10,000 inhabitants) ....n/a

Networked Readiness Index

Year (number of economies)	Rank
2006–2007 (122)	**67**
2005–2006 (115)	56
2004–2005 (104)	50

Global Competitiveness Index 2006–2007 (125)	81

Environment component		63
Market environment		**66**
1.01	Venture capital availability, 2006	45
1.02	Financial market sophistication, 2006	74
1.03	Technological readiness, 2006	67
1.04	State of cluster development, 2006	81
1.05	US utility patents, 2005	79
1.06	High-tech exports	n/a
1.07	Burden of government regulation, 2006	47
1.08	Extent and effect of taxation, 2006	16
1.09	Time required to start a business, 2006	111
1.10	No. of procedures required to start a business, 2006	79
1.11	Intensity of local competition, 2006	75
1.12	Freedom of the press, 2006	73
Political and regulatory environment		**45**
2.01	Effectiveness of law-making bodies, 2006	31
2.02	Laws relating to ICT, 2006	84
2.03	Judicial independence, 2006	25
2.04	Intellectual property protection, 2006	76
2.05	Efficiency of legal framework, 2006	31
2.06	Property rights, 2006	52
2.07	Quality of competition in the ISP sector, 2006	86
2.08	No. of procedures to enforce a contract, 2006	38
2.09	Time to enforce a contract, 2006	72
Infrastructure environment		**93**
3.01	Telephone lines, 2005	88
3.02	Secure Internet servers, 2004	91
3.03	Internet hosts, 2004	73
3.04	Electricity production	n/a
3.05	Availability of scientists and engineers, 2006	108
3.06	Quality of scientific research institutions, 2006	74
3.07	Tertiary enrollment, 2004	97

Readiness component		63
Individual readiness		**56**
4.01	Quality of math and science education, 2006	72
4.02	Quality of the educational system, 2006	56
4.03	Quality of public schools, 2006	49
4.04	Internet access in schools, 2006	83
4.05	Buyer sophistication, 2006	77
4.06	Residential telephone connection charge, 2005	44
4.07	Residential monthly telephone subscription, 2005	38
4.08	High-speed monthly broadband subscription, 2006	50
4.09	Lowest cost of broadband, 2006	61
4.10	Cost of mobile telephone call, 2005	58
Business readiness		**61**
5.01	Extent of staff training, 2006	68
5.02	Local availability of research and training, 2006	101
5.03	Quality of management schools, 2006	93
5.04	Company spending on R&D, 2006	87
5.05	University-industry research collaboration, 2006	86
5.06	Business telephone connection charge, 2005	45
5.07	Business monthly telephone subscription, 2005	39
5.08	Local supplier quality, 2006	92
5.09	Computer, comm., and other services imports	n/a
Government readiness		**83**
6.01	Government prioritization of ICT, 2006	65
6.02	Gov't. procurement of advanced tech products, 2006	63
6.03	Importance of ICT to gov't. vision of the future, 2006	66
6.04	E-participation index, 2005	82
6.05	E-government readiness index, 2005	77

153

Usage component		81
Individual usage		**64**
7.01	Mobile telephone subscribers, 2005	63
7.02	Personal computers, 2004	76
7.03	Broadband Internet subscribers	n/a
7.04	Internet users, 2005	99
7.05	Internet bandwidth	n/a
Business usage		**94**
8.01	Prevalence of foreign technology licensing, 2006	68
8.02	Firm-level technology absorption, 2006	91
8.03	Capacity for innovation, 2006	109
8.04	Availability of new telephone lines, 2006	92
8.05	Availability of mobile telephones, 2006	94
8.06	Extent of business Internet use, 2006	108
Government usage		**81**
9.01	Government success in ICT promotion, 2006	60
9.02	Availability of online services, 2006	83
9.03	ICT use and government efficiency, 2006	96
9.04	ICT pervasiveness, 2006	67

Brazil

Key indicators

Population (millions), 2005..186.4
GDP (PPP) per capita (US$), 2005..8,560.6
Internet users per 100 inhabitants, 2005....................................19.5
Internet bandwidth (Mbps/10,000 inhabitants), 2004................1.5

Networked Readiness Index

Year (number of economies)	Rank
2006–2007 (122) **53**
2005–2006 (115)..52
2004–2005 (104)..46

Global Competitiveness Index 2006–2007 (125)	66

Environment component	83
Market environment	**109**

1.01 Venture capital availability, 2006..95
1.02 Financial market sophistication, 2006...................................28
1.03 Technological readiness, 2006..58
1.04 State of cluster development, 2006.......................................34
1.05 US utility patents, 2005..50
1.06 High-tech exports, 2004...31
1.07 Burden of government regulation, 2006............................121
1.08 Extent and effect of taxation, 2006...................................122
1.09 Time required to start a business, 2006............................115
1.10 No. of procedures required to start a business, 2006.......113
1.11 Intensity of local competition, 2006.....................................40
1.12 Freedom of the press, 2006..35

Political and regulatory environment	**73**

2.01 Effectiveness of law-making bodies, 2006.......................109
2.02 Laws relating to ICT, 2006..48
2.03 Judicial independence, 2006...89
2.04 Intellectual property protection, 2006..................................63
2.05 Efficiency of legal framework, 2006......................................87
2.06 Property rights, 2006..62
2.07 Quality of competition in the ISP sector, 2006....................23
2.08 No. of procedures to enforce a contract, 2006....................97
2.09 Time to enforce a contract, 2006..91

Infrastructure environment	**58**

3.01 Telephone lines, 2004...55
3.02 Secure Internet servers, 2005..55
3.03 Internet hosts, 2004..38
3.04 Electricity production, 2003...62
3.05 Availability of scientists and engineers, 2006......................61
3.06 Quality of scientific research institutions, 2006..................36
3.07 Tertiary enrollment, 2003..73

Readiness component	50
Individual readiness	**72**

4.01 Quality of math and science education, 2006.....................98
4.02 Quality of the educational system, 2006...........................112
4.03 Quality of public schools, 2006..111
4.04 Internet access in schools, 2006..66
4.05 Buyer sophistication, 2006..58
4.06 Residential telephone connection charge, 2005.................37
4.07 Residential monthly telephone subscription, 2005.............86
4.08 High-speed monthly broadband subscription, 2006...........56
4.09 Lowest cost of broadband, 2006...33
4.10 Cost of mobile telephone call, 2005.....................................70

Business readiness	**42**

5.01 Extent of staff training, 2006..38
5.02 Local availability of research and training, 2006.................32
5.03 Quality of management schools, 2006..................................64
5.04 Company spending on R&D, 2006..30
5.05 University-industry research collaboration, 2006.................42
5.06 Business telephone connection charge, 2005.....................33
5.07 Business monthly telephone subscription, 2005................83
5.08 Local supplier quality, 2006..37
5.09 Computer, comm., and other services imports, 2004.........11

Government readiness	**35**

6.01 Government prioritization of ICT, 2006................................82
6.02 Gov't. procurement of advanced tech products, 2006........58
6.03 Importance of ICT to gov't. vision of the future, 2006........73
6.04 E-participation index, 2005..18
6.05 E-government readiness index, 2005....................................33

Usage component	45
Individual usage	**57**

7.01 Mobile telephone subscribers, 2005....................................64
7.02 Personal computers, 2004...54
7.03 Broadband Internet subscribers, 2005.................................50
7.04 Internet users, 2005...53
7.05 Internet bandwidth, 2004...45

Business usage	**32**

8.01 Prevalence of foreign technology licensing, 2006..............39
8.02 Firm-level technology absorption, 2006...............................47
8.03 Capacity for innovation, 2006...29
8.04 Availability of new telephone lines, 2006............................49
8.05 Availability of mobile telephones, 2006...............................53
8.06 Extent of business Internet use, 2006.................................23

Government usage	**30**

9.01 Government success in ICT promotion, 2006.......................57
9.02 Availability of online services, 2006.....................................23
9.03 ICT use and government efficiency, 2006............................19
9.04 ICT pervasiveness, 2006...55

Bulgaria

Key indicators

Population (millions), 2005..7.7
GDP (PPP) per capita (US$), 2005...9,223.3
Internet users per 100 inhabitants, 2005.................................20.6
Internet bandwidth (Mbps/10,000 inhabitants), 2004................0.8

Networked Readiness Index

Year (number of economies)	Rank
2006–2007 (122)	**72**
2005–2006 (115)	64
2004–2005 (104)	73

Global Competitiveness Index 2006–2007 (125)	72

Environment component	69
Market environment	**95**

1.01 Venture capital availability, 2006...54
1.02 Financial market sophistication, 2006.................................106
1.03 Technological readiness, 2006..89
1.04 State of cluster development, 2006.....................................112
1.05 US utility patents, 2005..51
1.06 High-tech exports, 2004...56
1.07 Burden of government regulation, 2006.............................90
1.08 Extent and effect of taxation, 2006......................................96
1.09 Time required to start a business, 2006.............................59
1.10 No. of procedures required to start a business, 2006.........49
1.11 Intensity of local competition, 2006.....................................98
1.12 Freedom of the press, 2006..83

Political and regulatory environment	**84**

2.01 Effectiveness of law-making bodies, 2006.........................98
2.02 Laws relating to ICT, 2006...36
2.03 Judicial independence, 2006..97
2.04 Intellectual property protection, 2006...................................96
2.05 Efficiency of legal framework, 2006....................................110
2.06 Property rights, 2006...89
2.07 Quality of competition in the ISP sector, 2006..................57
2.08 No. of procedures to enforce a contract, 2006..................69
2.09 Time to enforce a contract, 2006..59

Infrastructure environment	**44**

3.01 Telephone lines, 2005...37
3.02 Secure Internet servers, 2005..58
3.03 Internet hosts, 2004..44
3.04 Electricity production, 2003..37
3.05 Availability of scientists and engineers, 2006....................49
3.06 Quality of scientific research institutions, 2006.................67
3.07 Tertiary enrollment, 2004..41

Readiness component	68
Individual readiness	**59**

4.01 Quality of math and science education, 2006...................51
4.02 Quality of the educational system, 2006.............................82
4.03 Quality of public schools, 2006...52
4.04 Internet access in schools, 2006...61
4.05 Buyer sophistication, 2006..87
4.06 Residential telephone connection charge, 2005................59
4.07 Residential monthly telephone subscription, 2005............68
4.08 High-speed monthly broadband subscription, 2006...........54
4.09 Lowest cost of broadband, 2006..50
4.10 Cost of mobile telephone call, 2005.....................................71

Business readiness	**78**

5.01 Extent of staff training, 2006...111
5.02 Local availability of research and training, 2006................80
5.03 Quality of management schools, 2006.................................82
5.04 Company spending on R&D, 2006..96
5.05 University-industry research collaboration, 2006...............95
5.06 Business telephone connection charge, 2005....................53
5.07 Business monthly telephone subscription, 2005...............62
5.08 Local supplier quality, 2006..79
5.09 Computer, comm., and other services imports, 2004........73

Government readiness	**66**

6.01 Government prioritization of ICT, 2006...............................104
6.02 Gov't. procurement of advanced tech products, 2006......102
6.03 Importance of ICT to gov't. vision of the future, 2006........85
6.04 E-participation index, 2005...41
6.05 E-government readiness index, 2005....................................45

Usage component	76
Individual usage	**53**

7.01 Mobile telephone subscribers, 2005....................................38
7.02 Personal computers, 2004..66
7.03 Broadband Internet subscribers, 2004.................................88
7.04 Internet users, 2005..51
7.05 Internet bandwidth, 2004...55

Business usage	**93**

8.01 Prevalence of foreign technology licensing, 2006.............99
8.02 Firm-level technology absorption, 2006.............................114
8.03 Capacity for innovation, 2006...79
8.04 Availability of new telephone lines, 2006............................80
8.05 Availability of mobile telephones, 2006...............................85
8.06 Extent of business Internet use, 2006.................................85

Government usage	**77**

9.01 Government success in ICT promotion, 2006...................102
9.02 Availability of online services, 2006......................................69
9.03 ICT use and government efficiency, 2006..........................101
9.04 ICT pervasiveness, 2006...36

Burkina Faso

Key indicators

Population (millions), 2005..13.2
GDP (PPP) per capita (US$), 2005....................1,284.6
Internet users per 100 inhabitants, 2005.....................0.5
Internet bandwidth (Mbps/10,000 inhabitants), 2004.................0.0

Networked Readiness Index

Year (number of economies)	Rank
2006–2007 (122) **99**	
2005–2006 (115) ....—	
2004–2005 (104) ....—	

Global Competitiveness Index 2006–2007 (125)	116

Environment component	96
Market environment	**93**
1.01 Venture capital availability, 2006 108	
1.02 Financial market sophistication, 2006....................96	
1.03 Technological readiness, 2006...............................103	
1.04 State of cluster development, 2006 115	
1.05 US utility patents, 2005 79	
1.06 High-tech exports, 2004 67	
1.07 Burden of government regulation, 2006 25	
1.08 Extent and effect of taxation, 2006.......................56	
1.09 Time required to start a business, 2006.................63	
1.10 No. of procedures required to start a business, 2006.........36	
1.11 Intensity of local competition, 2006 86	
1.12 Freedom of the press, 2006...................................101	
Political and regulatory environment	**75**
2.01 Effectiveness of law-making bodies, 2006.............45	
2.02 Laws relating to ICT, 2006....................................100	
2.03 Judicial independence, 2006 83	
2.04 Intellectual property protection, 2006....................50	
2.05 Efficiency of legal framework, 2006......................82	
2.06 Property rights, 2006...68	
2.07 Quality of competition in the ISP sector, 2006..................94	
2.08 No. of procedures to enforce a contract, 2006.................91	
2.09 Time to enforce a contract, 2006 60	
Infrastructure environment	**104**
3.01 Telephone lines, 2005...112	
3.02 Secure Internet servers, 2005..............................109	
3.03 Internet hosts, 2004 111	
3.04 Electricity production ....n/a	
3.05 Availability of scientists and engineers, 2006..................103	
3.06 Quality of scientific research institutions, 2006 76	
3.07 Tertiary enrollment, 2004.....................................115	

Readiness component	111
Individual readiness	**117**
4.01 Quality of math and science education, 2006 76	
4.02 Quality of the educational system, 2006...........................113	
4.03 Quality of public schools, 2006 64	
4.04 Internet access in schools, 2006..........................119	
4.05 Buyer sophistication, 2006 117	
4.06 Residential telephone connection charge, 2005 103	
4.07 Residential monthly telephone subscription, 2005 108	
4.08 High-speed monthly broadband subscription, 2006...........100	
4.09 Lowest cost of broadband, 2006.........................101	
4.10 Cost of mobile telephone call, 2005....................100	
Business readiness	**99**
5.01 Extent of staff training, 2006...............................105	
5.02 Local availability of research and training, 2006 91	
5.03 Quality of management schools, 2006.................74	
5.04 Company spending on R&D, 2006 55	
5.05 University-industry research collaboration, 2006................75	
5.06 Business telephone connection charge, 2005 98	
5.07 Business monthly telephone subscription, 2005 103	
5.08 Local supplier quality, 2006 95	
5.09 Computer, comm., and other services imports ....n/a	
Government readiness	**96**
6.01 Government prioritization of ICT, 2006..................64	
6.02 Gov't. procurement of advanced tech products, 2006........38	
6.03 Importance of ICT to gov't. vision of the future, 2006........23	
6.04 E-participation index, 2005 101	
6.05 E-government readiness index, 2005...................119	

Usage component	85
Individual usage	**116**
7.01 Mobile telephone subscribers, 2005 116	
7.02 Personal computers, 2004.....................................116	
7.03 Broadband Internet subscribers, 2005 92	
7.04 Internet users, 2005 116	
7.05 Internet bandwidth, 2004 80	
Business usage	**90**
8.01 Prevalence of foreign technology licensing, 2006..............84	
8.02 Firm-level technology absorption, 2006...............71	
8.03 Capacity for innovation, 2006 87	
8.04 Availability of new telephone lines, 2006 93	
8.05 Availability of mobile telephones, 2006...............100	
8.06 Extent of business Internet use, 2006 92	
Government usage	**52**
9.01 Government success in ICT promotion, 2006....................13	
9.02 Availability of online services, 2006.....................85	
9.03 ICT use and government efficiency, 2006...........60	
9.04 ICT pervasiveness, 2006 54	

Burundi

Key indicators

Population (millions), 2005..7.5

GDP (PPP) per capita (US$), 2005..739.3

Internet users per 100 inhabitants, 2005......................................0.5

Internet bandwidth (Mbps/10,000 inhabitants) ....n/a

Networked Readiness Index

Year (number of economies)	Rank
2006–2007 (122) **121**	
2005–2006 (115) ....—	
2004–2005 (104) ....—	

Global Competitiveness Index 2006–2007 (125)   124

Environment component   120

Market environment   117

1.01	Venture capital availability, 2006 119	
1.02	Financial market sophistication, 2006..................................114	
1.03	Technological readiness, 2006..122	
1.04	State of cluster development, 2006 122	
1.05	US utility patents, 2005 79	
1.06	High-tech exports, 2004 83	
1.07	Burden of government regulation, 2006 88	
1.08	Extent and effect of taxation, 2006.....................................95	
1.09	Time required to start a business, 2006 79	
1.10	No. of procedures required to start a business, 2006.........79	
1.11	Intensity of local competition, 2006 116	
1.12	Freedom of the press, 2006...100	

Political and regulatory environment   117

2.01	Effectiveness of law-making bodies, 2006.........................92	
2.02	Laws relating to ICT, 2006...120	
2.03	Judicial independence, 2006 117	
2.04	Intellectual property protection, 2006................................120	
2.05	Efficiency of legal framework, 2006...................................111	
2.06	Property rights, 2006 117	
2.07	Quality of competition in the ISP sector, 2006..................113	
2.08	No. of procedures to enforce a contract, 2006 104	
2.09	Time to enforce a contract, 2006 49	

Infrastructure environment   116

3.01	Telephone lines, 2004...117	
3.02	Secure Internet servers, 2005 111	
3.03	Internet hosts, 2004 115	
3.04	Electricity production ....n/a	
3.05	Availability of scientists and engineers, 2006......................99	
3.06	Quality of scientific research institutions, 2006 119	
3.07	Tertiary enrollment, 2004..112	

Readiness component   121

Individual readiness   115

4.01	Quality of math and science education, 2006 74	
4.02	Quality of the educational system, 2006...........................104	
4.03	Quality of public schools, 2006 89	
4.04	Internet access in schools, 2006.......................................122	
4.05	Buyer sophistication, 2006 121	
4.06	Residential telephone connection charge, 2005 99	
4.07	Residential monthly telephone subscription, 2005 87	
4.08	High-speed monthly broadband subscription charge ....n/a	
4.09	Lowest cost of broadband, 2006..111	
4.10	Cost of mobile telephone call, 2005..................................107	

Business readiness   116

5.01	Extent of staff training, 2006...120	
5.02	Local availability of research and training, 2006 121	
5.03	Quality of management schools, 2006...............................115	
5.04	Company spending on R&D, 2006 115	
5.05	University-industry research collaboration, 2006...............115	
5.06	Business telephone connection charge, 2005 110	
5.07	Business monthly telephone subscription, 2005 70	
5.08	Local supplier quality, 2006 114	
5.09	Computer, comm., and other services imports ....n/a	

Government readiness   122

6.01	Government prioritization of ICT, 2006..............................122	
6.02	Gov't. procurement of advanced tech products, 2006......115	
6.03	Importance of ICT to gov't. vision of the future, 2006......120	
6.04	E-participation index, 2005 109	
6.05	E-government readiness index, 2005.................................116	

Usage component   119

Individual usage   117

7.01	Mobile telephone subscribers, 2005 120	
7.02	Personal computers, 2004..109	
7.03	Broadband Internet subscribers...n/a	
7.04	Internet users, 2005 114	
7.05	Internet bandwidth..n/a	

Business usage   121

8.01	Prevalence of foreign technology licensing, 2006.............120	
8.02	Firm-level technology absorption, 2006.............................105	
8.03	Capacity for innovation, 2006 120	
8.04	Availability of new telephone lines, 2006 105	
8.05	Availability of mobile telephones, 2006.............................121	
8.06	Extent of business Internet use, 2006 120	

Government usage   107

9.01	Government success in ICT promotion, 2006.....................80	
9.02	Availability of online services, 2006...................................109	
9.03	ICT use and government efficiency, 2006..........................111	
9.04	ICT pervasiveness, 2006 108	

Cambodia

Key indicators

Population (millions), 2005..14.1
GDP (PPP) per capita (US$), 2005.............................2,399.2
Internet users per 100 inhabitants, 2004.............................0.3
Internet bandwidth (Mbps/10,000 inhabitants)..........................n/a

Networked Readiness Index

Year (number of economies)	Rank
2006–2007 (122) **106**	
2005–2006 (115)...104	
2004–2005 (104) ....—	

Global Competitiveness Index 2006–2007 (125)	103

Environment component — 106

Market environment — 108

1.01 Venture capital availability, 2006..81
1.02 Financial market sophistication, 2006.................................115
1.03 Technological readiness, 2006...88
1.04 State of cluster development, 2006.......................................58
1.05 US utility patents, 2005 79
1.06 High-tech exports, 2004 86
1.07 Burden of government regulation, 2006 78
1.08 Extent and effect of taxation, 2006......................................28
1.09 Time required to start a business, 2006............................107
1.10 No. of procedures required to start a business, 2006.........61
1.11 Intensity of local competition, 2006 78
1.12 Freedom of the press, 2006...113

Political and regulatory environment — 92

2.01 Effectiveness of law-making bodies, 2006..........................58
2.02 Laws relating to ICT, 2006..110
2.03 Judicial independence, 2006 106
2.04 Intellectual property protection, 2006...............................105
2.05 Efficiency of legal framework, 2006.....................................92
2.06 Property rights, 2006..94
2.07 Quality of competition in the ISP sector, 2006.................81
2.08 No. of procedures to enforce a contract, 2006 60
2.09 Time to enforce a contract, 2006 48

Infrastructure environment — 120

3.01 Telephone lines, 2003...121
3.02 Secure Internet servers, 2005...116
3.03 Internet hosts, 2004 105
3.04 Electricity production ....n/a
3.05 Availability of scientists and engineers, 2006...................121
3.06 Quality of scientific research institutions, 2006 105
3.07 Tertiary enrollment, 2004...108

Readiness component — 103

Individual readiness — 105

4.01 Quality of math and science education, 2006...................112
4.02 Quality of the educational system, 2006.............................80
4.03 Quality of public schools, 2006 95
4.04 Internet access in schools, 2006...98
4.05 Buyer sophistication, 2006 74
4.06 Residential telephone connection charge...........................n/a
4.07 Residential monthly telephone subscription.......................n/a
4.08 High-speed monthly broadband subscription, 2006............97
4.09 Lowest cost of broadband, 2006...103
4.10 Cost of mobile telephone call, 2004....................................79

Business readiness — 102

5.01 Extent of staff training, 2006..88
5.02 Local availability of research and training, 2006.................95
5.03 Quality of management schools, 2006.............................104
5.04 Company spending on R&D, 2006 47
5.05 University-industry research collaboration, 2006................88
5.06 Business telephone connection charge.............................n/a
5.07 Business monthly telephone subscription.........................n/a
5.08 Local supplier quality, 2006 96
5.09 Computer, comm., and other services imports, 2004.........66

Government readiness — 73

6.01 Government prioritization of ICT, 2006...............................74
6.02 Gov't. procurement of advanced tech products, 2006........29
6.03 Importance of ICT to gov't. vision of the future, 2006........56
6.04 E-participation index, 2005 51
6.05 E-government readiness index, 2005...............................102

Usage component — 105

Individual usage — 111

7.01 Mobile telephone subscribers, 2005 109
7.02 Personal computers, 2004...115
7.03 Broadband Internet subscribers...n/a
7.04 Internet users, 2004 120
7.05 Internet bandwidth...n/a

Business usage — 104

8.01 Prevalence of foreign technology licensing, 2006............109
8.02 Firm-level technology absorption, 2006..............................79
8.03 Capacity for innovation, 2006 114
8.04 Availability of new telephone lines, 2006 100
8.05 Availability of mobile telephones, 2006............................109
8.06 Extent of business Internet use, 2006 93

Government usage — 95

9.01 Government success in ICT promotion, 2006...................89
9.02 Availability of online services, 2006...................................113
9.03 ICT use and government efficiency, 2006..........................37
9.04 ICT pervasiveness, 2006 105

Cameroon

Key indicators

Population (millions), 2005...16.3
GDP (PPP) per capita (US$), 2005..2,421.1
Internet users per 100 inhabitants, 2005......................................1.5
Internet bandwidth (Mbps/10,000 inhabitants) ....n/a

Networked Readiness Index

Year (number of economies)	Rank
2006–2007 (122)	**113**
2005–2006 (115)...	99
2004–2005 (104)	—

Global Competitiveness Index 2006–2007 (125)	108

Environment component	118
Market environment	**115**
1.01 Venture capital availability, 2006 115
1.02 Financial market sophistication, 2006.................................121
1.03 Technological readiness, 2006...107
1.04 State of cluster development, 2006 114
1.05 US utility patents, 2005 79
1.06 High-tech exports, 2004 91
1.07 Burden of government regulation, 2006 118
1.08 Extent and effect of taxation, 2006 111
1.09 Time required to start a business, 2006 71
1.10 No. of procedures required to start a business, 2006.........89
1.11 Intensity of local competition, 2006 102
1.12 Freedom of the press, 2006..81

Political and regulatory environment	**119**
2.01 Effectiveness of law-making bodies, 2006......................111
2.02 Laws relating to ICT, 2006...105
2.03 Judicial independence, 2006 109
2.04 Intellectual property protection, 2006................................67
2.05 Efficiency of legal framework, 2006..................................102
2.06 Property rights, 2006 103
2.07 Quality of competition in the ISP sector, 2006...............95
2.08 No. of procedures to enforce a contract, 2006 116
2.09 Time to enforce a contract, 2006 102

Infrastructure environment	**112**
3.01 Telephone lines, 2004..115
3.02 Secure Internet servers, 2005 112
3.03 Internet hosts, 2004 114
3.04 Electricity production, 2003 97
3.05 Availability of scientists and engineers, 2006......................72
3.06 Quality of scientific research institutions, 2006 107
3.07 Tertiary enrollment, 2004 100

Readiness component	104
Individual readiness	**102**
4.01 Quality of math and science education, 2006 80
4.02 Quality of the educational system, 2006............................87
4.03 Quality of public schools, 2006 83
4.04 Internet access in schools, 2006...116
4.05 Buyer sophistication, 2006 118
4.06 Residential telephone connection charge, 2005 90
4.07 Residential monthly telephone subscription, 2005 85
4.08 High-speed monthly broadband subscription, 2006 95
4.09 Lowest cost of broadband, 2006..97
4.10 Cost of mobile telephone call, 2004...................................97

Business readiness	**92**
5.01 Extent of staff training, 2006...110
5.02 Local availability of research and training, 2006 97
5.03 Quality of management schools, 2006..............................91
5.04 Company spending on R&D, 2006 103
5.05 University-industry research collaboration, 2006..............116
5.06 Business telephone connection charge, 2005 80
5.07 Business monthly telephone subscription, 2005 67
5.08 Local supplier quality, 2006 98
5.09 Computer, comm., and other services imports ....n/a

Government readiness	**115**
6.01 Government prioritization of ICT, 2006.............................114
6.02 Gov't. procurement of advanced tech products, 2006........96
6.03 Importance of ICT to gov't. vision of the future, 2006......109
6.04 E-participation index, 2005 92
6.05 E-government readiness index, 2005................................108

Usage component	107
Individual usage	**98**
7.01 Mobile telephone subscribers, 2005 97
7.02 Personal computers, 2004..103
7.03 Broadband Internet subscribers....................................n/a
7.04 Internet users, 2005..107
7.05 Internet bandwidth..n/a

Business usage	**109**
8.01 Prevalence of foreign technology licensing, 2006............103
8.02 Firm-level technology absorption, 2006...............................95
8.03 Capacity for innovation, 2006 110
8.04 Availability of new telephone lines, 2006 112
8.05 Availability of mobile telephones, 2006 97
8.06 Extent of business Internet use, 2006 119

Government usage	**105**
9.01 Government success in ICT promotion, 2006....................62
9.02 Availability of online services, 2006....................................114
9.03 ICT use and government efficiency, 2006.........................107
9.04 ICT pervasiveness, 2006 106

159

Canada

Key indicators

Population (millions), 2005..32.3

GDP (PPP) per capita (US$), 2005.......................................34,273.0

Internet users per 100 inhabitants, 2005.....................................52.1

Internet bandwidth (Mbps/10,000 inhabitants), 2004...............67.8

Networked Readiness Index

Year (number of economies)	Rank
2006–2007 (122) **11**	
2005–2006 (115)...6	
2004–2005 (104)...10	

Global Competitiveness Index 2006–2007 (125)	16

Environment component | 9

Market environment | 13

1.01 Venture capital availability, 2006...22
1.02 Financial market sophistication, 2006.......................................9
1.03 Technological readiness, 2006...12
1.04 State of cluster development, 2006..25
1.05 US utility patents, 2005..10
1.06 High-tech exports, 2004...26
1.07 Burden of government regulation, 2006.................................37
1.08 Extent and effect of taxation, 2006..65
1.09 Time required to start a business, 2006....................................2
1.10 No. of procedures required to start a business, 2006............1
1.11 Intensity of local competition, 2006.......................................10
1.12 Freedom of the press, 2006..10

Political and regulatory environment | 14

2.01 Effectiveness of law-making bodies, 2006.............................14
2.02 Laws relating to ICT, 2006..16
2.03 Judicial independence, 2006...18
2.04 Intellectual property protection, 2006....................................16
2.05 Efficiency of legal framework, 2006.......................................17
2.06 Property rights, 2006..20
2.07 Quality of competition in the ISP sector, 2006.....................10
2.08 No. of procedures to enforce a contract, 2006.......................5
2.09 Time to enforce a contract, 2006...38

Infrastructure environment | 8

3.01 Telephone lines, 2005..11
3.02 Secure Internet servers, 2005..3
3.03 Internet hosts, 2004...17
3.04 Electricity production, 2003..3
3.05 Availability of scientists and engineers, 2006.........................9
3.06 Quality of scientific research institutions, 2006...................11
3.07 Tertiary enrollment, 2002...23

Readiness component | 13

Individual readiness | 8

4.01 Quality of math and science education, 2006.......................22
4.02 Quality of the educational system, 2006................................14
4.03 Quality of public schools, 2006..9
4.04 Internet access in schools, 2006..12
4.05 Buyer sophistication, 2006..18
4.06 Residential telephone connection charge, 2005.....................8
4.07 Residential monthly telephone subscription, 2005...............21
4.08 High-speed monthly broadband subscription, 2006.............10
4.09 Lowest cost of broadband, 2006..14
4.10 Cost of mobile telephone call, 2005......................................14

Business readiness | 20

5.01 Extent of staff training, 2006...24
5.02 Local availability of research and training, 2006...................13
5.03 Quality of management schools, 2006.....................................4
5.04 Company spending on R&D, 2006..22
5.05 University-industry research collaboration, 2006..................14
5.06 Business telephone connection charge............................n/a
5.07 Business monthly telephone subscription, 2005...................33
5.08 Local supplier quality, 2006..13
5.09 Computer, comm., and other services imports, 2004.........28

Government readiness | 7

6.01 Government prioritization of ICT, 2006..................................30
6.02 Gov't. procurement of advanced tech products, 2006........36
6.03 Importance of ICT to gov't. vision of the future, 2006........55
6.04 E-participation index, 2005...4
6.05 E-government readiness index, 2005.......................................8

Usage component | 17

Individual usage | 13

7.01 Mobile telephone subscribers, 2005......................................59
7.02 Personal computers, 2004...5
7.03 Broadband Internet subscribers, 2005...................................10
7.04 Internet users, 2005..17
7.05 Internet bandwidth, 2004..10

Business usage | 16

8.01 Prevalence of foreign technology licensing, 2006..................9
8.02 Firm-level technology absorption, 2006.................................22
8.03 Capacity for innovation, 2006...19
8.04 Availability of new telephone lines, 2006...............................16
8.05 Availability of mobile telephones, 2006.................................40
8.06 Extent of business Internet use, 2006......................................9

Government usage | 24

9.01 Government success in ICT promotion, 2006........................38
9.02 Availability of online services, 2006.......................................14
9.03 ICT use and government efficiency, 2006...............................28
9.04 ICT pervasiveness, 2006...28

Chad

Key indicators

Population (millions), 2005...9.7
GDP (PPP) per capita (US$), 2005.......................................1,518.9
Internet users per 100 inhabitants, 2005....................................0.4
Internet bandwidth (Mbps/10,000 inhabitants), 2004.................0.0

Networked Readiness Index

Year (number of economies)	Rank
2006–2007 (122) **122**	
2005–2006 (115)..114	
2004–2005 (104)..104	

Global Competitiveness Index 2006–2007 (125)	123

Environment component	122
Market environment	**122**

1.01 Venture capital availability, 2006...99
1.02 Financial market sophistication, 2006..................................122
1.03 Technological readiness, 2006...121
1.04 State of cluster development, 2006.....................................118
1.05 US utility patents, 2005...79
1.06 High-tech exports...n/a
1.07 Burden of government regulation, 2006.............................107
1.08 Extent and effect of taxation, 2006....................................109
1.09 Time required to start a business, 2006..............................104
1.10 No. of procedures required to start a business, 2006........116
1.11 Intensity of local competition, 2006....................................119
1.12 Freedom of the press, 2006...119

Political and regulatory environment	**122**

2.01 Effectiveness of law-making bodies, 2006.........................117
2.02 Laws relating to ICT, 2006..121
2.03 Judicial independence, 2006...118
2.04 Intellectual property protection, 2006................................117
2.05 Efficiency of legal framework, 2006....................................119
2.06 Property rights, 2006...120
2.07 Quality of competition in the ISP sector, 2006...................120
2.08 No. of procedures to enforce a contract, 2006..................111
2.09 Time to enforce a contract, 2006.......................................101

Infrastructure environment	**117**

3.01 Telephone lines, 2004...122
3.02 Secure Internet servers..n/a
3.03 Internet hosts, 2004..120
3.04 Electricity production...n/a
3.05 Availability of scientists and engineers, 2006....................113
3.06 Quality of scientific research institutions, 2006.................118
3.07 Tertiary enrollment, 2004...116

Readiness component	122
Individual readiness	**122**

4.01 Quality of math and science education, 2006....................120
4.02 Quality of the educational system, 2006............................119
4.03 Quality of public schools, 2006..113
4.04 Internet access in schools, 2006..121
4.05 Buyer sophistication, 2006..122
4.06 Residential telephone connection charge, 2005...............106
4.07 Residential monthly telephone subscription, 2005...........107
4.08 High-speed monthly broadband subscription charge.........n/a
4.09 Lowest cost of broadband...n/a
4.10 Cost of mobile telephone call...n/a

Business readiness	**121**

5.01 Extent of staff training, 2006..121
5.02 Local availability of research and training, 2006................114
5.03 Quality of management schools, 2006...............................120
5.04 Company spending on R&D, 2006......................................117
5.05 University-industry research collaboration, 2006...............120
5.06 Business telephone connection charge, 2005...................102
5.07 Business monthly telephone subscription, 2005..............101
5.08 Local supplier quality, 2006..118
5.09 Computer, comm., and other services imports..................n/a

Government readiness	**121**

6.01 Government prioritization of ICT, 2006...............................121
6.02 Gov't. procurement of advanced tech products, 2006......105
6.03 Importance of ICT to gov't. vision of the future, 2006......119
6.04 E-participation index, 2005...109
6.05 E-government readiness index, 2005.................................117

Usage component	122
Individual usage	**120**

7.01 Mobile telephone subscribers, 2005..................................119
7.02 Personal computers, 2004...118
7.03 Broadband Internet subscribers..n/a
7.04 Internet users, 2005..118
7.05 Internet bandwidth, 2004..89

Business usage	**122**

8.01 Prevalence of foreign technology licensing, 2006.............122
8.02 Firm-level technology absorption, 2006.............................117
8.03 Capacity for innovation, 2006...108
8.04 Availability of new telephone lines, 2006...........................117
8.05 Availability of mobile telephones, 2006.............................113
8.06 Extent of business Internet use, 2006...............................122

Government usage	**115**

9.01 Government success in ICT promotion, 2006....................119
9.02 Availability of online services, 2006.....................................98
9.03 ICT use and government efficiency, 2006..........................112
9.04 ICT pervasiveness, 2006...111

Chile

Key indicators

Population (millions), 2005..16.3
GDP (PPP) per capita (US$), 2005.....................11,936.8
Internet users per 100 inhabitants, 2005....................18.0
Internet bandwidth (Mbps/10,000 inhabitants), 2004...............8.2

Networked Readiness Index

Year (number of economies)	Rank
2006–2007 (122) **31**	
2005–2006 (115)..29	
2004–2005 (104)..35	

Global Competitiveness Index 2006–2007 (125)	27

Environment component — 34

Market environment — 30

1.01 Venture capital availability, 2006.................................32
1.02 Financial market sophistication, 2006......................24
1.03 Technological readiness, 2006..................................26
1.04 State of cluster development, 2006 49
1.05 US utility patents, 2005 49
1.06 High-tech exports, 2004 72
1.07 Burden of government regulation, 2006 20
1.08 Extent and effect of taxation, 2006...........................30
1.09 Time required to start a business, 2006....................46
1.10 No. of procedures required to start a business, 2006.........49
1.11 Intensity of local competition, 2006 11
1.12 Freedom of the press, 2006.....................................31

Political and regulatory environment — 33

2.01 Effectiveness of law-making bodies, 2006.......................42
2.02 Laws relating to ICT, 2006..24
2.03 Judicial independence, 2006 56
2.04 Intellectual property protection, 2006........................45
2.05 Efficiency of legal framework, 2006...........................37
2.06 Property rights, 2006 30
2.07 Quality of competition in the ISP sector, 2006...................17
2.08 No. of procedures to enforce a contract, 2006 66
2.09 Time to enforce a contract, 2006 65

Infrastructure environment — 49

3.01 Telephone lines, 2005...59
3.02 Secure Internet servers, 2005 47
3.03 Internet hosts, 2004 41
3.04 Electricity production, 2003 52
3.05 Availability of scientists and engineers, 2006................33
3.06 Quality of scientific research institutions, 2006 48
3.07 Tertiary enrollment, 2004..39

Readiness component — 33

Individual readiness — 60

4.01 Quality of math and science education, 2006...................100
4.02 Quality of the educational system, 2006...........................75
4.03 Quality of public schools, 2006 102
4.04 Internet access in schools, 2006...............................36
4.05 Buyer sophistication, 2006 30
4.06 Residential telephone connection charge, 2005 36
4.07 Residential monthly telephone subscription, 2005 56
4.08 High-speed monthly broadband subscription, 2006............49
4.09 Lowest cost of broadband, 2006...............................42
4.10 Cost of mobile telephone call ....n/a

Business readiness — 32

5.01 Extent of staff training, 2006....................................34
5.02 Local availability of research and training, 2006 31
5.03 Quality of management schools, 2006...........................18
5.04 Company spending on R&D, 2006 48
5.05 University-industry research collaboration, 2006..............37
5.06 Business telephone connection charge, 2005 32
5.07 Business monthly telephone subscription, 2005 40
5.08 Local supplier quality, 2006 23
5.09 Computer, comm., and other services imports, 2004.........59

Government readiness — 18

6.01 Government prioritization of ICT, 2006.............................25
6.02 Gov't. procurement of advanced tech products, 2006........54
6.03 Importance of ICT to gov't. vision of the future, 2006........14
6.04 E-participation index, 2005 12
6.05 E-government readiness index, 2005................................22

Usage component — 33

Individual usage — 46

7.01 Mobile telephone subscribers, 2005 47
7.02 Personal computers, 2004...45
7.03 Broadband Internet subscribers, 2005 35
7.04 Internet users, 2005 56
7.05 Internet bandwidth, 2004 32

Business usage — 29

8.01 Prevalence of foreign technology licensing, 2006..............32
8.02 Firm-level technology absorption, 2006.......................33
8.03 Capacity for innovation, 2006 50
8.04 Availability of new telephone lines, 2006 20
8.05 Availability of mobile telephones, 2006 18
8.06 Extent of business Internet use, 2006 26

Government usage — 11

9.01 Government success in ICT promotion, 2006....................37
9.02 Availability of online services, 2006...............................9
9.03 ICT use and government efficiency, 2006.........................4
9.04 ICT pervasiveness, 2006 15

China

Key indicators

Population (millions), 2005..1315.8
GDP (PPP) per capita (US$), 2005...7,198.4
Internet users per 100 inhabitants, 2005................................8.4
Internet bandwidth (Mbps/10,000 inhabitants), 2004................0.6

Networked Readiness Index

Year (number of economies)	Rank
2006–2007 (122) **59**	
2005–2006 (115)..50	
2004–2005 (104)..41	

Global Competitiveness Index 2006–2007 (125)   54

Environment component   60
Market environment   61
1.01 Venture capital availability, 2006...89
1.02 Financial market sophistication, 2006....................................93
1.03 Technological readiness, 2006..69
1.04 State of cluster development, 2006..33
1.05 US utility patents, 2005..57
1.06 High-tech exports, 2004..9
1.07 Burden of government regulation, 2006..............................34
1.08 Extent and effect of taxation, 2006.......................................45
1.09 Time required to start a business, 2006...............................65
1.10 No. of procedures required to start a business, 2006.........95
1.11 Intensity of local competition, 2006....................................34
1.12 Freedom of the press, 2006...117

Political and regulatory environment   55
2.01 Effectiveness of law-making bodies, 2006...........................43
2.02 Laws relating to ICT, 2006...67
2.03 Judicial independence, 2006..77
2.04 Intellectual property protection, 2006..................................74
2.05 Efficiency of legal framework, 2006......................................74
2.06 Property rights, 2006..80
2.07 Quality of competition in the ISP sector, 2006...................70
2.08 No. of procedures to enforce a contract, 2006...................60
2.09 Time to enforce a contract, 2006...25

Infrastructure environment   74
3.01 Telephone lines, 2005..47
3.02 Secure Internet servers, 2005...97
3.03 Internet hosts, 2004...98
3.04 Electricity production, 2003...73
3.05 Availability of scientists and engineers, 2006.....................86
3.06 Quality of scientific research institutions, 2006.................63
3.07 Tertiary enrollment, 2004..78

Readiness component   58
Individual readiness   61
4.01 Quality of math and science education, 2006....................62
4.02 Quality of the educational system, 2006.............................86
4.03 Quality of public schools, 2006..54
4.04 Internet access in schools, 2006..49
4.05 Buyer sophistication, 2006..62
4.06 Residential telephone connection charge..........................n/a
4.07 Residential monthly telephone subscription, 2005............66
4.08 High-speed monthly broadband subscription, 2006..........45
4.09 Lowest cost of broadband, 2006..49
4.10 Cost of mobile telephone call, 2004....................................57

Business readiness   65
5.01 Extent of staff training, 2006..76
5.02 Local availability of research and training, 2006................46
5.03 Quality of management schools, 2006.................................92
5.04 Company spending on R&D, 2006..39
5.05 University-industry research collaboration, 2006................27
5.06 Business telephone connection charge..............................n/a
5.07 Business monthly telephone subscription, 2005................55
5.08 Local supplier quality, 2006..65
5.09 Computer, comm., and other services imports, 2004.........45

Government readiness   44
6.01 Government prioritization of ICT, 2006...............................90
6.02 Gov't. procurement of advanced tech products, 2006........21
6.03 Importance of ICT to gov't. vision of the future, 2006........36
6.04 E-participation index, 2005...49
6.05 E-government readiness index, 2005...................................56

Usage component   64
Individual usage   80
7.01 Mobile telephone subscribers, 2005....................................82
7.02 Personal computers, 2004..81
7.03 Broadband Internet subscribers, 2005.................................42
7.04 Internet users, 2005...76
7.05 Internet bandwidth, 2004..56

Business usage   71
8.01 Prevalence of foreign technology licensing, 2006..............88
8.02 Firm-level technology absorption, 2006...............................41
8.03 Capacity for innovation, 2006...43
8.04 Availability of new telephone lines, 2006...........................71
8.05 Availability of mobile telephones, 2006............................101
8.06 Extent of business Internet use, 2006.................................75

Government usage   46
9.01 Government success in ICT promotion, 2006....................64
9.02 Availability of online services, 2006....................................47
9.03 ICT use and government efficiency, 2006...........................44
9.04 ICT pervasiveness, 2006..60

Colombia

Key indicators

Population (millions), 2005..45.6
GDP (PPP) per capita (US$), 2005....................................7,326.1
Internet users per 100 inhabitants, 2005....................................10.4
Internet bandwidth (Mbps/10,000 inhabitants), 2004.................1.2

Networked Readiness Index

Year (number of economies)	Rank
2006–2007 (122) **64**
2005–2006 (115)..62
2004–2005 (104)..66

Global Competitiveness Index 2006–2007 (125)	65

Environment component	80
Market environment	**77**

1.01 Venture capital availability, 2006...71
1.02 Financial market sophistication, 2006...................................49
1.03 Technological readiness, 2006...64
1.04 State of cluster development, 2006.......................................67
1.05 US utility patents, 2005 64
1.06 High-tech exports, 2004 58
1.07 Burden of government regulation, 2006 95
1.08 Extent and effect of taxation, 2006.....................................107
1.09 Time required to start a business, 2006................................83
1.10 No. of procedures required to start a business, 2006.........95
1.11 Intensity of local competition, 2006 50
1.12 Freedom of the press, 2006..41

Political and regulatory environment	**79**

2.01 Effectiveness of law-making bodies, 2006...........................74
2.02 Laws relating to ICT, 2006..46
2.03 Judicial independence, 2006 65
2.04 Intellectual property protection, 2006...................................55
2.05 Efficiency of legal framework, 2006......................................59
2.06 Property rights, 2006 58
2.07 Quality of competition in the ISP sector, 2006....................55
2.08 No. of procedures to enforce a contract, 2006 78
2.09 Time to enforce a contract, 2006 114

Infrastructure environment	**73**

3.01 Telephone lines, 2005..67
3.02 Secure Internet servers, 2005 69
3.03 Internet hosts, 2004 54
3.04 Electricity production, 2003 76
3.05 Availability of scientists and engineers, 2006......................60
3.06 Quality of scientific research institutions, 2006 75
3.07 Tertiary enrollment, 2004...66

Readiness component	61
Individual readiness	**81**

4.01 Quality of math and science education, 2006.....................77
4.02 Quality of the educational system, 2006..............................55
4.03 Quality of public schools, 2006 66
4.04 Internet access in schools, 2006..73
4.05 Buyer sophistication, 2006 75
4.06 Residential telephone connection charge, 2004 85
4.07 Residential monthly telephone subscription, 2005 98
4.08 High-speed monthly broadband subscription, 2006............59
4.09 Lowest cost of broadband, 2006..77
4.10 Cost of mobile telephone call ....n/a

Business readiness	**58**

5.01 Extent of staff training, 2006..67
5.02 Local availability of research and training, 2006.................70
5.03 Quality of management schools, 2006..................................38
5.04 Company spending on R&D, 2006 69
5.05 University-industry research collaboration, 2006.................45
5.06 Business telephone connection charge, 2004 83
5.07 Business monthly telephone subscription, 2004 54
5.08 Local supplier quality, 2006 47
5.09 Computer, comm., and other services imports, 2004.........76

Government readiness	**34**

6.01 Government prioritization of ICT, 2006.................................60
6.02 Gov't. procurement of advanced tech products, 2006........60
6.03 Importance of ICT to gov't. vision of the future, 2006........70
6.04 E-participation index, 2005 12
6.05 E-government readiness index, 2005...................................53

Usage component	65
Individual usage	**71**

7.01 Mobile telephone subscribers, 2005 60
7.02 Personal computers, 2004...82
7.03 Broadband Internet subscribers, 2005 59
7.04 Internet users, 2005 70
7.05 Internet bandwidth, 2004 48

Business usage	**65**

8.01 Prevalence of foreign technology licensing, 2006...............75
8.02 Firm-level technology absorption, 2006...............................84
8.03 Capacity for innovation, 2006 55
8.04 Availability of new telephone lines, 2006 46
8.05 Availability of mobile telephones, 2006...............................48
8.06 Extent of business Internet use, 2006 71

Government usage	**66**

9.01 Government success in ICT promotion, 2006......................73
9.02 Availability of online services, 2006.....................................60
9.03 ICT use and government efficiency, 2006............................54
9.04 ICT pervasiveness, 2006 76

Costa Rica

Key indicators

Population (millions), 2005..4.3
GDP (PPP) per capita (US$), 2005......................................10,434.4
Internet users per 100 inhabitants, 2005................................25.4
Internet bandwidth (Mbps/10,000 inhabitants)........................n/a

Networked Readiness Index

Year (number of economies)	Rank
2006–2007 (122) **56**
2005–2006 (115)..69
2004–2005 (104)..61

Global Competitiveness Index 2006–2007 (125)	53

Environment component	53
Market environment	**54**
1.01 Venture capital availability, 2006 78
1.02 Financial market sophistication, 2006.................................55
1.03 Technological readiness, 2006..53
1.04 State of cluster development, 2006 74
1.05 US utility patents, 2005 43
1.06 High-tech exports, 2004 15
1.07 Burden of government regulation, 2006 97
1.08 Extent and effect of taxation, 2006....................................46
1.09 Time required to start a business, 2006............................105
1.10 No. of procedures required to start a business, 2006.........79
1.11 Intensity of local competition, 2006 48
1.12 Freedom of the press, 2006..25
Political and regulatory environment	**56**
2.01 Effectiveness of law-making bodies, 2006.......................108
2.02 Laws relating to ICT, 2006..47
2.03 Judicial independence, 2006 33
2.04 Intellectual property protection, 2006.................................48
2.05 Efficiency of legal framework, 2006....................................32
2.06 Property rights, 2006 59
2.07 Quality of competition in the ISP sector, 2006.................118
2.08 No. of procedures to enforce a contract, 2006 69
2.09 Time to enforce a contract, 2006 89
Infrastructure environment	**52**
3.01 Telephone lines, 2005..38
3.02 Secure Internet servers, 2005..30
3.03 Internet hosts, 2004 60
3.04 Electricity production, 2003 65
3.05 Availability of scientists and engineers, 2006....................37
3.06 Quality of scientific research institutions, 2006 38
3.07 Tertiary enrollment, 2004...69

Readiness component	53
Individual readiness	**53**
4.01 Quality of math and science education, 2006 67
4.02 Quality of the educational system, 2006............................40
4.03 Quality of public schools, 2006 50
4.04 Internet access in schools, 2006..81
4.05 Buyer sophistication, 2006 36
4.06 Residential telephone connection charge, 2004 53
4.07 Residential monthly telephone subscription, 2004 48
4.08 High-speed monthly broadband subscription, 2006...........79
4.09 Lowest cost of broadband, 2006...55
4.10 Cost of mobile telephone call, 2005....................................27
Business readiness	**36**
5.01 Extent of staff training, 2006...31
5.02 Local availability of research and training, 2006 40
5.03 Quality of management schools, 2006................................27
5.04 Company spending on R&D, 2006 33
5.05 University-industry research collaboration, 2006................39
5.06 Business telephone connection charge, 2004 46
5.07 Business monthly telephone subscription, 2004 38
5.08 Local supplier quality, 2006 36
5.09 Computer, comm., and other services imports, 2004.........63
Government readiness	**88**
6.01 Government prioritization of ICT, 2006...............................92
6.02 Gov't. procurement of advanced tech products, 2006........67
6.03 Importance of ICT to gov't. vision of the future, 2006........95
6.04 E-participation index, 2005 82
6.05 E-government readiness index, 2005...................................67

Usage component	69
Individual usage	**48**
7.01 Mobile telephone subscribers, 2005 87
7.02 Personal computers, 2004...33
7.03 Broadband Internet subscribers, 2004 60
7.04 Internet users, 2005 45
7.05 Internet bandwidth...n/a
Business usage	**83**
8.01 Prevalence of foreign technology licensing, 2006..............58
8.02 Firm-level technology absorption, 2006..............................57
8.03 Capacity for innovation, 2006 33
8.04 Availability of new telephone lines, 2006 102
8.05 Availability of mobile telephones, 2006............................119
8.06 Extent of business Internet use, 2006................................63
Government usage	**79**
9.01 Government success in ICT promotion, 2006.....................82
9.02 Availability of online services, 2006.....................................71
9.03 ICT use and government efficiency, 2006...........................90
9.04 ICT pervasiveness, 2006 61

165

Croatia

Key indicators

Population (millions), 2005..4.6
GDP (PPP) per capita (US$), 2005......................12,324.8
Internet users per 100 inhabitants, 2005....................31.9
Internet bandwidth (Mbps/10,000 inhabitants), 2004................3.2

Networked Readiness Index

Year (number of economies)	Rank
2006–2007 (122) **46**	
2005–2006 (115)..57	
2004–2005 (104)..58	

Global Competitiveness Index 2006–2007 (125)	51

Environment component	49
Market environment	**62**

1.01 Venture capital availability, 2006..65
1.02 Financial market sophistication, 2006.................................71
1.03 Technological readiness, 2006...78
1.04 State of cluster development, 2006.....................................13
1.05 US utility patents, 2005...32
1.06 High-tech exports, 2004...34
1.07 Burden of government regulation, 2006............................68
1.08 Extent and effect of taxation, 2006.....................................83
1.09 Time required to start a business, 2006............................85
1.10 No. of procedures required to start a business, 2006........61
1.11 Intensity of local competition, 2006...................................52
1.12 Freedom of the press, 2006...63

Political and regulatory environment	**54**

2.01 Effectiveness of law-making bodies, 2006.........................57
2.02 Laws relating to ICT, 2006...41
2.03 Judicial independence, 2006...80
2.04 Intellectual property protection, 2006................................56
2.05 Efficiency of legal framework, 2006....................................73
2.06 Property rights, 2006...79
2.07 Quality of competition in the ISP sector, 2006..................68
2.08 No. of procedures to enforce a contract, 2006.................23
2.09 Time to enforce a contract, 2006..76

Infrastructure environment	**39**

3.01 Telephone lines, 2005...30
3.02 Secure Internet servers, 2005...37
3.03 Internet hosts, 2004..45
3.04 Electricity production, 2003...54
3.05 Availability of scientists and engineers, 2006...................38
3.06 Quality of scientific research institutions, 2006................46
3.07 Tertiary enrollment, 2003...50

Readiness component	39
Individual readiness	**38**

4.01 Quality of math and science education, 2006....................31
4.02 Quality of the educational system, 2006............................53
4.03 Quality of public schools, 2006..32
4.04 Internet access in schools, 2006..46
4.05 Buyer sophistication, 2006...66
4.06 Residential telephone connection charge, 2005...............54
4.07 Residential monthly telephone subscription, 2005...........55
4.08 High-speed monthly broadband subscription, 2006..........44
4.09 Lowest cost of broadband, 2006...38
4.10 Cost of mobile telephone call, 2004...................................39

Business readiness	**40**

5.01 Extent of staff training, 2006...61
5.02 Local availability of research and training, 2006................34
5.03 Quality of management schools, 2006................................54
5.04 Company spending on R&D, 2006......................................52
5.05 University-industry research collaboration, 2006..............35
5.06 Business telephone connection charge, 2005...................47
5.07 Business monthly telephone subscription, 2005...............42
5.08 Local supplier quality, 2006...62
5.09 Computer, comm., and other services imports, 2004..........7

Government readiness	**50**

6.01 Government prioritization of ICT, 2006...............................49
6.02 Gov't. procurement of advanced tech products, 2006........69
6.03 Importance of ICT to gov't. vision of the future, 2006........71
6.04 E-participation index, 2005..51
6.05 E-government readiness index, 2005..................................47

Usage component	47
Individual usage	**45**

7.01 Mobile telephone subscribers, 2005..................................49
7.02 Personal computers, 2004..37
7.03 Broadband Internet subscribers, 2005...............................47
7.04 Internet users, 2005..35
7.05 Internet bandwidth, 2004...39

Business usage	**48**

8.01 Prevalence of foreign technology licensing, 2006.............30
8.02 Firm-level technology absorption, 2006..............................80
8.03 Capacity for innovation, 2006...53
8.04 Availability of new telephone lines, 2006...........................48
8.05 Availability of mobile telephones, 2006.............................55
8.06 Extent of business Internet use, 2006................................58

Government usage	**51**

9.01 Government success in ICT promotion, 2006.....................61
9.02 Availability of online services, 2006...................................64
9.03 ICT use and government efficiency, 2006...........................80
9.04 ICT pervasiveness, 2006..32

Cyprus

Key indicators

Population (millions), 2005...0.8
GDP (PPP) per capita (US$), 2005.....................................21,176.8
Internet users per 100 inhabitants, 2005....................................39.0
Internet bandwidth (Mbps/10,000 inhabitants), 2004.................3.7

Networked Readiness Index

Year (number of economies)	Rank
2006–2007 (122)	**43**
2005–2006 (115)	33
2004–2005 (104)	37

Global Competitiveness Index 2006–2007 (125)   46

Environment component   32
Market environment   42
1.01 Venture capital availability, 2006............................49
1.02 Financial market sophistication, 2006......................44
1.03 Technological readiness, 2006................................47
1.04 State of cluster development, 2006..........................52
1.05 US utility patents, 2005......................................26
1.06 High-tech exports, 2004......................................14
1.07 Burden of government regulation, 2006....................28
1.08 Extent and effect of taxation, 2006.........................19
1.09 Time required to start a business.........................n/a
1.10 No. of procedures required to start a business........n/a
1.11 Intensity of local competition, 2006.......................32
1.12 Freedom of the press, 2006.................................46

Political and regulatory environment   36
2.01 Effectiveness of law-making bodies, 2006................38
2.02 Laws relating to ICT, 2006...................................61
2.03 Judicial independence, 2006.................................30
2.04 Intellectual property protection, 2006....................37
2.05 Efficiency of legal framework, 2006........................28
2.06 Property rights, 2006...29
2.07 Quality of competition in the ISP sector, 2006..........40
2.08 No. of procedures to enforce a contract................n/a
2.09 Time to enforce a contract..................................n/a

Infrastructure environment   34
3.01 Telephone lines, 2005..16
3.02 Secure Internet servers, 2005..............................20
3.03 Internet hosts, 2004...42
3.04 Electricity production, 2003.................................42
3.05 Availability of scientists and engineers, 2006...........34
3.06 Quality of scientific research institutions, 2006.........77
3.07 Tertiary enrollment, 2004...................................55

Readiness component   49
Individual readiness   31
4.01 Quality of math and science education, 2006............30
4.02 Quality of the educational system, 2006..................26
4.03 Quality of public schools, 2006............................34
4.04 Internet access in schools, 2006...........................42
4.05 Buyer sophistication, 2006..................................39
4.06 Residential telephone connection charge, 2004.........31
4.07 Residential monthly telephone subscription, 2005......42
4.08 High-speed monthly broadband subscription, 2006.....34
4.09 Lowest cost of broadband, 2006...........................34
4.10 Cost of mobile telephone call, 2004........................8

Business readiness   52
5.01 Extent of staff training, 2006...............................70
5.02 Local availability of research and training, 2006.........65
5.03 Quality of management schools, 2006.....................49
5.04 Company spending on R&D, 2006..........................78
5.05 University-industry research collaboration, 2006........73
5.06 Business telephone connection charge, 2005............25
5.07 Business monthly telephone subscription, 2005.........29
5.08 Local supplier quality, 2006.................................49
5.09 Computer, comm., and other services imports, 2004...80

Government readiness   71
6.01 Government prioritization of ICT, 2006....................97
6.02 Gov't. procurement of advanced tech products, 2006...86
6.03 Importance of ICT to gov't. vision of the future, 2006...80
6.04 E-participation index, 2005.................................71
6.05 E-government readiness index, 2005......................37

Usage component   46
Individual usage   38
7.01 Mobile telephone subscribers, 2005.......................34
7.02 Personal computers, 2004...................................28
7.03 Broadband Internet subscribers, 2005....................40
7.04 Internet users, 2005..32
7.05 Internet bandwidth, 2004....................................36

Business usage   53
8.01 Prevalence of foreign technology licensing, 2006.......73
8.02 Firm-level technology absorption, 2006...................69
8.03 Capacity for innovation, 2006...............................81
8.04 Availability of new telephone lines, 2006.................24
8.05 Availability of mobile telephones, 2006...................41
8.06 Extent of business Internet use, 2006.....................47

Government usage   65
9.01 Government success in ICT promotion, 2006.............71
9.02 Availability of online services, 2006........................50
9.03 ICT use and government efficiency, 2006.................49
9.04 ICT pervasiveness, 2006.....................................85

167

Czech Republic

Key indicators

Population (millions), 2005 ....10.2
GDP (PPP) per capita (US$), 2005 ....18,340.5
Internet users per 100 inhabitants, 2005 ....27.0
Internet bandwidth (Mbps/10,000 inhabitants) ....n/a

Networked Readiness Index

Year (number of economies)	Rank
2006–2007 (122)	**34**
2005–2006 (115)	32
2004–2005 (104)	40

Global Competitiveness Index 2006–2007 (125)	29

Environment component — 36

Market environment — 43

1.01	Venture capital availability, 2006	60
1.02	Financial market sophistication, 2006	51
1.03	Technological readiness, 2006	33
1.04	State of cluster development, 2006	23
1.05	US utility patents, 2005	34
1.06	High-tech exports, 2004	24
1.07	Burden of government regulation, 2006	110
1.08	Extent and effect of taxation, 2006	87
1.09	Time required to start a business, 2006	38
1.10	No. of procedures required to start a business, 2006	61
1.11	Intensity of local competition, 2006	23
1.12	Freedom of the press, 2006	29

Political and regulatory environment — 52

2.01	Effectiveness of law-making bodies, 2006	65
2.02	Laws relating to ICT, 2006	44
2.03	Judicial independence, 2006	57
2.04	Intellectual property protection, 2006	52
2.05	Efficiency of legal framework, 2006	71
2.06	Property rights, 2006	64
2.07	Quality of competition in the ISP sector, 2006	39
2.08	No. of procedures to enforce a contract, 2006	16
2.09	Time to enforce a contract, 2006	103

Infrastructure environment — 32

3.01	Telephone lines, 2005	40
3.02	Secure Internet servers, 2005	36
3.03	Internet hosts, 2004	26
3.04	Electricity production, 2003	20
3.05	Availability of scientists and engineers, 2006	7
3.06	Quality of scientific research institutions, 2006	29
3.07	Tertiary enrollment, 2004	38

Readiness component — 32

Individual readiness — 28

4.01	Quality of math and science education, 2006	8
4.02	Quality of the educational system, 2006	30
4.03	Quality of public schools, 2006	11
4.04	Internet access in schools, 2006	27
4.05	Buyer sophistication, 2006	45
4.06	Residential telephone connection charge, 2004	62
4.07	Residential monthly telephone subscription, 2005	51
4.08	High-speed monthly broadband subscription, 2006	39
4.09	Lowest cost of broadband, 2006	29
4.10	Cost of mobile telephone call, 2004	40

Business readiness — 23

5.01	Extent of staff training, 2006	29
5.02	Local availability of research and training, 2006	20
5.03	Quality of management schools, 2006	36
5.04	Company spending on R&D, 2006	29
5.05	University-industry research collaboration, 2006	26
5.06	Business telephone connection charge, 2004	58
5.07	Business monthly telephone subscription, 2005	51
5.08	Local supplier quality, 2006	22
5.09	Computer, comm., and other services imports, 2004	14

Government readiness — 46

6.01	Government prioritization of ICT, 2006	77
6.02	Gov't. procurement of advanced tech products, 2006	53
6.03	Importance of ICT to gov't. vision of the future, 2006	92
6.04	E-participation index, 2005	46
6.05	E-government readiness index, 2005	29

Usage component — 37

Individual usage — 31

7.01	Mobile telephone subscribers, 2005	5
7.02	Personal computers, 2004	31
7.03	Broadband Internet subscribers, 2005	36
7.04	Internet users, 2005	42
7.05	Internet bandwidth	n/a

Business usage — 25

8.01	Prevalence of foreign technology licensing, 2006	37
8.02	Firm-level technology absorption, 2006	26
8.03	Capacity for innovation, 2006	27
8.04	Availability of new telephone lines, 2006	28
8.05	Availability of mobile telephones, 2006	17
8.06	Extent of business Internet use, 2006	20

Government usage — 87

9.01	Government success in ICT promotion, 2006	88
9.02	Availability of online services, 2006	78
9.03	ICT use and government efficiency, 2006	97
9.04	ICT pervasiveness, 2006	75

Denmark

Key indicators

Population (millions), 2005...5.4
GDP (PPP) per capita (US$), 2005.....................34,739.8
Internet users per 100 inhabitants, 2005.....................52.6
Internet bandwidth (Mbps/10,000 inhabitants), 2004.............348.3

Networked Readiness Index

Year (number of economies)	Rank
2006–2007 (122)	**1**
2005–2006 (115)	3
2004–2005 (104)	4

Global Competitiveness Index 2006–2007 (125)	4

Environment component	4
Market environment	**16**

1.01	Venture capital availability, 2006	10
1.02	Financial market sophistication, 2006	17
1.03	Technological readiness, 2006	10
1.04	State of cluster development, 2006	22
1.05	US utility patents, 2005	14
1.06	High-tech exports, 2004	25
1.07	Burden of government regulation, 2006	21
1.08	Extent and effect of taxation, 2006	114
1.09	Time required to start a business, 2006	3
1.10	No. of procedures required to start a business, 2006	4
1.11	Intensity of local competition, 2006	21
1.12	Freedom of the press, 2006	2

Political and regulatory environment	**1**

2.01	Effectiveness of law-making bodies, 2006	3
2.02	Laws relating to ICT, 2006	6
2.03	Judicial independence, 2006	7
2.04	Intellectual property protection, 2006	4
2.05	Efficiency of legal framework, 2006	1
2.06	Property rights, 2006	3
2.07	Quality of competition in the ISP sector, 2006	18
2.08	No. of procedures to enforce a contract, 2006	3
2.09	Time to enforce a contract, 2006	10

Infrastructure environment	**7**

3.01	Telephone lines, 2005	5
3.02	Secure Internet servers, 2005	9
3.03	Internet hosts, 2004	4
3.04	Electricity production, 2003	17
3.05	Availability of scientists and engineers, 2006	12
3.06	Quality of scientific research institutions, 2006	13
3.07	Tertiary enrollment, 2004	10

Readiness component	3
Individual readiness	**6**

4.01	Quality of math and science education, 2006	20
4.02	Quality of the educational system, 2006	5
4.03	Quality of public schools, 2006	10
4.04	Internet access in schools, 2006	6
4.05	Buyer sophistication, 2006	17
4.06	Residential telephone connection charge, 2005	26
4.07	Residential monthly telephone subscription, 2005	13
4.08	High-speed monthly broadband subscription, 2006	13
4.09	Lowest cost of broadband, 2006	22
4.10	Cost of mobile telephone call, 2004	7

Business readiness	**7**

5.01	Extent of staff training, 2006	2
5.02	Local availability of research and training, 2006	11
5.03	Quality of management schools, 2006	10
5.04	Company spending on R&D, 2006	8
5.05	University-industry research collaboration, 2006	15
5.06	Business telephone connection charge, 2005	22
5.07	Business monthly telephone subscription, 2005	11
5.08	Local supplier quality, 2006	11
5.09	Computer, comm., and other services imports, 2004	38

Government readiness	**2**

6.01	Government prioritization of ICT, 2006	11
6.02	Gov't. procurement of advanced tech products, 2006	15
6.03	Importance of ICT to gov't. vision of the future, 2006	9
6.04	E-participation index, 2005	7
6.05	E-government readiness index, 2005	2

169

Usage component	2
Individual usage	**3**

7.01	Mobile telephone subscribers, 2005	17
7.02	Personal computers, 2004	8
7.03	Broadband Internet subscribers, 2005	4
7.04	Internet users, 2005	16
7.05	Internet bandwidth, 2004	1

Business usage	**7**

8.01	Prevalence of foreign technology licensing, 2006	31
8.02	Firm-level technology absorption, 2006	14
8.03	Capacity for innovation, 2006	6
8.04	Availability of new telephone lines, 2006	8
8.05	Availability of mobile telephones, 2006	12
8.06	Extent of business Internet use, 2006	10

Government usage	**5**

9.01	Government success in ICT promotion, 2006	14
9.02	Availability of online services, 2006	4
9.03	ICT use and government efficiency, 2006	5
9.04	ICT pervasiveness, 2006	10

Dominican Republic

Key indicators

Population (millions), 2005...8.9
GDP (PPP) per capita (US$), 2005.........................7,626.8
Internet users per 100 inhabitants, 2005....................16.8
Internet bandwidth (Mbps/10,000 inhabitants) ....n/a

Networked Readiness Index

Year (number of economies)	Rank
2006–2007 (122) **66**	
2005–2006 (115)..89	
2004–2005 (104)..78	

Global Competitiveness Index 2006–2007 (125)	83

Environment component   70

Market environment   65

1.01 Venture capital availability, 2006..............................94
1.02 Financial market sophistication, 2006.......................77
1.03 Technological readiness, 2006..................................35
1.04 State of cluster development, 2006...........................86
1.05 US utility patents, 2005 62
1.06 High-tech exports...n/a
1.07 Burden of government regulation, 2006 47
1.08 Extent and effect of taxation, 2006...........................90
1.09 Time required to start a business, 2006...................101
1.10 No. of procedures required to start a business, 2006.........61
1.11 Intensity of local competition, 2006 88
1.12 Freedom of the press, 2006......................................48

Political and regulatory environment   61

2.01 Effectiveness of law-making bodies, 2006.........................106
2.02 Laws relating to ICT, 2006..56
2.03 Judicial independence, 2006 70
2.04 Intellectual property protection, 2006........................68
2.05 Efficiency of legal framework, 2006...........................79
2.06 Property rights, 2006...74
2.07 Quality of competition in the ISP sector, 2006...................47
2.08 No. of procedures to enforce a contract, 2006 51
2.09 Time to enforce a contract, 2006 63

Infrastructure environment   87

3.01 Telephone lines, 2005...81
3.02 Secure Internet servers, 2005...................................61
3.03 Internet hosts, 2004 47
3.04 Electricity production, 2003 71
3.05 Availability of scientists and engineers, 2006.............105
3.06 Quality of scientific research institutions, 2006 113
3.07 Tertiary enrollment, 2004..58

Readiness component   78

Individual readiness   79

4.01 Quality of math and science education, 2006...............116
4.02 Quality of the educational system, 2006.....................115
4.03 Quality of public schools, 2006 121
4.04 Internet access in schools, 2006...............................80
4.05 Buyer sophistication, 2006..72
4.06 Residential telephone connection charge, 2005 64
4.07 Residential monthly telephone subscription, 2005 93
4.08 High-speed monthly broadband subscription, 2006 58
4.09 Lowest cost of broadband, 2006................................63
4.10 Cost of mobile telephone call, 2005...........................60

Business readiness   86

5.01 Extent of staff training, 2006.....................................89
5.02 Local availability of research and training, 2006.............89
5.03 Quality of management schools, 2006........................78
5.04 Company spending on R&D, 2006 102
5.05 University-industry research collaboration, 2006............99
5.06 Business telephone connection charge, 2005 60
5.07 Business monthly telephone subscription, 2005 71
5.08 Local supplier quality, 2006 75
5.09 Computer, comm., and other services imports, 2004.........94

Government readiness   64

6.01 Government prioritization of ICT, 2006........................28
6.02 Gov't. procurement of advanced tech products, 2006........91
6.03 Importance of ICT to gov't. vision of the future, 2006........26
6.04 E-participation index, 2005 78
6.05 E-government readiness index, 2005..........................71

Usage component   53

Individual usage   62

7.01 Mobile telephone subscribers, 2005 72
7.02 Personal computers, 2004...74
7.03 Broadband Internet subscribers, 2005 58
7.04 Internet users, 2005 59
7.05 Internet bandwidth...n/a

Business usage   55

8.01 Prevalence of foreign technology licensing, 2006..............52
8.02 Firm-level technology absorption, 2006.......................64
8.03 Capacity for innovation, 2006 89
8.04 Availability of new telephone lines, 2006 34
8.05 Availability of mobile telephones, 2006.......................20
8.06 Extent of business Internet use, 2006 76

Government usage   43

9.01 Government success in ICT promotion, 2006................65
9.02 Availability of online services, 2006............................43
9.03 ICT use and government efficiency, 2006.....................32
9.04 ICT pervasiveness, 2006 76

Ecuador

Key indicators

Population (millions), 2005..13.2
GDP (PPP) per capita (US$), 2005........................4,316.2
Internet users per 100 inhabitants, 2005....................4.7
Internet bandwidth (Mbps/10,000 inhabitants), 2004.................0.4

Networked Readiness Index

Year (number of economies)	Rank
2006–2007 (122)	97
2005–2006 (115)	107
2004–2005 (104)	95

Global Competitiveness Index 2006–2007 (125)	90

Environment component	110
Market environment	**104**
1.01 Venture capital availability, 2006	112
1.02 Financial market sophistication, 2006	78
1.03 Technological readiness, 2006	81
1.04 State of cluster development, 2006	84
1.05 US utility patents, 2005	65
1.06 High-tech exports, 2004	69
1.07 Burden of government regulation, 2006	103
1.08 Extent and effect of taxation, 2006	79
1.09 Time required to start a business, 2006	99
1.10 No. of procedures required to start a business, 2006	106
1.11 Intensity of local competition, 2006	77
1.12 Freedom of the press, 2006	54
Political and regulatory environment	**110**
2.01 Effectiveness of law-making bodies, 2006	122
2.02 Laws relating to ICT, 2006	87
2.03 Judicial independence, 2006	113
2.04 Intellectual property protection, 2006	99
2.05 Efficiency of legal framework, 2006	117
2.06 Property rights, 2006	108
2.07 Quality of competition in the ISP sector, 2006	82
2.08 No. of procedures to enforce a contract, 2006	91
2.09 Time to enforce a contract, 2006	71
Infrastructure environment	**101**
3.01 Telephone lines, 2005	77
3.02 Secure Internet servers, 2005	71
3.03 Internet hosts, 2004	81
3.04 Electricity production, 2003	78
3.05 Availability of scientists and engineers, 2006	107
3.06 Quality of scientific research institutions, 2006	109
3.07 Tertiary enrollment	n/a

Readiness component	89
Individual readiness	**82**
4.01 Quality of math and science education, 2006	108
4.02 Quality of the educational system, 2006	117
4.03 Quality of public schools, 2006	117
4.04 Internet access in schools, 2006	94
4.05 Buyer sophistication, 2006	102
4.06 Residential telephone connection charge, 2005	56
4.07 Residential monthly telephone subscription, 2004	74
4.08 High-speed monthly broadband subscription, 2006	66
4.09 Lowest cost of broadband, 2006	83
4.10 Cost of mobile telephone call, 2005	81
Business readiness	**80**
5.01 Extent of staff training, 2006	103
5.02 Local availability of research and training, 2006	90
5.03 Quality of management schools, 2006	81
5.04 Company spending on R&D, 2006	93
5.05 University-industry research collaboration, 2006	90
5.06 Business telephone connection charge, 2004	63
5.07 Business monthly telephone subscription, 2004	76
5.08 Local supplier quality, 2006	88
5.09 Computer, comm., and other services imports, 2004	56
Government readiness	**113**
6.01 Government prioritization of ICT, 2006	119
6.02 Gov't. procurement of advanced tech products, 2006	108
6.03 Importance of ICT to gov't. vision of the future, 2006	110
6.04 E-participation index, 2005	78
6.05 E-government readiness index, 2005	78

Usage component	97
Individual usage	**81**
7.01 Mobile telephone subscribers, 2005	61
7.02 Personal computers, 2004	69
7.03 Broadband Internet subscribers, 2005	73
7.04 Internet users, 2005	94
7.05 Internet bandwidth, 2004	66
Business usage	**97**
8.01 Prevalence of foreign technology licensing, 2006	102
8.02 Firm-level technology absorption, 2006	99
8.03 Capacity for innovation, 2006	88
8.04 Availability of new telephone lines, 2006	104
8.05 Availability of mobile telephones, 2006	68
8.06 Extent of business Internet use, 2006	89
Government usage	**99**
9.01 Government success in ICT promotion, 2006	120
9.02 Availability of online services, 2006	70
9.03 ICT use and government efficiency, 2006	81
9.04 ICT pervasiveness, 2006	109

171

Egypt

Key indicators

Population (millions), 2005..74.0

GDP (PPP) per capita (US$), 2005.........................4,316.6

Internet users per 100 inhabitants, 2005.....................6.8

Internet bandwidth (Mbps/10,000 inhabitants), 2004.................0.2

Networked Readiness Index

Year (number of economies)	Rank
2006–2007 (122) **77**	
2005–2006 (115)...63	
2004–2005 (104)...57	

Global Competitiveness Index 2006–2007 (125)	63

Environment component	74
Market environment	**73**

1.01	Venture capital availability, 2006...........................85
1.02	Financial market sophistication, 2006....................76
1.03	Technological readiness, 2006................................66
1.04	State of cluster development, 2006........................59
1.05	US utility patents, 2005..71
1.06	High-tech exports, 2004...89
1.07	Burden of government regulation, 2006 72
1.08	Extent and effect of taxation, 2006.......................36
1.09	Time required to start a business, 2006................27
1.10	No. of procedures required to start a business, 2006.........61
1.11	Intensity of local competition, 2006......................68
1.12	Freedom of the press, 2006...................................99

Political and regulatory environment	**77**

2.01	Effectiveness of law-making bodies, 2006............64
2.02	Laws relating to ICT, 2006......................................80
2.03	Judicial independence, 2006 39
2.04	Intellectual property protection, 2006....................62
2.05	Efficiency of legal framework, 2006......................53
2.06	Property rights, 2006...57
2.07	Quality of competition in the ISP sector, 2006.....35
2.08	No. of procedures to enforce a contract, 2006.................113
2.09	Time to enforce a contract, 2006 108

Infrastructure environment	**72**

3.01	Telephone lines, 2005...72
3.02	Secure Internet servers, 2005................................92
3.03	Internet hosts, 2004..106
3.04	Electricity production, 2003 74
3.05	Availability of scientists and engineers, 2006.......40
3.06	Quality of scientific research institutions, 2006.................94
3.07	Tertiary enrollment, 2004.......................................59

Readiness component	82
Individual readiness	**85**

4.01	Quality of math and science education, 2006........93
4.02	Quality of the educational system, 2006.............103
4.03	Quality of public schools, 2006 112
4.04	Internet access in schools, 2006...........................57
4.05	Buyer sophistication, 2006 91
4.06	Residential telephone connection charge, 2005 97
4.07	Residential monthly telephone subscription, 2005 52
4.08	High-speed monthly broadband subscription, 2006...........85
4.09	Lowest cost of broadband, 2006...........................75
4.10	Cost of mobile telephone call, 2005......................61

Business readiness	**82**

5.01	Extent of staff training, 2006................................83
5.02	Local availability of research and training, 2006.................79
5.03	Quality of management schools, 2006...................88
5.04	Company spending on R&D, 2006...........................97
5.05	University-industry research collaboration, 2006.................93
5.06	Business telephone connection charge, 2005 101
5.07	Business monthly telephone subscription, 2005 47
5.08	Local supplier quality, 2006 56
5.09	Computer, comm., and other services imports, 2004.........36

Government readiness	**81**

6.01	Government prioritization of ICT, 2006..................79
6.02	Gov't. procurement of advanced tech products, 2006........82
6.03	Importance of ICT to gov't. vision of the future, 2006........44
6.04	E-participation index, 2005 71
6.05	E-government readiness index, 2005.....................82

Usage component	72
Individual usage	**93**

7.01	Mobile telephone subscribers, 2005.....................93
7.02	Personal computers, 2004.......................................85
7.03	Broadband Internet subscribers, 2005 76
7.04	Internet users, 2005..84
7.05	Internet bandwidth, 2004 70

Business usage	**51**

8.01	Prevalence of foreign technology licensing, 2006...............45
8.02	Firm-level technology absorption, 2006.................59
8.03	Capacity for innovation, 2006 83
8.04	Availability of new telephone lines, 2006 43
8.05	Availability of mobile telephones, 2006.................37
8.06	Extent of business Internet use, 2006 65

Government usage	**64**

9.01	Government success in ICT promotion, 2006....................49
9.02	Availability of online services, 2006.......................61
9.03	ICT use and government efficiency, 2006..............65
9.04	ICT pervasiveness, 2006 82

172

El Salvador

Key indicators

Population (millions), 2005..6.9
GDP (PPP) per capita (US$), 2005........................4,518.1
Internet users per 100 inhabitants, 2005....................9.3
Internet bandwidth (Mbps/10,000 inhabitants) ....n/a

Networked Readiness Index

Year (number of economies)	Rank
2006–2007 (122) **61**	
2005–2006 (115)..59	
2004—2005 (104) 70	

Global Competitiveness Index 2006–2007 (125)	61

Environment component	73
Market environment	**48**
1.01 Venture capital availability, 2006 84	
1.02 Financial market sophistication, 2006....................35	
1.03 Technological readiness, 2006...............................42	
1.04 State of cluster development, 2006 69	
1.05 US utility patents, 2005 66	
1.06 High-tech exports, 2004 64	
1.07 Burden of government regulation, 2006 41	
1.08 Extent and effect of taxation, 2006......................26	
1.09 Time required to start a business, 2006................44	
1.10 No. of procedures required to start a business, 2006.........61	
1.11 Intensity of local competition, 2006 39	
1.12 Freedom of the press, 2006..................................39	
Political and regulatory environment	**78**
2.01 Effectiveness of law-making bodies, 2006................105	
2.02 Laws relating to ICT, 2006....................................60	
2.03 Judicial independence, 2006 88	
2.04 Intellectual property protection, 2006....................60	
2.05 Efficiency of legal framework, 2006.......................90	
2.06 Property rights, 2006 67	
2.07 Quality of competition in the ISP sector, 2006....................37	
2.08 No. of procedures to enforce a contract, 2006 91	
2.09 Time to enforce a contract, 2006 92	
Infrastructure environment	**92**
3.01 Telephone lines, 2005...71	
3.02 Secure Internet servers, 2005 63	
3.03 Internet hosts, 2004 82	
3.04 Electricity production, 2003 87	
3.05 Availability of scientists and engineers, 2006....................101	
3.06 Quality of scientific research institutions, 2006 112	
3.07 Tertiary enrollment, 2004......................................80	

Readiness component	70
Individual readiness	**73**
4.01 Quality of math and science education, 2006 91	
4.02 Quality of the educational system, 2006..........................65	
4.03 Quality of public schools, 2006 77	
4.04 Internet access in schools, 2006...........................63	
4.05 Buyer sophistication, 2006 55	
4.06 Residential telephone connection charge, 2005 67	
4.07 Residential monthly telephone subscription, 2005 84	
4.08 High-speed monthly broadband subscription charge ....n/a	
4.09 Lowest cost of broadband, 2006...........................66	
4.10 Cost of mobile telephone call, 2005......................44	
Business readiness	**74**
5.01 Extent of staff training, 2006................................63	
5.02 Local availability of research and training, 2006 74	
5.03 Quality of management schools, 2006...............57	
5.04 Company spending on R&D, 2006 86	
5.05 University-industry research collaboration, 2006..............102	
5.06 Business telephone connection charge, 2005 62	
5.07 Business monthly telephone subscription, 2005 84	
5.08 Local supplier quality, 2006 61	
5.09 Computer, comm., and other services imports, 2004.........75	
Government readiness	**49**
6.01 Government prioritization of ICT, 2006...............34	
6.02 Gov't. procurement of advanced tech products, 2006........70	
6.03 Importance of ICT to gov't. vision of the future, 2006........22	
6.04 E-participation index, 2005 56	
6.05 E-government readiness index, 2005 70	

Usage component	51
Individual usage	**70**
7.01 Mobile telephone subscribers, 2005 78	
7.02 Personal computers, 2004.....................................75	
7.03 Broadband Internet subscribers, 2005 61	
7.04 Internet users, 2005 75	
7.05 Internet bandwidth...n/a	
Business usage	**46**
8.01 Prevalence of foreign technology licensing, 2006..............62	
8.02 Firm-level technology absorption, 2006...............60	
8.03 Capacity for innovation, 2006 67	
8.04 Availability of new telephone lines, 2006 23	
8.05 Availability of mobile telephones, 2006 27	
8.06 Extent of business Internet use, 2006 53	
Government usage	**34**
9.01 Government success in ICT promotion, 2006.....................52	
9.02 Availability of online services, 2006......................32	
9.03 ICT use and government efficiency, 2006...........................35	
9.04 ICT pervasiveness, 2006 42	

Estonia

Key indicators

Population (millions), 2005...1.3
GDP (PPP) per capita (US$), 2005.......................................16,414.0
Internet users per 100 inhabitants, 2005....................................51.9
Internet bandwidth (Mbps/10,000 inhabitants), 2004..............35.2

Networked Readiness Index

Year (number of economies)	Rank
2006–2007 (122) **20**	
2005–2006 (115)..23	
2004–2005 (104)..25	

Global Competitiveness Index 2006–2007 (125)	25

Environment component	25
Market environment	**25**

1.01	Venture capital availability, 2006...29
1.02	Financial market sophistication, 2006..................................29
1.03	Technological readiness, 2006...24
1.04	State of cluster development, 2006......................................65
1.05	US utility patents, 2005...30
1.06	High-tech exports, 2004..29
1.07	Burden of government regulation, 2006 9
1.08	Extent and effect of taxation, 2006.....................................13
1.09	Time required to start a business, 2006..............................65
1.10	No. of procedures required to start a business, 2006.........16
1.11	Intensity of local competition, 2006....................................31
1.12	Freedom of the press, 2006...11

Political and regulatory environment	**21**
2.01	Effectiveness of law-making bodies, 2006..........................26
2.02	Laws relating to ICT, 2006...1
2.03	Judicial independence, 2006..27
2.04	Intellectual property protection, 2006..................................32
2.05	Efficiency of legal framework, 2006.....................................24
2.06	Property rights, 2006..27
2.07	Quality of competition in the ISP sector, 2006....................13
2.08	No. of procedures to enforce a contract, 2006....................34
2.09	Time to enforce a contract, 2006..22

Infrastructure environment	**26**
3.01	Telephone lines, 2005..34
3.02	Secure Internet servers, 2005...26
3.03	Internet hosts, 2004...22
3.04	Electricity production, 2003..23
3.05	Availability of scientists and engineers, 2006.....................50
3.06	Quality of scientific research institutions, 2006.................28
3.07	Tertiary enrollment, 2004...18

Readiness component	23
Individual readiness	**26**

4.01	Quality of math and science education, 2006.....................18
4.02	Quality of the educational system, 2006.............................31
4.03	Quality of public schools, 2006...21
4.04	Internet access in schools, 2006...13
4.05	Buyer sophistication, 2006..37
4.06	Residential telephone connection charge, 2005.................43
4.07	Residential monthly telephone subscription, 2005.............41
4.08	High-speed monthly broadband subscription, 2006............32
4.09	Lowest cost of broadband, 2006..40
4.10	Cost of mobile telephone call, 2004....................................30

Business readiness	**25**
5.01	Extent of staff training, 2006...27
5.02	Local availability of research and training, 2006.................26
5.03	Quality of management schools, 2006.................................30
5.04	Company spending on R&D, 2006..32
5.05	University-industry research collaboration, 2006................28
5.06	Business telephone connection charge, 2005.....................38
5.07	Business monthly telephone subscription, 2005................32
5.08	Local supplier quality, 2006...32
5.09	Computer, comm., and other services imports, 2004.........41

Government readiness	**10**
6.01	Government prioritization of ICT, 2006..................................6
6.02	Gov't. procurement of advanced tech products, 2006........33
6.03	Importance of ICT to gov't. vision of the future, 2006........12
6.04	E-participation index, 2005..11
6.05	E-government readiness index, 2005...................................19

Usage component	12
Individual usage	**23**

7.01	Mobile telephone subscribers, 2005.....................................9
7.02	Personal computers, 2004..24
7.03	Broadband Internet subscribers, 2005................................21
7.04	Internet users, 2005...18
7.05	Internet bandwidth, 2004...16

Business usage	**22**
8.01	Prevalence of foreign technology licensing, 2006..............43
8.02	Firm-level technology absorption, 2006...............................19
8.03	Capacity for innovation, 2006..39
8.04	Availability of new telephone lines, 2006............................27
8.05	Availability of mobile telephones, 2006...............................13
8.06	Extent of business Internet use, 2006...................................3

Government usage	**2**
9.01	Government success in ICT promotion, 2006........................8
9.02	Availability of online services, 2006......................................1
9.03	ICT use and government efficiency, 2006..............................1
9.04	ICT pervasiveness, 2006..2

Ethiopia

Key indicators

Population (millions), 2005...77.4

GDP (PPP) per capita (US$), 2005..823.0

Internet users per 100 inhabitants, 2005.....................................0.2

Internet bandwidth (Mbps/10,000 inhabitants) ....n/a

Networked Readiness Index

Year (number of economies)	Rank
2006–2007 (122) **119**	
2005–2006 (115)..115	
2004–2005 (104)..102	

Global Competitiveness Index 2006–2007 (125)   120

Environment component	113
Market environment	**98**

1.01 Venture capital availability, 2006 118	
1.02 Financial market sophistication, 2006...............................119	
1.03 Technological readiness, 2006..119	
1.04 State of cluster development, 2006 95	
1.05 US utility patents, 2005 79	
1.06 High-tech exports...n/a	
1.07 Burden of government regulation, 2006 59	
1.08 Extent and effect of taxation, 2006...................................58	
1.09 Time required to start a business, 2006............................20	
1.10 No. of procedures required to start a business, 2006.........27	
1.11 Intensity of local competition, 2006 110	
1.12 Freedom of the press, 2006..121	

Political and regulatory environment	**112**
2.01 Effectiveness of law-making bodies, 2006.......................102	
2.02 Laws relating to ICT, 2006..108	
2.03 Judicial independence, 2006 111	
2.04 Intellectual property protection, 2006...............................107	
2.05 Efficiency of legal framework, 2006..................................105	
2.06 Property rights, 2006 97	
2.07 Quality of competition in the ISP sector, 2006..................121	
2.08 No. of procedures to enforce a contract, 2006 57	
2.09 Time to enforce a contract, 2006 97	

Infrastructure environment	**119**
3.01 Telephone lines, 2005..110	
3.02 Secure Internet servers, 2004 120	
3.03 Internet hosts, 2004 121	
3.04 Electricity production, 2003 104	
3.05 Availability of scientists and engineers, 2006...................116	
3.06 Quality of scientific research institutions, 2006 68	
3.07 Tertiary enrollment, 2004...111	

Readiness component	120
Individual readiness	**118**

4.01 Quality of math and science education, 2006 106	
4.02 Quality of the educational system, 2006...........................94	
4.03 Quality of public schools, 2006 101	
4.04 Internet access in schools, 2006.......................................108	
4.05 Buyer sophistication, 2006 112	
4.06 Residential telephone connection charge, 2004 111	
4.07 Residential monthly telephone subscription, 2004 101	
4.08 High-speed monthly broadband subscription, 2006..........106	
4.09 Lowest cost of broadband, 2006.......................................109	
4.10 Cost of mobile telephone call, 2005..................................98	

Business readiness	**122**
5.01 Extent of staff training, 2006...117	
5.02 Local availability of research and training, 2006 109	
5.03 Quality of management schools, 2006...............................109	
5.04 Company spending on R&D, 2006 116	
5.05 University-industry research collaboration, 2006...............111	
5.06 Business telephone connection charge, 2004 108	
5.07 Business monthly telephone subscription, 2004 108	
5.08 Local supplier quality, 2006 120	
5.09 Computer, comm., and other services imports, 2004.........53	

Government readiness	**111**
6.01 Government prioritization of ICT, 2006...............................98	
6.02 Gov't. procurement of advanced tech products, 2006......107	
6.03 Importance of ICT to gov't. vision of the future, 2006........48	
6.04 E-participation index, 2005 109	
6.05 E-government readiness index, 2005 118	

Usage component	108
Individual usage	**122**

7.01 Mobile telephone subscribers, 2005 122	
7.02 Personal computers, 2004..114	
7.03 Broadband Internet subscribers...n/a	
7.04 Internet users, 2005 122	
7.05 Internet bandwidth..n/a	

Business usage	**118**
8.01 Prevalence of foreign technology licensing, 2006.............111	
8.02 Firm-level technology absorption, 2006.............................119	
8.03 Capacity for innovation, 2006 113	
8.04 Availability of new telephone lines, 2006 106	
8.05 Availability of mobile telephones, 2006.............................117	
8.06 Extent of business Internet use, 2006 110	

Government usage	**92**
9.01 Government success in ICT promotion, 2006.....................79	
9.02 Availability of online services, 2006...................................88	
9.03 ICT use and government efficiency, 2006...........................76	
9.04 ICT pervasiveness, 2006 102	

175

Finland

Key indicators

Population (millions), 2005..5.2
GDP (PPP) per capita (US$), 2005.......................................31,207.7
Internet users per 100 inhabitants, 2005....................................53.3
Internet bandwidth (Mbps/10,000 inhabitants), 2004..............43.4

Networked Readiness Index

Year (number of economies)	Rank
2006–2007 (122) **4**
2005–2006 (115)...5
2004–2005 (104)...3

Global Competitiveness Index 2006–2007 (125)	2

Environment component	3
Market environment	**2**

1.01	Venture capital availability, 2006 4
1.02	Financial market sophistication, 2006...................................12
1.03	Technological readiness, 2006 1
1.04	State of cluster development, 2006 3
1.05	US utility patents, 2005 4
1.06	High-tech exports, 2004 17
1.07	Burden of government regulation, 2006 3
1.08	Extent and effect of taxation, 2006..98
1.09	Time required to start a business, 2006................................19
1.10	No. of procedures required to start a business, 2006...........4
1.11	Intensity of local competition, 2006 13
1.12	Freedom of the press, 2006..7

Political and regulatory environment	**9**

2.01	Effectiveness of law-making bodies, 2006............................5
2.02	Laws relating to ICT, 2006..10
2.03	Judicial independence, 2006 10
2.04	Intellectual property protection, 2006....................................2
2.05	Efficiency of legal framework, 2006..8
2.06	Property rights, 2006 8
2.07	Quality of competition in the ISP sector, 2006.....................9
2.08	No. of procedures to enforce a contract, 2006 41
2.09	Time to enforce a contract, 2006 15

Infrastructure environment	**4**

3.01	Telephone lines, 2005...32
3.02	Secure Internet servers, 2005 15
3.03	Internet hosts, 2004 5
3.04	Electricity production, 2003 6
3.05	Availability of scientists and engineers, 2006......................3
3.06	Quality of scientific research institutions, 2006 7
3.07	Tertiary enrollment, 2004..1

Readiness component	2
Individual readiness	**1**

4.01	Quality of math and science education, 2006 2
4.02	Quality of the educational system, 2006...............................1
4.03	Quality of public schools, 2006 1
4.04	Internet access in schools, 2006..3
4.05	Buyer sophistication, 2006 9
4.06	Residential telephone connection charge, 2005 25
4.07	Residential monthly telephone subscription, 2005 12
4.08	High-speed monthly broadband subscription, 2006 24
4.09	Lowest cost of broadband, 2006..2
4.10	Cost of mobile telephone call, 2005 11

Business readiness	**2**

5.01	Extent of staff training, 2006...8
5.02	Local availability of research and training, 2006 6
5.03	Quality of management schools, 2006....................................9
5.04	Company spending on R&D, 2006 6
5.05	University-industry research collaboration, 2006..................3
5.06	Business telephone connection charge, 2005 21
5.07	Business monthly telephone subscription, 2005 10
5.08	Local supplier quality, 2006 9
5.09	Computer, comm., and other services imports, 2004.........12

Government readiness	**8**

6.01	Government prioritization of ICT, 2006.................................14
6.02	Gov't. procurement of advanced tech products, 2006........13
6.03	Importance of ICT to gov't. vision of the future, 2006........16
6.04	E-participation index, 2005 15
6.05	E-government readiness index, 2005 9

Usage component	10
Individual usage	**14**

7.01	Mobile telephone subscribers, 2005 19
7.02	Personal computers, 2004...22
7.03	Broadband Internet subscribers, 2005 7
7.04	Internet users, 2005 15
7.05	Internet bandwidth, 2004 14

Business usage	**6**

8.01	Prevalence of foreign technology licensing, 2006...............33
8.02	Firm-level technology absorption, 2006.................................8
8.03	Capacity for innovation, 2006 4
8.04	Availability of new telephone lines, 2006 9
8.05	Availability of mobile telephones, 2006.................................6
8.06	Extent of business Internet use, 2006 15

Government usage	**13**

9.01	Government success in ICT promotion, 2006.....................16
9.02	Availability of online services, 2006......................................21
9.03	ICT use and government efficiency, 2006.............................18
9.04	ICT pervasiveness, 2006 5

France

Key indicators

Population (millions), 2005..60.5
GDP (PPP) per capita (US$), 2005.....................................29,187.2
Internet users per 100 inhabitants, 2005.....................................43.2
Internet bandwidth (Mbps/10,000 inhabitants), 2004..............33.1

Networked Readiness Index

Year (number of economies)	Rank
2006–2007 (122) **23**
2005–2006 (115)..22
2004–2005 (104)..20

Global Competitiveness Index 2006–2007 (125)	18

Environment component   21

Market environment   20

1.01   Venture capital availability, 2006...28
1.02   Financial market sophistication, 2006...................................14
1.03   Technological readiness, 2006..19
1.04   State of cluster development, 2006.....................................29
1.05   US utility patents, 2005...20
1.06   High-tech exports, 2004..22
1.07   Burden of government regulation, 2006..............................89
1.08   Extent and effect of taxation, 2006.....................................85
1.09   Time required to start a business, 2006.................................7
1.10   No. of procedures required to start a business, 2006.........27
1.11   Intensity of local competition, 2006.....................................14
1.12   Freedom of the press, 2006..17

Political and regulatory environment   19

2.01   Effectiveness of law-making bodies, 2006...........................29
2.02   Laws relating to ICT, 2006..13
2.03   Judicial independence, 2006...35
2.04   Intellectual property protection, 2006...................................11
2.05   Efficiency of legal framework, 2006.....................................27
2.06   Property rights, 2006...18
2.07   Quality of competition in the ISP sector, 2006....................16
2.08   No. of procedures to enforce a contract, 2006...................16
2.09   Time to enforce a contract, 2006...32

Infrastructure environment   21

3.01   Telephone lines, 2005..8
3.02   Secure Internet servers, 2005..28
3.03   Internet hosts, 2004..25
3.04   Electricity production, 2003...13
3.05   Availability of scientists and engineers, 2006........................5
3.06   Quality of scientific research institutions, 2006..................20
3.07   Tertiary enrollment, 2004..30

Readiness component   19

Individual readiness   12

4.01   Quality of math and science education, 2006.......................5
4.02   Quality of the educational system, 2006.............................27
4.03   Quality of public schools, 2006...15
4.04   Internet access in schools, 2006...24
4.05   Buyer sophistication, 2006...11
4.06   Residential telephone connection charge, 2005.................11
4.07   Residential monthly telephone subscription, 2005.............23
4.08   High-speed monthly broadband subscription, 2006.............7
4.09   Lowest cost of broadband, 2006..2
4.10   Cost of mobile telephone call, 2005....................................29

Business readiness   17

5.01   Extent of staff training, 2006...22
5.02   Local availability of research and training, 2006.................12
5.03   Quality of management schools, 2006...................................1
5.04   Company spending on R&D, 2006......................................14
5.05   University-industry research collaboration, 2006................29
5.06   Business telephone connection charge, 2005.....................10
5.07   Business monthly telephone subscription, 2005................14
5.08   Local supplier quality, 2006..12
5.09   Computer, comm., and other services imports, 2004.........26

Government readiness   23

6.01   Government prioritization of ICT, 2006...............................44
6.02   Gov't. procurement of advanced tech products, 2006..........9
6.03   Importance of ICT to gov't. vision of the future, 2006........33
6.04   E-participation index, 2005..24
6.05   E-government readiness index, 2005...................................23

Usage component   22

Individual usage   25

7.01   Mobile telephone subscribers, 2005....................................40
7.02   Personal computers, 2004..20
7.03   Broadband Internet subscribers, 2005................................18
7.04   Internet users, 2005..29
7.05   Internet bandwidth, 2004..17

Business usage   19

8.01   Prevalence of foreign technology licensing, 2006..............60
8.02   Firm-level technology absorption, 2006..............................37
8.03   Capacity for innovation, 2006...7
8.04   Availability of new telephone lines, 2006...............................6
8.05   Availability of mobile telephones, 2006..............................11
8.06   Extent of business Internet use, 2006.................................19

Government usage   27

9.01   Government success in ICT promotion, 2006.....................21
9.02   Availability of online services, 2006....................................22
9.03   ICT use and government efficiency, 2006...........................29
9.04   ICT pervasiveness, 2006..39

Georgia

Key indicators

Population (millions), 2005 ... 4.5
GDP (PPP) per capita (US$), 2005 ... 3,586.3
Internet users per 100 inhabitants, 2004 ... 3.9
Internet bandwidth (Mbps/10,000 inhabitants) ... n/a

Networked Readiness Index

Year (number of economies)	Rank
2006–2007 (122)	93
2005–2006 (115)	96
2004–2005 (104)	91

Global Competitiveness Index 2006–2007 (125) — 85

Environment component — 75
Market environment — 74
1.01 Venture capital availability, 2006 ... 97
1.02 Financial market sophistication, 2006 ... 90
1.03 Technological readiness, 2006 ... 92
1.04 State of cluster development, 2006 ... 113
1.05 US utility patents, 2005 ... 44
1.06 High-tech exports, 2004 ... 37
1.07 Burden of government regulation, 2006 ... 18
1.08 Extent and effect of taxation, 2006 ... 50
1.09 Time required to start a business, 2006 ... 20
1.10 No. of procedures required to start a business, 2006 ... 27
1.11 Intensity of local competition, 2006 ... 101
1.12 Freedom of the press, 2006 ... 78

Political and regulatory environment — 83
2.01 Effectiveness of law-making bodies, 2006 ... 83
2.02 Laws relating to ICT, 2006 ... 116
2.03 Judicial independence, 2006 ... 114
2.04 Intellectual property protection, 2006 ... 91
2.05 Efficiency of legal framework, 2006 ... 116
2.06 Property rights, 2006 ... 100
2.07 Quality of competition in the ISP sector, 2006 ... 45
2.08 No. of procedures to enforce a contract, 2006 ... 30
2.09 Time to enforce a contract, 2006 ... 24

Infrastructure environment — 71
3.01 Telephone lines, 2004 ... 69
3.02 Secure Internet servers, 2005 ... 68
3.03 Internet hosts, 2004 ... 72
3.04 Electricity production, 2003 ... 72
3.05 Availability of scientists and engineers, 2006 ... 79
3.06 Quality of scientific research institutions, 2006 ... 93
3.07 Tertiary enrollment, 2004 ... 40

Readiness component — 98
Individual readiness — 75
4.01 Quality of math and science education, 2006 ... 63
4.02 Quality of the educational system, 2006 ... 93
4.03 Quality of public schools, 2006 ... 74
4.04 Internet access in schools, 2006 ... 74
4.05 Buyer sophistication, 2006 ... 86
4.06 Residential telephone connection charge ... n/a
4.07 Residential monthly telephone subscription ... n/a
4.08 High-speed monthly broadband subscription, 2006 ... 63
4.09 Lowest cost of broadband, 2006 ... 71
4.10 Cost of mobile telephone call, 2005 ... 73

Business readiness — 117
5.01 Extent of staff training, 2006 ... 94
5.02 Local availability of research and training, 2006 ... 106
5.03 Quality of management schools, 2006 ... 106
5.04 Company spending on R&D, 2006 ... 107
5.05 University-industry research collaboration, 2006 ... 108
5.06 Business telephone connection charge ... n/a
5.07 Business monthly telephone subscription ... n/a
5.08 Local supplier quality, 2006 ... 116
5.09 Computer, comm., and other services imports, 2004 ... 91

Government readiness — 109
6.01 Government prioritization of ICT, 2006 ... 111
6.02 Gov't. procurement of advanced tech products, 2006 ... 106
6.03 Importance of ICT to gov't. vision of the future, 2006 ... 116
6.04 E-participation index, 2005 ... 101
6.05 E-government readiness index, 2005 ... 72

Usage component — 98
Individual usage — 72
7.01 Mobile telephone subscribers, 2005 ... 80
7.02 Personal computers, 2004 ... 79
7.03 Broadband Internet subscribers ... n/a
7.04 Internet users, 2004 ... 95
7.05 Internet bandwidth ... n/a

Business usage — 91
8.01 Prevalence of foreign technology licensing, 2006 ... 107
8.02 Firm-level technology absorption, 2006 ... 102
8.03 Capacity for innovation, 2006 ... 95
8.04 Availability of new telephone lines, 2006 ... 78
8.05 Availability of mobile telephones, 2006 ... 52
8.06 Extent of business Internet use, 2006 ... 98

Government usage — 108
9.01 Government success in ICT promotion, 2006 ... 104
9.02 Availability of online services, 2006 ... 118
9.03 ICT use and government efficiency, 2006 ... 89
9.04 ICT pervasiveness, 2006 ... 103

178

Germany

Key indicators

Population (millions), 2005..82.7
GDP (PPP) per capita (US$), 2005.....................30,579.4
Internet users per 100 inhabitants, 2005...................45.4
Internet bandwidth (Mbps/10,000 inhabitants), 2004.............68.6

Networked Readiness Index

179

Greece

Key indicators

Population (millions), 2005................................11.1
GDP (PPP) per capita (US$), 2005.....................22,391.6
Internet users per 100 inhabitants, 2005...................18.0
Internet bandwidth (Mbps/10,000 inhabitants), 2004.................5.9

Networked Readiness Index

Year (number of economies)	Rank
2006–2007 (122) **48**	
2005–2006 (115)...43	
2004–2005 (104)...42	

Global Competitiveness Index 2006–2007 (125)	47

Environment component — 33

Market environment — 71

1.01 Venture capital availability, 2006............................57
1.02 Financial market sophistication, 2006....................37
1.03 Technological readiness, 2006................................68
1.04 State of cluster development, 2006.........................75
1.05 US utility patents, 2005...36
1.06 High-tech exports, 2004..46
1.07 Burden of government regulation, 2006..................87
1.08 Extent and effect of taxation, 2006........................75
1.09 Time required to start a business, 2006..................74
1.10 No. of procedures required to start a business, 2006.......108
1.11 Intensity of local competition, 2006.......................64
1.12 Freedom of the press, 2006...................................22

Political and regulatory environment — 41

2.01 Effectiveness of law-making bodies, 2006..............41
2.02 Laws relating to ICT, 2006.....................................54
2.03 Judicial independence, 2006..................................46
2.04 Intellectual property protection, 2006....................38
2.05 Efficiency of legal framework, 2006.......................43
2.06 Property rights, 2006...37
2.07 Quality of competition in the ISP sector, 2006........60
2.08 No. of procedures to enforce a contract, 2006........23
2.09 Time to enforce a contract, 2006.........................100

Infrastructure environment — 25

3.01 Telephone lines, 2005...10
3.02 Secure Internet servers, 2005...............................40
3.03 Internet hosts, 2004..33
3.04 Electricity production, 2003...................................38
3.05 Availability of scientists and engineers, 2006.........17
3.06 Quality of scientific research institutions, 2006.......64
3.07 Tertiary enrollment, 2004..7

Readiness component — 51

Individual readiness — 58

4.01 Quality of math and science education, 2006..........46
4.02 Quality of the educational system, 2006.................59
4.03 Quality of public schools, 2006..............................53
4.04 Internet access in schools, 2006............................56
4.05 Buyer sophistication, 2006.....................................43
4.06 Residential telephone connection charge, 2005........9
4.07 Residential monthly telephone subscription, 2005....37
4.08 High-speed monthly broadband subscription, 2006...36
4.09 Lowest cost of broadband, 2006.............................41
4.10 Cost of mobile telephone call, 2005.......................32

Business readiness — 48

5.01 Extent of staff training, 2006.................................52
5.02 Local availability of research and training, 2006......59
5.03 Quality of management schools, 2006....................68
5.04 Company spending on R&D, 2006..........................71
5.05 University-industry research collaboration, 2006.....52
5.06 Business telephone connection charge, 2005...........7
5.07 Business monthly telephone subscription, 2005......27
5.08 Local supplier quality, 2006...................................44
5.09 Computer, comm., and other services imports, 2004.......69

Government readiness — 45

6.01 Government prioritization of ICT, 2006....................83
6.02 Gov't. procurement of advanced tech products, 2006.......73
6.03 Importance of ICT to gov't. vision of the future, 2006.......76
6.04 E-participation index, 2005.....................................56
6.05 E-government readiness index, 2005......................35

Usage component — 55

Individual usage — 47

7.01 Mobile telephone subscribers, 2005.......................29
7.02 Personal computers, 2004......................................56
7.03 Broadband Internet subscribers, 2005....................51
7.04 Internet users, 2005..55
7.05 Internet bandwidth, 2004.......................................34

Business usage — 64

8.01 Prevalence of foreign technology licensing, 2006....38
8.02 Firm-level technology absorption, 2006..................83
8.03 Capacity for innovation, 2006................................74
8.04 Availability of new telephone lines, 2006................53
8.05 Availability of mobile telephones, 2006..................30
8.06 Extent of business Internet use, 2006....................87

Government usage — 73

9.01 Government success in ICT promotion, 2006...........83
9.02 Availability of online services, 2006.......................63
9.03 ICT use and government efficiency, 2006................62
9.04 ICT pervasiveness, 2006..80

Guatemala

Key indicators

Population (millions), 2005..12.6
GDP (PPP) per capita (US$), 2005.........................4,154.8
Internet users per 100 inhabitants, 2005......................7.9
Internet bandwidth (Mbps/10,000 inhabitants), 2004................0.6

Networked Readiness Index

Year (number of economies)	Rank
2006–2007 (122) **79**	
2005–2006 (115)..98	
2004–2005 (104)..88	

Global Competitiveness Index 2006–2007 (125)   75

Environment component — 87

Market environment — 55
1.01 Venture capital availability, 2006...........................74
1.02 Financial market sophistication, 2006...................67
1.03 Technological readiness, 2006..............................46
1.04 State of cluster development, 2006......................57
1.05 US utility patents, 2005..72
1.06 High-tech exports, 2004.......................................54
1.07 Burden of government regulation, 2006...............38
1.08 Extent and effect of taxation, 2006......................37
1.09 Time required to start a business, 2006...............51
1.10 No. of procedures required to start a business, 2006.........95
1.11 Intensity of local competition, 2006....................56
1.12 Freedom of the press, 2006.................................34

Political and regulatory environment — 90
2.01 Effectiveness of law-making bodies, 2006.........112
2.02 Laws relating to ICT, 2006...................................81
2.03 Judicial independence, 2006................................78
2.04 Intellectual property protection, 2006..................72
2.05 Efficiency of legal framework, 2006.....................78
2.06 Property rights, 2006..72
2.07 Quality of competition in the ISP sector, 2006.....22
2.08 No. of procedures to enforce a contract, 2006.........109
2.09 Time to enforce a contract, 2006.........................79

Infrastructure environment — 97
3.01 Telephone lines, 2005..83
3.02 Secure Internet servers, 2005..............................62
3.03 Internet hosts, 2004...65
3.04 Electricity production, 2003..................................90
3.05 Availability of scientists and engineers, 2006.......90
3.06 Quality of scientific research institutions, 2006.......91
3.07 Tertiary enrollment, 2003......................................94

Readiness component — 85

Individual readiness — 92
4.01 Quality of math and science education, 2006...................109
4.02 Quality of the educational system, 2006...........................109
4.03 Quality of public schools, 2006..109
4.04 Internet access in schools, 2006..82
4.05 Buyer sophistication, 2006...76
4.06 Residential telephone connection charge, 2004...............107
4.07 Residential monthly telephone subscription, 2005.............77
4.08 High-speed monthly broadband subscription, 2006...........80
4.09 Lowest cost of broadband, 2006..85
4.10 Cost of mobile telephone call, 2004..................................23

Business readiness — 88
5.01 Extent of staff training, 2006..73
5.02 Local availability of research and training, 2006.................55
5.03 Quality of management schools, 2006................................56
5.04 Company spending on R&D, 2006......................................66
5.05 University-industry research collaboration, 2006................57
5.06 Business telephone connection charge, 2004...................103
5.07 Business monthly telephone subscription, 2004................61
5.08 Local supplier quality, 2006...52
5.09 Computer, comm., and other services imports, 2004.........97

Government readiness — 70
6.01 Government prioritization of ICT, 2006...............................96
6.02 Gov't. procurement of advanced tech products, 2006........83
6.03 Importance of ICT to gov't. vision of the future, 2006........61
6.04 E-participation index, 2005...36
6.05 E-government readiness index, 2005.................................83

Usage component — 61

Individual usage — 86
7.01 Mobile telephone subscribers, 2005..................................76
7.02 Personal computers, 2004..91
7.03 Broadband Internet subscribers, 2005...............................72
7.04 Internet users, 2005...79
7.05 Internet bandwidth, 2004...57

Business usage — 45
8.01 Prevalence of foreign technology licensing, 2006..............77
8.02 Firm-level technology absorption, 2006..............................61
8.03 Capacity for innovation, 2006...54
8.04 Availability of new telephone lines, 2006............................36
8.05 Availability of mobile telephones, 2006..............................23
8.06 Extent of business Internet use, 2006................................39

Government usage — 60
9.01 Government success in ICT promotion, 2006.....................99
9.02 Availability of online services, 2006...................................46
9.03 ICT use and government efficiency, 2006...........................41
9.04 ICT pervasiveness, 2006..68

181

Guyana

Key indicators

Population (millions), 2005...0.8
GDP (PPP) per capita (US$), 2005......................4,611.9
Internet users per 100 inhabitants, 2005................21.3
Internet bandwidth (Mbps/10,000 inhabitants) ....n/a

Networked Readiness Index

Year (number of economies)	Rank
2006–2007 (122) **98**	
2005–2006 (115)..111	
2004–2005 (104) ....—	

Global Competitiveness Index 2006–2007 (125)	111

Environment component	108
Market environment	**112**

1.01 Venture capital availability, 2006.........................114
1.02 Financial market sophistication, 2006.................118
1.03 Technological readiness, 2006.............................113
1.04 State of cluster development, 2006 93
1.05 US utility patents, 2005 79
1.06 High-tech exports, 2004 90
1.07 Burden of government regulation, 2006 100
1.08 Extent and effect of taxation, 2006....................105
1.09 Time required to start a business, 2006 86
1.10 No. of procedures required to start a business, 2006.........36
1.11 Intensity of local competition, 2006 104
1.12 Freedom of the press, 2006..................................95

Political and regulatory environment	**106**

2.01 Effectiveness of law-making bodies, 2006.........91
2.02 Laws relating to ICT, 2006...................................112
2.03 Judicial independence, 2006 90
2.04 Intellectual property protection, 2006................122
2.05 Efficiency of legal framework, 2006...................115
2.06 Property rights, 2006...110
2.07 Quality of competition in the ISP sector, 2006....84
2.08 No. of procedures to enforce a contract, 2006 ...57
2.09 Time to enforce a contract, 2006 95

Infrastructure environment	**94**

3.01 Telephone lines, 2005..70
3.02 Secure Internet servers, 2005 83
3.03 Internet hosts, 2004 77
3.04 Electricity production ....n/a
3.05 Availability of scientists and engineers, 2006...114
3.06 Quality of scientific research institutions, 2006 ..99
3.07 Tertiary enrollment, 2004......................................95

Readiness component	88
Individual readiness	**84**

4.01 Quality of math and science education, 2006....89
4.02 Quality of the educational system, 2006.............76
4.03 Quality of public schools, 2006 70
4.04 Internet access in schools, 2006.........................103
4.05 Buyer sophistication, 2006 94
4.06 Residential telephone connection charge, 2005 ...20
4.07 Residential monthly telephone subscription, 2005 ...80
4.08 High-speed monthly broadband subscription charge ....n/a
4.09 Lowest cost of broadband, 2006...........................82
4.10 Cost of mobile telephone call, 2005.....................82

Business readiness	**95**

5.01 Extent of staff training, 2006................................96
5.02 Local availability of research and training, 2006...111
5.03 Quality of management schools, 2006................100
5.04 Company spending on R&D, 2006 95
5.05 University-industry research collaboration, 2006...107
5.06 Business telephone connection charge, 2005 61
5.07 Business monthly telephone subscription, 2005 ...96
5.08 Local supplier quality, 2006..................................99
5.09 Computer, comm., and other services imports, 2004.........27

Government readiness	**103**

6.01 Government prioritization of ICT, 2006................80
6.02 Gov't. procurement of advanced tech products, 2006........95
6.03 Importance of ICT to gov't. vision of the future, 2006......114
6.04 E-participation index, 2005 92
6.05 E-government readiness index, 2005...................76

Usage component	106
Individual usage	**61**

7.01 Mobile telephone subscribers, 2005 74
7.02 Personal computers, 2004......................................83
7.03 Broadband Internet subscribers, 2005 67
7.04 Internet users, 2005 48
7.05 Internet bandwidth..n/a

Business usage	**114**

8.01 Prevalence of foreign technology licensing, 2006.............114
8.02 Firm-level technology absorption, 2006.............109
8.03 Capacity for innovation, 2006 99
8.04 Availability of new telephone lines, 2006 118
8.05 Availability of mobile telephones, 2006 106
8.06 Extent of business Internet use, 2006 91

Government usage	**111**

9.01 Government success in ICT promotion, 2006.....95
9.02 Availability of online services, 2006....................112
9.03 ICT use and government efficiency, 2006...........117
9.04 ICT pervasiveness, 2006 95

Honduras

Key indicators

Population (millions), 2005...7.2
GDP (PPP) per capita (US$), 2005.........................3,009.2
Internet users per 100 inhabitants, 2005.....................3.6
Internet bandwidth (Mbps/10,000 inhabitants) ....n/a

Networked Readiness Index

Year (number of economies)	Rank
2006–2007 (122) **94**	
2005–2006 (115)...100	
2004–2005 (104)..97	

Global Competitiveness Index 2006–2007 (125)   93

Environment component   98
Market environment   84

1.01   Venture capital availability, 2006 86
1.02   Financial market sophistication, 2006..................................86
1.03   Technological readiness, 2006..91
1.04   State of cluster development, 2006 63
1.05   US utility patents, 2005 79
1.06   High-tech exports...n/a
1.07   Burden of government regulation, 2006 79
1.08   Extent and effect of taxation, 2006 67
1.09   Time required to start a business, 2006.............................83
1.10   No. of procedures required to start a business, 2006.........95
1.11   Intensity of local competition, 2006 114
1.12   Freedom of the press, 2006..47

Political and regulatory environment   97

2.01   Effectiveness of law-making bodies, 2006..........................78
2.02   Laws relating to ICT, 2006...88
2.03   Judicial independence, 2006 102
2.04   Intellectual property protection, 2006................................93
2.05   Efficiency of legal framework, 2006...................................97
2.06   Property rights, 2006 102
2.07   Quality of competition in the ISP sector, 2006...................65
2.08   No. of procedures to enforce a contract, 2006 76
2.09   Time to enforce a contract, 2006 65

Infrastructure environment   109

3.01   Telephone lines, 2005...90
3.02   Secure Internet servers, 2005...70
3.03   Internet hosts, 2004 83
3.04   Electricity production, 2003 85
3.05   Availability of scientists and engineers, 2006...................110
3.06   Quality of scientific research institutions, 2006 114
3.07   Tertiary enrollment, 2004...84

Readiness component   90
Individual readiness   93

4.01   Quality of math and science education, 2006 114
4.02   Quality of the educational system, 2006...........................116
4.03   Quality of public schools, 2006 106
4.04   Internet access in schools, 2006..99
4.05   Buyer sophistication, 2006 106
4.06   Residential telephone connection charge, 2005 71
4.07   Residential monthly telephone subscription, 2005 64
4.08   High-speed monthly broadband subscription charge ....n/a
4.09   Lowest cost of broadband, 2006..86
4.10   Cost of mobile telephone call, 2005...................................85

Business readiness   91

5.01   Extent of staff training, 2006...92
5.02   Local availability of research and training, 2006 84
5.03   Quality of management schools, 2006...............................97
5.04   Company spending on R&D, 2006 112
5.05   University-industry research collaboration, 2006................91
5.06   Business telephone connection charge, 2005 74
5.07   Business monthly telephone subscription, 2005 74
5.08   Local supplier quality, 2006 89
5.09   Computer, comm., and other services imports, 2004.........64

Government readiness   95

6.01   Government prioritization of ICT, 2006..............................91
6.02   Gov't. procurement of advanced tech products, 2006......101
6.03   Importance of ICT to gov't. vision of the future, 2006......106
6.04   E-participation index, 2005 36
6.05   E-government readiness index, 2005 94

Usage component   103
Individual usage   96

7.01   Mobile telephone subscribers, 2005 94
7.02   Personal computers, 2004...94
7.03   Broadband Internet subscribers, 2004 97
7.04   Internet users, 2005 98
7.05   Internet bandwidth..n/a

Business usage   96

8.01   Prevalence of foreign technology licensing, 2006...............97
8.02   Firm-level technology absorption, 2006.............................100
8.03   Capacity for innovation, 2006 76
8.04   Availability of new telephone lines, 2006 110
8.05   Availability of mobile telephones, 2006.............................64
8.06   Extent of business Internet use, 2006 78

Government usage   98

9.01   Government success in ICT promotion, 2006...................110
9.02   Availability of online services, 2006...................................80
9.03   ICT use and government efficiency, 2006..........................94
9.04   ICT pervasiveness, 2006 99

183

Hong Kong SAR

Key indicators

Population (millions), 2005...7.0
GDP (PPP) per capita (US$), 2005.....................33,478.7
Internet users per 100 inhabitants, 2005....................50.1
Internet bandwidth (Mbps/10,000 inhabitants), 2004..............70.7

Networked Readiness Index

Year (number of economies)	Rank
2006–2007 (122) **12**	
2005–2006 (115)...11	
2004–2005 (104)..7	

Global Competitiveness Index 2006–2007 (125)   11

Environment component   16

Market environment   4

1.01	Venture capital availability, 2006..9
1.02	Financial market sophistication, 2006.....................................4
1.03	Technological readiness, 2006..22
1.04	State of cluster development, 2006 12
1.05	US utility patents, 2005 22
1.06	High-tech exports, 2004 7
1.07	Burden of government regulation, 2006 4
1.08	Extent and effect of taxation, 2006..2
1.09	Time required to start a business, 2006.............................12
1.10	No. of procedures required to start a business, 2006.........10
1.11	Intensity of local competition, 2006 6
1.12	Freedom of the press, 2006...28

Political and regulatory environment   12

2.01	Effectiveness of law-making bodies, 2006.......................35
2.02	Laws relating to ICT, 2006...19
2.03	Judicial independence, 2006 13
2.04	Intellectual property protection, 2006.................................20
2.05	Efficiency of legal framework, 2006......................................12
2.06	Property rights, 2006..14
2.07	Quality of competition in the ISP sector, 2006.....................8
2.08	No. of procedures to enforce a contract, 2006 4
2.09	Time to enforce a contract, 2006 12

Infrastructure environment   23

3.01	Telephone lines, 2005...12
3.02	Secure Internet servers, 2005..24
3.03	Internet hosts, 2004 14
3.04	Electricity production, 2003 39
3.05	Availability of scientists and engineers, 2006.....................27
3.06	Quality of scientific research institutions, 2006 24
3.07	Tertiary enrollment, 2004...61

Readiness component   10

Individual readiness   5

4.01	Quality of math and science education, 2006 6
4.02	Quality of the educational system, 2006..............................7
4.03	Quality of public schools, 2006 18
4.04	Internet access in schools, 2006..7
4.05	Buyer sophistication, 2006 4
4.06	Residential telephone connection charge, 2005 17
4.07	Residential monthly telephone subscription, 2005 19
4.08	High-speed monthly broadband subscription, 2006............23
4.09	Lowest cost of broadband, 2006..14
4.10	Cost of mobile telephone call, 2004......................................2

Business readiness   13

5.01	Extent of staff training, 2006..21
5.02	Local availability of research and training, 2006 18
5.03	Quality of management schools, 2006.................................13
5.04	Company spending on R&D, 2006 23
5.05	University-industry research collaboration, 2006................17
5.06	Business telephone connection charge, 2005 16
5.07	Business monthly telephone subscription, 2005 18
5.08	Local supplier quality, 2006 15
5.09	Computer, comm., and other services imports..................n/a

Government readiness   14

6.01	Government prioritization of ICT, 2006................................20
6.02	Gov't. procurement of advanced tech products, 2006........17
6.03	Importance of ICT to gov't. vision of the future, 2006........13
6.04	E-participation index...n/a
6.05	E-government readiness index ....n/a

Usage component   8

Individual usage   7

7.01	Mobile telephone subscribers, 2005 4
7.02	Personal computers, 2004...11
7.03	Broadband Internet subscribers, 2005 5
7.04	Internet users, 2005 19
7.05	Internet bandwidth, 2004 8

Business usage   18

8.01	Prevalence of foreign technology licensing, 2006..............12
8.02	Firm-level technology absorption, 2006...............................18
8.03	Capacity for innovation, 2006 22
8.04	Availability of new telephone lines, 2006 7
8.05	Availability of mobile telephones, 2006................................9
8.06	Extent of business Internet use, 2006 24

Government usage   9

9.01	Government success in ICT promotion, 2006.....................20
9.02	Availability of online services, 2006.....................................10
9.03	ICT use and government efficiency, 2006.............................9
9.04	ICT pervasiveness, 2006 18

Hungary

Key indicators

Population (millions), 2005..10.1

GDP (PPP) per capita (US$), 2005.....................................16,823.4

Internet users per 100 inhabitants, 2005.................................29.7

Internet bandwidth (Mbps/10,000 inhabitants), 2004................9.9

Networked Readiness Index

Year (number of economies)	Rank
2006–2007 (122) **33**	
2005–2006 (115)...38	
2004–2005 (104)...38	

Global Competitiveness Index 2006–2007 (125)	41

Environment component	28
Market environment	**29**

1.01 Venture capital availability, 2006...33
1.02 Financial market sophistication, 2006...................................34
1.03 Technological readiness, 2006..48
1.04 State of cluster development, 2006...6
1.05 US utility patents, 2005..29
1.06 High-tech exports, 2004...11
1.07 Burden of government regulation, 2006.................................77
1.08 Extent and effect of taxation, 2006.......................................77
1.09 Time required to start a business, 2006................................74
1.10 No. of procedures required to start a business, 2006.........16
1.11 Intensity of local competition, 2006.....................................30
1.12 Freedom of the press, 2006..40

Political and regulatory environment	**34**

2.01 Effectiveness of law-making bodies, 2006...........................56
2.02 Laws relating to ICT, 2006..40
2.03 Judicial independence, 2006...48
2.04 Intellectual property protection, 2006...................................33
2.05 Efficiency of legal framework, 2006......................................44
2.06 Property rights, 2006..32
2.07 Quality of competition in the ISP sector, 2006.....................72
2.08 No. of procedures to enforce a contract, 2006....................16
2.09 Time to enforce a contract, 2006...33

Infrastructure environment	**33**

3.01 Telephone lines, 2005...35
3.02 Secure Internet servers, 2005..41
3.03 Internet hosts, 2004...23
3.04 Electricity production, 2003..50
3.05 Availability of scientists and engineers, 2006......................30
3.06 Quality of scientific research institutions, 2006...................26
3.07 Tertiary enrollment, 2004..25

Readiness component	34
Individual readiness	**35**

4.01 Quality of math and science education, 2006.....................13
4.02 Quality of the educational system, 2006..............................42
4.03 Quality of public schools, 2006..26
4.04 Internet access in schools, 2006..29
4.05 Buyer sophistication, 2006...84
4.06 Residential telephone connection charge, 2005..................65
4.07 Residential monthly telephone subscription, 2005..............61
4.08 High-speed monthly broadband subscription, 2006...........38
4.09 Lowest cost of broadband, 2006..34
4.10 Cost of mobile telephone call, 2004.....................................36

Business readiness	**41**

5.01 Extent of staff training, 2006..53
5.02 Local availability of research and training, 2006..................36
5.03 Quality of management schools, 2006..................................33
5.04 Company spending on R&D, 2006..59
5.05 University-industry research collaboration, 2006.................30
5.06 Business telephone connection charge, 2004......................69
5.07 Business monthly telephone subscription, 2005..................52
5.08 Local supplier quality, 2006...58
5.09 Computer, comm., and other services imports, 2004.........10

Government readiness	**33**

6.01 Government prioritization of ICT, 2006.................................56
6.02 Gov't. procurement of advanced tech products, 2006........55
6.03 Importance of ICT to gov't. vision of the future, 2006........58
6.04 E-participation index, 2005...27
6.05 E-government readiness index, 2005....................................27

Usage component	40
Individual usage	**39**

7.01 Mobile telephone subscribers, 2005....................................25
7.02 Personal computers, 2004..44
7.03 Broadband Internet subscribers, 2005.................................34
7.04 Internet users, 2005...38
7.05 Internet bandwidth, 2004..29

Business usage	**30**

8.01 Prevalence of foreign technology licensing, 2006...............44
8.02 Firm-level technology absorption, 2006................................28
8.03 Capacity for innovation, 2006...32
8.04 Availability of new telephone lines, 2006.............................31
8.05 Availability of mobile telephones, 2006................................21
8.06 Extent of business Internet use, 2006..................................43

Government usage	**49**

9.01 Government success in ICT promotion, 2006.......................54
9.02 Availability of online services, 2006.....................................45
9.03 ICT use and government efficiency, 2006.............................69
9.04 ICT pervasiveness, 2006..52

Iceland

Key indicators

Population (millions), 2005..0.3
GDP (PPP) per capita (US$), 2005.........................35,115.1
Internet users per 100 inhabitants, 2005....................87.8
Internet bandwidth (Mbps/10,000 inhabitants), 2004.............42.3

Networked Readiness Index

Year (number of economies)	Rank
2006–2007 (122) **8**	
2005–2006 (115)..4	
2004–2005 (104)..2	

Global Competitiveness Index 2006–2007 (125)   14

Environment component   1

Market environment   10

1.01   Venture capital availability, 2006.............................12
1.02   Financial market sophistication, 2006....................16
1.03   Technological readiness, 2006...................................5
1.04   State of cluster development, 2006..........................47
1.05   US utility patents, 2005..13
1.06   High-tech exports, 2004..68
1.07   Burden of government regulation, 2006 1
1.08   Extent and effect of taxation, 2006...........................6
1.09   Time required to start a business, 2006....................3
1.10   No. of procedures required to start a business, 2006.........10
1.11   Intensity of local competition, 2006.........................24
1.12   Freedom of the press, 2006......................................20

Political and regulatory environment   3

2.01   Effectiveness of law-making bodies, 2006..................6
2.02   Laws relating to ICT, 2006.......................................17
2.03   Judicial independence, 2006....................................12
2.04   Intellectual property protection, 2006.......................7
2.05   Efficiency of legal framework, 2006...........................3
2.06   Property rights, 2006..2
2.07   Quality of competition in the ISP sector, 2006.........7
2.08   No. of procedures to enforce a contract, 2006..........1
2.09   Time to enforce a contract, 2006.............................39

Infrastructure environment   1

3.01   Telephone lines, 2005..4
3.02   Secure Internet servers, 2005....................................1
3.03   Internet hosts, 2004...2
3.04   Electricity production, 2003..1
3.05   Availability of scientists and engineers, 2006.........16
3.06   Quality of scientific research institutions, 2006........30
3.07   Tertiary enrollment, 2004...15

Readiness component   25

Individual readiness   10

4.01   Quality of math and science education, 2006.........33
4.02   Quality of the educational system, 2006...................3
4.03   Quality of public schools, 2006.................................6
4.04   Internet access in schools, 2006..............................1
4.05   Buyer sophistication, 2006.......................................27
4.06   Residential telephone connection charge...............n/a
4.07   Residential monthly telephone subscription...........n/a
4.08   High-speed monthly broadband subscription, 2006.............20
4.09   Lowest cost of broadband, 2006.............................18
4.10   Cost of mobile telephone call, 2005..........................6

Business readiness   34

5.01   Extent of staff training, 2006...................................11
5.02   Local availability of research and training, 2006.......19
5.03   Quality of management schools, 2006.....................16
5.04   Company spending on R&D, 2006..........................21
5.05   University-industry research collaboration, 2006......21
5.06   Business telephone connection charge...................n/a
5.07   Business monthly telephone subscription...............n/a
5.08   Local supplier quality, 2006.....................................19
5.09   Computer, comm., and other services imports, 2004.........72

Government readiness   25

6.01   Government prioritization of ICT, 2006....................15
6.02   Gov't. procurement of advanced tech products, 2006.........47
6.03   Importance of ICT to gov't. vision of the future, 2006..........8
6.04   E-participation index, 2005......................................62
6.05   E-government readiness index, 2005.......................15

Usage component   4

Individual usage   6

7.01   Mobile telephone subscribers, 2005........................11
7.02   Personal computers, 2004.......................................23
7.03   Broadband Internet subscribers, 2005......................1
7.04   Internet users, 2005...1
7.05   Internet bandwidth, 2004...15

Business usage   8

8.01   Prevalence of foreign technology licensing, 2006.............20
8.02   Firm-level technology absorption, 2006.....................1
8.03   Capacity for innovation, 2006..................................25
8.04   Availability of new telephone lines, 2006..................1
8.05   Availability of mobile telephones, 2006.....................4
8.06   Extent of business Internet use, 2006........................8

Government usage   3

9.01   Government success in ICT promotion, 2006.............9
9.02   Availability of online services, 2006...........................6
9.03   ICT use and government efficiency, 2006...................3
9.04   ICT pervasiveness, 2006..7

India

Key indicators

Population (millions), 2005..1103.4
GDP (PPP) per capita (US$), 2005..3,320.5
Internet users per 100 inhabitants, 2005....................................5.4
Internet bandwidth (Mbps/10,000 inhabitants), 2004................0.1

Networked Readiness Index

Year (number of economies)	Rank
2006–2007 (122)	**44**
2005–2006 (115)...	40
2004–2005 (104)...	39

Global Competitiveness Index 2006–2007 (125)	43

Environment component	46
Market environment	**28**

1.01 Venture capital availability, 2006..20
1.02 Financial market sophistication, 2006....................................32
1.03 Technological readiness, 2006...23
1.04 State of cluster development, 2006 11
1.05 US utility patents, 2005 54
1.06 High-tech exports, 2004 49
1.07 Burden of government regulation, 2006 66
1.08 Extent and effect of taxation, 2006......................................21
1.09 Time required to start a business, 2006................................65
1.10 No. of procedures required to start a business, 2006.........79
1.11 Intensity of local competition, 2006 4
1.12 Freedom of the press, 2006...12

Political and regulatory environment	**48**

2.01 Effectiveness of law-making bodies, 2006...........................20
2.02 Laws relating to ICT, 2006..31
2.03 Judicial independence, 2006 14
2.04 Intellectual property protection, 2006...................................34
2.05 Efficiency of legal framework, 2006......................................23
2.06 Property rights, 2006 25
2.07 Quality of competition in the ISP sector, 2006....................19
2.08 No. of procedures to enforce a contract, 2006 115
2.09 Time to enforce a contract, 2006 116

Infrastructure environment	**63**

3.01 Telephone lines, 2005..96
3.02 Secure Internet servers, 2005...89
3.03 Internet hosts, 2004 97
3.04 Electricity production, 2003 88
3.05 Availability of scientists and engineers, 2006........................4
3.06 Quality of scientific research institutions, 2006 14
3.07 Tertiary enrollment, 2004..90

Readiness component	37
Individual readiness	**52**

4.01 Quality of math and science education, 2006.......................7
4.02 Quality of the educational system, 2006.............................25
4.03 Quality of public schools, 2006 100
4.04 Internet access in schools, 2006...52
4.05 Buyer sophistication, 2006 26
4.06 Residential telephone connection charge, 2005 78
4.07 Residential monthly telephone subscription, 2005 88
4.08 High-speed monthly broadband subscription, 2006............65
4.09 Lowest cost of broadband, 2006..79
4.10 Cost of mobile telephone call, 2005....................................45

Business readiness	**28**

5.01 Extent of staff training, 2006..28
5.02 Local availability of research and training, 2006 28
5.03 Quality of management schools, 2006...................................3
5.04 Company spending on R&D, 2006 25
5.05 University-industry research collaboration, 2006.................34
5.06 Business telephone connection charge, 2005 67
5.07 Business monthly telephone subscription, 2005 93
5.08 Local supplier quality, 2006 28
5.09 Computer, comm., and other services imports.................n/a

Government readiness	**39**

6.01 Government prioritization of ICT, 2006................................11
6.02 Gov't. procurement of advanced tech products, 2006........40
6.03 Importance of ICT to gov't. vision of the future, 2006........20
6.04 E-participation index, 2005 56
6.05 E-government readiness index, 2005 75

Usage component	48
Individual usage	**107**

7.01 Mobile telephone subscribers, 2005 106
7.02 Personal computers, 2004...99
7.03 Broadband Internet subscribers, 2005 77
7.04 Internet users, 2005 89
7.05 Internet bandwidth, 2004 74

Business usage	**20**

8.01 Prevalence of foreign technology licensing, 2006.................1
8.02 Firm-level technology absorption, 2006...............................13
8.03 Capacity for innovation, 2006 28
8.04 Availability of new telephone lines, 2006 30
8.05 Availability of mobile telephones, 2006 25
8.06 Extent of business Internet use, 2006 31

Government usage	**36**

9.01 Government success in ICT promotion, 2006.....................17
9.02 Availability of online services, 2006.....................................40
9.03 ICT use and government efficiency, 2006...........................27
9.04 ICT pervasiveness, 2006 58

Indonesia

Key indicators

Population (millions), 2005..222.8
GDP (PPP) per capita (US$), 2005...4,459.1
Internet users per 100 inhabitants, 2005.................................7.2
Internet bandwidth (Mbps/10,000 inhabitants), 2004.................0.1

Networked Readiness Index

Year (number of economies)	Rank
2006–2007 (122) **62**	
2005–2006 (115)..68	
2004–2005 (104)..51	

Global Competitiveness Index 2006–2007 (125)	50

Environment component	56
Market environment	**40**

1.01	Venture capital availability, 2006..18
1.02	Financial market sophistication, 2006...................................83
1.03	Technological readiness, 2006...49
1.04	State of cluster development, 2006 14
1.05	US utility patents, 2005 75
1.06	High-tech exports, 2004 27
1.07	Burden of government regulation, 2006 5
1.08	Extent and effect of taxation, 2006 11
1.09	Time required to start a business, 2006.............................110
1.10	No. of procedures required to start a business, 2006.........89
1.11	Intensity of local competition, 2006 15
1.12	Freedom of the press, 2006..45

Political and regulatory environment	**68**

2.01	Effectiveness of law-making bodies, 2006...........................79
2.02	Laws relating to ICT, 2006...78
2.03	Judicial independence, 2006..91
2.04	Intellectual property protection, 2006...................................61
2.05	Efficiency of legal framework, 2006......................................75
2.06	Property rights, 2006..88
2.07	Quality of competition in the ISP sector, 2006....................26
2.08	No. of procedures to enforce a contract, 2006 69
2.09	Time to enforce a contract, 2006 78

Infrastructure environment	**76**

3.01	Telephone lines, 2005..94
3.02	Secure Internet servers, 2005 95
3.03	Internet hosts, 2004...86
3.04	Electricity production, 2003 92
3.05	Availability of scientists and engineers, 2006......................36
3.06	Quality of scientific research institutions, 2006 27
3.07	Tertiary enrollment, 2004...83

Readiness component	52
Individual readiness	**42**

4.01	Quality of math and science education, 2006 28
4.02	Quality of the educational system, 2006..............................23
4.03	Quality of public schools, 2006 46
4.04	Internet access in schools, 2006..75
4.05	Buyer sophistication, 2006 12
4.06	Residential telephone connection charge, 2005 73
4.07	Residential monthly telephone subscription, 2005 78
4.08	High-speed monthly broadband subscription, 2006...........69
4.09	Lowest cost of broadband, 2006...74
4.10	Cost of mobile telephone call, 2004.....................................43

Business readiness	**38**

5.01	Extent of staff training, 2006..40
5.02	Local availability of research and training, 2006 24
5.03	Quality of management schools, 2006..................................34
5.04	Company spending on R&D, 2006 26
5.05	University-industry research collaboration, 2006.................79
5.06	Business telephone connection charge, 2005 68
5.07	Business monthly telephone subscription, 2005 75
5.08	Local supplier quality, 2006 50
5.09	Computer, comm., and other services imports, 2004...........4

Government readiness	**92**

6.01	Government prioritization of ICT, 2006...............................118
6.02	Gov't. procurement of advanced tech products, 2006........23
6.03	Importance of ICT to gov't. vision of the future, 2006......112
6.04	E-participation index, 2005 34
6.05	E-government readiness index, 2005....................................81

Usage component	93
Individual usage	**90**

7.01	Mobile telephone subscribers, 2005 90
7.02	Personal computers, 2004..97
7.03	Broadband Internet subscribers...n/a
7.04	Internet users, 2005 81
7.05	Internet bandwidth, 2004 75

Business usage	**59**

8.01	Prevalence of foreign technology licensing, 2006.................4
8.02	Firm-level technology absorption, 2006...............................74
8.03	Capacity for innovation, 2006 59
8.04	Availability of new telephone lines, 2006 86
8.05	Availability of mobile telephones, 2006...............................73
8.06	Extent of business Internet use, 2006 48

Government usage	**113**

9.01	Government success in ICT promotion, 2006....................115
9.02	Availability of online services, 2006.....................................87
9.03	ICT use and government efficiency, 2006..........................115
9.04	ICT pervasiveness, 2006 117

188

Ireland

Key indicators

Population (millions), 2005..4.1
GDP (PPP) per capita (US$), 2005......................................40,609.8
Internet users per 100 inhabitants, 2005..................................27.6
Internet bandwidth (Mbps/10,000 inhabitants), 2004..............60.8

Networked Readiness Index

Year (number of economies)	Rank
2006–2007 (122) **21**	
2005–2006 (115)...20	
2004–2005 (104)...22	

Global Competitiveness Index 2006–2007 (125)   21

Environment component	18
Market environment	**14**

1.01 Venture capital availability, 2006..7
1.02 Financial market sophistication, 2006.....................................7
1.03 Technological readiness, 2006...37
1.04 State of cluster development, 2006.......................................35
1.05 US utility patents, 2005...23
1.06 High-tech exports, 2004..12
1.07 Burden of government regulation, 2006................................17
1.08 Extent and effect of taxation, 2006.......................................10
1.09 Time required to start a business, 2006................................27
1.10 No. of procedures required to start a business, 2006...........7
1.11 Intensity of local competition, 2006.....................................20
1.12 Freedom of the press, 2006...36

Political and regulatory environment	**18**

2.01 Effectiveness of law-making bodies, 2006...........................22
2.02 Laws relating to ICT, 2006...27
2.03 Judicial independence, 2006..11
2.04 Intellectual property protection, 2006...................................22
2.05 Efficiency of legal framework, 2006......................................22
2.06 Property rights, 2006..9
2.07 Quality of competition in the ISP sector, 2006....................62
2.08 No. of procedures to enforce a contract, 2006.....................7
2.09 Time to enforce a contract, 2006..14

Infrastructure environment	**18**

3.01 Telephone lines, 2005...19
3.02 Secure Internet servers, 2005..10
3.03 Internet hosts, 2004..24
3.04 Electricity production, 2003..30
3.05 Availability of scientists and engineers, 2006......................19
3.06 Quality of scientific research institutions, 2006..................15
3.07 Tertiary enrollment, 2004...27

Readiness component	20
Individual readiness	**9**

4.01 Quality of math and science education, 2006......................16
4.02 Quality of the educational system, 2006................................6
4.03 Quality of public schools, 2006...5
4.04 Internet access in schools, 2006...35
4.05 Buyer sophistication, 2006...8
4.06 Residential telephone connection charge, 2005..................23
4.07 Residential monthly telephone subscription, 2005.............32
4.08 High-speed monthly broadband subscription, 2006...........15
4.09 Lowest cost of broadband, 2006...23
4.10 Cost of mobile telephone call, 2005.....................................17

Business readiness	**6**

5.01 Extent of staff training, 2006...15
5.02 Local availability of research and training, 2006.................23
5.03 Quality of management schools, 2006..................................15
5.04 Company spending on R&D, 2006...15
5.05 University-industry research collaboration, 2006................19
5.06 Business telephone connection charge, 2005......................20
5.07 Business monthly telephone subscription, 2005.................21
5.08 Local supplier quality, 2006...16
5.09 Computer, comm., and other services imports, 2004...........3

Government readiness	**31**

6.01 Government prioritization of ICT, 2006.................................36
6.02 Gov't. procurement of advanced tech products, 2006........27
6.03 Importance of ICT to gov't. vision of the future, 2006........40
6.04 E-participation index, 2005..49
6.05 E-government readiness index, 2005....................................20

Usage component	24
Individual usage	**28**

7.01 Mobile telephone subscribers, 2005....................................15
7.02 Personal computers, 2004..19
7.03 Broadband Internet subscribers, 2005.................................33
7.04 Internet users, 2005..41
7.05 Internet bandwidth, 2004...12

Business usage	**28**

8.01 Prevalence of foreign technology licensing, 2006...............28
8.02 Firm-level technology absorption, 2006................................24
8.03 Capacity for innovation, 2006..21
8.04 Availability of new telephone lines, 2006.............................62
8.05 Availability of mobile telephones, 2006...............................62
8.06 Extent of business Internet use, 2006..................................25

Government usage	**16**

9.01 Government success in ICT promotion, 2006.......................42
9.02 Availability of online services, 2006..3
9.03 ICT use and government efficiency, 2006.............................16
9.04 ICT pervasiveness, 2006...22

189

Israel

Key indicators

Population (millions), 2005..6.7
GDP (PPP) per capita (US$), 2005.......................23,474.1
Internet users per 100 inhabitants, 2004...................46.6
Internet bandwidth (Mbps/10,000 inhabitants), 2004...............24.8

Networked Readiness Index

Year (number of economies)	Rank
2006–2007 (122) **18**	
2005–2006 (115)...19	
2004–2005 (104)...18	

Global Competitiveness Index 2006–2007 (125)   15

Environment component   19

Market environment   8

1.01   Venture capital availability, 2006 2
1.02   Financial market sophistication, 2006....................................15
1.03   Technological readiness, 2006..4
1.04   State of cluster development, 2006 21
1.05   US utility patents, 2005 5
1.06   High-tech exports, 2004 19
1.07   Burden of government regulation, 2006 22
1.08   Extent and effect of taxation, 2006.......................................57
1.09   Time required to start a business, 2006................................63
1.10   No. of procedures required to start a business, 2006.........10
1.11   Intensity of local competition, 2006 25
1.12   Freedom of the press, 2006..21

Political and regulatory environment   20

2.01   Effectiveness of law-making bodies, 2006...........................30
2.02   Laws relating to ICT, 2006..22
2.03   Judicial independence, 2006 3
2.04   Intellectual property protection, 2006....................................21
2.05   Efficiency of legal framework, 2006......................................21
2.06   Property rights, 2006...23
2.07   Quality of competition in the ISP sector, 2006.....................3
2.08   No. of procedures to enforce a contract, 2006 60
2.09   Time to enforce a contract, 2006 80

Infrastructure environment   19

3.01   Telephone lines, 2005...29
3.02   Secure Internet servers, 2005 23
3.03   Internet hosts, 2004 19
3.04   Electricity production, 2003 26
3.05   Availability of scientists and engineers, 2006.......................1
3.06   Quality of scientific research institutions, 2006 4
3.07   Tertiary enrollment, 2004..29

Readiness component   22

Individual readiness   22

4.01   Quality of math and science education, 2006 17
4.02   Quality of the educational system, 2006..............................22
4.03   Quality of public schools, 2006 24
4.04   Internet access in schools, 2006...16
4.05   Buyer sophistication, 2006 20
4.06   Residential telephone connection charge, 2004 27
4.07   Residential monthly telephone subscription, 2004 20
4.08   High-speed monthly broadband subscription, 2006 28
4.09   Lowest cost of broadband, 2006...30
4.10   Cost of mobile telephone call ....n/a

Business readiness   12

5.01   Extent of staff training, 2006...23
5.02   Local availability of research and training, 2006 10
5.03   Quality of management schools, 2006...................................14
5.04   Company spending on R&D, 2006 7
5.05   University-industry research collaboration, 2006...................6
5.06   Business telephone connection charge, 2004 23
5.07   Business monthly telephone subscription, 2004 12
5.08   Local supplier quality, 2006 21
5.09   Computer, comm., and other services imports, 2004.........23

Government readiness   29

6.01   Government prioritization of ICT, 2006..................................45
6.02   Gov't. procurement of advanced tech products, 2006........11
6.03   Importance of ICT to gov't. vision of the future, 2006........32
6.04   E-participation index, 2005 31
6.05   E-government readiness index, 2005 24

Usage component   14

Individual usage   12

7.01   Mobile telephone subscribers, 2005 6
7.02   Personal computers, 2004..4
7.03   Broadband Internet subscribers, 2005 13
7.04   Internet users, 2004 24
7.05   Internet bandwidth, 2004 22

Business usage   5

8.01   Prevalence of foreign technology licensing, 2006...............22
8.02   Firm-level technology absorption, 2006..................................4
8.03   Capacity for innovation, 2006 8
8.04   Availability of new telephone lines, 2006 10
8.05   Availability of mobile telephones, 2006.................................1
8.06   Extent of business Internet use, 2006 14

Government usage   28

9.01   Government success in ICT promotion, 2006......................23
9.02   Availability of online services, 2006.......................................26
9.03   ICT use and government efficiency, 2006..............................23
9.04   ICT pervasiveness, 2006 44

Italy

Key indicators

Population (millions), 2005..58.1

GDP (PPP) per capita (US$), 2005......................................28,533.8

Internet users per 100 inhabitants, 2005................................48.2

Internet bandwidth (Mbps/10,000 inhabitants) ....n/a

Networked Readiness Index

Year (number of economies)	Rank
2006–2007 (122) **38**	
2005–2006 (115)..42	
2004–2005 (104)..45	

Global Competitiveness Index 2006–2007 (125)	42

Environment component	51
Market environment	**58**

1.01 Venture capital availability, 2006..76
1.02 Financial market sophistication, 2006...................................46
1.03 Technological readiness, 2006..50
1.04 State of cluster development, 2006 85
1.05 US utility patents, 2005 25
1.06 High-tech exports, 2004 32
1.07 Burden of government regulation, 2006 119
1.08 Extent and effect of taxation, 2006 118
1.09 Time required to start a business, 2006..............................17
1.10 No. of procedures required to start a business, 2006.........49
1.11 Intensity of local competition, 2006 74
1.12 Freedom of the press, 2006..44

Political and regulatory environment	**74**

2.01 Effectiveness of law-making bodies, 2006..........................80
2.02 Laws relating to ICT, 2006..38
2.03 Judicial independence, 2006 67
2.04 Intellectual property protection, 2006...................................44
2.05 Efficiency of legal framework, 2006.....................................83
2.06 Property rights, 2006 43
2.07 Quality of competition in the ISP sector, 2006....................41
2.08 No. of procedures to enforce a contract, 2006 88
2.09 Time to enforce a contract, 2006 111

Infrastructure environment	**35**

3.01 Telephone lines, 2005..26
3.02 Secure Internet servers, 2005 35
3.03 Internet hosts, 2004 29
3.04 Electricity production, 2003 43
3.05 Availability of scientists and engineers, 2006.....................47
3.06 Quality of scientific research institutions, 2006 87
3.07 Tertiary enrollment, 2004..20

Readiness component	46
Individual readiness	**48**

4.01 Quality of math and science education, 2006 52
4.02 Quality of the educational system, 2006.............................71
4.03 Quality of public schools, 2006 42
4.04 Internet access in schools, 2006..48
4.05 Buyer sophistication, 2006 35
4.06 Residential telephone connection charge...........................n/a
4.07 Residential monthly telephone subscription, 2004 7
4.08 High-speed monthly broadband subscription, 2006...........11
4.09 Lowest cost of broadband, 2006..2
4.10 Cost of mobile telephone call, 2004....................................35

Business readiness	**46**

5.01 Extent of staff training, 2006..62
5.02 Local availability of research and training, 2006 27
5.03 Quality of management schools, 2006.................................47
5.04 Company spending on R&D, 2006 75
5.05 University-industry research collaboration, 2006................64
5.06 Business telephone connection charge...............................n/a
5.07 Business monthly telephone subscription, 2004 3
5.08 Local supplier quality, 2006 30
5.09 Computer, comm., and other services imports, 2004.........16

Government readiness	**47**

6.01 Government prioritization of ICT, 2006.............................101
6.02 Gov't. procurement of advanced tech products, 2006........92
6.03 Importance of ICT to gov't. vision of the future, 2006........81
6.04 E-participation index, 2005 44
6.05 E-government readiness index, 2005..................................25

Usage component	28
Individual usage	**19**

7.01 Mobile telephone subscribers, 2005 3
7.02 Personal computers, 2004...27
7.03 Broadband Internet subscribers, 2005 25
7.04 Internet users, 2005 22
7.05 Internet bandwidth..n/a

Business usage	**44**

8.01 Prevalence of foreign technology licensing, 2006...............63
8.02 Firm-level technology absorption, 2006...............................93
8.03 Capacity for innovation, 2006 20
8.04 Availability of new telephone lines, 2006 65
8.05 Availability of mobile telephones, 2006...............................44
8.06 Extent of business Internet use, 2006 40

Government usage	**41**

9.01 Government success in ICT promotion, 2006......................78
9.02 Availability of online services, 2006....................................58
9.03 ICT use and government efficiency, 2006............................12
9.04 ICT pervasiveness, 2006 59

Jamaica

Key indicators

Population (millions), 2005..2.7
GDP (PPP) per capita (US$), 2005........................4,381.0
Internet users per 100 inhabitants, 2004.....................39.9
Internet bandwidth (Mbps/10,000 inhabitants).............n/a

Networked Readiness Index

Year (number of economies)	Rank
2006–2007 (122) **45**	
2005–2006 (115)..54	
2004–2005 (104)..49	

Global Competitiveness Index 2006–2007 (125)	60

Environment component	47
Market environment	**36**

1.01 Venture capital availability, 2006................................87
1.02 Financial market sophistication, 2006.........................39
1.03 Technological readiness, 2006...................................40
1.04 State of cluster development, 2006.............................71
1.05 US utility patents, 2005..53
1.06 High-tech exports...n/a
1.07 Burden of government regulation, 2006......................85
1.08 Extent and effect of taxation, 2006............................89
1.09 Time required to start a business, 2006........................7
1.10 No. of procedures required to start a business, 2006.........16
1.11 Intensity of local competition, 2006...........................44
1.12 Freedom of the press, 2006......................................59

Political and regulatory environment	**40**

2.01 Effectiveness of law-making bodies, 2006...................46
2.02 Laws relating to ICT, 2006..68
2.03 Judicial independence, 2006.....................................47
2.04 Intellectual property protection, 2006........................59
2.05 Efficiency of legal framework, 2006............................58
2.06 Property rights, 2006...48
2.07 Quality of competition in the ISP sector, 2006.............38
2.08 No. of procedures to enforce a contract, 2006...............7
2.09 Time to enforce a contract, 2006...............................53

Infrastructure environment	**77**

3.01 Telephone lines, 2005..76
3.02 Secure Internet servers, 2005...................................54
3.03 Internet hosts, 2004..84
3.04 Electricity production, 2003......................................56
3.05 Availability of scientists and engineers, 2006..............88
3.06 Quality of scientific research institutions, 2006.............39
3.07 Tertiary enrollment, 2003...79

Readiness component	55
Individual readiness	**58**

4.01 Quality of math and science education, 2006...............87
4.02 Quality of the educational system, 2006.....................67
4.03 Quality of public schools, 2006.................................80
4.04 Internet access in schools, 2006................................64
4.05 Buyer sophistication, 2006.......................................47
4.06 Residential telephone connection charge, 2005............22
4.07 Residential monthly telephone subscription, 2005.........71
4.08 High-speed monthly broadband subscription, 2006........55
4.09 Lowest cost of broadband, 2006................................62
4.10 Cost of mobile telephone call, 2005...........................55

Business readiness	**50**

5.01 Extent of staff training, 2006....................................54
5.02 Local availability of research and training, 2006............56
5.03 Quality of management schools, 2006........................48
5.04 Company spending on R&D, 2006..............................40
5.05 University-industry research collaboration, 2006...........47
5.06 Business telephone connection charge, 2005...............26
5.07 Business monthly telephone subscription, 2005............85
5.08 Local supplier quality, 2006......................................59
5.09 Computer, comm., and other services imports, 2004........35

Government readiness	**48**

6.01 Government prioritization of ICT, 2006........................41
6.02 Gov't. procurement of advanced tech products, 2006........57
6.03 Importance of ICT to gov't. vision of the future, 2006........35
6.04 E-participation index, 2005.......................................70
6.05 E-government readiness index, 2005..........................58

Usage component	35
Individual usage	**29**

7.01 Mobile telephone subscribers, 2005...........................14
7.02 Personal computers, 2004..65
7.03 Broadband Internet subscribers................................n/a
7.04 Internet users, 2004..31
7.05 Internet bandwidth...n/a

Business usage	**52**

8.01 Prevalence of foreign technology licensing, 2006..........53
8.02 Firm-level technology absorption, 2006......................46
8.03 Capacity for innovation, 2006...................................69
8.04 Availability of new telephone lines, 2006....................64
8.05 Availability of mobile telephones, 2006......................32
8.06 Extent of business Internet use, 2006.........................52

Government usage	**42**

9.01 Government success in ICT promotion, 2006................51
9.02 Availability of online services, 2006...........................51
9.03 ICT use and government efficiency, 2006....................45
9.04 ICT pervasiveness, 2006..53

Japan

Key indicators

Population (millions), 2005...128.1
GDP (PPP) per capita (US$), 2005....................................30,615.4
Internet users per 100 inhabitants, 2005................................66.6
Internet bandwidth (Mbps/10,000 inhabitants), 2004..............10.4

Networked Readiness Index

Year (number of economies)	Rank
2006–2007 (122)	**14**
2005–2006 (115)	16
2004–2005 (104)	8

Global Competitiveness Index 2006–2007 (125)	7

Environment component | 12

Market environment 7

1.01 Venture capital availability, 2006 ... 23
1.02 Financial market sophistication, 2006 ... 18
1.03 Technological readiness, 2006 ... 2
1.04 State of cluster development, 2006 ... 27
1.05 US utility patents, 2005 ... 2
1.06 High-tech exports, 2004 ... 10
1.07 Burden of government regulation, 2006 ... 24
1.08 Extent and effect of taxation, 2006 ... 48
1.09 Time required to start a business, 2006 ... 37
1.10 No. of procedures required to start a business, 2006 ... 36
1.11 Intensity of local competition, 2006 ... 3
1.12 Freedom of the press, 2006 ... 27

Political and regulatory environment 15

2.01 Effectiveness of law-making bodies, 2006 ... 23
2.02 Laws relating to ICT, 2006 ... 25
2.03 Judicial independence, 2006 ... 22
2.04 Intellectual property protection, 2006 ... 12
2.05 Efficiency of legal framework, 2006 ... 19
2.06 Property rights, 2006 ... 15
2.07 Quality of competition in the ISP sector, 2006 ... 6
2.08 No. of procedures to enforce a contract, 2006 ... 13
2.09 Time to enforce a contract, 2006 ... 19

Infrastructure environment 14

3.01 Telephone lines, 2005 ... 23
3.02 Secure Internet servers, 2005 ... 18
3.03 Internet hosts, 2004 ... 12
3.04 Electricity production, 2003 ... 19
3.05 Availability of scientists and engineers, 2006 ... 2
3.06 Quality of scientific research institutions, 2006 ... 5
3.07 Tertiary enrollment, 2004 ... 31

Readiness component | 8

Individual readiness 14

4.01 Quality of math and science education, 2006 ... 14
4.02 Quality of the educational system, 2006 ... 19
4.03 Quality of public schools, 2006 ... 20
4.04 Internet access in schools, 2006 ... 18
4.05 Buyer sophistication, 2006 ... 1
4.06 Residential telephone connection charge, 2005 ... 52
4.07 Residential monthly telephone subscription, 2005 ... 16
4.08 High-speed monthly broadband subscription, 2006 ... 2
4.09 Lowest cost of broadband, 2006 ... 1
4.10 Cost of mobile telephone call ... n/a

Business readiness 5

5.01 Extent of staff training, 2006 ... 3
5.02 Local availability of research and training, 2006 ... 1
5.03 Quality of management schools, 2006 ... 59
5.04 Company spending on R&D, 2006 ... 2
5.05 University-industry research collaboration, 2006 ... 9
5.06 Business telephone connection charge, 2005 ... 44
5.07 Business monthly telephone subscription, 2005 ... 24
5.08 Local supplier quality, 2006 ... 2
5.09 Computer, comm., and other services imports, 2004 ... 34

Government readiness 11

6.01 Government prioritization of ICT, 2006 ... 8
6.02 Gov't. procurement of advanced tech products, 2006 ... 5
6.03 Importance of ICT to gov't. vision of the future, 2006 ... 19
6.04 E-participation index, 2005 ... 21
6.05 E-government readiness index, 2005 ... 14

Usage component | 20

Individual usage 22

7.01 Mobile telephone subscribers, 2005 ... 45
7.02 Personal computers, 2004 ... 17
7.03 Broadband Internet subscribers, 2005 ... 14
7.04 Internet users, 2005 ... 9
7.05 Internet bandwidth, 2004 ... 28

Business usage 1

8.01 Prevalence of foreign technology licensing, 2006 ... 10
8.02 Firm-level technology absorption, 2006 ... 2
8.03 Capacity for innovation, 2006 ... 2
8.04 Availability of new telephone lines, 2006 ... 2
8.05 Availability of mobile telephones, 2006 ... 5
8.06 Extent of business Internet use, 2006 ... 13

Government usage 35

9.01 Government success in ICT promotion, 2006 ... 12
9.02 Availability of online services, 2006 ... 41
9.03 ICT use and government efficiency, 2006 ... 57
9.04 ICT pervasiveness, 2006 ... 35

193

Jordan

Key indicators

Population (millions), 2005 ....5.7
GDP (PPP) per capita (US$), 2005 ....5,095.6
Internet users per 100 inhabitants, 2004 ....11.2
Internet bandwidth (Mbps/10,000 inhabitants), 2004 ....0.6

Networked Readiness Index

Year (number of economies)	Rank
2006–2007 (122)	**57**
2005–2006 (115)...	47
2004–2005 (104)...	44

Global Competitiveness Index 2006–2007 (125)	52

Environment component	52
Market environment	**63**

1.01 Venture capital availability, 2006 69
1.02 Financial market sophistication, 2006 62
1.03 Technological readiness, 2006 44
1.04 State of cluster development, 2006 76
1.05 US utility patents, 2005 79
1.06 High-tech exports, 2004 43
1.07 Burden of government regulation, 2006 23
1.08 Extent and effect of taxation, 2006 49
1.09 Time required to start a business, 2006 24
1.10 No. of procedures required to start a business, 2006.........79
1.11 Intensity of local competition, 2006 41
1.12 Freedom of the press, 2006 115

Political and regulatory environment	**39**

2.01 Effectiveness of law-making bodies, 2006................54
2.02 Laws relating to ICT, 2006.......................................64
2.03 Judicial independence, 2006 38
2.04 Intellectual property protection, 2006.......................42
2.05 Efficiency of legal framework, 2006 36
2.06 Property rights, 2006..46
2.07 Quality of competition in the ISP sector, 2006.........21
2.08 No. of procedures to enforce a contract, 2006 99
2.09 Time to enforce a contract, 2006 36

Infrastructure environment	**62**

3.01 Telephone lines, 2004..79
3.02 Secure Internet servers, 2005 72
3.03 Internet hosts, 2004 85
3.04 Electricity production, 2003 70
3.05 Availability of scientists and engineers, 2006............26
3.06 Quality of scientific research institutions, 2006 73
3.07 Tertiary enrollment, 2004...47

Readiness component	64
Individual readiness	**54**

4.01 Quality of math and science education, 2006...........56
4.02 Quality of the educational system, 2006...................44
4.03 Quality of public schools, 2006 63
4.04 Internet access in schools, 2006.............................40
4.05 Buyer sophistication, 2006 71
4.06 Residential telephone connection charge, 2005 69
4.07 Residential monthly telephone subscription, 2005 79
4.08 High-speed monthly broadband subscription, 2006...........42
4.09 Lowest cost of broadband, 2006.............................45
4.10 Cost of mobile telephone call, 2004 51

Business readiness	**85**

5.01 Extent of staff training, 2006...................................60
5.02 Local availability of research and training, 2006.........62
5.03 Quality of management schools, 2006......................76
5.04 Company spending on R&D, 2006 94
5.05 University-industry research collaboration, 2006.........84
5.06 Business telephone connection charge, 2005 71
5.07 Business monthly telephone subscription, 2005 88
5.08 Local supplier quality, 2006.....................................68
5.09 Computer, comm., and other services imports, 2004.........92

Government readiness	**60**

6.01 Government prioritization of ICT, 2006......................27
6.02 Gov't. procurement of advanced tech products, 2006.......84
6.03 Importance of ICT to gov't. vision of the future, 2006........25
6.04 E-participation index, 2005 82
6.05 E-government readiness index, 2005 65

Usage component	59
Individual usage	**84**

7.01 Mobile telephone subscribers, 2004 83
7.02 Personal computers, 2004..70
7.03 Broadband Internet subscribers, 2004 74
7.04 Internet users, 2004 67
7.05 Internet bandwidth, 2004 58

Business usage	**41**

8.01 Prevalence of foreign technology licensing, 2006..............25
8.02 Firm-level technology absorption, 2006.....................55
8.03 Capacity for innovation, 2006 75
8.04 Availability of new telephone lines, 2006 19
8.05 Availability of mobile telephones, 2006 15
8.06 Extent of business Internet use, 2006 67

Government usage	**61**

9.01 Government success in ICT promotion, 2006............31
9.02 Availability of online services, 2006..........................79
9.03 ICT use and government efficiency, 2006.................77
9.04 ICT pervasiveness, 2006 48

Kazakhstan

Key indicators

Population (millions), 2005..14.8
GDP (PPP) per capita (US$), 2005.......................................8,318.3
Internet users per 100 inhabitants, 2004.....................................2.7
Internet bandwidth (Mbps/10,000 inhabitants) ....n/a

Networked Readiness Index

Year (number of economies)	Rank
2006–2007 (122) **73**	
2005–2006 (115)...60	
2004–2005 (104) ....—	

Global Competitiveness Index 2006–2007 (125)   56

Environment component — 59

Market environment — 70

1.01 Venture capital availability, 2006..39
1.02 Financial market sophistication, 2006.....................................65
1.03 Technological readiness, 2006..80
1.04 State of cluster development, 2006 87
1.05 US utility patents, 2005 67
1.06 High-tech exports, 2004 77
1.07 Burden of government regulation, 2006 57
1.08 Extent and effect of taxation, 2006.......................................76
1.09 Time required to start a business, 2006.................................30
1.10 No. of procedures required to start a business, 2006.........27
1.11 Intensity of local competition, 2006 70
1.12 Freedom of the press, 2006...107

Political and regulatory environment — 65

2.01 Effectiveness of law-making bodies, 2006............................44
2.02 Laws relating to ICT, 2006...59
2.03 Judicial independence, 2006 93
2.04 Intellectual property protection, 2006....................................81
2.05 Efficiency of legal framework, 2006.......................................68
2.06 Property rights, 2006 83
2.07 Quality of competition in the ISP sector, 2006.....................83
2.08 No. of procedures to enforce a contract, 2006 78
2.09 Time to enforce a contract, 2006 7

Infrastructure environment — 60

3.01 Telephone lines, 2004...66
3.02 Secure Internet servers, 2005...87
3.03 Internet hosts, 2004 69
3.04 Electricity production, 2003 46
3.05 Availability of scientists and engineers, 2006.....................100
3.06 Quality of scientific research institutions, 2006 53
3.07 Tertiary enrollment, 2004...34

Readiness component — 80

Individual readiness — 95

4.01 Quality of math and science education, 2006 66
4.02 Quality of the educational system, 2006.............................50
4.03 Quality of public schools, 2006 62
4.04 Internet access in schools, 2006...47
4.05 Buyer sophistication, 2006 42
4.06 Residential telephone connection charge.........................n/a
4.07 Residential monthly telephone subscription......................n/a
4.08 High-speed monthly broadband subscription, 2006 94
4.09 Lowest cost of broadband, 2006...89
4.10 Cost of mobile telephone call ....n/a

Business readiness — 75

5.01 Extent of staff training, 2006..79
5.02 Local availability of research and training, 2006 67
5.03 Quality of management schools, 2006................................86
5.04 Company spending on R&D, 2006 57
5.05 University-industry research collaboration, 2006.................63
5.06 Business telephone connection charge...............................n/a
5.07 Business monthly telephone subscription..........................n/a
5.08 Local supplier quality, 2006 70
5.09 Computer, comm., and other services imports, 2004...........5

Government readiness — 43

6.01 Government prioritization of ICT, 2006................................39
6.02 Gov't. procurement of advanced tech products, 2006........56
6.03 Importance of ICT to gov't. vision of the future, 2006........37
6.04 E-participation index, 2005 46
6.05 E-government readiness index, 2005...................................63

Usage component — 71

Individual usage — 83

7.01 Mobile telephone subscribers, 2005 79
7.02 Personal computers ....n/a
7.03 Broadband Internet subscribers, 2004 89
7.04 Internet users, 2004 102
7.05 Internet bandwidth..n/a

Business usage — 75

8.01 Prevalence of foreign technology licensing, 2006..............83
8.02 Firm-level technology absorption, 2006..............................65
8.03 Capacity for innovation, 2006 62
8.04 Availability of new telephone lines, 2006 85
8.05 Availability of mobile telephones, 2006 80
8.06 Extent of business Internet use, 2006 70

Government usage — 47

9.01 Government success in ICT promotion, 2006.....................50
9.02 Availability of online services, 2006....................................44
9.03 ICT use and government efficiency, 2006...........................55
9.04 ICT pervasiveness, 2006 69

195

Kenya

Key indicators

Population (millions), 2005..34.3

GDP (PPP) per capita (US$), 2005.........................1,445.2

Internet users per 100 inhabitants, 2005.....................3.2

Internet bandwidth (Mbps/10,000 inhabitants), 2004.................0.0

Networked Readiness Index

Year (number of economies)	Rank
2006–2007 (122)	**95**
2005–2006 (115)..	91
2004–2005 (104)..	75

Global Competitiveness Index 2006–2007 (125)	94

Environment component	90
Market environment	**100**

1.01	Venture capital availability, 2006..	66
1.02	Financial market sophistication, 2006..................................	68
1.03	Technological readiness, 2006...	82
1.04	State of cluster development, 2006......................................	66
1.05	US utility patents, 2005..	59
1.06	High-tech exports, 2004..	74
1.07	Burden of government regulation, 2006..............................	83
1.08	Extent and effect of taxation, 2006....................................	104
1.09	Time required to start a business, 2006..............................	94
1.10	No. of procedures required to start a business, 2006.........	95
1.11	Intensity of local competition, 2006...................................	47
1.12	Freedom of the press, 2006..	114

Political and regulatory environment	**66**

2.01	Effectiveness of law-making bodies, 2006..........................	87
2.02	Laws relating to ICT, 2006...	76
2.03	Judicial independence, 2006...	84
2.04	Intellectual property protection, 2006.................................	84
2.05	Efficiency of legal framework, 2006....................................	91
2.06	Property rights, 2006...	81
2.07	Quality of competition in the ISP sector, 2006...................	58
2.08	No. of procedures to enforce a contract, 2006...................	34
2.09	Time to enforce a contract, 2006.......................................	40

Infrastructure environment	**90**

3.01	Telephone lines, 2005...	107
3.02	Secure Internet servers, 2005..	102
3.03	Internet hosts, 2004..	90
3.04	Electricity production, 2003..	99
3.05	Availability of scientists and engineers, 2006....................	56
3.06	Quality of scientific research institutions, 2006.................	31
3.07	Tertiary enrollment, 2004..	107

Readiness component	100
Individual readiness	**108**

4.01	Quality of math and science education, 2006.....................	81
4.02	Quality of the educational system, 2006............................	37
4.03	Quality of public schools, 2006..	81
4.04	Internet access in schools, 2006..	110
4.05	Buyer sophistication, 2006...	82
4.06	Residential telephone connection charge, 2005.................	88
4.07	Residential monthly telephone subscription, 2005...........	110
4.08	High-speed monthly broadband subscription, 2006...........	93
4.09	Lowest cost of broadband, 2006...	99
4.10	Cost of mobile telephone call, 2004..................................	101

Business readiness	**79**

5.01	Extent of staff training, 2006...	56
5.02	Local availability of research and training, 2006................	49
5.03	Quality of management schools, 2006................................	72
5.04	Company spending on R&D, 2006.......................................	34
5.05	University-industry research collaboration, 2006...............	50
5.06	Business telephone connection charge, 2005....................	78
5.07	Business monthly telephone subscription, 2005...............	105
5.08	Local supplier quality, 2006..	67
5.09	Computer, comm., and other services imports, 2004.........	60

Government readiness	**97**

6.01	Government prioritization of ICT, 2006...............................	103
6.02	Gov't. procurement of advanced tech products, 2006........	45
6.03	Importance of ICT to gov't. vision of the future, 2006........	77
6.04	E-participation index, 2005...	92
6.05	E-government readiness index, 2005..................................	97

Usage component	92
Individual usage	**105**

7.01	Mobile telephone subscribers, 2005...................................	99
7.02	Personal computers, 2004...	96
7.03	Broadband Internet subscribers, 2004................................	97
7.04	Internet users, 2005..	101
7.05	Internet bandwidth, 2004..	87

Business usage	**81**

8.01	Prevalence of foreign technology licensing, 2006..............	40
8.02	Firm-level technology absorption, 2006..............................	53
8.03	Capacity for innovation, 2006...	52
8.04	Availability of new telephone lines, 2006...........................	111
8.05	Availability of mobile telephones, 2006.............................	90
8.06	Extent of business Internet use, 2006................................	69

Government usage	**91**

9.01	Government success in ICT promotion, 2006......................	74
9.02	Availability of online services, 2006...................................	77
9.03	ICT use and government efficiency, 2006...........................	83
9.04	ICT pervasiveness, 2006...	113

196

Korea, Rep.

Key indicators

Population (millions), 2005...47.8
GDP (PPP) per capita (US$), 2005.....................................20,590.5
Internet users per 100 inhabitants, 2005...................................68.4
Internet bandwidth (Mbps/10,000 inhabitants), 2004..............14.8

Networked Readiness Index

Year (number of economies)	Rank
2006–2007 (122) **19**	
2005–2006 (115)...14	
2004–2005 (104)...24	

Global Competitiveness Index 2006–2007 (125)	24

Environment component	23
Market environment	**27**

1.01   Venture capital availability, 2006...68
1.02   Financial market sophistication, 2006.....................................42
1.03   Technological readiness, 2006..21
1.04   State of cluster development, 2006..31
1.05   US utility patents, 2005...9
1.06   High-tech exports, 2004...8
1.07   Burden of government regulation, 2006..................................49
1.08   Extent and effect of taxation, 2006..71
1.09   Time required to start a business, 2006..................................36
1.10   No. of procedures required to start a business, 2006.........89
1.11   Intensity of local competition, 2006..36
1.12   Freedom of the press, 2006..65

Political and regulatory environment	**26**

2.01   Effectiveness of law-making bodies, 2006............................65
2.02   Laws relating to ICT, 2006..11
2.03   Judicial independence, 2006..51
2.04   Intellectual property protection, 2006.....................................31
2.05   Efficiency of legal framework, 2006..47
2.06   Property rights, 2006...34
2.07   Quality of competition in the ISP sector, 2006.......................1
2.08   No. of procedures to enforce a contract, 2006.....................51
2.09   Time to enforce a contract, 2006...17

Infrastructure environment	**20**

3.01   Telephone lines, 2005...18
3.02   Secure Internet servers, 2005..49
3.03   Internet hosts, 2004..15
3.04   Electricity production, 2003...24
3.05   Availability of scientists and engineers, 2006.......................28
3.06   Quality of scientific research institutions, 2006.................22
3.07   Tertiary enrollment, 2004..2

Readiness component	9
Individual readiness	**23**

4.01   Quality of math and science education, 2006.....................23
4.02   Quality of the educational system, 2006...............................38
4.03   Quality of public schools, 2006..43
4.04   Internet access in schools, 2006..4
4.05   Buyer sophistication, 2006..21
4.06   Residential telephone connection charge, 2005.................30
4.07   Residential monthly telephone subscription, 2005...............9
4.08   High-speed monthly broadband subscription, 2006...........25
4.09   Lowest cost of broadband, 2006..2
4.10   Cost of mobile telephone call, 2005......................................16

Business readiness	**21**

5.01   Extent of staff training, 2006..18
5.02   Local availability of research and training, 2006.................29
5.03   Quality of management schools, 2006....................................53
5.04   Company spending on R&D, 2006..9
5.05   University-industry research collaboration, 2006................16
5.06   Business telephone connection charge, 2005.....................24
5.07   Business monthly telephone subscription, 2005...................7
5.08   Local supplier quality, 2006..26
5.09   Computer, comm., and other services imports, 2004.........24

Government readiness	**3**

6.01   Government prioritization of ICT, 2006...................................18
6.02   Gov't. procurement of advanced tech products, 2006........14
6.03   Importance of ICT to gov't. vision of the future, 2006........15
6.04   E-participation index, 2005...4
6.05   E-government readiness index, 2005.......................................5

Usage component	11
Individual usage	**11**

7.01   Mobile telephone subscribers, 2005......................................41
7.02   Personal computers, 2004...16
7.03   Broadband Internet subscribers, 2005.....................................2
7.04   Internet users, 2005..7
7.05   Internet bandwidth, 2004..25

Business usage	**11**

8.01   Prevalence of foreign technology licensing, 2006..............49
8.02   Firm-level technology absorption, 2006.................................11
8.03   Capacity for innovation, 2006...13
8.04   Availability of new telephone lines, 2006.............................22
8.05   Availability of mobile telephones, 2006................................28
8.06   Extent of business Internet use, 2006.....................................1

Government usage	**12**

9.01   Government success in ICT promotion, 2006.......................22
9.02   Availability of online services, 2006......................................20
9.03   ICT use and government efficiency, 2006.............................22
9.04   ICT pervasiveness, 2006...3

197

Kuwait

Key indicators

Population (millions), 2005..2.7
GDP (PPP) per capita (US$), 2005.......................................16,301.1
Internet users per 100 inhabitants, 2005....................................26.1
Internet bandwidth (Mbps/10,000 inhabitants), 2004...............1.1

Networked Readiness Index

Year (number of economies)	Rank
2006–2007 (122) **54**
2005–2006 (115)..46
2004–2005 (104) ....—

Global Competitiveness Index 2006–2007 (125)	44

Environment component	45
Market environment	**33**

1.01 Venture capital availability, 2006...27
1.02 Financial market sophistication, 2006....................................40
1.03 Technological readiness, 2006...43
1.04 State of cluster development, 2006..28
1.05 US utility patents, 2005 37
1.06 High-tech exports...n/a
1.07 Burden of government regulation, 2006 71
1.08 Extent and effect of taxation, 2006..4
1.09 Time required to start a business, 2006...............................65
1.10 No. of procedures required to start a business, 2006.........95
1.11 Intensity of local competition, 2006 63
1.12 Freedom of the press, 2006...93

Political and regulatory environment	**53**

2.01 Effectiveness of law-making bodies, 2006..........................37
2.02 Laws relating to ICT, 2006...89
2.03 Judicial independence, 2006 31
2.04 Intellectual property protection, 2006...................................58
2.05 Efficiency of legal framework, 2006......................................29
2.06 Property rights, 2006..54
2.07 Quality of competition in the ISP sector, 2006....................50
2.08 No. of procedures to enforce a contract, 2006.................111
2.09 Time to enforce a contract, 2006 42

Infrastructure environment	**48**

3.01 Telephone lines, 2005..62
3.02 Secure Internet servers, 2005...39
3.03 Internet hosts, 2004 75
3.04 Electricity production, 2003 4
3.05 Availability of scientists and engineers, 2006.....................63
3.06 Quality of scientific research institutions, 2006 57
3.07 Tertiary enrollment, 2004...74

Readiness component	62
Individual readiness	**41**

4.01 Quality of math and science education, 2006....................61
4.02 Quality of the educational system, 2006..............................61
4.03 Quality of public schools, 2006 56
4.04 Internet access in schools, 2006...37
4.05 Buyer sophistication, 2006 44
4.06 Residential telephone connection charge, 2004 40
4.07 Residential monthly telephone subscription, 2004 17
4.08 High-speed monthly broadband subscription, 2006............46
4.09 Lowest cost of broadband, 2006...52
4.10 Cost of mobile telephone call, 2004.....................................34

Business readiness	**57**

5.01 Extent of staff training, 2006..46
5.02 Local availability of research and training, 2006 53
5.03 Quality of management schools, 2006..................................60
5.04 Company spending on R&D, 2006...81
5.05 University-industry research collaboration, 2006................85
5.06 Business telephone connection charge, 2004 54
5.07 Business monthly telephone subscription, 2004 36
5.08 Local supplier quality, 2006 38
5.09 Computer, comm., and other services imports, 2004.........98

Government readiness	**101**

6.01 Government prioritization of ICT, 2006...............................105
6.02 Gov't. procurement of advanced tech products, 2006........98
6.03 Importance of ICT to gov't. vision of the future, 2006........99
6.04 E-participation index, 2005 109
6.05 E-government readiness index, 2005...................................68

Usage component	60
Individual usage	**44**

7.01 Mobile telephone subscribers, 2005 32
7.02 Personal computers, 2004...39
7.03 Broadband Internet subscribers, 2004 57
7.04 Internet users, 2005 43
7.05 Internet bandwidth, 2004 50

Business usage	**50**

8.01 Prevalence of foreign technology licensing, 2006..............47
8.02 Firm-level technology absorption, 2006..............................39
8.03 Capacity for innovation, 2006 107
8.04 Availability of new telephone lines, 2006 45
8.05 Availability of mobile telephones, 2006...............................19
8.06 Extent of business Internet use, 2006 68

Government usage	**96**

9.01 Government success in ICT promotion, 2006.....................86
9.02 Availability of online services, 2006....................................106
9.03 ICT use and government efficiency, 2006............................98
9.04 ICT pervasiveness, 2006 81

Kyrgyz Republic

Key indicators

Population (millions), 2005...5.3
GDP (PPP) per capita (US$), 2005.......................................2,087.7
Internet users per 100 inhabitants, 2005.................................5.3
Internet bandwidth (Mbps/10,000 inhabitants), 2004................0.1

Networked Readiness Index

Year (number of economies)	Rank
2006–2007 (122) **105**	
2005–2006 (115)..103	
2004–2005 (104) ....—	

Global Competitiveness Index 2006–2007 (125)   107

Environment component	103
Market environment	**102**

1.01 Venture capital availability, 2006...88
1.02 Financial market sophistication, 2006.................................113
1.03 Technological readiness, 2006...120
1.04 State of cluster development, 2006 100
1.05 US utility patents, 2005 79
1.06 High-tech exports..n/a
1.07 Burden of government regulation, 2006 117
1.08 Extent and effect of taxation, 2006....................................120
1.09 Time required to start a business, 2006.............................33
1.10 No. of procedures required to start a business, 2006........36
1.11 Intensity of local competition, 2006 118
1.12 Freedom of the press, 2006..91

Political and regulatory environment	**105**

2.01 Effectiveness of law-making bodies, 2006...........................73
2.02 Laws relating to ICT, 2006...118
2.03 Judicial independence, 2006 115
2.04 Intellectual property protection, 2006................................100
2.05 Efficiency of legal framework, 2006...................................114
2.06 Property rights, 2006 116
2.07 Quality of competition in the ISP sector, 2006....................88
2.08 No. of procedures to enforce a contract, 2006 101
2.09 Time to enforce a contract, 2006 3

Infrastructure environment	**81**

3.01 Telephone lines, 2005...85
3.02 Secure Internet servers, 2005...90
3.03 Internet hosts, 2004 74
3.04 Electricity production, 2003 55
3.05 Availability of scientists and engineers, 2006...................109
3.06 Quality of scientific research institutions, 2006 98
3.07 Tertiary enrollment, 2004...45

Readiness component	94
Individual readiness	**88**

4.01 Quality of math and science education, 2006 84
4.02 Quality of the educational system, 2006............................77
4.03 Quality of public schools, 2006 79
4.04 Internet access in schools, 2006...90
4.05 Buyer sophistication, 2006 104
4.06 Residential telephone connection charge, 2004 83
4.07 Residential monthly telephone subscription, 2004 83
4.08 High-speed monthly broadband subscription, 2006...........89
4.09 Lowest cost of broadband, 2006...96
4.10 Cost of mobile telephone call, 2004....................................84

Business readiness	**110**

5.01 Extent of staff training, 2006..118
5.02 Local availability of research and training, 2006 103
5.03 Quality of management schools, 2006...............................110
5.04 Company spending on R&D, 2006 109
5.05 University-industry research collaboration, 2006...............104
5.06 Business telephone connection charge, 2004 99
5.07 Business monthly telephone subscription, 2004 81
5.08 Local supplier quality, 2006 105
5.09 Computer, comm., and other services imports, 2004.........61

Government readiness	**100**

6.01 Government prioritization of ICT, 2006................................87
6.02 Gov't. procurement of advanced tech products, 2006......119
6.03 Importance of ICT to gov't. vision of the future, 2006......117
6.04 E-participation index, 2005 56
6.05 E-government readiness index, 2005...................................69

199

Usage component	118
Individual usage	**104**

7.01 Mobile telephone subscribers, 2005 103
7.02 Personal computers, 2004..93
7.03 Broadband Internet subscribers, 2005 84
7.04 Internet users, 2005 92
7.05 Internet bandwidth, 2004 78

Business usage	**113**

8.01 Prevalence of foreign technology licensing, 2006.............117
8.02 Firm-level technology absorption, 2006..............................118
8.03 Capacity for innovation, 2006 82
8.04 Availability of new telephone lines, 2006 99
8.05 Availability of mobile telephones, 2006 107
8.06 Extent of business Internet use, 2006 114

Government usage	**116**

9.01 Government success in ICT promotion, 2006...................117
9.02 Availability of online services, 2006...................................111
9.03 ICT use and government efficiency, 2006..........................102
9.04 ICT pervasiveness, 2006 118

Latvia

Key indicators

Population (millions), 2005...2.3
GDP (PPP) per capita (US$), 2005.......................................12,666.1
Internet users per 100 inhabitants, 2005...................................44.6
Internet bandwidth (Mbps/10,000 inhabitants), 2004...............9.8

Networked Readiness Index

Year (number of economies)	Rank
2006–2007 (122) **42**	
2005–2006 (115)..51	
2004–2005 (104)..56	

Global Competitiveness Index 2006–2007 (125)   36

Environment component	39
Market environment	**41**

1.01   Venture capital availability, 2006..43
1.02   Financial market sophistication, 2006......................................61
1.03   Technological readiness, 2006...59
1.04   State of cluster development, 2006..45
1.05   US utility patents, 2005...41
1.06   High-tech exports, 2004...50
1.07   Burden of government regulation, 2006...................................36
1.08   Extent and effect of taxation, 2006..38
1.09   Time required to start a business, 2006....................................20
1.10   No. of procedures required to start a business, 2006.........10
1.11   Intensity of local competition, 2006...55
1.12   Freedom of the press, 2006...43

Political and regulatory environment	**42**

2.01   Effectiveness of law-making bodies, 2006...............................53
2.02   Laws relating to ICT, 2006..63
2.03   Judicial independence, 2006...59
2.04   Intellectual property protection, 2006.......................................64
2.05   Efficiency of legal framework, 2006..61
2.06   Property rights, 2006...51
2.07   Quality of competition in the ISP sector, 2006.........................44
2.08   No. of procedures to enforce a contract, 2006..................16
2.09   Time to enforce a contract, 2006...18

Infrastructure environment	**38**

3.01   Telephone lines, 2005...39
3.02   Secure Internet servers, 2005..38
3.03   Internet hosts, 2004..32
3.04   Electricity production, 2003...68
3.05   Availability of scientists and engineers, 2006..........................97
3.06   Quality of scientific research institutions, 2006.................61
3.07   Tertiary enrollment, 2004..9

Readiness component	42
Individual readiness	**34**

4.01   Quality of math and science education, 2006..................35
4.02   Quality of the educational system, 2006..............................35
4.03   Quality of public schools, 2006..30
4.04   Internet access in schools, 2006..33
4.05   Buyer sophistication, 2006...64
4.06   Residential telephone connection charge, 2005................51
4.07   Residential monthly telephone subscription, 2005.............44
4.08   High-speed monthly broadband subscription, 2006............72
4.09   Lowest cost of broadband, 2006..64
4.10   Cost of mobile telephone call, 2005..5

Business readiness	**45**

5.01   Extent of staff training, 2006..42
5.02   Local availability of research and training, 2006.................50
5.03   Quality of management schools, 2006.....................................39
5.04   Company spending on R&D, 2006...50
5.05   University-industry research collaboration, 2006................51
5.06   Business telephone connection charge, 2005....................43
5.07   Business monthly telephone subscription, 2005................49
5.08   Local supplier quality, 2006...45
5.09   Computer, comm., and other services imports, 2004.........52

Government readiness	**63**

6.01   Government prioritization of ICT, 2006....................................88
6.02   Gov't. procurement of advanced tech products, 2006........93
6.03   Importance of ICT to gov't. vision of the future, 2006........83
6.04   E-participation index, 2005...51
6.05   E-government readiness index, 2005.......................................32

Usage component	43
Individual usage	**33**

7.01   Mobile telephone subscribers, 2005..37
7.02   Personal computers, 2004..32
7.03   Broadband Internet subscribers, 2005.....................................27
7.04   Internet users, 2005..28
7.05   Internet bandwidth, 2004..30

Business usage	**47**

8.01   Prevalence of foreign technology licensing, 2006..............80
8.02   Firm-level technology absorption, 2006...................................51
8.03   Capacity for innovation, 2006...42
8.04   Availability of new telephone lines, 2006.................................57
8.05   Availability of mobile telephones, 2006....................................58
8.06   Extent of business Internet use, 2006......................................36

Government usage	**68**

9.01   Government success in ICT promotion, 2006....................96
9.02   Availability of online services, 2006...73
9.03   ICT use and government efficiency, 2006................................70
9.04   ICT pervasiveness, 2006..38

Lesotho

Key indicators

Population (millions), 2005 ....1.8
GDP (PPP) per capita (US$), 2005 ....2,113.4
Internet users per 100 inhabitants, 2004 ....2.4
Internet bandwidth (Mbps/10,000 inhabitants) ....n/a

Networked Readiness Index

Year (number of economies)	Rank
2006–2007 (122)	**116**
2005–2006 (115)	—
2004–2005 (104)	—

Global Competitiveness Index 2006–2007 (125)   112

Environment component   109

Market environment   96

1.01	Venture capital availability, 2006	113
1.02	Financial market sophistication, 2006	117
1.03	Technological readiness, 2006	109
1.04	State of cluster development, 2006	91
1.05	US utility patents, 2005	79
1.06	High-tech exports	n/a
1.07	Burden of government regulation, 2006	101
1.08	Extent and effect of taxation, 2006	69
1.09	Time required to start a business, 2006	101
1.10	No. of procedures required to start a business, 2006	36
1.11	Intensity of local competition, 2006	91
1.12	Freedom of the press, 2006	80

Political and regulatory environment   109

2.01	Effectiveness of law-making bodies, 2006	70
2.02	Laws relating to ICT, 2006	106
2.03	Judicial independence, 2006	54
2.04	Intellectual property protection, 2006	106
2.05	Efficiency of legal framework, 2006	57
2.06	Property rights, 2006	109
2.07	Quality of competition in the ISP sector, 2006	110
2.08	No. of procedures to enforce a contract, 2006	116
2.09	Time to enforce a contract, 2006	98

Infrastructure environment   107

3.01	Telephone lines, 2005	102
3.02	Secure Internet servers	n/a
3.03	Internet hosts, 2004	103
3.04	Electricity production	n/a
3.05	Availability of scientists and engineers, 2006	115
3.06	Quality of scientific research institutions, 2006	110
3.07	Tertiary enrollment, 2003	109

Readiness component   117

Individual readiness   111

4.01	Quality of math and science education, 2006	111
4.02	Quality of the educational system, 2006	84
4.03	Quality of public schools, 2006	88
4.04	Internet access in schools, 2006	118
4.05	Buyer sophistication, 2006	107
4.06	Residential telephone connection charge, 2005	96
4.07	Residential monthly telephone subscription, 2005	111
4.08	High-speed monthly broadband subscription charge	n/a
4.09	Lowest cost of broadband	n/a
4.10	Cost of mobile telephone call, 2004	88

Business readiness   120

5.01	Extent of staff training, 2006	91
5.02	Local availability of research and training, 2006	115
5.03	Quality of management schools, 2006	119
5.04	Company spending on R&D, 2006	118
5.05	University-industry research collaboration, 2006	117
5.06	Business telephone connection charge, 2005	89
5.07	Business monthly telephone subscription, 2005	106
5.08	Local supplier quality, 2006	119
5.09	Computer, comm., and other services imports, 2004	99

Government readiness   110

6.01	Government prioritization of ICT, 2006	100
6.02	Gov't. procurement of advanced tech products, 2006	110
6.03	Importance of ICT to gov't. vision of the future, 2006	107
6.04	E-participation index, 2005	101
6.05	E-government readiness index, 2005	93

Usage component   116

Individual usage   100

7.01	Mobile telephone subscribers, 2005	98
7.02	Personal computers	n/a
7.03	Broadband Internet subscribers, 2005	94
7.04	Internet users, 2004	103
7.05	Internet bandwidth	n/a

Business usage   115

8.01	Prevalence of foreign technology licensing, 2006	104
8.02	Firm-level technology absorption, 2006	112
8.03	Capacity for innovation, 2006	112
8.04	Availability of new telephone lines, 2006	114
8.05	Availability of mobile telephones, 2006	110
8.06	Extent of business Internet use, 2006	113

Government usage   114

9.01	Government success in ICT promotion, 2006	106
9.02	Availability of online services, 2006	103
9.03	ICT use and government efficiency, 2006	119
9.04	ICT pervasiveness, 2006	115

Lithuania

Key indicators

Population (millions), 2005...3.4
GDP (PPP) per capita (US$), 2005.......................14,158.4
Internet users per 100 inhabitants, 2004....................35.7
Internet bandwidth (Mbps/10,000 inhabitants), 2004...............1.9

Networked Readiness Index

Year (number of economies)	Rank
2006–2007 (122) **39**	
2005–2006 (115)..44	
2004–2005 (104)..43	

Global Competitiveness Index 2006–2007 (125)	40

Environment component	41
Market environment	**50**

1.01 Venture capital availability, 2006..........................46
1.02 Financial market sophistication, 2006....................58
1.03 Technological readiness, 2006..............................70
1.04 State of cluster development, 2006 80
1.05 US utility patents, 2005 40
1.06 High-tech exports, 2004 51
1.07 Burden of government regulation, 2006 58
1.08 Extent and effect of taxation, 2006......................61
1.09 Time required to start a business, 2006................44
1.10 No. of procedures required to start a business, 2006.........27
1.11 Intensity of local competition, 2006 37
1.12 Freedom of the press, 2006................................32

Political and regulatory environment	**43**

2.01 Effectiveness of law-making bodies, 2006.............68
2.02 Laws relating to ICT, 2006.................................43
2.03 Judicial independence, 2006 75
2.04 Intellectual property protection, 2006...................70
2.05 Efficiency of legal framework, 2006 67
2.06 Property rights, 2006 49
2.07 Quality of competition in the ISP sector, 2006.........33
2.08 No. of procedures to enforce a contract, 2006........30
2.09 Time to enforce a contract, 2006 4

Infrastructure environment	**36**

3.01 Telephone lines, 2004..54
3.02 Secure Internet servers, 2005.............................44
3.03 Internet hosts, 2004 30
3.04 Electricity production, 2003 36
3.05 Availability of scientists and engineers, 2006...........48
3.06 Quality of scientific research institutions, 2006 44
3.07 Tertiary enrollment, 2004...................................12

Readiness component	43
Individual readiness	**37**

4.01 Quality of math and science education, 2006 26
4.02 Quality of the educational system, 2006...........................45
4.03 Quality of public schools, 2006 40
4.04 Internet access in schools, 2006.....................................31
4.05 Buyer sophistication, 2006 63
4.06 Residential telephone connection charge, 2005 61
4.07 Residential monthly telephone subscription, 2004 58
4.08 High-speed monthly broadband subscription, 2006...........31
4.09 Lowest cost of broadband, 2006.....................................27
4.10 Cost of mobile telephone call, 2004.................................48

Business readiness	**47**

5.01 Extent of staff training, 2006...43
5.02 Local availability of research and training, 2006 48
5.03 Quality of management schools, 2006.............................50
5.04 Company spending on R&D, 2006 49
5.05 University-industry research collaboration, 2006................55
5.06 Business telephone connection charge, 2005 52
5.07 Business monthly telephone subscription, 2004 41
5.08 Local supplier quality, 2006 48
5.09 Computer, comm., and other services imports, 2004.........84

Government readiness	**55**

6.01 Government prioritization of ICT, 2006.............................46
6.02 Gov't. procurement of advanced tech products, 2006........71
6.03 Importance of ICT to gov't. vision of the future, 2006........79
6.04 E-participation index, 2005 68
6.05 E-government readiness index, 2005...............................40

Usage component	34
Individual usage	**34**

7.01 Mobile telephone subscribers, 2005 2
7.02 Personal computers, 2004...42
7.03 Broadband Internet subscribers, 2005 32
7.04 Internet users, 2004 33
7.05 Internet bandwidth, 2004 43

Business usage	**39**

8.01 Prevalence of foreign technology licensing, 2006...............72
8.02 Firm-level technology absorption, 2006............................45
8.03 Capacity for innovation, 2006 44
8.04 Availability of new telephone lines, 2006 52
8.05 Availability of mobile telephones, 2006............................42
8.06 Extent of business Internet use, 2006 29

Government usage	**38**

9.01 Government success in ICT promotion, 2006....................55
9.02 Availability of online services, 2006.................................31
9.03 ICT use and government efficiency, 2006..........................56
9.04 ICT pervasiveness, 2006 41

Luxembourg

Key indicators

Population (millions), 2005..0.5

GDP (PPP) per capita (US$), 2005.....................................69,799.6

Internet users per 100 inhabitants, 2005................................67.7

Internet bandwidth (Mbps/10,000 inhabitants), 2004.............32.4

Networked Readiness Index

Year (number of economies)	Rank
2006–2007 (122) **25**	
2005–2006 (115)..26	
2004–2005 (104)..17	

Global Competitiveness Index 2006–2007 (125)	22

Environment component — 22

Market environment — 24

1.01   Venture capital availability, 2006...8
1.02   Financial market sophistication, 2006......................................3
1.03   Technological readiness, 2006..41
1.04   State of cluster development, 2006 47
1.05   US utility patents, 2005 11
1.06   High-tech exports, 2004 48
1.07   Burden of government regulation, 2006 14
1.08   Extent and effect of taxation, 2006...9
1.09   Time required to start a business ....n/a
1.10   No. of procedures required to start a business ....n/a
1.11   Intensity of local competition, 2006 62
1.12   Freedom of the press, 2006...33

Political and regulatory environment — 22

2.01   Effectiveness of law-making bodies, 2006..............................9
2.02   Laws relating to ICT, 2006..21
2.03   Judicial independence, 2006 23
2.04   Intellectual property protection, 2006...................................15
2.05   Efficiency of legal framework, 2006......................................16
2.06   Property rights, 2006 12
2.07   Quality of competition in the ISP sector, 2006...................49
2.08   No. of procedures to enforce a contract ....n/a
2.09   Time to enforce a contract...n/a

Infrastructure environment — 22

3.01   Telephone lines, 2005..14
3.02   Secure Internet servers, 2005 6
3.03   Internet hosts, 2004 16
3.04   Electricity production, 2003 31
3.05   Availability of scientists and engineers, 2006....................76
3.06   Quality of scientific research institutions, 2006 56
3.07   Tertiary enrollment, 2004..88

Readiness component — 26

Individual readiness — 18

4.01   Quality of math and science education, 2006....................32
4.02   Quality of the educational system, 2006.............................36
4.03   Quality of public schools, 2006 27
4.04   Internet access in schools, 2006...15
4.05   Buyer sophistication, 2006 5
4.06   Residential telephone connection charge, 2005 2
4.07   Residential monthly telephone subscription, 2005 8
4.08   High-speed monthly broadband subscription, 2006 8
4.09   Lowest cost of broadband, 2006...20
4.10   Cost of mobile telephone call, 2004....................................3

Business readiness — 29

5.01   Extent of staff training, 2006..14
5.02   Local availability of research and training, 2006.................51
5.03   Quality of management schools, 2006.................................79
5.04   Company spending on R&D, 2006 20
5.05   University-industry research collaboration, 2006................43
5.06   Business telephone connection charge, 2005 3
5.07   Business monthly telephone subscription, 2005 4
5.08   Local supplier quality, 2006 27
5.09   Computer, comm., and other services imports, 2004.........49

Government readiness — 32

6.01   Government prioritization of ICT, 2006................................38
6.02   Gov't. procurement of advanced tech products, 2006..........7
6.03   Importance of ICT to gov't. vision of the future, 2006........27
6.04   E-participation index, 2005 60
6.05   E-government readiness index, 2005....................................28

Usage component — 19

Individual usage — 9

7.01   Mobile telephone subscribers, 2005 1
7.02   Personal computers, 2004..10
7.03   Broadband Internet subscribers, 2005 19
7.04   Internet users, 2005 8
7.05   Internet bandwidth, 2004 19

Business usage — 26

8.01   Prevalence of foreign technology licensing, 2006...............57
8.02   Firm-level technology absorption, 2006...............................34
8.03   Capacity for innovation, 2006 16
8.04   Availability of new telephone lines, 2006 25
8.05   Availability of mobile telephones, 2006 33
8.06   Extent of business Internet use, 2006.................................33

Government usage — 31

9.01   Government success in ICT promotion, 2006....................35
9.02   Availability of online services, 2006....................................35
9.03   ICT use and government efficiency, 2006............................39
9.04   ICT pervasiveness, 2006 19

203

Macedonia, FYR

Key indicators

Population (millions), 2005..2.0
GDP (PPP) per capita (US$), 2005..7,748.1
Internet users per 100 inhabitants, 2005....................................7.9
Internet bandwidth (Mbps/10,000 inhabitants)........................n/a

Networked Readiness Index

Year (number of economies)	Rank
2006–2007 (122) **81**	
2005–2006 (115)..82	
2004–2005 (104)..85	

Global Competitiveness Index 2006–2007 (125)	80

Environment component	78
Market environment	**86**

1.01 Venture capital availability, 2006..44
1.02 Financial market sophistication, 2006.....................................87
1.03 Technological readiness, 2006...98
1.04 State of cluster development, 2006..105
1.05 US utility patents, 2005...79
1.06 High-tech exports, 2004...66
1.07 Burden of government regulation, 2006..................................69
1.08 Extent and effect of taxation, 2006...86
1.09 Time required to start a business, 2006..................................24
1.10 No. of procedures required to start a business, 2006.........61
1.11 Intensity of local competition, 2006.......................................89
1.12 Freedom of the press, 2006...76

Political and regulatory environment	**96**

2.01 Effectiveness of law-making bodies, 2006..............................76
2.02 Laws relating to ICT, 2006...94
2.03 Judicial independence, 2006..105
2.04 Intellectual property protection, 2006...................................103
2.05 Efficiency of legal framework, 2006..94
2.06 Property rights, 2006...105
2.07 Quality of competition in the ISP sector, 2006....................112
2.08 No. of procedures to enforce a contract, 2006...................41
2.09 Time to enforce a contract, 2006...41

Infrastructure environment	**53**

3.01 Telephone lines, 2005..49
3.02 Secure Internet servers, 2005...93
3.03 Internet hosts, 2004...67
3.04 Electricity production...n/a
3.05 Availability of scientists and engineers, 2006........................51
3.06 Quality of scientific research institutions, 2006..................83
3.07 Tertiary enrollment, 2004...65

Readiness component	71
Individual readiness	**66**

4.01 Quality of math and science education, 2006....................40
4.02 Quality of the educational system, 2006...........................43
4.03 Quality of public schools, 2006..44
4.04 Internet access in schools, 2006...84
4.05 Buyer sophistication, 2006...98
4.06 Residential telephone connection charge, 2005................50
4.07 Residential monthly telephone subscription, 2005............82
4.08 High-speed monthly broadband subscription, 2006...........84
4.09 Lowest cost of broadband, 2006...59
4.10 Cost of mobile telephone call, 2005....................................78

Business readiness	**70**

5.01 Extent of staff training, 2006..69
5.02 Local availability of research and training, 2006................87
5.03 Quality of management schools, 2006................................85
5.04 Company spending on R&D, 2006.....................................100
5.05 University-industry research collaboration, 2006................58
5.06 Business telephone connection charge, 2005...................42
5.07 Business monthly telephone subscription, 2005...............77
5.08 Local supplier quality, 2006...86
5.09 Computer, comm., and other services imports, 2004.........17

Government readiness	**91**

6.01 Government prioritization of ICT, 2006..............................109
6.02 Gov't. procurement of advanced tech products, 2006........94
6.03 Importance of ICT to gov't. vision of the future, 2006........94
6.04 E-participation index, 2005..62
6.05 E-government readiness index, 2005..................................66

Usage component	89
Individual usage	**56**

7.01 Mobile telephone subscribers, 2005...................................51
7.02 Personal computers, 2004...63
7.03 Broadband Internet subscribers, 2005................................62
7.04 Internet users, 2005...80
7.05 Internet bandwidth...n/a

Business usage	**92**

8.01 Prevalence of foreign technology licensing, 2006...............93
8.02 Firm-level technology absorption, 2006.............................115
8.03 Capacity for innovation, 2006..66
8.04 Availability of new telephone lines, 2006............................69
8.05 Availability of mobile telephones, 2006..............................79
8.06 Extent of business Internet use, 2006................................115

Government usage	**103**

9.01 Government success in ICT promotion, 2006....................103
9.02 Availability of online services, 2006....................................91
9.03 ICT use and government efficiency, 2006..........................109
9.04 ICT pervasiveness, 2006...93

Madagascar

Key indicators

Population (millions), 2005...18.6

GDP (PPP) per capita (US$), 2005...........................908.5

Internet users per 100 inhabitants, 2005...................0.5

Internet bandwidth (Mbps/10,000 inhabitants), 2004...............0.0

Networked Readiness Index

Year (number of economies)	Rank
2006–2007 (122) **102**	
2005–2006 (115)..102	
2004–2005 (104)...87	

Global Competitiveness Index 2006–2007 (125)	109

Environment component	97
Market environment	**94**

1.01	Venture capital availability, 2006..80
1.02	Financial market sophistication, 2006..............................103
1.03	Technological readiness, 2006..90
1.04	State of cluster development, 2006.....................................92
1.05	US utility patents, 2005 79
1.06	High-tech exports, 2004 92
1.07	Burden of government regulation, 2006 91
1.08	Extent and effect of taxation, 2006....................................68
1.09	Time required to start a business, 2006.............................33
1.10	No. of procedures required to start a business, 2006.........61
1.11	Intensity of local competition, 2006 111
1.12	Freedom of the press, 2006..94

Political and regulatory environment	**86**

2.01	Effectiveness of law-making bodies, 2006..........................77
2.02	Laws relating to ICT, 2006..95
2.03	Judicial independence, 2006 85
2.04	Intellectual property protection, 2006................................69
2.05	Efficiency of legal framework, 2006.....................................81
2.06	Property rights, 2006...98
2.07	Quality of competition in the ISP sector, 2006....................91
2.08	No. of procedures to enforce a contract, 2006 51
2.09	Time to enforce a contract, 2006 83

Infrastructure environment	**99**

3.01	Telephone lines, 2005..119
3.02	Secure Internet servers, 2005...106
3.03	Internet hosts, 2004 107
3.04	Electricity production ....n/a
3.05	Availability of scientists and engineers, 2006...................68
3.06	Quality of scientific research institutions, 2006 92
3.07	Tertiary enrollment, 2004...110

Readiness component	105
Individual readiness	**104**

4.01	Quality of math and science education, 2006 79
4.02	Quality of the educational system, 2006............................89
4.03	Quality of public schools, 2006 99
4.04	Internet access in schools, 2006.......................................107
4.05	Buyer sophistication, 2006 119
4.06	Residential telephone connection charge, 2005 68
4.07	Residential monthly telephone subscription, 2004 114
4.08	High-speed monthly broadband subscription charge ....n/a
4.09	Lowest cost of broadband, 2006...91
4.10	Cost of mobile telephone call, 2004..................................103

Business readiness	**106**

5.01	Extent of staff training, 2006...100
5.02	Local availability of research and training, 2006 99
5.03	Quality of management schools, 2006................................73
5.04	Company spending on R&D, 2006 61
5.05	University-industry research collaboration, 2006................82
5.06	Business telephone connection charge, 2004 96
5.07	Business monthly telephone subscription, 2004 112
5.08	Local supplier quality, 2006 106
5.09	Computer, comm., and other services imports.................n/a

Government readiness	**90**

6.01	Government prioritization of ICT, 2006...............................61
6.02	Gov't. procurement of advanced tech products, 2006........65
6.03	Importance of ICT to gov't. vision of the future, 2006........50
6.04	E-participation index, 2005 82
6.05	E-government readiness index, 2005 107

Usage component	100
Individual usage	**118**

7.01	Mobile telephone subscribers, 2005 118
7.02	Personal computers, 2004...108
7.03	Broadband Internet subscribers, 2004 97
7.04	Internet users, 2005 113
7.05	Internet bandwidth, 2004 85

Business usage	**102**

8.01	Prevalence of foreign technology licensing, 2006.............108
8.02	Firm-level technology absorption, 2006..............................67
8.03	Capacity for innovation, 2006 78
8.04	Availability of new telephone lines, 2006 108
8.05	Availability of mobile telephones, 2006............................105
8.06	Extent of business Internet use, 2006...............................99

Government usage	**78**

9.01	Government success in ICT promotion, 2006....................26
9.02	Availability of online services, 2006...................................102
9.03	ICT use and government efficiency, 2006..........................52
9.04	ICT pervasiveness, 2006 86

205

Malawi

Key indicators

Population (millions), 2005..12.9
GDP (PPP) per capita (US$), 2005.............................596.1
Internet users per 100 inhabitants, 2005.....................0.4
Internet bandwidth (Mbps/10,000 inhabitants), 2004.................0.0

Networked Readiness Index

Year (number of economies)	Rank
2006–2007 (122) **111**	
2005–2006 (115) ....—	
2004–2005 (104)..93	

Global Competitiveness Index 2006–2007 (125)	117

Environment component	100
Market environment	**105**

1.01   Venture capital availability, 2006...106
1.02   Financial market sophistication, 2006...............................110
1.03   Technological readiness, 2006...105
1.04   State of cluster development, 2006 104
1.05   US utility patents, 2005 79
1.06   High-tech exports, 2004 78
1.07   Burden of government regulation, 2006 33
1.08   Extent and effect of taxation, 2006......................................84
1.09   Time required to start a business, 2006 71
1.10   No. of procedures required to start a business, 2006.........61
1.11   Intensity of local competition, 2006 113
1.12   Freedom of the press, 2006...75

Political and regulatory environment	**71**

2.01   Effectiveness of law-making bodies, 2006..........................81
2.02   Laws relating to ICT, 2006...102
2.03   Judicial independence, 2006 43
2.04   Intellectual property protection, 2006...............................95
2.05   Efficiency of legal framework, 2006......................................63
2.06   Property rights, 2006..73
2.07   Quality of competition in the ISP sector, 2006....................78
2.08   No. of procedures to enforce a contract, 2006 88
2.09   Time to enforce a contract, 2006 35

Infrastructure environment	**114**

3.01   Telephone lines, 2005..109
3.02   Secure Internet servers, 2005..107
3.03   Internet hosts, 2004 118
3.04   Electricity production ....n/a
3.05   Availability of scientists and engineers, 2006....................119
3.06   Quality of scientific research institutions, 2006 80
3.07   Tertiary enrollment, 2004...120

Readiness component	109
Individual readiness	**112**

4.01   Quality of math and science education, 2006 118
4.02   Quality of the educational system, 2006............................96
4.03   Quality of public schools, 2006 119
4.04   Internet access in schools, 2006..105
4.05   Buyer sophistication, 2006 100
4.06   Residential telephone connection charge, 2005 91
4.07   Residential monthly telephone subscription, 2005 94
4.08   High-speed monthly broadband subscription charge ....n/a
4.09   Lowest cost of broadband...n/a
4.10   Cost of mobile telephone call, 2004....................................106

Business readiness	**96**

5.01   Extent of staff training, 2006...86
5.02   Local availability of research and training, 2006................105
5.03   Quality of management schools, 2006................................118
5.04   Company spending on R&D, 2006 84
5.05   University-industry research collaboration, 2006................95
5.06   Business telephone connection charge, 2005 81
5.07   Business monthly telephone subscription, 2005 80
5.08   Local supplier quality, 2006 112
5.09   Computer, comm., and other services imports ....n/a

Government readiness	**108**

6.01   Government prioritization of ICT, 2006...............................88
6.02   Gov't. procurement of advanced tech products, 2006........87
6.03   Importance of ICT to gov't. vision of the future, 2006......111
6.04   E-participation index, 2005 101
6.05   E-government readiness index, 2005................................104

Usage component	113
Individual usage	**119**

7.01   Mobile telephone subscribers, 2005 117
7.02   Personal computers, 2004...119
7.03   Broadband Internet subscribers, 2005 91
7.04   Internet users, 2005 119
7.05   Internet bandwidth, 2004 90

Business usage	**111**

8.01   Prevalence of foreign technology licensing, 2006...............94
8.02   Firm-level technology absorption, 2006.............................107
8.03   Capacity for innovation, 2006 104
8.04   Availability of new telephone lines, 2006 116
8.05   Availability of mobile telephones, 2006.............................103
8.06   Extent of business Internet use, 2006 102

Government usage	**106**

9.01   Government success in ICT promotion, 2006.....................98
9.02   Availability of online services, 2006.....................................93
9.03   ICT use and government efficiency, 2006..........................113
9.04   ICT pervasiveness, 2006 107

Malaysia

Key indicators

Population (millions), 2005..25.3
GDP (PPP) per capita (US$), 2005.....................11,201.1
Internet users per 100 inhabitants, 2005...................42.4
Internet bandwidth (Mbps/10,000 inhabitants), 2004.................1.2

Networked Readiness Index

Year (number of economies)	Rank
2006–2007 (122) **26**	
2005–2006 (115)..24	
2004–2005 (104)..27	

Global Competitiveness Index 2006–2007 (125)   26

Environment component   26

Market environment   17

1.01 Venture capital availability, 2006..........................19
1.02 Financial market sophistication, 2006....................31
1.03 Technological readiness, 2006...............................18
1.04 State of cluster development, 2006 5
1.05 US utility patents, 2005 31
1.06 High-tech exports, 2004 3
1.07 Burden of government regulation, 2006 7
1.08 Extent and effect of taxation, 2006......................12
1.09 Time required to start a business, 2006...............51
1.10 No. of procedures required to start a business, 2006.........49
1.11 Intensity of local competition, 2006 16
1.12 Freedom of the press, 2006..................................112

Political and regulatory environment   23

2.01 Effectiveness of law-making bodies, 2006.............7
2.02 Laws relating to ICT, 2006....................................12
2.03 Judicial independence, 2006 24
2.04 Intellectual property protection, 2006..................23
2.05 Efficiency of legal framework, 2006......................18
2.06 Property rights, 2006 24
2.07 Quality of competition in the ISP sector, 2006....32
2.08 No. of procedures to enforce a contract, 2006 ...60
2.09 Time to enforce a contract, 2006 61

Infrastructure environment   50

3.01 Telephone lines, 2005...68
3.02 Secure Internet servers, 2005 53
3.03 Internet hosts, 2004 53
3.04 Electricity production, 2003 51
3.05 Availability of scientists and engineers, 2006......24
3.06 Quality of scientific research institutions, 2006 17
3.07 Tertiary enrollment, 2003......................................60

Readiness component   18

Individual readiness   17

4.01 Quality of math and science education, 2006 12
4.02 Quality of the educational system, 2006...........10
4.03 Quality of public schools, 2006 17
4.04 Internet access in schools, 2006........................30
4.05 Buyer sophistication, 2006 25
4.06 Residential telephone connection charge, 2005 18
4.07 Residential monthly telephone subscription, 2005 59
4.08 High-speed monthly broadband subscription, 2006............35
4.09 Lowest cost of broadband, 2006.........................39
4.10 Cost of mobile telephone call, 2004...................42

Business readiness   16

5.01 Extent of staff training, 2006...............................17
5.02 Local availability of research and training, 2006 22
5.03 Quality of management schools, 2006................22
5.04 Company spending on R&D, 2006 10
5.05 University-industry research collaboration, 2006................12
5.06 Business telephone connection charge, 2005 17
5.07 Business monthly telephone subscription, 2005 53
5.08 Local supplier quality, 2006 24
5.09 Computer, comm., and other services imports.................n/a

Government readiness   21

6.01 Government prioritization of ICT, 2006................2
6.02 Gov't. procurement of advanced tech products, 2006..........2
6.03 Importance of ICT to gov't. vision of the future, 2006.........3
6.04 E-participation index, 2005 51
6.05 E-government readiness index, 2005..................43

Usage component   27

Individual usage   42

7.01 Mobile telephone subscribers, 2005 44
7.02 Personal computers, 2004...................................35
7.03 Broadband Internet subscribers, 2005 48
7.04 Internet users, 2005 30
7.05 Internet bandwidth, 2004 47

Business usage   23

8.01 Prevalence of foreign technology licensing, 2006................6
8.02 Firm-level technology absorption, 2006..............15
8.03 Capacity for innovation, 2006 23
8.04 Availability of new telephone lines, 2006 47
8.05 Availability of mobile telephones, 2006 49
8.06 Extent of business Internet use, 2006 27

Government usage   6

9.01 Government success in ICT promotion, 2006......3
9.02 Availability of online services, 2006....................16
9.03 ICT use and government efficiency, 2006...........11
9.04 ICT pervasiveness, 2006 17

Mali

Key indicators

Population (millions), 2005..13.5
GDP (PPP) per capita (US$), 2005.......................................1,154.0
Internet users per 100 inhabitants, 2005...................................0.5
Internet bandwidth (Mbps/10,000 inhabitants), 2004................0.0

Networked Readiness Index

Year (number of economies)	Rank
2006–2007 (122)	**101**
2005–2006 (115)	95
2004–2005 (104)	92

Global Competitiveness Index 2006–2007 (125)	118

Environment component	93
Market environment	**101**

1.01	Venture capital availability, 2006	96
1.02	Financial market sophistication, 2006	100
1.03	Technological readiness, 2006	101
1.04	State of cluster development, 2006	119
1.05	US utility patents, 2005	79
1.06	High-tech exports	n/a
1.07	Burden of government regulation, 2006	59
1.08	Extent and effect of taxation, 2006	82
1.09	Time required to start a business, 2006	78
1.10	No. of procedures required to start a business, 2006	95
1.11	Intensity of local competition, 2006	112
1.12	Freedom of the press, 2006	66

Political and regulatory environment	**70**

2.01	Effectiveness of law-making bodies, 2006	34
2.02	Laws relating to ICT, 2006	103
2.03	Judicial independence, 2006	64
2.04	Intellectual property protection, 2006	66
2.05	Efficiency of legal framework, 2006	76
2.06	Property rights, 2006	76
2.07	Quality of competition in the ISP sector, 2006	63
2.08	No. of procedures to enforce a contract, 2006	46
2.09	Time to enforce a contract, 2006	105

Infrastructure environment	**98**

3.01	Telephone lines, 2005	113
3.02	Secure Internet servers, 2005	115
3.03	Internet hosts, 2004	110
3.04	Electricity production	n/a
3.05	Availability of scientists and engineers, 2006	71
3.06	Quality of scientific research institutions, 2006	86
3.07	Tertiary enrollment, 2004	114

Readiness component	115
Individual readiness	**119**

4.01	Quality of math and science education, 2006	99
4.02	Quality of the educational system, 2006	107
4.03	Quality of public schools, 2006	105
4.04	Internet access in schools, 2006	93
4.05	Buyer sophistication, 2006	113
4.06	Residential telephone connection charge, 2005	98
4.07	Residential monthly telephone subscription, 2005	113
4.08	High-speed monthly broadband subscription charge	n/a
4.09	Lowest cost of broadband	n/a
4.10	Cost of mobile telephone call, 2004	104

Business readiness	**112**

5.01	Extent of staff training, 2006	115
5.02	Local availability of research and training, 2006	93
5.03	Quality of management schools, 2006	87
5.04	Company spending on R&D, 2006	89
5.05	University-industry research collaboration, 2006	114
5.06	Business telephone connection charge, 2005	91
5.07	Business monthly telephone subscription, 2005	110
5.08	Local supplier quality, 2006	103
5.09	Computer, comm., and other services imports	n/a

Government readiness	**87**

6.01	Government prioritization of ICT, 2006	31
6.02	Gov't. procurement of advanced tech products, 2006	28
6.03	Importance of ICT to gov't. vision of the future, 2006	17
6.04	E-participation index, 2005	109
6.05	E-government readiness index, 2005	120

Usage component	83
Individual usage	**114**

7.01	Mobile telephone subscribers, 2005	108
7.02	Personal computers, 2004	113
7.03	Broadband Internet subscribers, 2005	97
7.04	Internet users, 2005	115
7.05	Internet bandwidth, 2004	86

Business usage	**87**

8.01	Prevalence of foreign technology licensing, 2006	101
8.02	Firm-level technology absorption, 2006	88
8.03	Capacity for innovation, 2006	64
8.04	Availability of new telephone lines, 2006	87
8.05	Availability of mobile telephones, 2006	91
8.06	Extent of business Internet use, 2006	83

Government usage	**45**

9.01	Government success in ICT promotion, 2006	11
9.02	Availability of online services, 2006	89
9.03	ICT use and government efficiency, 2006	47
9.04	ICT pervasiveness, 2006	49

Malta

Key indicators

Population (millions), 2005..0.4
GDP (PPP) per capita (US$), 2005.....................19,739.1
Internet users per 100 inhabitants, 2005...................31.7
Internet bandwidth (Mbps/10,000 inhabitants), 2004.............19.4

Networked Readiness Index

Year (number of economies)	Rank
2006–2007 (122)	**27**
2005–2006 (115)	30
2004–2005 (104)	28

Global Competitiveness Index 2006–2007 (125)	39

Environment component · 29

Market environment · 38

1.01 Venture capital availability, 2006...48
1.02 Financial market sophistication, 2006....................................48
1.03 Technological readiness, 2006..29
1.04 State of cluster development, 2006......................................101
1.05 US utility patents, 2005...33
1.06 High-tech exports, 2004..4
1.07 Burden of government regulation, 2006................................80
1.08 Extent and effect of taxation, 2006.......................................59
1.09 Time required to start a business......................................n/a
1.10 No. of procedures required to start a business..................n/a
1.11 Intensity of local competition, 2006.....................................29
1.12 Freedom of the press, 2006...53

Political and regulatory environment · 29

2.01 Effectiveness of law-making bodies, 2006..........................27
2.02 Laws relating to ICT, 2006...23
2.03 Judicial independence, 2006...26
2.04 Intellectual property protection, 2006...................................43
2.05 Efficiency of legal framework, 2006......................................39
2.06 Property rights, 2006..41
2.07 Quality of competition in the ISP sector, 2006....................30
2.08 No. of procedures to enforce a contract, 2006....................69
2.09 Time to enforce a contract, 2006..57

Infrastructure environment · 31

3.01 Telephone lines, 2005..15
3.02 Secure Internet servers, 2005...11
3.03 Internet hosts, 2004...39
3.04 Electricity production, 2003...35
3.05 Availability of scientists and engineers, 2006......................57
3.06 Quality of scientific research institutions, 2006..................88
3.07 Tertiary enrollment, 2004...67

Readiness component · 28

Individual readiness · 32

4.01 Quality of math and science education, 2006......................44
4.02 Quality of the educational system, 2006..............................28
4.03 Quality of public schools, 2006...35
4.04 Internet access in schools, 2006..26
4.05 Buyer sophistication, 2006..52
4.06 Residential telephone connection charge, 2005..................35
4.07 Residential monthly telephone subscription, 2005..............25
4.08 High-speed monthly broadband subscription, 2006...............1
4.09 Lowest cost of broadband, 2006...13
4.10 Cost of mobile telephone call, 2005.....................................52

Business readiness · 53

5.01 Extent of staff training, 2006...44
5.02 Local availability of research and training, 2006..................88
5.03 Quality of management schools, 2006..................................44
5.04 Company spending on R&D, 2006...64
5.05 University-industry research collaboration, 2006.................76
5.06 Business telephone connection charge, 2005......................48
5.07 Business monthly telephone subscription, 2005..................43
5.08 Local supplier quality, 2006...53
5.09 Computer, comm., and other services imports, 2004.........78

Government readiness · 16

6.01 Government prioritization of ICT, 2006....................................9
6.02 Gov't. procurement of advanced tech products, 2006........61
6.03 Importance of ICT to gov't. vision of the future, 2006.........2
6.04 E-participation index, 2005..19
6.05 E-government readiness index, 2005....................................21

Usage component · 29

Individual usage · 37

7.01 Mobile telephone subscribers, 2005.....................................39
7.02 Personal computers, 2004..43
7.03 Broadband Internet subscribers, 2005..................................28
7.04 Internet users, 2005...36
7.05 Internet bandwidth, 2004...24

Business usage · 40

8.01 Prevalence of foreign technology licensing, 2006...............36
8.02 Firm-level technology absorption, 2006................................40
8.03 Capacity for innovation, 2006..65
8.04 Availability of new telephone lines, 2006.............................40
8.05 Availability of mobile telephones, 2006................................67
8.06 Extent of business Internet use, 2006..................................34

Government usage · 4

9.01 Government success in ICT promotion, 2006.........................2
9.02 Availability of online services, 2006..5
9.03 ICT use and government efficiency, 2006...............................8
9.04 ICT pervasiveness, 2006..16

209

Mauritania

Key indicators

Population (millions), 2005..3.1
GDP (PPP) per capita (US$), 2005...2,535.2
Internet users per 100 inhabitants, 2005.......................................0.7
Internet bandwidth (Mbps/10,000 inhabitants) ....n/a

Networked Readiness Index

Year (number of economies)	Rank
2006–2007 (122) **87**	
2005–2006 (115) ....—	
2004–2005 (104) ....—	

Global Competitiveness Index 2006–2007 (125)	114

Environment component	101
Market environment	**83**

1.01 Venture capital availability, 2006..110
1.02 Financial market sophistication, 2006.....................................104
1.03 Technological readiness, 2006..116
1.04 State of cluster development, 2006...89
1.05 US utility patents, 2005 79
1.06 High-tech exports..n/a
1.07 Burden of government regulation, 2006 6
1.08 Extent and effect of taxation, 2006 13
1.09 Time required to start a business, 2006............................106
1.10 No. of procedures required to start a business, 2006.........79
1.11 Intensity of local competition, 2006 117
1.12 Freedom of the press, 2006..70

Political and regulatory environment	**94**

2.01 Effectiveness of law-making bodies, 2006.........................107
2.02 Laws relating to ICT, 2006...113
2.03 Judicial independence, 2006 58
2.04 Intellectual property protection, 2006....................................77
2.05 Efficiency of legal framework, 2006 51
2.06 Property rights, 2006 91
2.07 Quality of competition in the ISP sector, 2006...................116
2.08 No. of procedures to enforce a contract, 2006 88
2.09 Time to enforce a contract, 2006 47

Infrastructure environment	**118**

3.01 Telephone lines, 2005..104
3.02 Secure Internet servers, 2005..98
3.03 Internet hosts, 2004 116
3.04 Electricity production ....n/a
3.05 Availability of scientists and engineers, 2006.......................77
3.06 Quality of scientific research institutions, 2006 122
3.07 Tertiary enrollment, 2004..102

Readiness component	101
Individual readiness	**109**

4.01 Quality of math and science education, 2006 94
4.02 Quality of the educational system, 2006............................108
4.03 Quality of public schools, 2006 104
4.04 Internet access in schools, 2006...115
4.05 Buyer sophistication, 2006 116
4.06 Residential telephone connection charge, 2005 75
4.07 Residential monthly telephone subscription, 2005 102
4.08 High-speed monthly broadband subscription, 2006..........104
4.09 Lowest cost of broadband, 2006...108
4.10 Cost of mobile telephone call, 2005.....................................89

Business readiness	**103**

5.01 Extent of staff training, 2006...77
5.02 Local availability of research and training, 2006 119
5.03 Quality of management schools, 2006..................................121
5.04 Company spending on R&D, 2006 122
5.05 University-industry research collaboration, 2006.................121
5.06 Business telephone connection charge, 2005 65
5.07 Business monthly telephone subscription, 2005 94
5.08 Local supplier quality, 2006 107
5.09 Computer, comm., and other services imports ....n/a

Government readiness	**51**

6.01 Government prioritization of ICT, 2006....................................3
6.02 Gov't. procurement of advanced tech products, 2006........19
6.03 Importance of ICT to gov't. vision of the future, 2006..........5
6.04 E-participation index, 2005 92
6.05 E-government readiness index, 2005................................115

Usage component	49
Individual usage	**95**

7.01 Mobile telephone subscribers, 2005 89
7.02 Personal computers, 2004...95
7.03 Broadband Internet subscribers, 2005 90
7.04 Internet users, 2005 112
7.05 Internet bandwidth..n/a

Business usage	**57**

8.01 Prevalence of foreign technology licensing, 2006...............66
8.02 Firm-level technology absorption, 2006................................16
8.03 Capacity for innovation, 2006 77
8.04 Availability of new telephone lines, 2006 60
8.05 Availability of mobile telephones, 2006................................72
8.06 Extent of business Internet use, 2006 72

Government usage	**21**

9.01 Government success in ICT promotion, 2006........................5
9.02 Availability of online services, 2006......................................57
9.03 ICT use and government efficiency, 2006.............................6
9.04 ICT pervasiveness, 2006 11

Mauritius

Key indicators

Population (millions), 2005..1.2
GDP (PPP) per capita (US$), 2005.......................................12,894.6
Internet users per 100 inhabitants, 2004..................................14.6
Internet bandwidth (Mbps/10,000 inhabitants)..........................n/a

Networked Readiness Index

Year (number of economies)	Rank
2006–2007 (122)	**51**
2005–2006 (115)...	45
2004–2005 (104)...	47

Global Competitiveness Index 2006–2007 (125)	55

Environment component — 48

Market environment — 51

1.01 Venture capital availability, 2006.......................................46
1.02 Financial market sophistication, 2006..................................43
1.03 Technological readiness, 2006..54
1.04 State of cluster development, 2006.....................................78
1.05 US utility patents, 2005...79
1.06 High-tech exports, 2004..55
1.07 Burden of government regulation, 2006............................112
1.08 Extent and effect of taxation, 2006....................................17
1.09 Time required to start a business, 2006..............................86
1.10 No. of procedures required to start a business, 2006.........16
1.11 Intensity of local competition, 2006...................................84
1.12 Freedom of the press, 2006...55

Political and regulatory environment — 46

2.01 Effectiveness of law-making bodies, 2006..........................24
2.02 Laws relating to ICT, 2006..35
2.03 Judicial independence, 2006...42
2.04 Intellectual property protection, 2006.................................39
2.05 Efficiency of legal framework, 2006....................................33
2.06 Property rights, 2006..28
2.07 Quality of competition in the ISP sector, 2006..................100
2.08 No. of procedures to enforce a contract, 2006...................78
2.09 Time to enforce a contract, 2006..93

Infrastructure environment — 61

3.01 Telephone lines, 2005..43
3.02 Secure Internet servers, 2005..50
3.03 Internet hosts, 2004...56
3.04 Electricity production..n/a
3.05 Availability of scientists and engineers, 2006.....................86
3.06 Quality of scientific research institutions, 2006..................70
3.07 Tertiary enrollment, 2004...82

Readiness component — 48

Individual readiness — 49

4.01 Quality of math and science education, 2006.....................59
4.02 Quality of the educational system, 2006.............................64
4.03 Quality of public schools, 2006..45
4.04 Internet access in schools, 2006..62
4.05 Buyer sophistication, 2006...49
4.06 Residential telephone connection charge, 2005.................45
4.07 Residential monthly telephone subscription, 2005.............34
4.08 High-speed monthly broadband subscription, 2006...........71
4.09 Lowest cost of broadband, 2006..59
4.10 Cost of mobile telephone call, 2005...................................18

Business readiness — 54

5.01 Extent of staff training, 2006...33
5.02 Local availability of research and training, 2006.................85
5.03 Quality of management schools, 2006................................69
5.04 Company spending on R&D, 2006......................................73
5.05 University-industry research collaboration, 2006................68
5.06 Business telephone connection charge, 2005.....................57
5.07 Business monthly telephone subscription, 2005.................45
5.08 Local supplier quality, 2006...46
5.09 Computer, comm., and other services imports, 2004.........67

Government readiness — 41

6.01 Government prioritization of ICT, 2006...............................17
6.02 Gov't. procurement of advanced tech products, 2006........39
6.03 Importance of ICT to gov't. vision of the future, 2006........47
6.04 E-participation index, 2005..62
6.05 E-government readiness index, 2005.................................51

Usage component — 58

Individual usage — 49

7.01 Mobile telephone subscribers, 2005..................................55
7.02 Personal computers, 2004..41
7.03 Broadband Internet subscribers, 2004...............................71
7.04 Internet users, 2004...63
7.05 Internet bandwidth...n/a

Business usage — 67

8.01 Prevalence of foreign technology licensing, 2006..............50
8.02 Firm-level technology absorption, 2006..............................73
8.03 Capacity for innovation, 2006..73
8.04 Availability of new telephone lines, 2006............................50
8.05 Availability of mobile telephones, 2006..............................51
8.06 Extent of business Internet use, 2006..............................101

Government usage — 70

9.01 Government success in ICT promotion, 2006.....................27
9.02 Availability of online services, 2006...................................75
9.03 ICT use and government efficiency, 2006...........................95
9.04 ICT pervasiveness, 2006..73

211

Mexico

Key indicators

Population (millions), 2005...107.0
GDP (PPP) per capita (US$), 2005.....................10,185.7
Internet users per 100 inhabitants, 2005...................17.4
Internet bandwidth (Mbps/10,000 inhabitants), 2004.................1.1

Networked Readiness Index

Year (number of economies)	Rank
2006–2007 (122)	49
2005–2006 (115)..	55
2004–2005 (104)..	60

Global Competitiveness Index 2006–2007 (125)   58

Environment component — 54

Market environment — 44
1.01 Venture capital availability, 2006...67
1.02 Financial market sophistication, 2006...................................38
1.03 Technological readiness, 2006...57
1.04 State of cluster development, 2006.......................................43
1.05 US utility patents, 2005 42
1.06 High-tech exports, 2004 16
1.07 Burden of government regulation, 2006 92
1.08 Extent and effect of taxation, 2006......................................73
1.09 Time required to start a business, 2006................................46
1.10 No. of procedures required to start a business, 2006.........36
1.11 Intensity of local competition, 2006 57
1.12 Freedom of the press, 2006...38

Political and regulatory environment — 60
2.01 Effectiveness of law-making bodies, 2006........................101
2.02 Laws relating to ICT, 2006...42
2.03 Judicial independence, 2006 66
2.04 Intellectual property protection, 2006..................................54
2.05 Efficiency of legal framework, 2006......................................77
2.06 Property rights, 2006...60
2.07 Quality of competition in the ISP sector, 2006...................69
2.08 No. of procedures to enforce a contract, 2006 78
2.09 Time to enforce a contract, 2006 53

Infrastructure environment — 69
3.01 Telephone lines, 2005...64
3.02 Secure Internet servers, 2005 59
3.03 Internet hosts, 2004 40
3.04 Electricity production, 2003 61
3.05 Availability of scientists and engineers, 2006.....................85
3.06 Quality of scientific research institutions, 2006 54
3.07 Tertiary enrollment, 2004...71

Readiness component — 41

Individual readiness — 67
4.01 Quality of math and science education, 2006...................101
4.02 Quality of the educational system, 2006..............................81
4.03 Quality of public schools, 2006 90
4.04 Internet access in schools, 2006..53
4.05 Buyer sophistication, 2006...59
4.06 Residential telephone connection charge, 2005 63
4.07 Residential monthly telephone subscription, 2005 69
4.08 High-speed monthly broadband subscription, 2006...........77
4.09 Lowest cost of broadband, 2006...46
4.10 Cost of mobile telephone call, 2005.....................................31

Business readiness — 51
5.01 Extent of staff training, 2006..47
5.02 Local availability of research and training, 2006.................47
5.03 Quality of management schools, 2006..................................43
5.04 Company spending on R&D, 2006...60
5.05 University-industry research collaboration, 2006................40
5.06 Business telephone connection charge, 2005 59
5.07 Business monthly telephone subscription, 2005 56
5.08 Local supplier quality, 2006...51
5.09 Computer, comm., and other services imports, 2004.........90

Government readiness — 24
6.01 Government prioritization of ICT, 2006................................57
6.02 Gov't. procurement of advanced tech products, 2006........77
6.03 Importance of ICT to gov't. vision of the future, 2006........43
6.04 E-participation index, 2005 7
6.05 E-government readiness index, 2005...................................31

Usage component — 50

Individual usage — 59
7.01 Mobile telephone subscribers, 2005 65
7.02 Personal computers, 2004..53
7.03 Broadband Internet subscribers, 2005 46
7.04 Internet users, 2005 58
7.05 Internet bandwidth, 2004 51

Business usage — 54
8.01 Prevalence of foreign technology licensing, 2006..............42
8.02 Firm-level technology absorption, 2006...............................75
8.03 Capacity for innovation, 2006 56
8.04 Availability of new telephone lines, 2006 51
8.05 Availability of mobile telephones, 2006...............................63
8.06 Extent of business Internet use, 2006 56

Government usage — 37
9.01 Government success in ICT promotion, 2006......................67
9.02 Availability of online services, 2006.....................................33
9.03 ICT use and government efficiency, 2006.............................31
9.04 ICT pervasiveness, 2006 47

Moldova

Key indicators

Population (millions), 2005...4.2
GDP (PPP) per capita (US$), 2005...2,526.7
Internet users per 100 inhabitants, 2004.....................................9.5
Internet bandwidth (Mbps/10,000 inhabitants), 2004.................0.4

Networked Readiness Index

Year (number of economies)	Rank
2006–2007 (122)	**92**
2005–2006 (115)	94
2004–2005 (104)	—

Global Competitiveness Index 2006–2007 (125)   86

Environment component — 95

Market environment — 111

1.01 Venture capital availability, 2006 93
1.02 Financial market sophistication, 2006 99
1.03 Technological readiness, 2006 115
1.04 State of cluster development, 2006 116
1.05 US utility patents, 2005 61
1.06 High-tech exports, 2004 62
1.07 Burden of government regulation, 2006 98
1.08 Extent and effect of taxation, 2006 103
1.09 Time required to start a business, 2006 51
1.10 No. of procedures required to start a business, 2006.........61
1.11 Intensity of local competition, 2006 79
1.12 Freedom of the press, 2006 98

Political and regulatory environment — 93

2.01 Effectiveness of law-making bodies, 2006 71
2.02 Laws relating to ICT, 2006 79
2.03 Judicial independence, 2006 110
2.04 Intellectual property protection, 2006 89
2.05 Efficiency of legal framework, 2006 108
2.06 Property rights, 2006 101
2.07 Quality of competition in the ISP sector, 2006 75
2.08 No. of procedures to enforce a contract, 2006 78
2.09 Time to enforce a contract, 2006 29

Infrastructure environment — 75

3.01 Telephone lines, 2005 58
3.02 Secure Internet servers, 2005 73
3.03 Internet hosts, 2004 58
3.04 Electricity production, 2003 81
3.05 Availability of scientists and engineers, 2006 98
3.06 Quality of scientific research institutions, 2006 102
3.07 Tertiary enrollment, 2004 52

Readiness component — 91

Individual readiness — 90

4.01 Quality of math and science education, 2006 58
4.02 Quality of the educational system, 2006 68
4.03 Quality of public schools, 2006 75
4.04 Internet access in schools, 2006 87
4.05 Buyer sophistication, 2006 99
4.06 Residential telephone connection charge, 2005 81
4.07 Residential monthly telephone subscription, 2005 72
4.08 High-speed monthly broadband subscription, 2006 92
4.09 Lowest cost of broadband, 2006...95
4.10 Cost of mobile telephone call, 2004 90

Business readiness — 90

5.01 Extent of staff training, 2006 101
5.02 Local availability of research and training, 2006 94
5.03 Quality of management schools, 2006 101
5.04 Company spending on R&D, 2006 106
5.05 University-industry research collaboration, 2006 100
5.06 Business telephone connection charge, 2005 72
5.07 Business monthly telephone subscription, 2005 72
5.08 Local supplier quality, 2006 102
5.09 Computer, comm., and other services imports, 2004.........48

Government readiness — 102

6.01 Government prioritization of ICT, 2006 71
6.02 Gov't. procurement of advanced tech products, 2006......113
6.03 Importance of ICT to gov't. vision of the future, 2006........82
6.04 E-participation index, 2005 109
6.05 E-government readiness index, 2005 89

Usage component — 96

Individual usage — 88

7.01 Mobile telephone subscribers, 2005 86
7.02 Personal computers, 2004..88
7.03 Broadband Internet subscribers, 2005 69
7.04 Internet users, 2004 73
7.05 Internet bandwidth, 2004 63

Business usage — 98

8.01 Prevalence of foreign technology licensing, 2006.............116
8.02 Firm-level technology absorption, 2006 98
8.03 Capacity for innovation, 2006 60
8.04 Availability of new telephone lines, 2006 83
8.05 Availability of mobile telephones, 2006 89
8.06 Extent of business Internet use, 2006 118

Government usage — 93

9.01 Government success in ICT promotion, 2006 81
9.02 Availability of online services, 2006 99
9.03 ICT use and government efficiency, 2006 105
9.04 ICT pervasiveness, 2006 57

213

Mongolia

Key indicators

Population (millions), 2005	2.6
GDP (PPP) per capita (US$), 2005	2,175.3
Internet users per 100 inhabitants, 2005	10.1
Internet bandwidth (Mbps/10,000 inhabitants)	n/a

Networked Readiness Index

Year (number of economies)	Rank
2006–2007 (122)	**90**
2005–2006 (115)	92
2004–2005 (104)	—

Global Competitiveness Index 2006–2007 (125)	92

Environment component	85
Market environment	**88**
1.01 Venture capital availability, 2006	116
1.02 Financial market sophistication, 2006	95
1.03 Technological readiness, 2006	99
1.04 State of cluster development, 2006	108
1.05 US utility patents, 2005	79
1.06 High-tech exports	n/a
1.07 Burden of government regulation, 2006	82
1.08 Extent and effect of taxation, 2006	113
1.09 Time required to start a business, 2006	30
1.10 No. of procedures required to start a business, 2006	36
1.11 Intensity of local competition, 2006	87
1.12 Freedom of the press, 2006	87
Political and regulatory environment	**91**
2.01 Effectiveness of law-making bodies, 2006	96
2.02 Laws relating to ICT, 2006	99
2.03 Judicial independence, 2006	95
2.04 Intellectual property protection, 2006	108
2.05 Efficiency of legal framework, 2006	112
2.06 Property rights, 2006	78
2.07 Quality of competition in the ISP sector, 2006	87
2.08 No. of procedures to enforce a contract, 2006	51
2.09 Time to enforce a contract, 2006	30
Infrastructure environment	**64**
3.01 Telephone lines, 2005	93
3.02 Secure Internet servers, 2005	75
3.03 Internet hosts, 2004	104
3.04 Electricity production	n/a
3.05 Availability of scientists and engineers, 2006	53
3.06 Quality of scientific research institutions, 2006	81
3.07 Tertiary enrollment, 2004	49

Readiness component	93
Individual readiness	**97**
4.01 Quality of math and science education, 2006	60
4.02 Quality of the educational system, 2006	95
4.03 Quality of public schools, 2006	84
4.04 Internet access in schools, 2006	89
4.05 Buyer sophistication, 2006	88
4.06 Residential telephone connection charge, 2004	95
4.07 Residential monthly telephone subscription	n/a
4.08 High-speed monthly broadband subscription, 2006	91
4.09 Lowest cost of broadband, 2006	94
4.10 Cost of mobile telephone call, 2004	94
Business readiness	**105**
5.01 Extent of staff training, 2006	84
5.02 Local availability of research and training, 2006	78
5.03 Quality of management schools, 2006	114
5.04 Company spending on R&D, 2006	88
5.05 University-industry research collaboration, 2006	71
5.06 Business telephone connection charge, 2004	97
5.07 Business monthly telephone subscription	n/a
5.08 Local supplier quality, 2006	115
5.09 Computer, comm., and other services imports, 2004	87
Government readiness	**61**
6.01 Government prioritization of ICT, 2006	32
6.02 Gov't. procurement of advanced tech products, 2006	111
6.03 Importance of ICT to gov't. vision of the future, 2006	34
6.04 E-participation index, 2005	41
6.05 E-government readiness index, 2005	79

Usage component	91
Individual usage	**73**
7.01 Mobile telephone subscribers, 2005	91
7.02 Personal computers, 2004	49
7.03 Broadband Internet subscribers, 2005	82
7.04 Internet users, 2005	71
7.05 Internet bandwidth	n/a
Business usage	**105**
8.01 Prevalence of foreign technology licensing, 2006	98
8.02 Firm-level technology absorption, 2006	108
8.03 Capacity for innovation, 2006	96
8.04 Availability of new telephone lines, 2006	98
8.05 Availability of mobile telephones, 2006	108
8.06 Extent of business Internet use, 2006	97
Government usage	**85**
9.01 Government success in ICT promotion, 2006	58
9.02 Availability of online services, 2006	104
9.03 ICT use and government efficiency, 2006	100
9.04 ICT pervasiveness, 2006	51

Morocco

Key indicators

Population (millions), 2005...31.5
GDP (PPP) per capita (US$), 2005...............................4,503.2
Internet users per 100 inhabitants, 2005............................15.2
Internet bandwidth (Mbps/10,000 inhabitants), 2004.................0.4

Networked Readiness Index

Year (number of economies)	Rank
2006–2007 (122) **76**	
2005–2006 (115)..77	
2004–2005 (104)..54	

Global Competitiveness Index 2006–2007 (125)   70

Environment component	76
Market environment	**68**

1.01 Venture capital availability, 2006............................90
1.02 Financial market sophistication, 2006.......................82
1.03 Technological readiness, 2006.................................71
1.04 State of cluster development, 2006..........................88
1.05 US utility patents, 2005..76
1.06 High-tech exports, 2004.......................................36
1.07 Burden of government regulation, 2006....................46
1.08 Extent and effect of taxation, 2006.........................62
1.09 Time required to start a business, 2006....................15
1.10 No. of procedures required to start a business, 2006.........16
1.11 Intensity of local competition, 2006.........................71
1.12 Freedom of the press, 2006..................................111

Political and regulatory environment	**69**

2.01 Effectiveness of law-making bodies, 2006.................62
2.02 Laws relating to ICT, 2006....................................74
2.03 Judicial independence, 2006..................................74
2.04 Intellectual property protection, 2006......................53
2.05 Efficiency of legal framework, 2006.........................54
2.06 Property rights, 2006...56
2.07 Quality of competition in the ISP sector, 2006............97
2.08 No. of procedures to enforce a contract, 2006............97
2.09 Time to enforce a contract, 2006............................89

Infrastructure environment	**84**

3.01 Telephone lines, 2005...97
3.02 Secure Internet servers, 2005................................88
3.03 Internet hosts, 2004..96
3.04 Electricity production, 2003....................................86
3.05 Availability of scientists and engineers, 2006.............20
3.06 Quality of scientific research institutions, 2006..........82
3.07 Tertiary enrollment, 2004......................................91

Readiness component	81
Individual readiness	**78**

4.01 Quality of math and science education, 2006..............49
4.02 Quality of the educational system, 2006...................90
4.03 Quality of public schools, 2006...............................67
4.04 Internet access in schools, 2006.............................72
4.05 Buyer sophistication, 2006....................................93
4.06 Residential telephone connection charge, 2004...........80
4.07 Residential monthly telephone subscription, 2004........100
4.08 High-speed monthly broadband subscription, 2006........73
4.09 Lowest cost of broadband, 2006.............................53
4.10 Cost of mobile telephone call, 2004.........................87

Business readiness	**84**

5.01 Extent of staff training, 2006.................................85
5.02 Local availability of research and training, 2006..........58
5.03 Quality of management schools, 2006.......................26
5.04 Company spending on R&D, 2006...........................75
5.05 University-industry research collaboration, 2006..........66
5.06 Business telephone connection charge, 2004..............88
5.07 Business monthly telephone subscription, 2004...........99
5.08 Local supplier quality, 2006...................................78
5.09 Computer, comm., and other services imports, 2004.........47

Government readiness	**85**

6.01 Government prioritization of ICT, 2006......................52
6.02 Gov't. procurement of advanced tech products, 2006........64
6.03 Importance of ICT to gov't. vision of the future, 2006........29
6.04 E-participation index, 2005....................................92
6.05 E-government readiness index, 2005........................105

Usage component	67
Individual usage	**74**

7.01 Mobile telephone subscribers, 2005.........................70
7.02 Personal computers, 2004.....................................90
7.03 Broadband Internet subscribers, 2005......................56
7.04 Internet users, 2005..62
7.05 Internet bandwidth, 2004......................................64

Business usage	**70**

8.01 Prevalence of foreign technology licensing, 2006.........78
8.02 Firm-level technology absorption, 2006.....................42
8.03 Capacity for innovation, 2006.................................92
8.04 Availability of new telephone lines, 2006...................44
8.05 Availability of mobile telephones, 2006.....................69
8.06 Extent of business Internet use, 2006......................107

Government usage	**55**

9.01 Government success in ICT promotion, 2006...............41
9.02 Availability of online services, 2006.........................81
9.03 ICT use and government efficiency, 2006...................43
9.04 ICT pervasiveness, 2006.......................................50

Mozambique

Key indicators

Population (millions), 2005..19.8
GDP (PPP) per capita (US$), 2005...1,378.9
Internet users per 100 inhabitants, 2004....................................0.7
Internet bandwidth (Mbps/10,000 inhabitants) ....n/a

Networked Readiness Index

Year (number of economies)	Rank
2006–2007 (122)	**115**
2005–2006 (115)..	101
2004–2005 (104)..	96

Global Competitiveness Index 2006–2007 (125)	121

Environment component	119
Market environment	**118**

1.01	Venture capital availability, 2006 121
1.02	Financial market sophistication, 2006...................................112
1.03	Technological readiness, 2006..111
1.04	State of cluster development, 2006 117
1.05	US utility patents, 2005 79
1.06	High-tech exports...n/a
1.07	Burden of government regulation, 2006 114
1.08	Extent and effect of taxation, 2006..94
1.09	Time required to start a business, 2006...............................112
1.10	No. of procedures required to start a business, 2006...........95
1.11	Intensity of local competition, 2006 108
1.12	Freedom of the press, 2006...87

Political and regulatory environment	**113**
2.01	Effectiveness of law-making bodies, 2006.............................89
2.02	Laws relating to ICT, 2006..114
2.03	Judicial independence, 2006 100
2.04	Intellectual property protection, 2006..................................102
2.05	Efficiency of legal framework, 2006.....................................104
2.06	Property rights, 2006..84
2.07	Quality of competition in the ISP sector, 2006......................92
2.08	No. of procedures to enforce a contract, 2006.....................85
2.09	Time to enforce a contract, 2006 108

Infrastructure environment	**121**
3.01	Telephone lines, 2004..118
3.02	Secure Internet servers, 2005...108
3.03	Internet hosts, 2004 89
3.04	Electricity production, 2003 89
3.05	Availability of scientists and engineers, 2006....................117
3.06	Quality of scientific research institutions, 2006 95
3.07	Tertiary enrollment, 2004..118

Readiness component	113
Individual readiness	**116**

4.01	Quality of math and science education, 2006 96
4.02	Quality of the educational system, 2006............................111
4.03	Quality of public schools, 2006 103
4.04	Internet access in schools, 2006..106
4.05	Buyer sophistication, 2006 111
4.06	Residential telephone connection charge, 2005 86
4.07	Residential monthly telephone subscription, 2005 115
4.08	High-speed monthly broadband subscription, 2006..........102
4.09	Lowest cost of broadband, 2006..106
4.10	Cost of mobile telephone call, 2004...................................92

Business readiness	**107**
5.01	Extent of staff training, 2006..78
5.02	Local availability of research and training, 2006 110
5.03	Quality of management schools, 2006...............................116
5.04	Company spending on R&D, 2006 104
5.05	University-industry research collaboration, 2006................94
5.06	Business telephone connection charge, 2005 75
5.07	Business monthly telephone subscription, 2005 113
5.08	Local supplier quality, 2006 111
5.09	Computer, comm., and other services imports, 2004.........37

Government readiness	**89**
6.01	Government prioritization of ICT, 2006................................81
6.02	Gov't. procurement of advanced tech products, 2006........99
6.03	Importance of ICT to gov't. vision of the future, 2006........84
6.04	E-participation index, 2005 30
6.05	E-government readiness index, 2005..............................109

Usage component	110
Individual usage	**110**

7.01	Mobile telephone subscribers, 2005 112
7.02	Personal computers, 2004...107
7.03	Broadband Internet subscribers..n/a
7.04	Internet users, 2004 111
7.05	Internet bandwidth...n/a

Business usage	**101**
8.01	Prevalence of foreign technology licensing, 2006.............106
8.02	Firm-level technology absorption, 2006.............................116
8.03	Capacity for innovation, 2006 97
8.04	Availability of new telephone lines, 2006 84
8.05	Availability of mobile telephones, 2006..............................86
8.06	Extent of business Internet use, 2006 112

Government usage	**112**
9.01	Government success in ICT promotion, 2006......................76
9.02	Availability of online services, 2006...................................117
9.03	ICT use and government efficiency, 2006...........................114
9.04	ICT pervasiveness, 2006 112

Namibia

Key indicators

Population (millions), 2005..2.0
GDP (PPP) per capita (US$), 2005.....................7,478.4
Internet users per 100 inhabitants, 2004.....................3.7
Internet bandwidth (Mbps/10,000 inhabitants), 2004................0.0

Networked Readiness Index

Year (number of economies)	Rank
2006–2007 (122) **85**	
2005–2006 (115)..78	
2004–2005 (104)..55	

Global Competitiveness Index 2006–2007 (125)	84

Environment component	67
Market environment	**64**

1.01   Venture capital availability, 2006 50
1.02   Financial market sophistication, 2006...................................60
1.03   Technological readiness, 2006...65
1.04   State of cluster development, 2006 94
1.05   US utility patents, 2005 79
1.06   High-tech exports...n/a
1.07   Burden of government regulation, 2006 84
1.08   Extent and effect of taxation, 2006 42
1.09   Time required to start a business, 2006.............................108
1.10   No. of procedures required to start a business, 2006.........61
1.11   Intensity of local competition, 2006 60
1.12   Freedom of the press, 2006 61

Political and regulatory environment	**38**

2.01   Effectiveness of law-making bodies, 2006.........................51
2.02   Laws relating to ICT, 2006..75
2.03   Judicial independence, 2006 28
2.04   Intellectual property protection, 2006...................................40
2.05   Efficiency of legal framework, 2006......................................38
2.06   Property rights, 2006 31
2.07   Quality of competition in the ISP sector, 2006...................99
2.08   No. of procedures to enforce a contract, 2006 60
2.09   Time to enforce a contract, 2006 21

Infrastructure environment	**115**

3.01   Telephone lines, 2004...91
3.02   Secure Internet servers, 2005 60
3.03   Internet hosts, 2004 68
3.04   Electricity production, 2003 82
3.05   Availability of scientists and engineers, 2006...................118
3.06   Quality of scientific research institutions, 2006 101
3.07   Tertiary enrollment, 2003..98

Readiness component	86
Individual readiness	**91**

4.01   Quality of math and science education, 2006...................113
4.02   Quality of the educational system, 2006............................106
4.03   Quality of public schools, 2006 97
4.04   Internet access in schools, 2006..77
4.05   Buyer sophistication, 2006 65
4.06   Residential telephone connection charge, 2005 57
4.07   Residential monthly telephone subscription, 2005 73
4.08   High-speed monthly broadband subscription charge ....n/a
4.09   Lowest cost of broadband...n/a
4.10   Cost of mobile telephone call, 2005......................................75

Business readiness	**72**

5.01   Extent of staff training, 2006...64
5.02   Local availability of research and training, 2006 117
5.03   Quality of management schools, 2006...............................117
5.04   Company spending on R&D, 2006 80
5.05   University-industry research collaboration, 2006................78
5.06   Business telephone connection charge, 2005 56
5.07   Business monthly telephone subscription, 2005 63
5.08   Local supplier quality, 2006 72
5.09   Computer, comm., and other services imports, 2004.........33

Government readiness	**106**

6.01   Government prioritization of ICT, 2006..............................106
6.02   Gov't. procurement of advanced tech products, 2006........81
6.03   Importance of ICT to gov't. vision of the future, 2006......102
6.04   E-participation index, 2005 109
6.05   E-government readiness index, 2005 91

Usage component	94
Individual usage	**87**

7.01   Mobile telephone subscribers, 2005 88
7.02   Personal computers, 2004..52
7.03   Broadband Internet subscribers, 2004 97
7.04   Internet users, 2004 97
7.05   Internet bandwidth, 2004 82

Business usage	**78**

8.01   Prevalence of foreign technology licensing, 2006..............65
8.02   Firm-level technology absorption, 2006...............................89
8.03   Capacity for innovation, 2006 100
8.04   Availability of new telephone lines, 2006 75
8.05   Availability of mobile telephones, 2006 82
8.06   Extent of business Internet use, 2006 73

Government usage	**102**

9.01   Government success in ICT promotion, 2006....................105
9.02   Availability of online services, 2006......................................97
9.03   ICT use and government efficiency, 2006...........................106
9.04   ICT pervasiveness, 2006 88

Nepal

Key indicators

Population (millions), 2005..27.1
GDP (PPP) per capita (US$), 2005.......................................1,675.2
Internet users per 100 inhabitants, 2005....................................0.4
Internet bandwidth (Mbps/10,000 inhabitants).........................n/a

Networked Readiness Index

Year (number of economies)	Rank
2006–2007 (122)	**108**
2005–2006 (115)	—
2004–2005 (104)	—

Global Competitiveness Index 2006–2007 (125)	110

Environment component	91
Market environment	**78**

1.01 Venture capital availability, 2006 107
1.02 Financial market sophistication, 2006.............................107
1.03 Technological readiness, 2006 106
1.04 State of cluster development, 2006 72
1.05 US utility patents, 2005 79
1.06 High-tech exports...n/a
1.07 Burden of government regulation, 2006 73
1.08 Extent and effect of taxation, 2006.................................44
1.09 Time required to start a business, 2006............................56
1.10 No. of procedures required to start a business, 2006.........27
1.11 Intensity of local competition, 2006 85
1.12 Freedom of the press, 2006...97

Political and regulatory environment	**76**

2.01 Effectiveness of law-making bodies, 2006........................97
2.02 Laws relating to ICT, 2006..104
2.03 Judicial independence, 2006 52
2.04 Intellectual property protection, 2006.............................113
2.05 Efficiency of legal framework, 2006.................................85
2.06 Property rights, 2006 77
2.07 Quality of competition in the ISP sector, 2006...................48
2.08 No. of procedures to enforce a contract, 2006 46
2.09 Time to enforce a contract, 2006 81

Infrastructure environment	**110**

3.01 Telephone lines, 2005...103
3.02 Secure Internet servers, 2005 96
3.03 Internet hosts, 2004 100
3.04 Electricity production, 2003 102
3.05 Availability of scientists and engineers, 2006.....................95
3.06 Quality of scientific research institutions, 2006 85
3.07 Tertiary enrollment, 2004...99

Readiness component	112
Individual readiness	**106**

4.01 Quality of math and science education, 2006 90
4.02 Quality of the educational system, 2006..........................100
4.03 Quality of public schools, 2006 120
4.04 Internet access in schools, 2006.....................................91
4.05 Buyer sophistication, 2006 105
4.06 Residential telephone connection charge, 2005 94
4.07 Residential monthly telephone subscription, 2005 104
4.08 High-speed monthly broadband subscription charge ....n/a
4.09 Lowest cost of broadband, 2006.....................................107
4.10 Cost of mobile telephone call, 2005................................76

Business readiness	**111**

5.01 Extent of staff training, 2006...116
5.02 Local availability of research and training, 2006 112
5.03 Quality of management schools, 2006.............................103
5.04 Company spending on R&D, 2006 101
5.05 University-industry research collaboration, 2006...............113
5.06 Business telephone connection charge, 2005 87
5.07 Business monthly telephone subscription, 2005 98
5.08 Local supplier quality, 2006 100
5.09 Computer, comm., and other services imports, 2004.........82

Government readiness	**107**

6.01 Government prioritization of ICT, 2006..............................70
6.02 Gov't. procurement of advanced tech products, 2006......118
6.03 Importance of ICT to gov't. vision of the future, 2006........98
6.04 E-participation index, 2005 71
6.05 E-government readiness index, 2005 100

Usage component	109
Individual usage	**121**

7.01 Mobile telephone subscribers, 2005 121
7.02 Personal computers, 2004...110
7.03 Broadband Internet subscribers.....................................n/a
7.04 Internet users, 2005 117
7.05 Internet bandwidth..n/a

Business usage	**106**

8.01 Prevalence of foreign technology licensing, 2006..............96
8.02 Firm-level technology absorption, 2006............................101
8.03 Capacity for innovation, 2006 111
8.04 Availability of new telephone lines, 2006 101
8.05 Availability of mobile telephones, 2006...........................118
8.06 Extent of business Internet use, 2006 84

Government usage	**104**

9.01 Government success in ICT promotion, 2006....................84
9.02 Availability of online services, 2006................................108
9.03 ICT use and government efficiency, 2006.........................85
9.04 ICT pervasiveness, 2006 116

Netherlands

Key indicators

Population (millions), 2005...16.3

GDP (PPP) per capita (US$), 2005.....................................30,861.5

Internet users per 100 inhabitants, 2005.....................74.0

Internet bandwidth (Mbps/10,000 inhabitants), 2004.............206.2

Networked Readiness Index

Year (number of economies)	Rank
2006–2007 (122) **6**	
2005–2006 (115)..12	
2004–2005 (104)..16	

Global Competitiveness Index 2006–2007 (125)	9

Environment component — 11

Market environment — 12

1.01 Venture capital availability, 2006..3
1.02 Financial market sophistication, 2006......................................6
1.03 Technological readiness, 2006...14
1.04 State of cluster development, 2006 18
1.05 US utility patents, 2005 15
1.06 High-tech exports, 2004 18
1.07 Burden of government regulation, 2006 45
1.08 Extent and effect of taxation, 2006 39
1.09 Time required to start a business, 2006............................11
1.10 No. of procedures required to start a business, 2006.........16
1.11 Intensity of local competition, 2006 9
1.12 Freedom of the press, 2006..3

Political and regulatory environment — 5

2.01 Effectiveness of law-making bodies, 2006.........................13
2.02 Laws relating to ICT, 2006..9
2.03 Judicial independence, 2006 2
2.04 Intellectual property protection, 2006...................................5
2.05 Efficiency of legal framework, 2006......................................4
2.06 Property rights, 2006 5
2.07 Quality of competition in the ISP sector, 2006.................2
2.08 No. of procedures to enforce a contract, 2006 23
2.09 Time to enforce a contract, 2006 51

Infrastructure environment — 11

3.01 Telephone lines, 2005..20
3.02 Secure Internet servers, 2005..13
3.03 Internet hosts, 2004 3
3.04 Electricity production, 2003 33
3.05 Availability of scientists and engineers, 2006.....................31
3.06 Quality of scientific research institutions, 2006 12
3.07 Tertiary enrollment, 2004..26

Readiness component — 12

Individual readiness — 15

4.01 Quality of math and science education, 2006....................15
4.02 Quality of the educational system, 2006.............................18
4.03 Quality of public schools, 2006 7
4.04 Internet access in schools, 2006...8
4.05 Buyer sophistication, 2006 6
4.06 Residential telephone connection charge...........................n/a
4.07 Residential monthly telephone subscription, 2004 31
4.08 High-speed monthly broadband subscription, 2006...............4
4.09 Lowest cost of broadband, 2006..2
4.10 Cost of mobile telephone call ....n/a

Business readiness — 10

5.01 Extent of staff training, 2006..6
5.02 Local availability of research and training, 2006 7
5.03 Quality of management schools, 2006.................................12
5.04 Company spending on R&D, 2006 13
5.05 University-industry research collaboration, 2006.................13
5.06 Business telephone connection charge...............................n/a
5.07 Business monthly telephone subscription, 2004 13
5.08 Local supplier quality, 2006 8
5.09 Computer, comm., and other services imports, 2004...........8

Government readiness — 12

6.01 Government prioritization of ICT, 2006................................26
6.02 Gov't. procurement of advanced tech products, 2006........16
6.03 Importance of ICT to gov't. vision of the future, 2006........46
6.04 E-participation index, 2005 10
6.05 E-government readiness index, 2005....................................12

Usage component — 3

Individual usage — 1

7.01 Mobile telephone subscribers, 2005 21
7.02 Personal computers, 2004..7
7.03 Broadband Internet subscribers, 2005 3
7.04 Internet users, 2005 3
7.05 Internet bandwidth, 2004 2

Business usage — 9

8.01 Prevalence of foreign technology licensing, 2006.................3
8.02 Firm-level technology absorption, 2006................................27
8.03 Capacity for innovation, 2006 11
8.04 Availability of new telephone lines, 2006............................13
8.05 Availability of mobile telephones, 2006...............................7
8.06 Extent of business Internet use, 2006 5

Government usage — 18

9.01 Government success in ICT promotion, 2006....................34
9.02 Availability of online services, 2006......................................19
9.03 ICT use and government efficiency, 2006...........................24
9.04 ICT pervasiveness, 2006 6

New Zealand

Key indicators

Population (millions), 2005...4.0
GDP (PPP) per capita (US$), 2005........................24,797.2
Internet users per 100 inhabitants, 2005........................68.4
Internet bandwidth (Mbps/10,000 inhabitants), 2004..............11.7

Networked Readiness Index

Year (number of economies)	Rank
2006–2007 (122)	22
2005–2006 (115)..	21
2004–2005 (104)..	21

Global Competitiveness Index 2006–2007 (125)	23

Environment component	15
Market environment	**21**

1.01	Venture capital availability, 2006.............................	14
1.02	Financial market sophistication, 2006...................................	26
1.03	Technological readiness, 2006..................................	31
1.04	State of cluster development, 2006	50
1.05	US utility patents, 2005	24
1.06	High-tech exports, 2004	35
1.07	Burden of government regulation, 2006	30
1.08	Extent and effect of taxation, 2006.............................	60
1.09	Time required to start a business, 2006	15
1.10	No. of procedures required to start a business, 2006...........1
1.11	Intensity of local competition, 2006	16
1.12	Freedom of the press, 2006................................	13

Political and regulatory environment	**16**

2.01	Effectiveness of law-making bodies, 2006.......................	10
2.02	Laws relating to ICT, 2006...	18
2.03	Judicial independence, 2006	4
2.04	Intellectual property protection, 2006........................	13
2.05	Efficiency of legal framework, 2006...........................	13
2.06	Property rights, 2006	17
2.07	Quality of competition in the ISP sector, 2006................	101
2.08	No. of procedures to enforce a contract, 2006	46
2.09	Time to enforce a contract, 2006	1

Infrastructure environment	**9**

3.01	Telephone lines, 2005..	28
3.02	Secure Internet servers, 2005...................................	5
3.03	Internet hosts, 2004	9
3.04	Electricity production, 2003	12
3.05	Availability of scientists and engineers, 2006......................	43
3.06	Quality of scientific research institutions, 2006	19
3.07	Tertiary enrollment, 2004......................................	3

Readiness component	24
Individual readiness	**16**

4.01	Quality of math and science education, 2006	25
4.02	Quality of the educational system, 2006.............................	21
4.03	Quality of public schools, 2006	16
4.04	Internet access in schools, 2006...............................	21
4.05	Buyer sophistication, 2006	22
4.06	Residential telephone connection charge, 2005	5
4.07	Residential monthly telephone subscription, 2005	49
4.08	High-speed monthly broadband subscription, 2006.............	19
4.09	Lowest cost of broadband, 2006................................	25
4.10	Cost of mobile telephone call, 2004...........................	47

Business readiness	**24**

5.01	Extent of staff training, 2006....................................	25
5.02	Local availability of research and training, 2006	25
5.03	Quality of management schools, 2006..........................	25
5.04	Company spending on R&D, 2006	35
5.05	University-industry research collaboration, 2006.................	23
5.06	Business telephone connection charge, 2005	6
5.07	Business monthly telephone subscription, 2005	44
5.08	Local supplier quality, 2006	20
5.09	Computer, comm., and other services imports, 2004.........	57

Government readiness	**15**

6.01	Government prioritization of ICT, 2006............................	51
6.02	Gov't. procurement of advanced tech products, 2006........	47
6.03	Importance of ICT to gov't. vision of the future, 2006........	65
6.04	E-participation index, 2005	6
6.05	E-government readiness index, 2005	13

Usage component	25
Individual usage	**26**

7.01	Mobile telephone subscribers, 2005	33
7.02	Personal computers, 2004.......................................	21
7.03	Broadband Internet subscribers, 2005	31
7.04	Internet users, 2005	6
7.05	Internet bandwidth, 2004	26

Business usage	**27**

8.01	Prevalence of foreign technology licensing, 2006...............	14
8.02	Firm-level technology absorption, 2006..............................	32
8.03	Capacity for innovation, 2006	26
8.04	Availability of new telephone lines, 2006	39
8.05	Availability of mobile telephones, 2006	84
8.06	Extent of business Internet use, 2006	22

Government usage	**33**

9.01	Government success in ICT promotion, 2006.....................	72
9.02	Availability of online services, 2006...................................	18
9.03	ICT use and government efficiency, 2006..........................	47
9.04	ICT pervasiveness, 2006	31

Nicaragua

Key indicators

Population (millions), 2005..5.5
GDP (PPP) per capita (US$), 2005.....................................3,636.0
Internet users per 100 inhabitants, 2004.............................2.2
Internet bandwidth (Mbps/10,000 inhabitants) ....n/a

Networked Readiness Index

Year (number of economies)	Rank
2006–2007 (122) **103**
2005–2006 (115)..112
2004–2005 (104)..103

Global Competitiveness Index 2006–2007 (125)   95

Environment component	99
Market environment	**76**

1.01 Venture capital availability, 2006...104
1.02 Financial market sophistication, 2006...................................80
1.03 Technological readiness, 2006..83
1.04 State of cluster development, 2006.......................................82
1.05 US utility patents, 2005 79
1.06 High-tech exports, 2004 75
1.07 Burden of government regulation, 2006 65
1.08 Extent and effect of taxation, 2006.....................................108
1.09 Time required to start a business, 2006................................76
1.10 No. of procedures required to start a business, 2006.........16
1.11 Intensity of local competition, 2006 97
1.12 Freedom of the press, 2006...18

Political and regulatory environment	**104**

2.01 Effectiveness of law-making bodies, 2006........................120
2.02 Laws relating to ICT, 2006..82
2.03 Judicial independence, 2006...121
2.04 Intellectual property protection, 2006..................................90
2.05 Efficiency of legal framework, 2006...................................120
2.06 Property rights, 2006..107
2.07 Quality of competition in the ISP sector, 2006....................96
2.08 No. of procedures to enforce a contract, 2006 13
2.09 Time to enforce a contract, 2006 69

Infrastructure environment	**105**

3.01 Telephone lines, 2004...99
3.02 Secure Internet servers, 2005..80
3.03 Internet hosts, 2004 66
3.04 Electricity production, 2003 93
3.05 Availability of scientists and engineers, 2006....................104
3.06 Quality of scientific research institutions, 2006 111
3.07 Tertiary enrollment, 2003...81

Readiness component	106
Individual readiness	**99**

4.01 Quality of math and science education, 2006 105
4.02 Quality of the educational system, 2006............................102
4.03 Quality of public schools, 2006 107
4.04 Internet access in schools, 2006...95
4.05 Buyer sophistication, 2006 94
4.06 Residential telephone connection charge, 2005 105
4.07 Residential monthly telephone subscription, 2005 95
4.08 High-speed monthly broadband subscription, 2006............83
4.09 Lowest cost of broadband, 2006..88
4.10 Cost of mobile telephone call, 2005....................................96

Business readiness	**119**

5.01 Extent of staff training, 2006...104
5.02 Local availability of research and training, 2006.................81
5.03 Quality of management schools, 2006.................................65
5.04 Company spending on R&D, 2006 105
5.05 University-industry research collaboration, 2006................98
5.06 Business telephone connection charge, 2005 106
5.07 Business monthly telephone subscription, 2005 107
5.08 Local supplier quality, 2006 101
5.09 Computer, comm., and other services imports, 2004.........81

Government readiness	**93**

6.01 Government prioritization of ICT, 2006................................67
6.02 Gov't. procurement of advanced tech products, 2006......100
6.03 Importance of ICT to gov't. vision of the future, 2006........74
6.04 E-participation index, 2005 68
6.05 E-government readiness index, 2005..................................92

Usage component	99
Individual usage	**99**

7.01 Mobile telephone subscribers, 2004 100
7.02 Personal computers, 2004..84
7.03 Broadband Internet subscribers, 2004 80
7.04 Internet users, 2004 104
7.05 Internet bandwidth...n/a

Business usage	**103**

8.01 Prevalence of foreign technology licensing, 2006.............113
8.02 Firm-level technology absorption, 2006.............................111
8.03 Capacity for innovation, 2006 93
8.04 Availability of new telephone lines, 2006 103
8.05 Availability of mobile telephones, 2006...............................76
8.06 Extent of business Internet use, 2006 95

Government usage	**82**

9.01 Government success in ICT promotion, 2006.....................94
9.02 Availability of online services, 2006....................................76
9.03 ICT use and government efficiency, 2006...........................82
9.04 ICT pervasiveness, 2006 71

221

Nigeria

Key indicators

Population (millions), 2005...131.5

GDP (PPP) per capita (US$), 2005........................1,188.4

Internet users per 100 inhabitants, 2005.....................3.8

Internet bandwidth (Mbps/10,000 inhabitants) ....n/a

Networked Readiness Index

Year (number of economies)	Rank
2006–2007 (122)	**88**
2005–2006 (115)...	90
2004–2005 (104)...	86

Global Competitiveness Index 2006–2007 (125)	101

Environment component — 71

Market environment — 59

1.01 Venture capital availability, 2006....................................72
1.02 Financial market sophistication, 2006...................................79
1.03 Technological readiness, 2006...85
1.04 State of cluster development, 2006 37
1.05 US utility patents, 2005 79
1.06 High-tech exports...n/a
1.07 Burden of government regulation, 2006 43
1.08 Extent and effect of taxation, 2006..................................23
1.09 Time required to start a business, 2006.............................79
1.10 No. of procedures required to start a business, 2006.........49
1.11 Intensity of local competition, 2006 103
1.12 Freedom of the press, 2006...92

Political and regulatory environment — 62

2.01 Effectiveness of law-making bodies, 2006.........................63
2.02 Laws relating to ICT, 2006...69
2.03 Judicial independence, 2006 81
2.04 Intellectual property protection, 2006................................87
2.05 Efficiency of legal framework, 2006...................................80
2.06 Property rights, 2006...86
2.07 Quality of competition in the ISP sector, 2006...................67
2.08 No. of procedures to enforce a contract, 2006 27
2.09 Time to enforce a contract, 2006 62

Infrastructure environment — 95

3.01 Telephone lines, 2005...106
3.02 Secure Internet servers, 2005..104
3.03 Internet hosts, 2004 117
3.04 Electricity production, 2003 98
3.05 Availability of scientists and engineers, 2006....................83
3.06 Quality of scientific research institutions, 2006 45
3.07 Tertiary enrollment, 2004..92

Readiness component — 95

Individual readiness — 103

4.01 Quality of math and science education, 2006...................102
4.02 Quality of the educational system, 2006............................70
4.03 Quality of public schools, 2006 108
4.04 Internet access in schools, 2006.......................................96
4.05 Buyer sophistication, 2006 83
4.06 Residential telephone connection charge, 2004 104
4.07 Residential monthly telephone subscription, 2004 103
4.08 High-speed monthly broadband subscription charge ....n/a
4.09 Lowest cost of broadband...n/a
4.10 Cost of mobile telephone call, 2004..................................80

Business readiness — 87

5.01 Extent of staff training, 2006..65
5.02 Local availability of research and training, 2006 60
5.03 Quality of management schools, 2006...............................94
5.04 Company spending on R&D, 2006 38
5.05 University-industry research collaboration, 2006................49
5.06 Business telephone connection charge, 2004 100
5.07 Business monthly telephone subscription, 2004 95
5.08 Local supplier quality, 2006 93
5.09 Computer, comm., and other services imports, 2004.........13

Government readiness — 75

6.01 Government prioritization of ICT, 2006...............................63
6.02 Gov't. procurement of advanced tech products, 2006........20
6.03 Importance of ICT to gov't. vision of the future, 2006........42
6.04 E-participation index, 2005 71
6.05 E-government readiness index, 2005...............................106

Usage component — 82

Individual usage — 101

7.01 Mobile telephone subscribers, 2005 96
7.02 Personal computers, 2004..106
7.03 Broadband Internet subscribers, 2005 96
7.04 Internet users, 2005 96
7.05 Internet bandwidth...n/a

Business usage — 82

8.01 Prevalence of foreign technology licensing, 2006...............59
8.02 Firm-level technology absorption, 2006.............................78
8.03 Capacity for innovation, 2006 58
8.04 Availability of new telephone lines, 2006 91
8.05 Availability of mobile telephones, 2006.............................95
8.06 Extent of business Internet use, 2006 81

Government usage — 67

9.01 Government success in ICT promotion, 2006....................48
9.02 Availability of online services, 2006...................................55
9.03 ICT use and government efficiency, 2006..........................61
9.04 ICT pervasiveness, 2006 91

Norway

Key indicators

Population (millions), 2005..4.6

GDP (PPP) per capita (US$), 2005.....................................42,364.2

Internet users per 100 inhabitants, 2005....................................73.6

Internet bandwidth (Mbps/10,000 inhabitants), 2004...............94.5

Networked Readiness Index

Year (number of economies)	Rank
2006–2007 (122) **10**	
2005–2006 (115)..13	
2004–2005 (104)..13	

Global Competitiveness Index 2006–2007 (125)	12

Environment component	7
Market environment	**18**

1.01 Venture capital availability, 2006...6
1.02 Financial market sophistication, 2006...................................20
1.03 Technological readiness, 2006..9
1.04 State of cluster development, 2006.......................................30
1.05 US utility patents, 2005...19
1.06 High-tech exports, 2004..44
1.07 Burden of government regulation, 2006...............................19
1.08 Extent and effect of taxation, 2006......................................51
1.09 Time required to start a business, 2006................................17
1.10 No. of procedures required to start a business, 2006...........7
1.11 Intensity of local competition, 2006.....................................19
1.12 Freedom of the press, 2006..4

Political and regulatory environment	**4**

2.01 Effectiveness of law-making bodies, 2006.............................8
2.02 Laws relating to ICT, 2006..4
2.03 Judicial independence, 2006...5
2.04 Intellectual property protection, 2006..................................14
2.05 Efficiency of legal framework, 2006.......................................5
2.06 Property rights, 2006...13
2.07 Quality of competition in the ISP sector, 2006....................15
2.08 No. of procedures to enforce a contract, 2006......................1
2.09 Time to enforce a contract, 2006...23

Infrastructure environment	**6**

3.01 Telephone lines, 2005...21
3.02 Secure Internet servers, 2005..14
3.03 Internet hosts, 2004..7
3.04 Electricity production, 2003..2
3.05 Availability of scientists and engineers, 2006......................25
3.06 Quality of scientific research institutions, 2006..................18
3.07 Tertiary enrollment, 2004..6

Readiness component	21
Individual readiness	**20**

4.01 Quality of math and science education, 2006......................54
4.02 Quality of the educational system, 2006..............................17
4.03 Quality of public schools, 2006...19
4.04 Internet access in schools, 2006..22
4.05 Buyer sophistication, 2006..23
4.06 Residential telephone connection charge, 2005..................16
4.07 Residential monthly telephone subscription, 2005..............10
4.08 High-speed monthly broadband subscription, 2006...............9
4.09 Lowest cost of broadband, 2006...14
4.10 Cost of mobile telephone call, 2004.......................................4

Business readiness	**19**

5.01 Extent of staff training, 2006..10
5.02 Local availability of research and training, 2006..................15
5.03 Quality of management schools, 2006...................................23
5.04 Company spending on R&D, 2006...19
5.05 University-industry research collaboration, 2006.................18
5.06 Business telephone connection charge, 2005......................15
5.07 Business monthly telephone subscription, 2005....................8
5.08 Local supplier quality, 2006..17
5.09 Computer, comm., and other services imports, 2004.........58

Government readiness	**20**

6.01 Government prioritization of ICT, 2006..................................23
6.02 Gov't. procurement of advanced tech products, 2006........34
6.03 Importance of ICT to gov't. vision of the future, 2006........21
6.04 E-participation index, 2005..26
6.05 E-government readiness index, 2005....................................10

Usage component	7
Individual usage	**5**

7.01 Mobile telephone subscribers, 2005.....................................13
7.02 Personal computers, 2004...14
7.03 Broadband Internet subscribers, 2005....................................8
7.04 Internet users, 2005...4
7.05 Internet bandwidth, 2004...7

Business usage	**12**

8.01 Prevalence of foreign technology licensing, 2006................24
8.02 Firm-level technology absorption, 2006.................................12
8.03 Capacity for innovation, 2006..15
8.04 Availability of new telephone lines, 2006..............................14
8.05 Availability of mobile telephones, 2006................................22
8.06 Extent of business Internet use, 2006...................................11

Government usage	**15**

9.01 Government success in ICT promotion, 2006........................25
9.02 Availability of online services, 2006......................................11
9.03 ICT use and government efficiency, 2006..............................21
9.04 ICT pervasiveness, 2006..14

Pakistan

Key indicators

Population (millions), 2005...157.9
GDP (PPP) per capita (US$), 2005.........................2,652.5
Internet users per 100 inhabitants, 2005.....................6.8
Internet bandwidth (Mbps/10,000 inhabitants), 2004.................0.1

Networked Readiness Index

Year (number of economies)	Rank
2006–2007 (122) **84**	
2005–2006 (115)...67	
2004–2005 (104)...63	

Global Competitiveness Index 2006–2007 (125)	91

Environment component	94
Market environment	**67**

1.01 Venture capital availability, 2006............................61
1.02 Financial market sophistication, 2006...................53
1.03 Technological readiness, 2006..............................77
1.04 State of cluster development, 2006.......................55
1.05 US utility patents, 2005..78
1.06 High-tech exports, 2004.......................................63
1.07 Burden of government regulation, 2006................54
1.08 Extent and effect of taxation, 2006.......................33
1.09 Time required to start a business, 2006................38
1.10 No. of procedures required to start a business, 2006.........79
1.11 Intensity of local competition, 2006......................73
1.12 Freedom of the press, 2006..................................84

Political and regulatory environment	**100**

2.01 Effectiveness of law-making bodies, 2006............59
2.02 Laws relating to ICT, 2006....................................65
2.03 Judicial independence, 2006.................................79
2.04 Intellectual property protection, 2006...................79
2.05 Efficiency of legal framework, 2006......................89
2.06 Property rights, 2006..93
2.07 Quality of competition in the ISP sector, 2006......36
2.08 No. of procedures to enforce a contract, 2006......113
2.09 Time to enforce a contract, 2006.........................106

Infrastructure environment	**102**

3.01 Telephone lines, 2005...100
3.02 Secure Internet servers, 2005.............................101
3.03 Internet hosts, 2004..94
3.04 Electricity production, 2003...................................91
3.05 Availability of scientists and engineers, 2006........78
3.06 Quality of scientific research institutions, 2006......62
3.07 Tertiary enrollment, 2004....................................104

Readiness component	83
Individual readiness	**94**

4.01 Quality of math and science education, 2006........85
4.02 Quality of the educational system, 2006...............73
4.03 Quality of public schools, 2006............................78
4.04 Internet access in schools, 2006..........................51
4.05 Buyer sophistication, 2006...................................67
4.06 Residential telephone connection charge, 2005.....58
4.07 Residential monthly telephone subscription, 2005.....89
4.08 High-speed monthly broadband subscription, 2006.....96
4.09 Lowest cost of broadband, 2006.........................100
4.10 Cost of mobile telephone call, 2005.....................64

Business readiness	**64**

5.01 Extent of staff training, 2006................................90
5.02 Local availability of research and training, 2006.....83
5.03 Quality of management schools, 2006...................71
5.04 Company spending on R&D, 2006.........................51
5.05 University-industry research collaboration, 2006.....61
5.06 Business telephone connection charge, 2005........51
5.07 Business monthly telephone subscription, 2005.....73
5.08 Local supplier quality, 2006..................................66
5.09 Computer, comm., and other services imports, 2004.........43

Government readiness	**80**

6.01 Government prioritization of ICT, 2006..................62
6.02 Gov't. procurement of advanced tech products, 2006........47
6.03 Importance of ICT to gov't. vision of the future, 2006........53
6.04 E-participation index, 2005...................................62
6.05 E-government readiness index, 2005...................103

Usage component	79
Individual usage	**102**

7.01 Mobile telephone subscribers, 2005...................105
7.02 Personal computers...n/a
7.03 Broadband Internet subscribers, 2005..................86
7.04 Internet users, 2005...83
7.05 Internet bandwidth, 2004......................................79

Business usage	**73**

8.01 Prevalence of foreign technology licensing, 2006....81
8.02 Firm-level technology absorption, 2006.................85
8.03 Capacity for innovation, 2006...............................38
8.04 Availability of new telephone lines, 2006...............89
8.05 Availability of mobile telephones, 2006................102
8.06 Extent of business Internet use, 2006...................50

Government usage	**63**

9.01 Government success in ICT promotion, 2006.........46
9.02 Availability of online services, 2006......................59
9.03 ICT use and government efficiency, 2006..............63
9.04 ICT pervasiveness, 2006......................................84

224

Panama

Key indicators

Population (millions), 2005..3.2
GDP (PPP) per capita (US$), 2005.....................................7,282.8
Internet users per 100 inhabitants, 2005....................................6.4
Internet bandwidth (Mbps/10,000 inhabitants), 2004................2.9

Networked Readiness Index

Year (number of economies)	Rank
2006–2007 (122)	**65**
2005–2006 (115)	66
2004–2005 (104)	69

Global Competitiveness Index 2006–2007 (125)	57

Environment component	58
Market environment	**45**
1.01 Venture capital availability, 2006	35
1.02 Financial market sophistication, 2006	27
1.03 Technological readiness, 2006	28
1.04 State of cluster development, 2006	42
1.05 US utility patents, 2005	79
1.06 High-tech exports, 2004	94
1.07 Burden of government regulation, 2006	74
1.08 Extent and effect of taxation, 2006	91
1.09 Time required to start a business, 2006	27
1.10 No. of procedures required to start a business, 2006	27
1.11 Intensity of local competition, 2006	61
1.12 Freedom of the press, 2006	56
Political and regulatory environment	**80**
2.01 Effectiveness of law-making bodies, 2006	104
2.02 Laws relating to ICT, 2006	55
2.03 Judicial independence, 2006	103
2.04 Intellectual property protection, 2006	49
2.05 Efficiency of legal framework, 2006	95
2.06 Property rights, 2006	44
2.07 Quality of competition in the ISP sector, 2006	34
2.08 No. of procedures to enforce a contract, 2006	102
2.09 Time to enforce a contract, 2006	96
Infrastructure environment	**66**
3.01 Telephone lines, 2005	73
3.02 Secure Internet servers, 2005	32
3.03 Internet hosts, 2004	63
3.04 Electricity production, 2003	67
3.05 Availability of scientists and engineers, 2006	102
3.06 Quality of scientific research institutions, 2006	90
3.07 Tertiary enrollment, 2004	37

Readiness component	67
Individual readiness	**62**
4.01 Quality of math and science education, 2006	104
4.02 Quality of the educational system, 2006	104
4.03 Quality of public schools, 2006	94
4.04 Internet access in schools, 2006	58
4.05 Buyer sophistication, 2006	33
4.06 Residential telephone connection charge, 2005	48
4.07 Residential monthly telephone subscription, 2005	43
4.08 High-speed monthly broadband subscription, 2006	62
4.09 Lowest cost of broadband, 2006	73
4.10 Cost of mobile telephone call, 2005	49
Business readiness	**67**
5.01 Extent of staff training, 2006	58
5.02 Local availability of research and training, 2006	73
5.03 Quality of management schools, 2006	83
5.04 Company spending on R&D, 2006	74
5.05 University-industry research collaboration, 2006	83
5.06 Business telephone connection charge, 2005	40
5.07 Business monthly telephone subscription, 2005	65
5.08 Local supplier quality, 2006	54
5.09 Computer, comm., and other services imports, 2004	85
Government readiness	**78**
6.01 Government prioritization of ICT, 2006	110
6.02 Gov't. procurement of advanced tech products, 2006	89
6.03 Importance of ICT to gov't. vision of the future, 2006	93
6.04 E-participation index, 2005	36
6.05 E-government readiness index, 2005	62

Usage component	74
Individual usage	**82**
7.01 Mobile telephone subscribers, 2005	67
7.02 Personal computers, 2004	80
7.03 Broadband Internet subscribers, 2005	64
7.04 Internet users, 2005	85
7.05 Internet bandwidth, 2004	41
Business usage	**49**
8.01 Prevalence of foreign technology licensing, 2006	48
8.02 Firm-level technology absorption, 2006	50
8.03 Capacity for innovation, 2006	91
8.04 Availability of new telephone lines, 2006	55
8.05 Availability of mobile telephones, 2006	56
8.06 Extent of business Internet use, 2006	38
Government usage	**89**
9.01 Government success in ICT promotion, 2006	112
9.02 Availability of online services, 2006	67
9.03 ICT use and government efficiency, 2006	86
9.04 ICT pervasiveness, 2006	90

225

Paraguay

Key indicators

Population (millions), 2005..(2
GDP (PPP) per capita (US$), 2005.........................4,887.7
Internet users per 100 inhabitants, 2005......................3.2
Internet bandwidth (Mbps/10,000 inhabitants), 2004.................0.3

Networked Readiness Index

Year (number of economies)	Rank
2006–2007 (122) **114**	
2005–2006 (115)..113	
2004–2005 (104)...98	

Global Competitiveness Index 2006–2007 (125)   106

Environment component	116
Market environment	**114**

1.01 Venture capital availability, 2006...........................117
1.02 Financial market sophistication, 2006..................101
1.03 Technological readiness, 2006.............................100
1.04 State of cluster development, 2006 107
1.05 US utility patents, 2005 79
1.06 High-tech exports, 2004 71
1.07 Burden of government regulation, 2006 81
1.08 Extent and effect of taxation, 2006....................29
1.09 Time required to start a business, 2006.............103
1.10 No. of procedures required to start a business, 2006.......113
1.11 Intensity of local competition, 2006 106
1.12 Freedom of the press, 2006....................................42

Political and regulatory environment	**120**

2.01 Effectiveness of law-making bodies, 2006.........118
2.02 Laws relating to ICT, 2006...................................115
2.03 Judicial independence, 2006 120
2.04 Intellectual property protection, 2006..................114
2.05 Efficiency of legal framework, 2006....................121
2.06 Property rights, 2006 115
2.07 Quality of competition in the ISP sector, 2006..................105
2.08 No. of procedures to enforce a contract, 2006 103
2.09 Time to enforce a contract, 2006 64

Infrastructure environment	**96**

3.01 Telephone lines, 2005...95
3.02 Secure Internet servers, 2005...............................86
3.03 Internet hosts, 2004 71
3.04 Electricity production, 2003 16
3.05 Availability of scientists and engineers, 2006....................120
3.06 Quality of scientific research institutions, 2006 121
3.07 Tertiary enrollment, 2003..70

Readiness component	107
Individual readiness	**100**

4.01 Quality of math and science education, 2006...................121
4.02 Quality of the educational system, 2006..........................122
4.03 Quality of public schools, 2006 118
4.04 Internet access in schools, 2006.........................114
4.05 Buyer sophistication, 2006 109
4.06 Residential telephone connection charge, 2005 101
4.07 Residential monthly telephone subscription, 2005 76
4.08 High-speed monthly broadband subscription charge ....n/a
4.09 Lowest cost of broadband, 2006..........................76
4.10 Cost of mobile telephone call, 2005....................68

Business readiness	**109**

5.01 Extent of staff training, 2006...............................112
5.02 Local availability of research and training, 2006 107
5.03 Quality of management schools, 2006...............107
5.04 Company spending on R&D, 2006 120
5.05 University-industry research collaboration, 2006..............118
5.06 Business telephone connection charge, 2005 93
5.07 Business monthly telephone subscription, 2005 69
5.08 Local supplier quality, 2006 104
5.09 Computer, comm., and other services imports, 2004.........93

Government readiness	**117**

6.01 Government prioritization of ICT, 2006.............117
6.02 Gov't. procurement of advanced tech products, 2006......117
6.03 Importance of ICT to gov't. vision of the future, 2006......118
6.04 E-participation index, 2005 101
6.05 E-government readiness index, 2005.................88

Usage component	114
Individual usage	**89**

7.01 Mobile telephone subscribers, 2005 81
7.02 Personal computers, 2004.....................................67
7.03 Broadband Internet subscribers, 2005 79
7.04 Internet users, 2005 100
7.05 Internet bandwidth, 2004 67

Business usage	**110**

8.01 Prevalence of foreign technology licensing, 2006.............118
8.02 Firm-level technology absorption, 2006..............121
8.03 Capacity for innovation, 2006 101
8.04 Availability of new telephone lines, 2006 109
8.05 Availability of mobile telephones, 2006 59
8.06 Extent of business Internet use, 2006 111

Government usage	**118**

9.01 Government success in ICT promotion, 2006...................122
9.02 Availability of online services, 2006....................107
9.03 ICT use and government efficiency, 2006.........................108
9.04 ICT pervasiveness, 2006 114

Peru

Key indicators

Population (millions), 2005..28.0

GDP (PPP) per capita (US$), 2005....................................5,983.2

Internet users per 100 inhabitants, 2005....................................16.4

Internet bandwidth (Mbps/10,000 inhabitants), 2004................2.0

Networked Readiness Index

Year (number of economies)	Rank
2006–007 (122) **78**	
2005–006 (115)...85	
2004–005 (104)...90	

Global Competitiveness Index 2006–007 (125)   74

Environment component	88
Market environment	**72**

1.01 Venture capital availability, 2006..79
1.02 Financial market sophistication, 2006.................................50
1.03 Technological readiness, 2006...61
1.04 State of cluster development, 2006......................................64
1.05 US utility patents, 2005...68
1.06 High-tech exports, 2004..79
1.07 Burden of government regulation, 2006 109
1.08 Extent and effect of taxation, 2006.....................................80
1.09 Time required to start a business, 2006...........................100
1.10 No. of procedures required to start a business, 2006.........61
1.11 Intensity of local competition, 2006 58
1.12 Freedom of the press, 2006...16

Political and regulatory environment	**98**

2.01 Effectiveness of law-making bodies, 2006.......................119
2.02 Laws relating to ICT, 2006..66
2.03 Judicial independence, 2006 116
2.04 Intellectual property protection, 2006.................................94
2.05 Efficiency of legal framework, 2006..................................109
2.06 Property rights, 2006...104
2.07 Quality of competition in the ISP sector, 2006...................61
2.08 No. of procedures to enforce a contract, 2006 75
2.09 Time to enforce a contract, 2006 27

Infrastructure environment	**82**

3.01 Telephone lines, 2005..86
3.02 Secure Internet servers, 2005..65
3.03 Internet hosts, 2004 55
3.04 Electricity production, 2003...80
3.05 Availability of scientists and engineers, 2006....................73
3.06 Quality of scientific research institutions, 2006 106
3.07 Tertiary enrollment, 2004..57

Readiness component	79
Individual readiness	**89**

4.01 Quality of math and science education, 2006...................122
4.02 Quality of the educational system, 2006...........................121
4.03 Quality of public schools, 2006 122
4.04 Internet access in schools, 2006...59
4.05 Buyer sophistication, 2006 73
4.06 Residential telephone connection charge, 2004 82
4.07 Residential monthly telephone subscription, 2004 97
4.08 High-speed monthly broadband subscription, 2006...........60
4.09 Lowest cost of broadband, 2006...68
4.10 Cost of mobile telephone call, 2005............................. 77

Business readiness	**71**

5.01 Extent of staff training, 2006...66
5.02 Local availability of research and training, 2006................63
5.03 Quality of management schools, 2006................................45
5.04 Company spending on R&D, 2006 67
5.05 University-industry research collaboration, 2006................97
5.06 Business telephone connection charge, 2004 73
5.07 Business monthly telephone subscription, 2004 86
5.08 Local supplier quality, 2006 43
5.09 Computer, comm., and other services imports, 2004.........54

Government readiness	**69**

6.01 Government prioritization of ICT, 2006...............................84
6.02 Gov't. procurement of advanced tech products, 2006......104
6.03 Importance of ICT to gov't. vision of the future, 2006........86
6.04 E-participation index, 2005 36
6.05 E-government readiness index, 2005..................................55

Usage component	66
Individual usage	**78**

7.01 Mobile telephone subscribers, 2005 92
7.02 Personal computers, 2004..59
7.03 Broadband Internet subscribers, 2005 53
7.04 Internet users, 2005 60
7.05 Internet bandwidth, 2004 42

Business usage	**60**

8.01 Prevalence of foreign technology licensing, 2006..............76
8.02 Firm-level technology absorption, 2006...............................76
8.03 Capacity for innovation, 2006 57
8.04 Availability of new telephone lines, 2006...........................41
8.05 Availability of mobile telephones, 2006..............................66
8.06 Extent of business Internet use, 2006 46

Government usage	**69**

9.01 Government success in ICT promotion, 2006...................101
9.02 Availability of online services, 2006....................................42
9.03 ICT use and government efficiency, 2006...........................46
9.04 ICT pervasiveness, 2006 83

227

Philippines

Key indicators

Population (millions), 2005...83.1
GDP (PPP) per capita (US$), 2005.......................4,922.8
Internet users per 100 inhabitants, 2004....................5.3
Internet bandwidth (Mbps/10,000 inhabitants), 2004.................0.4

Networked Readiness Index

Year (number of economies)	Rank
2006–2007 (122) **69**
2005–2006 (115)...70
2004–2005 (104)...67

Global Competitiveness Index 2006–2007 (125)	71

Environment component	61
Market environment	**47**

1.01 Venture capital availability, 2006.............................77
1.02 Financial market sophistication, 2006.....................66
1.03 Technological readiness, 2006................................63
1.04 State of cluster development, 2006.........................46
1.05 US utility patents, 2005.......................................63
1.06 High-tech exports, 2004.......................................5
1.07 Burden of government regulation, 2006.................106
1.08 Extent and effect of taxation, 2006........................40
1.09 Time required to start a business, 2006.................89
1.10 No. of procedures required to start a business, 2006.........79
1.11 Intensity of local competition, 2006.....................49
1.12 Freedom of the press, 2006.................................72

Political and regulatory environment	**59**

2.01 Effectiveness of law-making bodies, 2006....................100
2.02 Laws relating to ICT, 2006...................................53
2.03 Judicial independence, 2006.................................76
2.04 Intellectual property protection, 2006....................83
2.05 Efficiency of legal framework, 2006.......................84
2.06 Property rights, 2006...70
2.07 Quality of competition in the ISP sector, 2006..............31
2.08 No. of procedures to enforce a contract, 2006 34
2.09 Time to enforce a contract, 2006 86

Infrastructure environment	**85**

3.01 Telephone lines, 2005..98
3.02 Secure Internet servers, 2005...............................76
3.03 Internet hosts, 2004..78
3.04 Electricity production, 2003..................................84
3.05 Availability of scientists and engineers, 2006.............84
3.06 Quality of scientific research institutions, 2006 78
3.07 Tertiary enrollment, 2004.....................................63

Readiness component	77
Individual readiness	**80**

4.01 Quality of math and science education, 2006...................107
4.02 Quality of the educational system, 2006.....................60
4.03 Quality of public schools, 2006 91
4.04 Internet access in schools, 2006............................68
4.05 Buyer sophistication, 2006 46
4.06 Residential telephone connection charge, 2005 77
4.07 Residential monthly telephone subscription, 2005 105
4.08 High-speed monthly broadband subscription, 2006.............81
4.09 Lowest cost of broadband, 2006.............................54
4.10 Cost of mobile telephone call, 2004........................74

Business readiness	**89**

5.01 Extent of staff training, 2006................................35
5.02 Local availability of research and training, 2006.................75
5.03 Quality of management schools, 2006.......................46
5.04 Company spending on R&D, 2006 56
5.05 University-industry research collaboration, 2006.............67
5.06 Business telephone connection charge, 2005 77
5.07 Business monthly telephone subscription, 2005 111
5.08 Local supplier quality, 2006 63
5.09 Computer, comm., and other services imports, 2004.........71

Government readiness	**42**

6.01 Government prioritization of ICT, 2006.....................95
6.02 Gov't. procurement of advanced tech products, 2006........88
6.03 Importance of ICT to gov't. vision of the future, 2006........88
6.04 E-participation index, 2005 19
6.05 E-government readiness index, 2005.......................41

Usage component	70
Individual usage	**77**

7.01 Mobile telephone subscribers, 2005 69
7.02 Personal computers, 2004.....................................77
7.03 Broadband Internet subscribers..............................n/a
7.04 Internet users, 2004 91
7.05 Internet bandwidth, 2004......................................65

Business usage	**56**

8.01 Prevalence of foreign technology licensing, 2006...............56
8.02 Firm-level technology absorption, 2006.....................48
8.03 Capacity for innovation, 2006 63
8.04 Availability of new telephone lines, 2006 70
8.05 Availability of mobile telephones, 2006....................36
8.06 Extent of business Internet use, 2006 51

Government usage	**76**

9.01 Government success in ICT promotion, 2006...................68
9.02 Availability of online services, 2006........................68
9.03 ICT use and government efficiency, 2006...................51
9.04 ICT pervasiveness, 2006 96

228

Poland

Key indicators

Population (millions), 2005...38.5

GDP (PPP) per capita (US$), 2005........................12,994.2

Internet users per 100 inhabitants, 2005....................26.0

Internet bandwidth (Mbps/10,000 inhabitants), 2004...............5.5

Networked Readiness Index

Year (number of economies)	Rank
2006–2007 (122) **58**	
2005–2006 (115)...53	
2004–2005 (104)...72	

Global Competitiveness Index 2006–2007 (125)	48

Environment component	64
Market environment	**69**
1.01 Venture capital availability, 2006.........................36	
1.02 Financial market sophistication, 2006...................56	
1.03 Technological readiness, 2006.............................76	
1.04 State of cluster development, 2006 79	
1.05 US utility patents, 2005 47	
1.06 High-tech exports, 2004 52	
1.07 Burden of government regulation, 2006 55	
1.08 Extent and effect of taxation, 2006......................63	
1.09 Time required to start a business, 2006...............56	
1.10 No. of procedures required to start a business, 2006.........61	
1.11 Intensity of local competition, 2006 96	
1.12 Freedom of the press, 2006.................................79	
Political and regulatory environment	**95**
2.01 Effectiveness of law-making bodies, 2006...........93	
2.02 Laws relating to ICT, 2006..................................62	
2.03 Judicial independence, 2006 69	
2.04 Intellectual property protection, 2006..................57	
2.05 Efficiency of legal framework, 2006.....................65	
2.06 Property rights, 2006 90	
2.07 Quality of competition in the ISP sector, 2006.....79	
2.08 No. of procedures to enforce a contract, 2006 91	
2.09 Time to enforce a contract, 2006 107	
Infrastructure environment	**40**
3.01 Telephone lines, 2005..42	
3.02 Secure Internet servers, 2005.............................43	
3.03 Internet hosts, 2004 48	
3.04 Electricity production, 2003 47	
3.05 Availability of scientists and engineers, 2006.......75	
3.06 Quality of scientific research institutions, 2006 58	
3.07 Tertiary enrollment, 2004.....................................22	

Readiness component	56
Individual readiness	**51**
4.01 Quality of math and science education, 2006.......53	
4.02 Quality of the educational system, 2006..............34	
4.03 Quality of public schools, 2006 38	
4.04 Internet access in schools, 2006..........................50	
4.05 Buyer sophistication, 2006 60	
4.06 Residential telephone connection charge...........n/a	
4.07 Residential monthly telephone subscription, 2004 70	
4.08 High-speed monthly broadband subscription, 2006...........48	
4.09 Lowest cost of broadband, 2006...........................31	
4.10 Cost of mobile telephone call, 2005.....................41	
Business readiness	**62**
5.01 Extent of staff training, 2006..............................48	
5.02 Local availability of research and training, 2006.................39	
5.03 Quality of management schools, 2006..................55	
5.04 Company spending on R&D, 2006.........................31	
5.05 University-industry research collaboration, 2006................38	
5.06 Business telephone connection charge................n/a	
5.07 Business monthly telephone subscription..............n/a	
5.08 Local supplier quality, 2006 60	
5.09 Computer, comm., and other services imports, 2004.........22	
Government readiness	**59**
6.01 Government prioritization of ICT, 2006...............115	
6.02 Gov't. procurement of advanced tech products, 2006........76	
6.03 Importance of ICT to gov't. vision of the future, 2006........90	
6.04 E-participation index, 2005 29	
6.05 E-government readiness index, 2005 37	

229

Usage component	63
Individual usage	**43**
7.01 Mobile telephone subscribers, 2005 43	
7.02 Personal computers, 2004....................................36	
7.03 Broadband Internet subscribers, 2005 39	
7.04 Internet users, 2005 44	
7.05 Internet bandwidth, 2004 35	
Business usage	**77**
8.01 Prevalence of foreign technology licensing, 2006...............85	
8.02 Firm-level technology absorption, 2006...............82	
8.03 Capacity for innovation, 2006 30	
8.04 Availability of new telephone lines, 2006 94	
8.05 Availability of mobile telephones, 2006............114	
8.06 Extent of business Internet use, 2006 41	
Government usage	**88**
9.01 Government success in ICT promotion, 2006...................108	
9.02 Availability of online services, 2006....................65	
9.03 ICT use and government efficiency, 2006...........91	
9.04 ICT pervasiveness, 2006 87	

Portugal

Key indicators

Population (millions), 2005..10.5

GDP (PPP) per capita (US$), 2005......................................19,334.6

Internet users per 100 inhabitants, 2005...................................28.0

Internet bandwidth (Mbps/10,000 inhabitants), 2004................8.3

Networked Readiness Index

Year (number of economies)	Rank
2006–2007 (122) **28**	
2005–2006 (115)..27	
2004–2005 (104)..30	

Global Competitiveness Index 2006–2007 (125)	34

Environment component	27
Market environment	**32**

1.01	Venture capital availability, 2006	33
1.02	Financial market sophistication, 2006	25
1.03	Technological readiness, 2006	52
1.04	State of cluster development, 2006	54
1.05	US utility patents, 2005	39
1.06	High-tech exports, 2004	33
1.07	Burden of government regulation, 2006	44
1.08	Extent and effect of taxation, 2006	55
1.09	Time required to start a business, 2006	7
1.10	No. of procedures required to start a business, 2006	36
1.11	Intensity of local competition, 2006	45
1.12	Freedom of the press, 2006	9

Political and regulatory environment	**27**

2.01	Effectiveness of law-making bodies, 2006	36
2.02	Laws relating to ICT, 2006	31
2.03	Judicial independence, 2006	19
2.04	Intellectual property protection, 2006	24
2.05	Efficiency of legal framework, 2006	45
2.06	Property rights, 2006	33
2.07	Quality of competition in the ISP sector, 2006	43
2.08	No. of procedures to enforce a contract, 2006	30
2.09	Time to enforce a contract, 2006	70

Infrastructure environment	**28**

3.01	Telephone lines, 2005	33
3.02	Secure Internet servers, 2005	31
3.03	Internet hosts, 2004	21
3.04	Electricity production, 2003	44
3.05	Availability of scientists and engineers, 2006	32
3.06	Quality of scientific research institutions, 2006	35
3.07	Tertiary enrollment, 2004	28

Readiness component	30
Individual readiness	**40**

4.01	Quality of math and science education, 2006	83
4.02	Quality of the educational system, 2006	57
4.03	Quality of public schools, 2006	41
4.04	Internet access in schools, 2006	34
4.05	Buyer sophistication, 2006	51
4.06	Residential telephone connection charge, 2005	41
4.07	Residential monthly telephone subscription, 2005	50
4.08	High-speed monthly broadband subscription, 2006	33
4.09	Lowest cost of broadband, 2006	21
4.10	Cost of mobile telephone call, 2004	19

Business readiness	**37**

5.01	Extent of staff training, 2006	55
5.02	Local availability of research and training, 2006	38
5.03	Quality of management schools, 2006	32
5.04	Company spending on R&D, 2006	53
5.05	University-industry research collaboration, 2006	33
5.06	Business telephone connection charge, 2005	31
5.07	Business monthly telephone subscription, 2005	34
5.08	Local supplier quality, 2006	41
5.09	Computer, comm., and other services imports, 2004	31

Government readiness	**26**

6.01	Government prioritization of ICT, 2006	5
6.02	Gov't. procurement of advanced tech products, 2006	26
6.03	Importance of ICT to gov't. vision of the future, 2006	7
6.04	E-participation index, 2005	46
6.05	E-government readiness index, 2005	30

Usage component	31
Individual usage	**35**

7.01	Mobile telephone subscribers, 2005	8
7.02	Personal computers, 2004	46
7.03	Broadband Internet subscribers, 2005	26
7.04	Internet users, 2005	40
7.05	Internet bandwidth, 2004	31

Business usage	**36**

8.01	Prevalence of foreign technology licensing, 2006	13
8.02	Firm-level technology absorption, 2006	63
8.03	Capacity for innovation, 2006	40
8.04	Availability of new telephone lines, 2006	38
8.05	Availability of mobile telephones, 2006	43
8.06	Extent of business Internet use, 2006	37

Government usage	**29**

9.01	Government success in ICT promotion, 2006	19
9.02	Availability of online services, 2006	30
9.03	ICT use and government efficiency, 2006	15
9.04	ICT pervasiveness, 2006	43

Qatar

Key indicators

Population (millions), 2005..0.8
GDP (PPP) per capita (US$), 2005......................................31,397.3
Internet users per 100 inhabitants, 2005...................................28.2
Internet bandwidth (Mbps/10,000 inhabitants), 2004................6.2

Networked Readiness Index

Year (number of economies)	Rank
2006–2007 (122)	**36**
2005–2006 (115)...	39
2004–2005 (104)	—

Global Competitiveness Index 2006–2007 (125)	38

Environment component	44
Market environment	**53**
1.01 Venture capital availability, 2006..37	
1.02 Financial market sophistication, 2006....................................47	
1.03 Technological readiness, 2006..27	
1.04 State of cluster development, 2006...56	
1.05 US utility patents, 2005...79	
1.06 High-tech exports, 2004..88	
1.07 Burden of government regulation, 2006..................................16	
1.08 Extent and effect of taxation, 2006...5	
1.09 Time required to start a business...n/a	
1.10 No. of procedures required to start a business...................n/a	
1.11 Intensity of local competition, 2006.......................................67	
1.12 Freedom of the press, 2006..90	
Political and regulatory environment	**37**
2.01 Effectiveness of law-making bodies, 2006............................32	
2.02 Laws relating to ICT, 2006..39	
2.03 Judicial independence, 2006...20	
2.04 Intellectual property protection, 2006....................................27	
2.05 Efficiency of legal framework, 2006.......................................26	
2.06 Property rights, 2006...38	
2.07 Quality of competition in the ISP sector, 2006....................111	
2.08 No. of procedures to enforce a contract............................n/a	
2.09 Time to enforce a contract...n/a	
Infrastructure environment	**45**
3.01 Telephone lines, 2005...48	
3.02 Secure Internet servers, 2005..48	
3.03 Internet hosts, 2004...88	
3.04 Electricity production, 2003...5	
3.05 Availability of scientists and engineers, 2006......................80	
3.06 Quality of scientific research institutions, 2006..................49	
3.07 Tertiary enrollment, 2004..77	

Readiness component	36
Individual readiness	**36**
4.01 Quality of math and science education, 2006.......................38	
4.02 Quality of the educational system, 2006...............................20	
4.03 Quality of public schools, 2006..39	
4.04 Internet access in schools, 2006..39	
4.05 Buyer sophistication, 2006..48	
4.06 Residential telephone connection charge, 2005.....................7	
4.07 Residential monthly telephone subscription, 2005................5	
4.08 High-speed monthly broadband subscription, 2006.............26	
4.09 Lowest cost of broadband..n/a	
4.10 Cost of mobile telephone call, 2004......................................12	
Business readiness	**43**
5.01 Extent of staff training, 2006..57	
5.02 Local availability of research and training, 2006..................57	
5.03 Quality of management schools, 2006...................................40	
5.04 Company spending on R&D, 2006..42	
5.05 University-industry research collaboration, 2006.................60	
5.06 Business telephone connection charge, 2005........................5	
5.07 Business monthly telephone subscription, 2005..................30	
5.08 Local supplier quality, 2006..80	
5.09 Computer, comm., and other services imports..................n/a	
Government readiness	**36**
6.01 Government prioritization of ICT, 2006..................................10	
6.02 Gov't. procurement of advanced tech products, 2006........24	
6.03 Importance of ICT to gov't. vision of the future, 2006........11	
6.04 E-participation index, 2005...82	
6.05 E-government readiness index, 2005.....................................61	

Usage component	36
Individual usage	**41**
7.01 Mobile telephone subscribers, 2005......................................26	
7.02 Personal computers, 2004...38	
7.03 Broadband Internet subscribers, 2005..................................38	
7.04 Internet users, 2005...39	
7.05 Internet bandwidth, 2004..33	
Business usage	**62**
8.01 Prevalence of foreign technology licensing, 2006...............21	
8.02 Firm-level technology absorption, 2006................................43	
8.03 Capacity for innovation, 2006...61	
8.04 Availability of new telephone lines, 2006..............................68	
8.05 Availability of mobile telephones, 2006................................99	
8.06 Extent of business Internet use, 2006...................................66	
Government usage	**17**
9.01 Government success in ICT promotion, 2006.......................10	
9.02 Availability of online services, 2006.......................................24	
9.03 ICT use and government efficiency, 2006..............................13	
9.04 ICT pervasiveness, 2006...34	

231

Romania

Key indicators

Population (millions), 2005..21.7
GDP (PPP) per capita (US$), 2005.......................................8,785.0
Internet users per 100 inhabitants, 2004....................................20.8
Internet bandwidth (Mbps/10,000 inhabitants), 2004................1.9

Networked Readiness Index

Year (number of economies)	Rank
2006–2007 (122)...	**55**
2005–2006 (115)...	58
2004–2005 (104)...	53

Global Competitiveness Index 2006–2007 (125)	68

Environment component	62
Market environment	**52**

1.01 Venture capital availability, 2006..70
1.02 Financial market sophistication, 2006...................................85
1.03 Technological readiness, 2006...74
1.04 State of cluster development, 2006..10
1.05 US utility patents, 2005...56
1.06 High-tech exports, 2004...45
1.07 Burden of government regulation, 2006...............................51
1.08 Extent and effect of taxation, 2006.....................................106
1.09 Time required to start a business, 2006................................12
1.10 No. of procedures required to start a business, 2006.........10
1.11 Intensity of local competition, 2006.....................................59
1.12 Freedom of the press, 2006...64

Political and regulatory environment	**87**

2.01 Effectiveness of law-making bodies, 2006..........................99
2.02 Laws relating to ICT, 2006...57
2.03 Judicial independence, 2006..86
2.04 Intellectual property protection, 2006..................................80
2.05 Efficiency of legal framework, 2006......................................86
2.06 Property rights, 2006..75
2.07 Quality of competition in the ISP sector, 2006....................85
2.08 No. of procedures to enforce a contract, 2006...................99
2.09 Time to enforce a contract, 2006...33

Infrastructure environment	**55**

3.01 Telephone lines, 2005..60
3.02 Secure Internet servers, 2005...64
3.03 Internet hosts, 2004...62
3.04 Electricity production, 2003...58
3.05 Availability of scientists and engineers, 2006.....................41
3.06 Quality of scientific research institutions, 2006.................66
3.07 Tertiary enrollment, 2004...44

Readiness component	44
Individual readiness	**39**

4.01 Quality of math and science education, 2006.....................11
4.02 Quality of the educational system, 2006.............................51
4.03 Quality of public schools, 2006...36
4.04 Internet access in schools, 2006...65
4.05 Buyer sophistication, 2006...69
4.06 Residential telephone connection charge, 2005.................15
4.07 Residential monthly telephone subscription, 2005.............53
4.08 High-speed monthly broadband subscription, 2006...........43
4.09 Lowest cost of broadband, 2006...44
4.10 Cost of mobile telephone call, 2005.....................................53

Business readiness	**49**

5.01 Extent of staff training, 2006...81
5.02 Local availability of research and training, 2006................44
5.03 Quality of management schools, 2006..................................70
5.04 Company spending on R&D, 2006...70
5.05 University-industry research collaboration, 2006................77
5.06 Business telephone connection charge, 2005.....................14
5.07 Business monthly telephone subscription, 2005................50
5.08 Local supplier quality, 2006...69
5.09 Computer, comm., and other services imports, 2004.........19

Government readiness	**45**

6.01 Government prioritization of ICT, 2006.................................99
6.02 Gov't. procurement of advanced tech products, 2006........74
6.03 Importance of ICT to gov't. vision of the future, 2006........72
6.04 E-participation index, 2005..31
6.05 E-government readiness index, 2005....................................44

Usage component	57
Individual usage	**50**

7.01 Mobile telephone subscribers, 2005....................................52
7.02 Personal computers, 2004..50
7.03 Broadband Internet subscribers, 2005.................................37
7.04 Internet users, 2004..49
7.05 Internet bandwidth, 2004...44

Business usage	**79**

8.01 Prevalence of foreign technology licensing, 2006..............67
8.02 Firm-level technology absorption, 2006...............................72
8.03 Capacity for innovation, 2006..84
8.04 Availability of new telephone lines, 2006............................88
8.05 Availability of mobile telephones, 2006...............................83
8.06 Extent of business Internet use, 2006..................................80

Government usage	**54**

9.01 Government success in ICT promotion, 2006......................63
9.02 Availability of online services, 2006.....................................66
9.03 ICT use and government efficiency, 2006.............................78
9.04 ICT pervasiveness, 2006..37

Russian Federation

Key indicators

Population (millions), 2005..143.2

GDP (PPP) per capita (US$), 2005....................................11,041.1

Internet users per 100 inhabitants, 2005..................................15.2

Internet bandwidth (Mbps/10,000 inhabitants), 2004................1.0

Networked Readiness Index

Year (number of economies)	Rank
2006–2007 (122)	**70**
2005–2006 (115)	72
2004–2005 (104)	62

Global Competitiveness Index 2006–2007 (125)	62

Environment component	57
Market environment	**82**

1.01	Venture capital availability, 2006	62
1.02	Financial market sophistication, 2006	84
1.03	Technological readiness, 2006	84
1.04	State of cluster development, 2006	73
1.05	US utility patents, 2005	38
1.06	High-tech exports, 2004	57
1.07	Burden of government regulation, 2006	113
1.08	Extent and effect of taxation, 2006	92
1.09	Time required to start a business, 2006	49
1.10	No. of procedures required to start a business, 2006	27
1.11	Intensity of local competition, 2006	65
1.12	Freedom of the press, 2006	95

Political and regulatory environment	**89**

2.01	Effectiveness of law-making bodies, 2006	85
2.02	Laws relating to ICT, 2006	86
2.03	Judicial independence, 2006	107
2.04	Intellectual property protection, 2006	110
2.05	Efficiency of legal framework, 2006	103
2.06	Property rights, 2006	112
2.07	Quality of competition in the ISP sector, 2006	74
2.08	No. of procedures to enforce a contract, 2006	60
2.09	Time to enforce a contract, 2006	5

Infrastructure environment	**37**

3.01	Telephone lines, 2005	44
3.02	Secure Internet servers, 2005	77
3.03	Internet hosts, 2004	51
3.04	Electricity production, 2003	29
3.05	Availability of scientists and engineers, 2006	46
3.06	Quality of scientific research institutions, 2006	32
3.07	Tertiary enrollment, 2004	14

Readiness component	75
Individual readiness	**63**

4.01	Quality of math and science education, 2006	43
4.02	Quality of the educational system, 2006	54
4.03	Quality of public schools, 2006	51
4.04	Internet access in schools, 2006	54
4.05	Buyer sophistication, 2006	53
4.06	Residential telephone connection charge	n/a
4.07	Residential monthly telephone subscription	n/a
4.08	High-speed monthly broadband subscription, 2006	74
4.09	Lowest cost of broadband, 2006	84
4.10	Cost of mobile telephone call, 2004	46

Business readiness	**93**

5.01	Extent of staff training, 2006	98
5.02	Local availability of research and training, 2006	69
5.03	Quality of management schools, 2006	84
5.04	Company spending on R&D, 2006	44
5.05	University-industry research collaboration, 2006	54
5.06	Business telephone connection charge	n/a
5.07	Business monthly telephone subscription	n/a
5.08	Local supplier quality, 2006	71
5.09	Computer, comm., and other services imports, 2004	39

Government readiness	**76**

6.01	Government prioritization of ICT, 2006	75
6.02	Gov't. procurement of advanced tech products, 2006	78
6.03	Importance of ICT to gov't. vision of the future, 2006	105
6.04	E-participation index, 2005	60
6.05	E-government readiness index, 2005	50

Usage component	73
Individual usage	**51**

7.01	Mobile telephone subscribers, 2005	36
7.02	Personal computers, 2004	55
7.03	Broadband Internet subscribers, 2005	54
7.04	Internet users, 2005	61
7.05	Internet bandwidth, 2004	53

Business usage	**74**

8.01	Prevalence of foreign technology licensing, 2006	105
8.02	Firm-level technology absorption, 2006	81
8.03	Capacity for innovation, 2006	49
8.04	Availability of new telephone lines, 2006	90
8.05	Availability of mobile telephones, 2006	70
8.06	Extent of business Internet use, 2006	42

Government usage	**94**

9.01	Government success in ICT promotion, 2006	97
9.02	Availability of online services, 2006	84
9.03	ICT use and government efficiency, 2006	87
9.04	ICT pervasiveness, 2006	94

Serbia and Montenegro

Key indicators

Population (millions), 2005..10.5
GDP (PPP) per capita (US$), 2005.......................................5,347.8
Internet users per 100 inhabitants, 2004....................................18.6
Internet bandwidth (Mbps/10,000 inhabitants), 2004................0.9

Networked Readiness Index

Year (number of economies)	Rank
2006–2007 (122)	**74**
2005–2006 (115)..	80
2004–2005 (104)..	79

Global Competitiveness Index 2006–2007 (125)	87

Environment component	84
Market environment	**97**

1.01	Venture capital availability, 2006..	82
1.02	Financial market sophistication, 2006.................................	108
1.03	Technological readiness, 2006..	118
1.04	State of cluster development, 2006.....................................	111
1.05	US utility patents, 2005..	79
1.06	High-tech exports...	n/a
1.07	Burden of government regulation, 2006............................	120
1.08	Extent and effect of taxation, 2006....................................	97
1.09	Time required to start a business, 2006...............................	33
1.10	No. of procedures required to start a business, 2006.........	94
1.11	Intensity of local competition, 2006....................................	93
1.12	Freedom of the press, 2006..	68

Political and regulatory environment	**102**

2.01	Effectiveness of law-making bodies, 2006..........................	61
2.02	Laws relating to ICT, 2006..	70
2.03	Judicial independence, 2006...	101
2.04	Intellectual property protection, 2006..................................	112
2.05	Efficiency of legal framework, 2006.....................................	100
2.06	Property rights, 2006..	87
2.07	Quality of competition in the ISP sector, 2006...................	89
2.08	No. of procedures to enforce a contract, 2006..................	91
2.09	Time to enforce a contract, 2006..	81

Infrastructure environment	**46**

3.01	Telephone lines, 2004..	36
3.02	Secure Internet servers, 2005...	79
3.03	Internet hosts, 2004..	57
3.04	Electricity production, 2003...	45
3.05	Availability of scientists and engineers, 2006.....................	39
3.06	Quality of scientific research institutions, 2006.................	43
3.07	Tertiary enrollment, 2001...	53

Readiness component	69
Individual readiness	**55**

4.01	Quality of math and science education, 2006....................	24
4.02	Quality of the educational system, 2006............................	46
4.03	Quality of public schools, 2006	37
4.04	Internet access in schools, 2006..	79
4.05	Buyer sophistication, 2006...	103
4.06	Residential telephone connection charge, 2005	74
4.07	Residential monthly telephone subscription, 2005	6
4.08	High-speed monthly broadband subscription, 2006...........	64
4.09	Lowest cost of broadband, 2006..	72
4.10	Cost of mobile telephone call, 2005..................................	59

Business readiness	**60**

5.01	Extent of staff training, 2006...	119
5.02	Local availability of research and training, 2006.................	52
5.03	Quality of management schools, 2006...............................	67
5.04	Company spending on R&D, 2006.....................................	77
5.05	University-industry research collaboration, 2006................	62
5.06	Business telephone connection charge, 2005	79
5.07	Business monthly telephone subscription, 2005	2
5.08	Local supplier quality, 2006...	87
5.09	Computer, comm., and other services imports..................	n/a

Government readiness	**99**

6.01	Government prioritization of ICT, 2006...............................	86
6.02	Gov't. procurement of advanced tech products, 2006........	51
6.03	Importance of ICT to gov't. vision of the future, 2006........	57
6.04	E-participation index, 2005	82
6.05	E-government readiness index, 2005.................................	112

Usage component	77
Individual usage	**52**

7.01	Mobile telephone subscribers, 2005	50
7.02	Personal computers, 2004...	72
7.03	Broadband Internet subscribers..	n/a
7.04	Internet users, 2004	54
7.05	Internet bandwidth, 2004	54

Business usage	**99**

8.01	Prevalence of foreign technology licensing, 2006..............	95
8.02	Firm-level technology absorption, 2006.............................	104
8.03	Capacity for innovation, 2006	105
8.04	Availability of new telephone lines, 2006	95
8.05	Availability of mobile telephones, 2006..............................	93
8.06	Extent of business Internet use, 2006	94

Government usage	**74**

9.01	Government success in ICT promotion, 2006.....................	85
9.02	Availability of online services, 2006....................................	92
9.03	ICT use and government efficiency, 2006...........................	79
9.04	ICT pervasiveness, 2006	33

Singapore

Key indicators

Population (millions), 2005..4.3
GDP (PPP) per capita (US$), 2005......................................28,368.1
Internet users per 100 inhabitants, 2004....................................57.9
Internet bandwidth (Mbps/10,000 inhabitants), 2004...............57.3

Networked Readiness Index

Year (number of economies)	Rank
2006–2007 (122) **3**	
2005–2006 (115)..2	
2004–2005 (104)..1	

Global Competitiveness Index 2006–2007 (125)   5

Environment component	13
Market environment	**6**

1.01 Venture capital availability, 2006..13
1.02 Financial market sophistication, 2006....................................13
1.03 Technological readiness, 2006...11
1.04 State of cluster development, 2006 7
1.05 US utility patents, 2005 12
1.06 High-tech exports, 2004 1
1.07 Burden of government regulation, 2006 2
1.08 Extent and effect of taxation, 2006...7
1.09 Time required to start a business, 2006....................................6
1.10 No. of procedures required to start a business, 2006.........16
1.11 Intensity of local competition, 2006 26
1.12 Freedom of the press, 2006..116

Political and regulatory environment	**11**

2.01 Effectiveness of law-making bodies, 2006............................1
2.02 Laws relating to ICT, 2006...2
2.03 Judicial independence, 2006 29
2.04 Intellectual property protection, 2006......................................9
2.05 Efficiency of legal framework, 2006.......................................14
2.06 Property rights, 2006..11
2.07 Quality of competition in the ISP sector, 2006....................29
2.08 No. of procedures to enforce a contract, 2006 51
2.09 Time to enforce a contract, 2006 2

Infrastructure environment	**16**

3.01 Telephone lines, 2005...25
3.02 Secure Internet servers, 2005...17
3.03 Internet hosts, 2004 13
3.04 Electricity production, 2003 18
3.05 Availability of scientists and engineers, 2006.....................15
3.06 Quality of scientific research institutions, 2006 10
3.07 Tertiary enrollment, 2004..35

Readiness component	1
Individual readiness	**2**

4.01 Quality of math and science education, 2006.......................1
4.02 Quality of the educational system, 2006................................2
4.03 Quality of public schools, 2006 2
4.04 Internet access in schools, 2006..5
4.05 Buyer sophistication, 2006 19
4.06 Residential telephone connection charge, 2005 1
4.07 Residential monthly telephone subscription, 2005 4
4.08 High-speed monthly broadband subscription, 2006...........27
4.09 Lowest cost of broadband, 2006...2
4.10 Cost of mobile telephone call, 2005.....................................10

Business readiness	**15**

5.01 Extent of staff training, 2006..12
5.02 Local availability of research and training, 2006.................17
5.03 Quality of management schools, 2006....................................8
5.04 Company spending on R&D, 2006..11
5.05 University-industry research collaboration, 2006.................8
5.06 Business telephone connection charge, 2005......................2
5.07 Business monthly telephone subscription, 2005 5
5.08 Local supplier quality, 2006..25
5.09 Computer, comm., and other services imports, 2004.........32

Government readiness	**1**

6.01 Government prioritization of ICT, 2006...................................1
6.02 Gov't. procurement of advanced tech products, 2006..........1
6.03 Importance of ICT to gov't. vision of the future, 2006..........1
6.04 E-participation index, 2005 2
6.05 E-government readiness index, 2005....................................7

235

Usage component	5
Individual usage	**10**

7.01 Mobile telephone subscribers, 2005 10
7.02 Personal computers, 2004..9
7.03 Broadband Internet subscribers, 2005 17
7.04 Internet users, 2004 13
7.05 Internet bandwidth, 2004 13

Business usage	**13**

8.01 Prevalence of foreign technology licensing, 2006................2
8.02 Firm-level technology absorption, 2006.................................7
8.03 Capacity for innovation, 2006 24
8.04 Availability of new telephone lines, 2006 5
8.05 Availability of mobile telephones, 2006 8
8.06 Extent of business Internet use, 2006 21

Government usage	**1**

9.01 Government success in ICT promotion, 2006.......................1
9.02 Availability of online services, 2006.......................................2
9.03 ICT use and government efficiency, 2006..............................2
9.04 ICT pervasiveness, 2006 1

Slovak Republic

Key indicators

Population (millions), 2005..5.4
GDP (PPP) per capita (US$), 2005.......................................16,040.7
Internet users per 100 inhabitants, 2005...................................46.3
Internet bandwidth (Mbps/10,000 inhabitants), 2004..............22.9

Networked Readiness Index

Year (number of economies)	Rank
2006–2007 (122) **41**	
2005–2006 (115)..41	
2004–2005 (104)..48	

Global Competitiveness Index 2006–2007 (125)	37

Environment component	43
Market environment	**35**

1.01 Venture capital availability, 2006...40
1.02 Financial market sophistication, 2006.....................................54
1.03 Technological readiness, 2006..51
1.04 State of cluster development, 2006...51
1.05 US utility patents, 2005...79
1.06 High-tech exports, 2004...39
1.07 Burden of government regulation, 2006...................................52
1.08 Extent and effect of taxation, 2006..8
1.09 Time required to start a business, 2006...................................43
1.10 No. of procedures required to start a business, 2006.........49
1.11 Intensity of local competition, 2006...53
1.12 Freedom of the press, 2006..23

Political and regulatory environment	**47**

2.01 Effectiveness of law-making bodies, 2006..........................52
2.02 Laws relating to ICT, 2006...45
2.03 Judicial independence, 2006..68
2.04 Intellectual property protection, 2006....................................47
2.05 Efficiency of legal framework, 2006..66
2.06 Property rights, 2006...45
2.07 Quality of competition in the ISP sector, 2006....................52
2.08 No. of procedures to enforce a contract, 2006.................41
2.09 Time to enforce a contract, 2006...77

Infrastructure environment	**43**

3.01 Telephone lines, 2005..57
3.02 Secure Internet servers, 2005...51
3.03 Internet hosts, 2004...36
3.04 Electricity production, 2003..34
3.05 Availability of scientists and engineers, 2006......................23
3.06 Quality of scientific research institutions, 2006 71
3.07 Tertiary enrollment, 2004...54

Readiness component	40
Individual readiness	**43**

4.01 Quality of math and science education, 2006....................21
4.02 Quality of the educational system, 2006..............................49
4.03 Quality of public schools, 2006...28
4.04 Internet access in schools, 2006...38
4.05 Buyer sophistication, 2006..78
4.06 Residential telephone connection charge, 2004.................34
4.07 Residential monthly telephone subscription, 2004.............57
4.08 High-speed monthly broadband subscription, 2006...........57
4.09 Lowest cost of broadband, 2006...36
4.10 Cost of mobile telephone call...n/a

Business readiness	**35**

5.01 Extent of staff training, 2006...45
5.02 Local availability of research and training, 2006................45
5.03 Quality of management schools, 2006...................................58
5.04 Company spending on R&D, 2006...45
5.05 University-industry research collaboration, 2006................31
5.06 Business telephone connection charge, 2004....................30
5.07 Business monthly telephone subscription, 2004...............46
5.08 Local supplier quality, 2006...42
5.09 Computer, comm., and other services imports..................n/a

Government readiness	**53**

6.01 Government prioritization of ICT, 2006...............................76
6.02 Gov't. procurement of advanced tech products, 2006........85
6.03 Importance of ICT to gov't. vision of the future, 2006........62
6.04 E-participation index, 2005...51
6.05 E-government readiness index, 2005...................................36

Usage component	38
Individual usage	**36**

7.01 Mobile telephone subscribers, 2005....................................35
7.02 Personal computers, 2004...29
7.03 Broadband Internet subscribers, 2005.................................43
7.04 Internet users, 2005...25
7.05 Internet bandwidth, 2004...23

Business usage	**34**

8.01 Prevalence of foreign technology licensing, 2006..............41
8.02 Firm-level technology absorption, 2006................................31
8.03 Capacity for innovation, 2006..48
8.04 Availability of new telephone lines, 2006............................21
8.05 Availability of mobile telephones, 2006................................26
8.06 Extent of business Internet use, 2006..................................35

Government usage	**58**

9.01 Government success in ICT promotion, 2006....................77
9.02 Availability of online services, 2006.....................................74
9.03 ICT use and government efficiency, 2006.........................71
9.04 ICT pervasiveness, 2006...23

Slovenia

Key indicators

Population (millions), 2005..2.0
GDP (PPP) per capita (US$), 2005......................................21,808.4
Internet users per 100 inhabitants, 2005....................................55.4
Internet bandwidth (Mbps/10,000 inhabitants) ....n/a

Networked Readiness Index

Year (number of economies)	Rank
2006–2007 (122)	**30**
2005–2006 (115)...	35
2004–2005 (104)...	32

Global Competitiveness Index 2006–2007 (125)	33

Environment component	42
Market environment	**57**
1.01 Venture capital availability, 2006..	51
1.02 Financial market sophistication, 2006....................................	52
1.03 Technological readiness, 2006..	45
1.04 State of cluster development, 2006	26
1.05 US utility patents, 2005	28
1.06 High-tech exports, 2004	38
1.07 Burden of government regulation, 2006	70
1.08 Extent and effect of taxation, 2006.....................................	100
1.09 Time required to start a business, 2006..............................	96
1.10 No. of procedures required to start a business, 2006..........	49
1.11 Intensity of local competition, 2006	46
1.12 Freedom of the press, 2006..	67

Political and regulatory environment	**51**
2.01 Effectiveness of law-making bodies, 2006..........................	50
2.02 Laws relating to ICT, 2006..	29
2.03 Judicial independence, 2006	44
2.04 Intellectual property protection, 2006.................................	35
2.05 Efficiency of legal framework, 2006....................................	46
2.06 Property rights, 2006	50
2.07 Quality of competition in the ISP sector, 2006....................	42
2.08 No. of procedures to enforce a contract, 2006	34
2.09 Time to enforce a contract, 2006	115

Infrastructure environment	**30**
3.01 Telephone lines, 2005...	31
3.02 Secure Internet servers, 2005..	29
3.03 Internet hosts, 2004	31
3.04 Electricity production, 2003	27
3.05 Availability of scientists and engineers, 2006.....................	94
3.06 Quality of scientific research institutions, 2006	41
3.07 Tertiary enrollment, 2004..	11

Readiness component	31
Individual readiness	**30**
4.01 Quality of math and science education, 2006	39
4.02 Quality of the educational system, 2006.............................	52
4.03 Quality of public schools, 2006	33
4.04 Internet access in schools, 2006...	19
4.05 Buyer sophistication, 2006	31
4.06 Residential telephone connection charge, 2004	39
4.07 Residential monthly telephone subscription, 2004	36
4.08 High-speed monthly broadband subscription, 2006	29
4.09 Lowest cost of broadband, 2006...	23
4.10 Cost of mobile telephone call, 2005....................................	9

Business readiness	**27**
5.01 Extent of staff training, 2006..	32
5.02 Local availability of research and training, 2006	37
5.03 Quality of management schools, 2006.................................	42
5.04 Company spending on R&D, 2006.......................................	27
5.05 University-industry research collaboration, 2006.................	36
5.06 Business telephone connection charge, 2004	36
5.07 Business monthly telephone subscription, 2004	26
5.08 Local supplier quality, 2006	34
5.09 Computer, comm., and other services imports, 2004.........	20

Government readiness	**37**
6.01 Government prioritization of ICT, 2006................................	59
6.02 Gov't. procurement of advanced tech products, 2006........	80
6.03 Importance of ICT to gov't. vision of the future, 2006........	49
6.04 E-participation index, 2005	45
6.05 E-government readiness index, 2005...................................	26

237

Usage component	26
Individual usage	**24**
7.01 Mobile telephone subscribers, 2005	31
7.02 Personal computers, 2004...	25
7.03 Broadband Internet subscribers, 2005	30
7.04 Internet users, 2005	14
7.05 Internet bandwidth..	n/a

Business usage	**33**
8.01 Prevalence of foreign technology licensing, 2006...............	69
8.02 Firm-level technology absorption, 2006..............................	66
8.03 Capacity for innovation, 2006	18
8.04 Availability of new telephone lines, 2006	42
8.05 Availability of mobile telephones, 2006..............................	35
8.06 Extent of business Internet use, 2006	30

Government usage	**40**
9.01 Government success in ICT promotion, 2006......................	53
9.02 Availability of online services, 2006....................................	37
9.03 ICT use and government efficiency, 2006............................	66
9.04 ICT pervasiveness, 2006	25

South Africa

Key indicators

Population (millions), 2005..47.4
GDP (PPP) per capita (US$), 2005.....................12,160.6
Internet users per 100 inhabitants, 2005...................10.8
Internet bandwidth (Mbps/10,000 inhabitants), 2004...............0.2

Networked Readiness Index

Year (number of economies)	Rank
2006–2007 (122) **47**	
2005–2006 (115)...37	
2004–2005 (104)...34	

Global Competitiveness Index 2006–2007 (125)	45

Environment component	40
Market environment	**34**

1.01	Venture capital availability, 2006	38
1.02	Financial market sophistication, 2006	19
1.03	Technological readiness, 2006	38
1.04	State of cluster development, 2006	39
1.05	US utility patents, 2005	35
1.06	High-tech exports, 2004	47
1.07	Burden of government regulation, 2006	62
1.08	Extent and effect of taxation, 2006	27
1.09	Time required to start a business, 2006	65
1.10	No. of procedures required to start a business, 2006	49
1.11	Intensity of local competition, 2006	38
1.12	Freedom of the press, 2006	24

Political and regulatory environment	**25**

2.01	Effectiveness of law-making bodies, 2006	17
2.02	Laws relating to ICT, 2006	28
2.03	Judicial independence, 2006	21
2.04	Intellectual property protection, 2006	25
2.05	Efficiency of legal framework, 2006	15
2.06	Property rights, 2006	22
2.07	Quality of competition in the ISP sector, 2006	90
2.08	No. of procedures to enforce a contract, 2006	38
2.09	Time to enforce a contract, 2006	86

Infrastructure environment	**70**

3.01	Telephone lines, 2005	82
3.02	Secure Internet servers, 2005	46
3.03	Internet hosts, 2004	46
3.04	Electricity production, 2003	40
3.05	Availability of scientists and engineers, 2006	92
3.06	Quality of scientific research institutions, 2006	25
3.07	Tertiary enrollment, 2003	85

Readiness component	45
Individual readiness	**70**

4.01	Quality of math and science education, 2006	115
4.02	Quality of the educational system, 2006	97
4.03	Quality of public schools, 2006	87
4.04	Internet access in schools, 2006	78
4.05	Buyer sophistication, 2006	34
4.06	Residential telephone connection charge, 2005	47
4.07	Residential monthly telephone subscription, 2005	75
4.08	High-speed monthly broadband subscription, 2006	61
4.09	Lowest cost of broadband, 2006	70
4.10	Cost of mobile telephone call, 2005	62

Business readiness	**31**

5.01	Extent of staff training, 2006	26
5.02	Local availability of research and training, 2006	30
5.03	Quality of management schools, 2006	19
5.04	Company spending on R&D, 2006	24
5.05	University-industry research collaboration, 2006	22
5.06	Business telephone connection charge, 2005	39
5.07	Business monthly telephone subscription, 2005	68
5.08	Local supplier quality, 2006	29
5.09	Computer, comm., and other services imports, 2004	89

Government readiness	**40**

6.01	Government prioritization of ICT, 2006	47
6.02	Gov't. procurement of advanced tech products, 2006	32
6.03	Importance of ICT to gov't. vision of the future, 2006	67
6.04	E-participation index, 2005	33
6.05	E-government readiness index, 2005	57

Usage component	54
Individual usage	**60**

7.01	Mobile telephone subscribers, 2005	46
7.02	Personal computers, 2004	61
7.03	Broadband Internet subscribers, 2005	66
7.04	Internet users, 2005	69
7.05	Internet bandwidth, 2004	71

Business usage	**43**

8.01	Prevalence of foreign technology licensing, 2006	7
8.02	Firm-level technology absorption, 2006	30
8.03	Capacity for innovation, 2006	37
8.04	Availability of new telephone lines, 2006	77
8.05	Availability of mobile telephones, 2006	76
8.06	Extent of business Internet use, 2006	49

Government usage	**59**

9.01	Government success in ICT promotion, 2006	59
9.02	Availability of online services, 2006	56
9.03	ICT use and government efficiency, 2006	75
9.04	ICT pervasiveness, 2006	64

238

Spain

Key indicators

Population (millions), 2005..43.1
GDP (PPP) per capita (US$), 2005......................................26,320.3
Internet users per 100 inhabitants, 2005..................................35.4
Internet bandwidth (Mbps/10,000 inhabitants), 2004..............27.9

Networked Readiness Index

Year (number of economies)	Rank
2006–2007 (122)	**32**
2005–2006 (115)..	31
2004–2005 (104)..	29

Global Competitiveness Index 2006–2007 (125)	28

Environment component — 30

Market environment — 39

1.01 Venture capital availability, 2006..30
1.02 Financial market sophistication, 2006.....................................22
1.03 Technological readiness, 2006...32
1.04 State of cluster development, 2006 35
1.05 US utility patents, 2005 27
1.06 High-tech exports, 2004 40
1.07 Burden of government regulation, 2006 67
1.08 Extent and effect of taxation, 2006..64
1.09 Time required to start a business, 2006................................88
1.10 No. of procedures required to start a business, 2006.........61
1.11 Intensity of local competition, 2006 33
1.12 Freedom of the press, 2006...30

Political and regulatory environment — 35

2.01 Effectiveness of law-making bodies, 2006.........................33
2.02 Laws relating to ICT, 2006..33
2.03 Judicial independence, 2006 63
2.04 Intellectual property protection, 2006.....................................29
2.05 Efficiency of legal framework, 2006..48
2.06 Property rights, 2006 26
2.07 Quality of competition in the ISP sector, 2006....................51
2.08 No. of procedures to enforce a contract, 2006 27
2.09 Time to enforce a contract, 2006 74

Infrastructure environment — 29

3.01 Telephone lines, 2005..27
3.02 Secure Internet servers, 2005...27
3.03 Internet hosts, 2004 37
3.04 Electricity production, 2003 32
3.05 Availability of scientists and engineers, 2006......................42
3.06 Quality of scientific research institutions, 2006 47
3.07 Tertiary enrollment, 2004..16

Readiness component — 38

Individual readiness — 44

4.01 Quality of math and science education, 2006....................82
4.02 Quality of the educational system, 2006...........................66
4.03 Quality of public schools, 2006 60
4.04 Internet access in schools, 2006..44
4.05 Buyer sophistication, 2006 29
4.06 Residential telephone connection charge, 2005 19
4.07 Residential monthly telephone subscription, 2005 27
4.08 High-speed monthly broadband subscription, 2006...........21
4.09 Lowest cost of broadband, 2006...32
4.10 Cost of mobile telephone call, 2005....................................20

Business readiness — 26

5.01 Extent of staff training, 2006...41
5.02 Local availability of research and training, 2006.................35
5.03 Quality of management schools, 2006................................11
5.04 Company spending on R&D, 2006 46
5.05 University-industry research collaboration, 2006................44
5.06 Business telephone connection charge, 2005 18
5.07 Business monthly telephone subscription, 2005 16
5.08 Local supplier quality, 2006 31
5.09 Computer, comm., and other services imports, 2004.........15

Government readiness — 57

6.01 Government prioritization of ICT, 2006...............................58
6.02 Gov't. procurement of advanced tech products, 2006........52
6.03 Importance of ICT to gov't. vision of the future, 2006........75
6.04 E-participation index, 2005 71
6.05 E-government readiness index, 2005...................................39

Usage component — 32

Individual usage — 30

7.01 Mobile telephone subscribers, 2005 22
7.02 Personal computers, 2004..30
7.03 Broadband Internet subscribers, 2005 24
7.04 Internet users, 2005 34
7.05 Internet bandwidth, 2004 21

Business usage — 35

8.01 Prevalence of foreign technology licensing, 2006...............11
8.02 Firm-level technology absorption, 2006..............................56
8.03 Capacity for innovation, 2006 34
8.04 Availability of new telephone lines, 2006 54
8.05 Availability of mobile telephones, 2006...............................34
8.06 Extent of business Internet use, 2006 44

Government usage — 39

9.01 Government success in ICT promotion, 2006....................86
9.02 Availability of online services, 2006.....................................36
9.03 ICT use and government efficiency, 2006...........................36
9.04 ICT pervasiveness, 2006 29

Sri Lanka

Key indicators

Population (millions), 2005..20.7
GDP (PPP) per capita (US$), 2005.........................4,383.7
Internet users per 100 inhabitants, 2004.....................1.4
Internet bandwidth (Mbps/10,000 inhabitants), 2004...............0.2

Networked Readiness Index

Year (number of economies)	Rank
2006–2007 (122) **86**	
2005–2006 (115)...83	
2004–2005 (104)...71	

Global Competitiveness Index 2006–2007 (125)	79

Environment component — 81

Market environment — 81

1.01 Venture capital availability, 2006...........................53
1.02 Financial market sophistication, 2006....................72
1.03 Technological readiness, 2006................................79
1.04 State of cluster development, 2006........................68
1.05 US utility patents, 2005 74
1.06 High-tech exports, 2004 65
1.07 Burden of government regulation, 2006 86
1.08 Extent and effect of taxation, 2006........................72
1.09 Time required to start a business, 2006...............96
1.10 No. of procedures required to start a business, 2006.........36
1.11 Intensity of local competition, 2006 69
1.12 Freedom of the press, 2006....................................85

Political and regulatory environment — 58

2.01 Effectiveness of law-making bodies, 2006...........75
2.02 Laws relating to ICT, 2006......................................71
2.03 Judicial independence, 2006 71
2.04 Intellectual property protection, 2006...................65
2.05 Efficiency of legal framework, 2006.......................64
2.06 Property rights, 2006 63
2.07 Quality of competition in the ISP sector, 2006...................53
2.08 No. of procedures to enforce a contract, 2006 13
2.09 Time to enforce a contract, 2006 104

Infrastructure environment — 89

3.01 Telephone lines, 2005..92
3.02 Secure Internet servers, 2005.................................81
3.03 Internet hosts, 2004 101
3.04 Electricity production, 2003 96
3.05 Availability of scientists and engineers, 2006.....67
3.06 Quality of scientific research institutions, 2006 42
3.07 Tertiary enrollment, 2004......................................106

Readiness component — 87

Individual readiness — 86

4.01 Quality of math and science education, 2006.....................64
4.02 Quality of the educational system, 2006...............69
4.03 Quality of public schools, 2006 57
4.04 Internet access in schools, 2006...........................86
4.05 Buyer sophistication, 2006 61
4.06 Residential telephone connection charge, 2005 108
4.07 Residential monthly telephone subscription, 2005 90
4.08 High-speed monthly broadband subscription, 2006............75
4.09 Lowest cost of broadband, 2006...........................67
4.10 Cost of mobile telephone call, 2005......................66

Business readiness — 101

5.01 Extent of staff training, 2006..................................75
5.02 Local availability of research and training, 2006.................82
5.03 Quality of management schools, 2006...................66
5.04 Company spending on R&D, 2006 63
5.05 University-industry research collaboration, 2006.................53
5.06 Business telephone connection charge, 2005 104
5.07 Business monthly telephone subscription, 2005 102
5.08 Local supplier quality, 2006 73
5.09 Computer, comm., and other services imports, 2004.........79

Government readiness — 79

6.01 Government prioritization of ICT, 2006.................53
6.02 Gov't. procurement of advanced tech products, 2006........66
6.03 Importance of ICT to gov't. vision of the future, 2006........64
6.04 E-participation index, 2005 82
6.05 E-government readiness index, 2005.....................80

Usage component — 86

Individual usage — 103

7.01 Mobile telephone subscribers, 2005 95
7.02 Personal computers, 2004.......................................87
7.03 Broadband Internet subscribers, 2005 83
7.04 Internet users, 2004 108
7.05 Internet bandwidth, 2004 73

Business usage — 68

8.01 Prevalence of foreign technology licensing, 2006..............74
8.02 Firm-level technology absorption, 2006................87
8.03 Capacity for innovation, 2006 46
8.04 Availability of new telephone lines, 2006 67
8.05 Availability of mobile telephones, 2006 71
8.06 Extent of business Internet use, 2006 77

Government usage — 90

9.01 Government success in ICT promotion, 2006....................69
9.02 Availability of online services, 2006......................82
9.03 ICT use and government efficiency, 2006............88
9.04 ICT pervasiveness, 2006 98

Suriname

Key indicators

Population (millions), 2005..0.4
GDP (PPP) per capita (US$), 2005......................................5,682.8
Internet users per 100 inhabitants, 2005...................................7.1
Internet bandwidth (Mbps/10,000 inhabitants), 2004................1.0

Networked Readiness Index

Year (number of economies)	Rank
2006–2007 (122)	**110**
2005–2006 (115)	—
2004–2005 (104)	—

Global Competitiveness Index 2006–2007 (125)	100

Environment component	112
Market environment	**116**

1.01	Venture capital availability, 2006	120
1.02	Financial market sophistication, 2006	88
1.03	Technological readiness, 2006	110
1.04	State of cluster development, 2006	103
1.05	US utility patents, 2005	79
1.06	High-tech exports	n/a
1.07	Burden of government regulation, 2006	94
1.08	Extent and effect of taxation, 2006	110
1.09	Time required to start a business, 2006	116
1.10	No. of procedures required to start a business, 2006	95
1.11	Intensity of local competition, 2006	81
1.12	Freedom of the press, 2006	51

Political and regulatory environment	**114**

2.01	Effectiveness of law-making bodies, 2006	114
2.02	Laws relating to ICT, 2006	122
2.03	Judicial independence, 2006	41
2.04	Intellectual property protection, 2006	121
2.05	Efficiency of legal framework, 2006	49
2.06	Property rights, 2006	92
2.07	Quality of competition in the ISP sector, 2006	122
2.08	No. of procedures to enforce a contract, 2006	51
2.09	Time to enforce a contract, 2006	112

Infrastructure environment	**83**

3.01	Telephone lines, 2005	65
3.02	Secure Internet servers, 2005	56
3.03	Internet hosts, 2004	91
3.04	Electricity production	n/a
3.05	Availability of scientists and engineers, 2006	106
3.06	Quality of scientific research institutions, 2006	96
3.07	Tertiary enrollment, 2005	87

Readiness component	92
Individual readiness	**83**

4.01	Quality of math and science education, 2006	75
4.02	Quality of the educational system, 2006	114
4.03	Quality of public schools, 2006	71
4.04	Internet access in schools, 2006	109
4.05	Buyer sophistication, 2006	108
4.06	Residential telephone connection charge, 2004	93
4.07	Residential monthly telephone subscription, 2004	28
4.08	High-speed monthly broadband subscription, 2006	78
4.09	Lowest cost of broadband, 2006	87
4.10	Cost of mobile telephone call, 2005	67

Business readiness	**83**

5.01	Extent of staff training, 2006	108
5.02	Local availability of research and training, 2006	113
5.03	Quality of management schools, 2006	95
5.04	Company spending on R&D, 2006	98
5.05	University-industry research collaboration, 2006	103
5.06	Business telephone connection charge, 2004	85
5.07	Business monthly telephone subscription, 2004	17
5.08	Local supplier quality, 2006	108
5.09	Computer, comm., and other services imports, 2004	18

Government readiness	**120**

6.01	Government prioritization of ICT, 2006	120
6.02	Gov't. procurement of advanced tech products, 2006	114
6.03	Importance of ICT to gov't. vision of the future, 2006	122
6.04	E-participation index, 2005	109
6.05	E-government readiness index, 2005	90

Usage component	120
Individual usage	**75**

7.01	Mobile telephone subscribers, 2005	58
7.02	Personal computers, 2003	78
7.03	Broadband Internet subscribers, 2005	70
7.04	Internet users, 2005	82
7.05	Internet bandwidth, 2004	52

Business usage	**120**

8.01	Prevalence of foreign technology licensing, 2006	119
8.02	Firm-level technology absorption, 2006	110
8.03	Capacity for innovation, 2006	90
8.04	Availability of new telephone lines, 2006	119
8.05	Availability of mobile telephones, 2006	115
8.06	Extent of business Internet use, 2006	106

Government usage	**122**

9.01	Government success in ICT promotion, 2006	121
9.02	Availability of online services, 2006	120
9.03	ICT use and government efficiency, 2006	116
9.04	ICT pervasiveness, 2006	122

241

Sweden

Key indicators

Population (millions), 2005..9.0
GDP (PPP) per capita (US$), 2005......................................29,926.0
Internet users per 100 inhabitants, 2005....................................76.2
Internet bandwidth (Mbps/10,000 inhabitants), 2004.............174.9

Networked Readiness Index

Year (number of economies)	Rank
2006–2007 (122) **2**	
2005–2006 (115)..8	
2004–2005 (104)..6	

Global Competitiveness Index 2006–2007 (125)	3

Environment component	5
Market environment	**9**

1.01 Venture capital availability, 2006..11
1.02 Financial market sophistication, 2006..8
1.03 Technological readiness, 2006...3
1.04 State of cluster development, 2006 17
1.05 US utility patents, 2005 7
1.06 High-tech exports, 2004 23
1.07 Burden of government regulation, 2006 42
1.08 Extent and effect of taxation, 2006..115
1.09 Time required to start a business, 2006...................................20
1.10 No. of procedures required to start a business, 2006...........4
1.11 Intensity of local competition, 2006 7
1.12 Freedom of the press, 2006..5

Political and regulatory environment	**10**

2.01 Effectiveness of law-making bodies, 2006...........................19
2.02 Laws relating to ICT, 2006..14
2.03 Judicial independence, 2006 17
2.04 Intellectual property protection, 2006.......................................8
2.05 Efficiency of legal framework, 2006 6
2.06 Property rights, 2006 16
2.07 Quality of competition in the ISP sector, 2006....................12
2.08 No. of procedures to enforce a contract, 2006 9
2.09 Time to enforce a contract, 2006 11

Infrastructure environment	**3**

3.01 Telephone lines, 2004..1
3.02 Secure Internet servers, 2005..12
3.03 Internet hosts, 2004 10
3.04 Electricity production, 2003 7
3.05 Availability of scientists and engineers, 2006........................8
3.06 Quality of scientific research institutions, 2006 8
3.07 Tertiary enrollment, 2004...4

Readiness component	11
Individual readiness	**21**

4.01 Quality of math and science education, 2006.....................37
4.02 Quality of the educational system, 2006............................24
4.03 Quality of public schools, 2006 23
4.04 Internet access in schools, 2006..2
4.05 Buyer sophistication, 2006 16
4.06 Residential telephone connection charge, 2005 24
4.07 Residential monthly telephone subscription, 2005 15
4.08 High-speed monthly broadband subscription, 2006...........16
4.09 Lowest cost of broadband, 2006..2
4.10 Cost of mobile telephone call ....n/a

Business readiness	**11**

5.01 Extent of staff training, 2006...3
5.02 Local availability of research and training, 2006 8
5.03 Quality of management schools, 2006....................................21
5.04 Company spending on R&D, 2006 5
5.05 University-industry research collaboration, 2006..................2
5.06 Business telephone connection charge...............................n/a
5.07 Business monthly telephone subscription.........................n/a
5.08 Local supplier quality, 2006 6
5.09 Computer, comm., and other services imports, 2004...........9

Government readiness	**9**

6.01 Government prioritization of ICT, 2006...............................22
6.02 Gov't. procurement of advanced tech products, 2006........18
6.03 Importance of ICT to gov't. vision of the future, 2006........24
6.04 E-participation index, 2005 14
6.05 E-government readiness index, 2005 3

Usage component	1
Individual usage	**2**

7.01 Mobile telephone subscribers, 2005 24
7.02 Personal computers, 2004...3
7.03 Broadband Internet subscribers, 2005 9
7.04 Internet users, 2005 2
7.05 Internet bandwidth, 2004 3

Business usage	**2**

8.01 Prevalence of foreign technology licensing, 2006..............17
8.02 Firm-level technology absorption, 2006.................................3
8.03 Capacity for innovation, 2006 3
8.04 Availability of new telephone lines, 2006 11
8.05 Availability of mobile telephones, 2006 2
8.06 Extent of business Internet use, 2006 4

Government usage	**7**

9.01 Government success in ICT promotion, 2006....................18
9.02 Availability of online services, 2006...7
9.03 ICT use and government efficiency, 2006...........................10
9.04 ICT pervasiveness, 2006 9

Switzerland

Key indicators

Population (millions), 2005...7.3
GDP (PPP) per capita (US$), 2005......................32,570.9
Internet users per 100 inhabitants, 2005..................49.6
Internet bandwidth (Mbps/10,000 inhabitants), 2004.............96.4

Networked Readiness Index

Year (number of economies)	Rank
2006–2007 (122)	5
2005–2006 (115)	9
2004–2005 (104)	9

Global Competitiveness Index 2006–2007 (125)	1

Environment component — 6

Market environment — 3

1.01 Venture capital availability, 2006 21
1.02 Financial market sophistication, 2006......................2
1.03 Technological readiness, 2006................................8
1.04 State of cluster development, 2006.........................15
1.05 US utility patents, 2005 6
1.06 High-tech exports, 2004 13
1.07 Burden of government regulation, 2006 13
1.08 Extent and effect of taxation, 2006........................15
1.09 Time required to start a business, 2006..................30
1.10 No. of procedures required to start a business, 2006.........16
1.11 Intensity of local competition, 2006 35
1.12 Freedom of the press, 2006...................................6

Political and regulatory environment — 8

2.01 Effectiveness of law-making bodies, 2006................11
2.02 Laws relating to ICT, 2006....................................8
2.03 Judicial independence, 2006..................................9
2.04 Intellectual property protection, 2006.....................3
2.05 Efficiency of legal framework, 2006.........................7
2.06 Property rights, 2006 4
2.07 Quality of competition in the ISP sector, 2006.........24
2.08 No. of procedures to enforce a contract, 2006 23
2.09 Time to enforce a contract, 2006 13

Infrastructure environment — 10

3.01 Telephone lines, 2005...2
3.02 Secure Internet servers, 2005 7
3.03 Internet hosts, 2004 18
3.04 Electricity production, 2003 15
3.05 Availability of scientists and engineers, 2006...........6
3.06 Quality of scientific research institutions, 2006 1
3.07 Tertiary enrollment, 2004....................................36

Readiness component — 5

Individual readiness — 3

4.01 Quality of math and science education, 2006.............4
4.02 Quality of the educational system, 2006..................4
4.03 Quality of public schools, 2006 3
4.04 Internet access in schools, 2006...........................10
4.05 Buyer sophistication, 2006 2
4.06 Residential telephone connection charge, 2005 13
4.07 Residential monthly telephone subscription, 2005 11
4.08 High-speed monthly broadband subscription, 2006.....5
4.09 Lowest cost of broadband, 2006...........................14
4.10 Cost of mobile telephone call, 2004.......................15

Business readiness — 1

5.01 Extent of staff training, 2006................................1
5.02 Local availability of research and training, 2006.........4
5.03 Quality of management schools, 2006......................2
5.04 Company spending on R&D, 2006 1
5.05 University-industry research collaboration, 2006.........1
5.06 Business telephone connection charge, 2005 1
5.07 Business monthly telephone subscription, 2005 9
5.08 Local supplier quality, 2006 3
5.09 Computer, comm., and other services imports, 2004.........30

Government readiness — 19

6.01 Government prioritization of ICT, 2006....................21
6.02 Gov't. procurement of advanced tech products, 2006.........6
6.03 Importance of ICT to gov't. vision of the future, 2006........45
6.04 E-participation index, 2005..................................22
6.05 E-government readiness index, 2005 17

Usage component — 6

Individual usage — 4

7.01 Mobile telephone subscribers, 2005 27
7.02 Personal computers, 2004.....................................1
7.03 Broadband Internet subscribers, 2005 6
7.04 Internet users, 2005 20
7.05 Internet bandwidth, 2004 6

Business usage — 4

8.01 Prevalence of foreign technology licensing, 2006.......16
8.02 Firm-level technology absorption, 2006....................6
8.03 Capacity for innovation, 2006 5
8.04 Availability of new telephone lines, 2006 4
8.05 Availability of mobile telephones, 2006 16
8.06 Extent of business Internet use, 2006 7

Government usage — 19

9.01 Government success in ICT promotion, 2006.............30
9.02 Availability of online services, 2006.......................27
9.03 ICT use and government efficiency, 2006.................30
9.04 ICT pervasiveness, 2006 4

243

Taiwan, China

Key indicators

Population (millions), 2005...22.6
GDP (PPP) per capita (US$), 2005......................27,721.1
Internet users per 100 inhabitants, 2005...................58.0
Internet bandwidth (Mbps/10,000 inhabitants), 2004..............31.4

Networked Readiness Index

Year (number of economies)	Rank
2006–2007 (122) **13**
2005–2006 (115)...7
2004–2005 (104)..15

Global Competitiveness Index 2006–2007 (125)   13

Environment component   17

Market environment   5

1.01   Venture capital availability, 2006..............................24
1.02   Financial market sophistication, 2006....................33
1.03   Technological readiness, 2006................................16
1.04   State of cluster development, 2006 1
1.05   US utility patents, 2005 3
1.06   High-tech exports, 2004 2
1.07   Burden of government regulation, 2006 12
1.08   Extent and effect of taxation, 2006........................20
1.09   Time required to start a business, 2006..............89
1.10   No. of procedures required to start a business, 2006.........36
1.11   Intensity of local competition, 2006 22
1.12   Freedom of the press, 2006....................................52

Political and regulatory environment   31

2.01   Effectiveness of law-making bodies, 2006..........49
2.02   Laws relating to ICT, 2006......................................26
2.03   Judicial independence, 2006 53
2.04   Intellectual property protection, 2006...................26
2.05   Efficiency of legal framework, 2006.......................41
2.06   Property rights, 2006...39
2.07   Quality of competition in the ISP sector, 2006.....20
2.08   No. of procedures to enforce a contract, 2006...........46
2.09   Time to enforce a contract, 2006 73

Infrastructure environment   12

3.01   Telephone lines, 2005..7
3.02   Secure Internet servers, 2006 22
3.03   Internet hosts, 2004 11
3.04   Electricity production, 2003 14
3.05   Availability of scientists and engineers, 2006......14
3.06   Quality of scientific research institutions, 2006.........21
3.07   Tertiary enrollment, 2004..8

Readiness component   7

Individual readiness   7

4.01   Quality of math and science education, 2006.........10
4.02   Quality of the educational system, 2006.................9
4.03   Quality of public schools, 2006 13
4.04   Internet access in schools, 2006...........................20
4.05   Buyer sophistication, 2006 24
4.06   Residential telephone connection charge, 2005 42
4.07   Residential monthly telephone subscription, 2005 1
4.08   High-speed monthly broadband subscription, 2006............17
4.09   Lowest cost of broadband, 2006...........................10
4.10   Cost of mobile telephone call, 2004......................25

Business readiness   18

5.01   Extent of staff training, 2006.................................19
5.02   Local availability of research and training, 2006.................21
5.03   Quality of management schools, 2006...................24
5.04   Company spending on R&D, 2006 12
5.05   University-industry research collaboration, 2006............7
5.06   Business telephone connection charge, 2005 37
5.07   Business monthly telephone subscription, 2005 15
5.08   Local supplier quality, 2006 14
5.09   Computer, comm., and other services imports, 2004.........25

Government readiness   6

6.01   Government prioritization of ICT, 2006...................4
6.02   Gov't. procurement of advanced tech products, 2006..........3
6.03   Importance of ICT to gov't. vision of the future, 2006........17
6.04   E-participation index..n/a
6.05   E-government readiness index ....n/a

Usage component   13

Individual usage   17

7.01   Mobile telephone subscribers, 2005 20
7.02   Personal computers, 2004......................................18
7.03   Broadband Internet subscribers, 2005 11
7.04   Internet users, 2005 12
7.05   Internet bandwidth, 2004 20

Business usage   17

8.01   Prevalence of foreign technology licensing, 2006.................5
8.02   Firm-level technology absorption, 2006.................5
8.03   Capacity for innovation, 2006 17
8.04   Availability of new telephone lines, 2006 33
8.05   Availability of mobile telephones, 2006.................47
8.06   Extent of business Internet use, 2006 18

Government usage   8

9.01   Government success in ICT promotion, 2006.....................6
9.02   Availability of online services, 2006......................17
9.03   ICT use and government efficiency, 2006...............17
9.04   ICT pervasiveness, 2006 13

Tanzania

Key indicators

Population (millions), 2005..38.3
GDP (PPP) per capita (US$), 2005............................723.3
Internet users per 100 inhabitants, 2004......................0.9
Internet bandwidth (Mbps/10,000 inhabitants) ....n/a

Networked Readiness Index

Year (number of economies)	Rank
2006–2007 (122) **91**	
2005–2006 (115)...84	
2004–2005 (104)...83	

Global Competitiveness Index 2006–2007 (125)   104

Environment component	72
Market environment	**85**

1.01 Venture capital availability, 2006 58
1.02 Financial market sophistication, 2006........................89
1.03 Technological readiness, 2006...................................75
1.04 State of cluster development, 2006 95
1.05 US utility patents, 2005 79
1.06 High-tech exports, 2004 85
1.07 Burden of government regulation, 2006 29
1.08 Extent and effect of taxation, 2006 52
1.09 Time required to start a business, 2006 51
1.10 No. of procedures required to start a business, 2006.........95
1.11 Intensity of local competition, 2006 83
1.12 Freedom of the press, 2006 77

Political and regulatory environment	**49**

2.01 Effectiveness of law-making bodies, 2006..................16
2.02 Laws relating to ICT, 2006...73
2.03 Judicial independence, 2006 55
2.04 Intellectual property protection, 2006.........................78
2.05 Efficiency of legal framework, 2006............................62
2.06 Property rights, 2006 82
2.07 Quality of competition in the ISP sector, 2006...........80
2.08 No. of procedures to enforce a contract, 2006 16
2.09 Time to enforce a contract, 2006 44

Infrastructure environment	**100**

3.01 Telephone lines, 2004..116
3.02 Secure Internet servers, 2004 118
3.03 Internet hosts, 2004 95
3.04 Electricity production, 2003 103
3.05 Availability of scientists and engineers, 2006..............69
3.06 Quality of scientific research institutions, 2006 40
3.07 Tertiary enrollment, 2004...117

Readiness component	102
Individual readiness	**107**

4.01 Quality of math and science education, 2006 103
4.02 Quality of the educational system, 2006............................83
4.03 Quality of public schools, 2006 95
4.04 Internet access in schools, 2006..97
4.05 Buyer sophistication, 2006 81
4.06 Residential telephone connection charge, 2005 102
4.07 Residential monthly telephone subscription, 2005 109
4.08 High-speed monthly broadband subscription charge ....n/a
4.09 Lowest cost of broadband ....n/a
4.10 Cost of mobile telephone call, 2004...................................91

Business readiness	**98**

5.01 Extent of staff training, 2006...87
5.02 Local availability of research and training, 2006 54
5.03 Quality of management schools, 2006.................................96
5.04 Company spending on R&D, 2006 41
5.05 University-industry research collaboration, 2006.................41
5.06 Business telephone connection charge, 2005 95
5.07 Business monthly telephone subscription, 2005 104
5.08 Local supplier quality, 2006 90
5.09 Computer, comm., and other services imports, 2004.........65

Government readiness	**77**

6.01 Government prioritization of ICT, 2006...............................29
6.02 Gov't. procurement of advanced tech products, 2006........37
6.03 Importance of ICT to gov't. vision of the future, 2006........51
6.04 E-participation index, 2005 92
6.05 E-government readiness index, 2005 101

Usage component	87
Individual usage	**112**

7.01 Mobile telephone subscribers, 2004 115
7.02 Personal computers, 2004..105
7.03 Broadband Internet subscribers..n/a
7.04 Internet users, 2004 110
7.05 Internet bandwidth..n/a

Business usage	**85**

8.01 Prevalence of foreign technology licensing, 2006..............91
8.02 Firm-level technology absorption, 2006..............................70
8.03 Capacity for innovation, 2006 98
8.04 Availability of new telephone lines, 2006 76
8.05 Availability of mobile telephones, 2006..............................81
8.06 Extent of business Internet use, 2006 86

Government usage	**75**

9.01 Government success in ICT promotion, 2006.....................29
9.02 Availability of online services, 2006....................................94
9.03 ICT use and government efficiency, 2006...........................68
9.04 ICT pervasiveness, 2006 79

245

Thailand

Key indicators

Population (millions), 2005..64.2
GDP (PPP) per capita (US$), 2005..8,367.9
Internet users per 100 inhabitants, 2005..................................11.0
Internet bandwidth (Mbps/10,000 inhabitants), 2004.................0.5

Networked Readiness Index

Year (number of economies)	Rank
2006–2007 (122) **37**
2005–2006 (115)...34
2004–2005 (104)...36

Global Competitiveness Index 2006–2007 (125)	35

Environment component — 38

Market environment — 31

1.01	Venture capital availability, 2006	42
1.02	Financial market sophistication, 2006	41
1.03	Technological readiness, 2006	36
1.04	State of cluster development, 2006	40
1.05	US utility patents, 2005	60
1.06	High-tech exports	n/a
1.07	Burden of government regulation, 2006	15
1.08	Extent and effect of taxation, 2006	22
1.09	Time required to start a business, 2006	61
1.10	No. of procedures required to start a business, 2006	36
1.11	Intensity of local competition, 2006	42
1.12	Freedom of the press, 2006	86

Political and regulatory environment — 32

2.01	Effectiveness of law-making bodies, 2006	40
2.02	Laws relating to ICT, 2006	37
2.03	Judicial independence, 2006	45
2.04	Intellectual property protection, 2006	41
2.05	Efficiency of legal framework, 2006	40
2.06	Property rights, 2006	40
2.07	Quality of competition in the ISP sector, 2006	28
2.08	No. of procedures to enforce a contract, 2006	38
2.09	Time to enforce a contract, 2006	56

Infrastructure environment — 59

3.01	Telephone lines, 2005	80
3.02	Secure Internet servers, 2005	66
3.03	Internet hosts, 2004	52
3.04	Electricity production, 2003	64
3.05	Availability of scientists and engineers, 2006	45
3.06	Quality of scientific research institutions, 2006	37
3.07	Tertiary enrollment, 2004	42

Readiness component — 35

Individual readiness — 47

4.01	Quality of math and science education, 2006	47
4.02	Quality of the educational system, 2006	41
4.03	Quality of public schools, 2006	58
4.04	Internet access in schools, 2006	41
4.05	Buyer sophistication, 2006	41
4.06	Residential telephone connection charge, 2005	76
4.07	Residential monthly telephone subscription, 2005	45
4.08	High-speed monthly broadband subscription, 2006	47
4.09	Lowest cost of broadband, 2006	47
4.10	Cost of mobile telephone call, 2005	56

Business readiness — 39

5.01	Extent of staff training, 2006	30
5.02	Local availability of research and training, 2006	66
5.03	Quality of management schools, 2006	35
5.04	Company spending on R&D, 2006	37
5.05	University-industry research collaboration, 2006	24
5.06	Business telephone connection charge, 2005	66
5.07	Business monthly telephone subscription, 2005	35
5.08	Local supplier quality, 2006	40
5.09	Computer, comm., and other services imports, 2004	50

Government readiness — 30

6.01	Government prioritization of ICT, 2006	19
6.02	Gov't. procurement of advanced tech products, 2006	25
6.03	Importance of ICT to gov't. vision of the future, 2006	10
6.04	E-participation index, 2005	41
6.05	E-government readiness index, 2005	46

Usage component — 42

Individual usage — 67

7.01	Mobile telephone subscribers, 2004	66
7.02	Personal computers, 2004	68
7.03	Broadband Internet subscribers	n/a
7.04	Internet users, 2005	68
7.05	Internet bandwidth, 2004	60

Business usage — 38

8.01	Prevalence of foreign technology licensing, 2006	18
8.02	Firm-level technology absorption, 2006	29
8.03	Capacity for innovation, 2006	51
8.04	Availability of new telephone lines, 2006	56
8.05	Availability of mobile telephones, 2006	76
8.06	Extent of business Internet use, 2006	28

Government usage — 20

9.01	Government success in ICT promotion, 2006	15
9.02	Availability of online services, 2006	28
9.03	ICT use and government efficiency, 2006	14
9.04	ICT pervasiveness, 2006	30

246

Trinidad and Tobago

Key indicators

Population (millions), 2005..1.3
GDP (PPP) per capita (US$), 2005....................................14,257.5
Internet users per 100 inhabitants, 2004...................................12.2
Internet bandwidth (Mbps/10,000 inhabitants), 2004.................1.4

Networked Readiness Index

Year (number of economies)	Rank
2006–2007 (122) **68**	
2005–2006 (115)...74	
2004–2005 (104)...59	

Global Competitiveness Index 2006–2007 (125)	67

Environment component	68
Market environment	**46**

1.01	Venture capital availability, 2006 41
1.02	Financial market sophistication, 2006...................................57
1.03	Technological readiness, 2006...62
1.04	State of cluster development, 2006 70
1.05	US utility patents, 2005 79
1.06	High-tech exports..n/a
1.07	Burden of government regulation, 2006 64
1.08	Extent and effect of taxation, 2006......................................24
1.09	Time required to start a business, 2006..............................79
1.10	No. of procedures required to start a business, 2006.........49
1.11	Intensity of local competition, 2006 72
1.12	Freedom of the press, 2006..60

Political and regulatory environment	**103**

2.01	Effectiveness of law-making bodies, 2006..........................82
2.02	Laws relating to ICT, 2006...97
2.03	Judicial independence, 2006 49
2.04	Intellectual property protection, 2006..................................75
2.05	Efficiency of legal framework, 2006.....................................50
2.06	Property rights, 2006 65
2.07	Quality of competition in the ISP sector, 2006.................117
2.08	No. of procedures to enforce a contract, 2006 78
2.09	Time to enforce a contract, 2006 113

Infrastructure environment	**65**

3.01	Telephone lines, 2005...53
3.02	Secure Internet servers, 2005 45
3.03	Internet hosts, 2004 43
3.04	Electricity production, 2003 41
3.05	Availability of scientists and engineers, 2006.....................54
3.06	Quality of scientific research institutions, 2006 65
3.07	Tertiary enrollment, 2004...89

Readiness component	54
Individual readiness	**46**

4.01	Quality of math and science education, 2006 48
4.02	Quality of the educational system, 2006.............................48
4.03	Quality of public schools, 2006 55
4.04	Internet access in schools, 2006..76
4.05	Buyer sophistication, 2006 50
4.06	Residential telephone connection charge, 2004 3
4.07	Residential monthly telephone subscription, 2004 18
4.08	High-speed monthly broadband subscription, 2006...........40
4.09	Lowest cost of broadband, 2006...69
4.10	Cost of mobile telephone call, 2005 33

Business readiness	**44**

5.01	Extent of staff training, 2006..51
5.02	Local availability of research and training, 2006 72
5.03	Quality of management schools, 2006.................................41
5.04	Company spending on R&D, 2006 65
5.05	University-industry research collaboration, 2006................56
5.06	Business telephone connection charge, 2004 11
5.07	Business monthly telephone subscription, 2004 58
5.08	Local supplier quality, 2006 55
5.09	Computer, comm., and other services imports ....n/a

Government readiness	**84**

6.01	Government prioritization of ICT, 2006...............................85
6.02	Gov't. procurement of advanced tech products, 2006........50
6.03	Importance of ICT to gov't. vision of the future, 2006........96
6.04	E-participation index, 2005 71
6.05	E-government readiness index, 2005 64

Usage component	90
Individual usage	**63**

7.01	Mobile telephone subscribers, 2005 53
7.02	Personal computers, 2003..62
7.03	Broadband Internet subscribers, 2005 55
7.04	Internet users, 2004 66
7.05	Internet bandwidth, 2004 46

Business usage	**84**

8.01	Prevalence of foreign technology licensing, 2006..............35
8.02	Firm-level technology absorption, 2006...............................54
8.03	Capacity for innovation, 2006 103
8.04	Availability of new telephone lines, 2006 97
8.05	Availability of mobile telephones, 2006 104
8.06	Extent of business Internet use, 2006 64

Government usage	**110**

9.01	Government success in ICT promotion, 2006.....................75
9.02	Availability of online services, 2006...................................115
9.03	ICT use and government efficiency, 2006..........................118
9.04	ICT pervasiveness, 2006 101

247

Tunisia

Key indicators

Population (millions), 2005...10.1
GDP (PPP) per capita (US$), 2005.........................8,254.8
Internet users per 100 inhabitants, 2005.....................9.5
Internet bandwidth (Mbps/10,000 inhabitants), 2004.................0.4

Networked Readiness Index

Year (number of economies)	Rank
2006–2007 (122) **35**	
2005–2006 (115)...36	
2004–2005 (104)...31	

Global Competitiveness Index 2006–2007 (125)	30

Environment component — 37

Market environment — 37

1.01 Venture capital availability, 2006...31
1.02 Financial market sophistication, 2006...................................59
1.03 Technological readiness, 2006...30
1.04 State of cluster development, 2006 32
1.05 US utility patents, 2005 69
1.06 High-tech exports, 2004 42
1.07 Burden of government regulation, 2006 11
1.08 Extent and effect of taxation, 2006 18
1.09 Time required to start a business, 2006 12
1.10 No. of procedures required to start a business, 2006.........61
1.11 Intensity of local competition, 2006 43
1.12 Freedom of the press, 2006...103

Political and regulatory environment — 28

2.01 Effectiveness of law-making bodies, 2006.........................18
2.02 Laws relating to ICT, 2006...49
2.03 Judicial independence, 2006 34
2.04 Intellectual property protection, 2006...............................30
2.05 Efficiency of legal framework, 2006......................................30
2.06 Property rights, 2006..35
2.07 Quality of competition in the ISP sector, 2006...................56
2.08 No. of procedures to enforce a contract, 2006...................16
2.09 Time to enforce a contract, 2006 67

Infrastructure environment — 57

3.01 Telephone lines, 2005..78
3.02 Secure Internet servers, 2005...82
3.03 Internet hosts, 2004 109
3.04 Electricity production, 2003 75
3.05 Availability of scientists and engineers, 2006....................10
3.06 Quality of scientific research institutions, 2006 33
3.07 Tertiary enrollment, 2004...64

Readiness component — 29

Individual readiness — 24

4.01 Quality of math and science education, 2006 9
4.02 Quality of the educational system, 2006...........................11
4.03 Quality of public schools, 2006 14
4.04 Internet access in schools, 2006...32
4.05 Buyer sophistication, 2006 28
4.06 Residential telephone connection charge, 2005 38
4.07 Residential monthly telephone subscription, 2005 39
4.08 High-speed monthly broadband subscription, 2006............76
4.09 Lowest cost of broadband, 2006...78
4.10 Cost of mobile telephone call, 2005.....................................50

Business readiness — 33

5.01 Extent of staff training, 2006...36
5.02 Local availability of research and training, 2006.................33
5.03 Quality of management schools, 2006.................................20
5.04 Company spending on R&D, 2006.......................................36
5.05 University-industry research collaboration, 2006.................32
5.06 Business telephone connection charge, 2005 34
5.07 Business monthly telephone subscription, 2005 28
5.08 Local supplier quality, 2006 33
5.09 Computer, comm., and other services imports, 2004.........70

Government readiness — 38

6.01 Government prioritization of ICT, 2006...............................13
6.02 Gov't. procurement of advanced tech products, 2006..........4
6.03 Importance of ICT to gov't. vision of the future, 2006..........6
6.04 E-participation index, 2005 109
6.05 E-government readiness index, 2005 96

Usage component — 44

Individual usage — 69

7.01 Mobile telephone subscribers, 2005 57
7.02 Personal computers, 2004..73
7.03 Broadband Internet subscribers, 2005 75
7.04 Internet users, 2005 74
7.05 Internet bandwidth, 2004 62

Business usage — 37

8.01 Prevalence of foreign technology licensing, 2006...............34
8.02 Firm-level technology absorption, 2006...............................36
8.03 Capacity for innovation, 2006 31
8.04 Availability of new telephone lines, 2006 32
8.05 Availability of mobile telephones, 2006...............................60
8.06 Extent of business Internet use, 2006 59

Government usage — 23

9.01 Government success in ICT promotion, 2006........................4
9.02 Availability of online services, 2006......................................39
9.03 ICT use and government efficiency, 2006...........................25
9.04 ICT pervasiveness, 2006 12

Turkey

Key indicators

Population (millions), 2005...73.2
GDP (PPP) per capita (US$), 2005........................7,949.9
Internet users per 100 inhabitants, 2005....................21.9
Internet bandwidth (Mbps/10,000 inhabitants), 2004.............1.2

Networked Readiness Index

Year (number of economies)	Rank
2006–2007 (122) **52**	
2005–2006 (115)...48	
2004–2005 (104)...52	

Global Competitiveness Index 2006–2007 (125)	59

Environment component	50
Market environment	**49**

1.01 Venture capital availability, 2006.............................75
1.02 Financial market sophistication, 2006....................36
1.03 Technological readiness, 2006.................................55
1.04 State of cluster development, 2006.........................44
1.05 US utility patents, 2005..70
1.06 High-tech exports, 2004 60
1.07 Burden of government regulation, 2006 63
1.08 Extent and effect of taxation, 2006.........................81
1.09 Time required to start a business, 2006...................10
1.10 No. of procedures required to start a business, 2006.........36
1.11 Intensity of local competition, 2006 27
1.12 Freedom of the press, 2006.....................................62

Political and regulatory environment	**50**

2.01 Effectiveness of law-making bodies, 2006..........................39
2.02 Laws relating to ICT, 2006.......................................52
2.03 Judicial independence, 2006 50
2.04 Intellectual property protection, 2006.....................71
2.05 Efficiency of legal framework, 2006.........................56
2.06 Property rights, 2006 53
2.07 Quality of competition in the ISP sector, 2006....................54
2.08 No. of procedures to enforce a contract, 2006 69
2.09 Time to enforce a contract, 2006 55

Infrastructure environment	**56**

3.01 Telephone lines, 2005..50
3.02 Secure Internet servers, 2005 52
3.03 Internet hosts, 2004 49
3.04 Electricity production, 2003 63
3.05 Availability of scientists and engineers, 2006.....................44
3.06 Quality of scientific research institutions, 2006 55
3.07 Tertiary enrollment, 2004...62

Readiness component	57
Individual readiness	**64**

4.01 Quality of math and science education, 2006....................57
4.02 Quality of the educational system, 2006.........................72
4.03 Quality of public schools, 2006 76
4.04 Internet access in schools, 2006.............................55
4.05 Buyer sophistication, 2006 54
4.06 Residential telephone connection charge, 2004 6
4.07 Residential monthly telephone subscription, 2004 67
4.08 High-speed monthly broadband subscription, 2006............70
4.09 Lowest cost of broadband, 2006............................65
4.10 Cost of mobile telephone call ....n/a

Business readiness	**56**

5.01 Extent of staff training, 2006...................................39
5.02 Local availability of research and training, 2006.................41
5.03 Quality of management schools, 2006....................61
5.04 Company spending on R&D, 2006 62
5.05 University-industry research collaboration, 2006................46
5.06 Business telephone connection charge, 2004 4
5.07 Business monthly telephone subscription, 2004 92
5.08 Local supplier quality, 2006 39
5.09 Computer, comm., and other services imports, 2004.........68

Government readiness	**52**

6.01 Government prioritization of ICT, 2006.............................77
6.02 Gov't. procurement of advanced tech products, 2006........62
6.03 Importance of ICT to gov't. vision of the future, 2006........69
6.04 E-participation index, 2005 34
6.05 E-government readiness index, 2005................................59

249

Usage component	52
Individual usage	**54**

7.01 Mobile telephone subscribers, 2005 54
7.02 Personal computers, 2004..71
7.03 Broadband Internet subscribers, 2005 45
7.04 Internet users, 2005...46
7.05 Internet bandwidth, 2004 49

Business usage	**42**

8.01 Prevalence of foreign technology licensing, 2006.............46
8.02 Firm-level technology absorption, 2006...................25
8.03 Capacity for innovation, 2006 47
8.04 Availability of new telephone lines, 2006...............58
8.05 Availability of mobile telephones, 2006..................50
8.06 Extent of business Internet use, 2006....................61

Government usage	**53**

9.01 Government success in ICT promotion, 2006....................70
9.02 Availability of online services, 2006........................52
9.03 ICT use and government efficiency, 2006.........................42
9.04 ICT pervasiveness, 2006 62

Uganda

Key indicators

Population (millions), 2005..28.8

GDP (PPP) per capita (US$), 2005.................................1,501.3

Internet users per 100 inhabitants, 2005...................................1.7

Internet bandwidth (Mbps/10,000 inhabitants), 2004.................0.0

Networked Readiness Index

Year (number of economies)	Rank
2006–2007 (122) **100**	
2005–2006 (115)...79	
2004–2005 (104)...77	

Global Competitiveness Index 2006–2007 (125)	113

Environment component	89
Market environment	**107**

1.01 Venture capital availability, 2006...73
1.02 Financial market sophistication, 2006................................109
1.03 Technological readiness, 2006..87
1.04 State of cluster development, 2006.......................................61
1.05 US utility patents, 2005..79
1.06 High-tech exports, 2004...59
1.07 Burden of government regulation, 2006.............................39
1.08 Extent and effect of taxation, 2006....................................102
1.09 Time required to start a business, 2006...............................51
1.10 No. of procedures required to start a business, 2006........113
1.11 Intensity of local competition, 2006....................................51
1.12 Freedom of the press, 2006..104

Political and regulatory environment	**63**

2.01 Effectiveness of law-making bodies, 2006..........................60
2.02 Laws relating to ICT, 2006...92
2.03 Judicial independence, 2006..60
2.04 Intellectual property protection, 2006................................104
2.05 Efficiency of legal framework, 2006.....................................70
2.06 Property rights, 2006..99
2.07 Quality of competition in the ISP sector, 2006....................76
2.08 No. of procedures to enforce a contract, 2006.....................9
2.09 Time to enforce a contract, 2006...68

Infrastructure environment	**88**

3.01 Telephone lines, 2005..120
3.02 Secure Internet servers, 2005..117
3.03 Internet hosts, 2004...102
3.04 Electricity production...n/a
3.05 Availability of scientists and engineers, 2006......................82
3.06 Quality of scientific research institutions, 2006.................34
3.07 Tertiary enrollment, 2004...103

Readiness component	116
Individual readiness	**121**

4.01 Quality of math and science education, 2006.....................95
4.02 Quality of the educational system, 2006.............................63
4.03 Quality of public schools, 2006...98
4.04 Internet access in schools, 2006.......................................112
4.05 Buyer sophistication, 2006..96
4.06 Residential telephone connection charge, 2005...............110
4.07 Residential monthly telephone subscription, 2005...........112
4.08 High-speed monthly broadband subscription, 2006..........105
4.09 Lowest cost of broadband, 2006.......................................110
4.10 Cost of mobile telephone call, 2004..................................102

Business readiness	**113**

5.01 Extent of staff training, 2006...74
5.02 Local availability of research and training, 2006.................61
5.03 Quality of management schools, 2006................................99
5.04 Company spending on R&D, 2006......................................72
5.05 University-industry research collaboration, 2006................59
5.06 Business telephone connection charge, 2005...................107
5.07 Business monthly telephone subscription, 2005..............109
5.08 Local supplier quality, 2006...110
5.09 Computer, comm., and other services imports, 2004.........40

Government readiness	**82**

6.01 Government prioritization of ICT, 2006...............................66
6.02 Gov't. procurement of advanced tech products, 2006........46
6.03 Importance of ICT to gov't. vision of the future, 2006........30
6.04 E-participation index, 2005...82
6.05 E-government readiness index, 2005..................................99

Usage component	80
Individual usage	**113**

7.01 Mobile telephone subscribers, 2005.................................114
7.02 Personal computers, 2004...111
7.03 Broadband Internet subscribers...n/a
7.04 Internet users, 2005..106
7.05 Internet bandwidth, 2004..83

Business usage	**76**

8.01 Prevalence of foreign technology licensing, 2006...............51
8.02 Firm-level technology absorption, 2006..............................92
8.03 Capacity for innovation, 2006...86
8.04 Availability of new telephone lines, 2006............................74
8.05 Availability of mobile telephones, 2006...............................87
8.06 Extent of business Internet use, 2006.................................74

Government usage	**56**

9.01 Government success in ICT promotion, 2006......................40
9.02 Availability of online services, 2006....................................49
9.03 ICT use and government efficiency, 2006............................73
9.04 ICT pervasiveness, 2006...72

Ukraine

Key indicators

Population (millions), 2005..46.5

GDP (PPP) per capita (US$), 2005...7,212.7

Internet users per 100 inhabitants, 2005....................................9.8

Internet bandwidth (Mbps/10,000 inhabitants), 2004................0.2

Networked Readiness Index

Year (number of economies)	Rank
2006–2007 (122)	**75**
2005–2006 (115)...	76
2004–2005 (104)...	82

Global Competitiveness Index 2006–2007 (125)	78

Environment component	66
Market environment	**75**

1.01 Venture capital availability, 2006..59
1.02 Financial market sophistication, 2006...................................81
1.03 Technological readiness, 2006..94
1.04 State of cluster development, 2006.......................................83
1.05 US utility patents, 2005 52
1.06 High-tech exports...n/a
1.07 Burden of government regulation, 2006 96
1.08 Extent and effect of taxation, 2006.....................................99
1.09 Time required to start a business, 2006..............................61
1.10 No. of procedures required to start a business, 2006.........61
1.11 Intensity of local competition, 2006 80
1.12 Freedom of the press, 2006..71

Political and regulatory environment	**88**

2.01 Effectiveness of law-making bodies, 2006..........................94
2.02 Laws relating to ICT, 2006...91
2.03 Judicial independence, 2006 96
2.04 Intellectual property protection, 2006..................................97
2.05 Efficiency of legal framework, 2006.....................................96
2.06 Property rights, 2006...111
2.07 Quality of competition in the ISP sector, 2006..................108
2.08 No. of procedures to enforce a contract, 2006 46
2.09 Time to enforce a contract, 2006 7

Infrastructure environment	**41**

3.01 Telephone lines, 2004..51
3.02 Secure Internet servers, 2005..85
3.03 Internet hosts, 2004 59
3.04 Electricity production, 2003 48
3.05 Availability of scientists and engineers, 2006......................70
3.06 Quality of scientific research institutions, 2006 51
3.07 Tertiary enrollment, 2004..17

Readiness component	65
Individual readiness	**57**

4.01 Quality of math and science education, 2006.....................50
4.02 Quality of the educational system, 2006.............................47
4.03 Quality of public schools, 2006 48
4.04 Internet access in schools, 2006...88
4.05 Buyer sophistication, 2006 68
4.06 Residential telephone connection charge, 2005 66
4.07 Residential monthly telephone subscription, 2005 54
4.08 High-speed monthly broadband subscription, 2006............53
4.09 Lowest cost of broadband, 2006...51
4.10 Cost of mobile telephone call, 2005....................................72

Business readiness	**77**

5.01 Extent of staff training, 2006...99
5.02 Local availability of research and training, 2006.................86
5.03 Quality of management schools, 2006.................................89
5.04 Company spending on R&D, 2006 82
5.05 University-industry research collaboration, 2006.................69
5.06 Business telephone connection charge, 2005 86
5.07 Business monthly telephone subscription, 2005 48
5.08 Local supplier quality, 2006 76
5.09 Computer, comm., and other services imports, 2004.........42

Government readiness	**68**

6.01 Government prioritization of ICT, 2006..............................107
6.02 Gov't. procurement of advanced tech products, 2006........89
6.03 Importance of ICT to gov't. vision of the future, 2006......113
6.04 E-participation index, 2005 28
6.05 E-government readiness index, 2005...................................48

Usage component	95
Individual usage	**76**

7.01 Mobile telephone subscribers, 2005 75
7.02 Personal computers, 2004..86
7.03 Broadband Internet subscribers..n/a
7.04 Internet users, 2005 72
7.05 Internet bandwidth, 2004 72

Business usage	**95**

8.01 Prevalence of foreign technology licensing, 2006.............112
8.02 Firm-level technology absorption, 2006..............................94
8.03 Capacity for innovation, 2006 45
8.04 Availability of new telephone lines, 2006 96
8.05 Availability of mobile telephones, 2006...............................96
8.06 Extent of business Internet use, 2006 88

Government usage	**100**

9.01 Government success in ICT promotion, 2006....................113
9.02 Availability of online services, 2006.....................................90
9.03 ICT use and government efficiency, 2006............................92
9.04 ICT pervasiveness, 2006 100

251

United Arab Emirates

Key indicators

Population (millions), 2005...4.5
GDP (PPP) per capita (US$), 2005.....................27,957.1
Internet users per 100 inhabitants, 2005....................31.1
Internet bandwidth (Mbps/10,000 inhabitants), 2004................3.5

Networked Readiness Index

Year (number of economies)	Rank
2006–2007 (122) **29**
2005–2006 (115)..28
2004–2005 (104)..23

Global Competitiveness Index 2006–2007 (125)	32

Environment component	35
Market environment	**26**

1.01   Venture capital availability, 2006...........................17
1.02   Financial market sophistication, 2006....................45
1.03   Technological readiness, 2006................................15
1.04   State of cluster development, 2006 9
1.05   US utility patents, 2005 44
1.06   High-tech exports..n/a
1.07   Burden of government regulation, 2006 8
1.08   Extent and effect of taxation, 2006 3
1.09   Time required to start a business, 2006.................98
1.10   No. of procedures required to start a business, 2006.........89
1.11   Intensity of local competition, 2006 28
1.12   Freedom of the press, 2006..................................102

Political and regulatory environment	**44**

2.01   Effectiveness of law-making bodies, 2006.............28
2.02   Laws relating to ICT, 2006.....................................34
2.03   Judicial independence, 2006 40
2.04   Intellectual property protection, 2006....................28
2.05   Efficiency of legal framework, 2006.......................34
2.06   Property rights, 2006...42
2.07   Quality of competition in the ISP sector, 2006.................114
2.08   No. of procedures to enforce a contract, 2006 69
2.09   Time to enforce a contract, 2006 88

Infrastructure environment	**51**

3.01   Telephone lines, 2005..45
3.02   Secure Internet servers, 2005................................33
3.03   Internet hosts, 2004 50
3.04   Electricity production, 2003 9
3.05   Availability of scientists and engineers, 2006.................80
3.06   Quality of scientific research institutions, 2006 60
3.07   Tertiary enrollment, 2003.......................................72

Readiness component	27
Individual readiness	**33**

4.01   Quality of math and science education, 2006.................41
4.02   Quality of the educational system, 2006...............32
4.03   Quality of public schools, 2006 47
4.04   Internet access in schools, 2006...........................25
4.05   Buyer sophistication, 2006 32
4.06   Residential telephone connection charge, 2004 10
4.07   Residential monthly telephone subscription, 2005 2
4.08   High-speed monthly broadband subscription, 2006............30
4.09   Lowest cost of broadband ....n/a
4.10   Cost of mobile telephone call, 2004........................1

Business readiness	**30**

5.01   Extent of staff training, 2006.................................37
5.02   Local availability of research and training, 2006.................42
5.03   Quality of management schools, 2006...................52
5.04   Company spending on R&D, 2006 42
5.05   University-industry research collaboration, 2006................48
5.06   Business telephone connection charge, 2005 8
5.07   Business monthly telephone subscription, 2005 1
5.08   Local supplier quality, 2006 35
5.09   Computer, comm., and other services imports.................n/a

Government readiness	**28**

6.01   Government prioritization of ICT, 2006....................7
6.02   Gov't. procurement of advanced tech products, 2006........12
6.03   Importance of ICT to gov't. vision of the future, 2006..........4
6.04   E-participation index, 2005 62
6.05   E-government readiness index, 2005 42

Usage component	30
Individual usage	**40**

7.01   Mobile telephone subscribers, 2005 16
7.02   Personal computers, 2004......................................34
7.03   Broadband Internet subscribers, 2005 41
7.04   Internet users, 2005 37
7.05   Internet bandwidth, 2004 37

Business usage	**31**

8.01   Prevalence of foreign technology licensing, 2006...............15
8.02   Firm-level technology absorption, 2006.................21
8.03   Capacity for innovation, 2006 72
8.04   Availability of new telephone lines, 2006 18
8.05   Availability of mobile telephones, 2006.................29
8.06   Extent of business Internet use, 2006 45

Government usage	**10**

9.01   Government success in ICT promotion, 2006.................7
9.02   Availability of online services, 2006......................25
9.03   ICT use and government efficiency, 2006...................7
9.04   ICT pervasiveness, 2006 24

United Kingdom

Key indicators

Population (millions), 2005...59.7
GDP (PPP) per capita (US$), 2005....................................30,436.0
Internet users per 100 inhabitants, 2005.............................47.8
Internet bandwidth (Mbps/10,000 inhabitants), 2004.............130.7

Networked Readiness Index

Year (number of economies)	Rank
2006–2007 (122) **9**
2005–2006 (115)...10
2004–2005 (104)...12

Global Competitiveness Index 2006–2007 (125)   10

Environment component — 10
Market environment — 11

1.01 Venture capital availability, 2006.............................5
1.02 Financial market sophistication, 2006........................1
1.03 Technological readiness, 2006.................................13
1.04 State of cluster development, 2006..........................4
1.05 US utility patents, 2005...17
1.06 High-tech exports, 2004...21
1.07 Burden of government regulation, 2006...................40
1.08 Extent and effect of taxation, 2006..........................25
1.09 Time required to start a business, 2006....................24
1.10 No. of procedures required to start a business, 2006.........16
1.11 Intensity of local competition, 2006.........................2
1.12 Freedom of the press, 2006....................................15

Political and regulatory environment — 2

2.01 Effectiveness of law-making bodies, 2006.................2
2.02 Laws relating to ICT, 2006......................................5
2.03 Judicial independence, 2006...................................8
2.04 Intellectual property protection, 2006.......................6
2.05 Efficiency of legal framework, 2006.........................10
2.06 Property rights, 2006...6
2.07 Quality of competition in the ISP sector, 2006...........5
2.08 No. of procedures to enforce a contract, 2006...........9
2.09 Time to enforce a contract, 2006............................16

Infrastructure environment — 13

3.01 Telephone lines, 2005..13
3.02 Secure Internet servers, 2005.................................8
3.03 Internet hosts, 2004..20
3.04 Electricity production, 2003....................................28
3.05 Availability of scientists and engineers, 2006............22
3.06 Quality of scientific research institutions, 2006.........3
3.07 Tertiary enrollment, 2004.......................................24

Readiness component — 6
Individual readiness — 27

4.01 Quality of math and science education, 2006............36
4.02 Quality of the educational system, 2006...................29
4.03 Quality of public schools, 2006...............................29
4.04 Internet access in schools, 2006.............................11
4.05 Buyer sophistication, 2006......................................3
4.06 Residential telephone connection charge, 2005.........29
4.07 Residential monthly telephone subscription, 2005......26
4.08 High-speed monthly broadband subscription, 2006.....12
4.09 Lowest cost of broadband, 2006..............................10
4.10 Cost of mobile telephone call...............................n/a

Business readiness — 14

5.01 Extent of staff training, 2006..................................16
5.02 Local availability of research and training, 2006.........3
5.03 Quality of management schools, 2006......................5
5.04 Company spending on R&D, 2006...........................16
5.05 University-industry research collaboration, 2006........10
5.06 Business telephone connection charge, 2005............29
5.07 Business monthly telephone subscription, 2005.........23
5.08 Local supplier quality, 2006....................................7
5.09 Computer, comm., and other services imports, 2004.........46

Government readiness — 4

6.01 Government prioritization of ICT, 2006.....................16
6.02 Gov't. procurement of advanced tech products, 2006........31
6.03 Importance of ICT to gov't. vision of the future, 2006........28
6.04 E-participation index, 2005.....................................1
6.05 E-government readiness index, 2005........................4

Usage component — 9
Individual usage — 8

7.01 Mobile telephone subscribers, 2005........................7
7.02 Personal computers, 2004......................................12
7.03 Broadband Internet subscribers, 2005.....................16
7.04 Internet users, 2005..23
7.05 Internet bandwidth, 2004.......................................4

Business usage — 10

8.01 Prevalence of foreign technology licensing, 2006.......19
8.02 Firm-level technology absorption, 2006....................23
8.03 Capacity for innovation, 2006..................................12
8.04 Availability of new telephone lines, 2006..................15
8.05 Availability of mobile telephones, 2006....................14
8.06 Extent of business Internet use, 2006......................2

Government usage — 25

9.01 Government success in ICT promotion, 2006.............56
9.02 Availability of online services, 2006.........................8
9.03 ICT use and government efficiency, 2006..................33
9.04 ICT pervasiveness, 2006..20

United States

Key indicators

Population (millions), 2005..298.2

GDP (PPP) per capita (US$), 2005.....................................41,399.4

Internet users per 100 inhabitants, 2004...............................63.0

Internet bandwidth (Mbps/10,000 inhabitants), 2004..............33.1

Networked Readiness Index

Year (number of economies)	Rank
2006–2007 (122) **7**
2005–2006 (115)...1
2004–2005 (104)...5

Global Competitiveness Index 2006–2007 (125)	6

Environment component	2
Market environment	**1**

1.01	Venture capital availability, 2006.............................1
1.02	Financial market sophistication, 2006.....................5
1.03	Technological readiness, 2006.................................7
1.04	State of cluster development, 2006.........................2
1.05	US utility patents, 2005...1
1.06	High-tech exports, 2004...6
1.07	Burden of government regulation, 2006.................26
1.08	Extent and effect of taxation, 2006.......................31
1.09	Time required to start a business, 2006..................3
1.10	No. of procedures required to start a business, 2006.........10
1.11	Intensity of local competition, 2006........................5
1.12	Freedom of the press, 2006...................................26

Political and regulatory environment	**17**
2.01	Effectiveness of law-making bodies, 2006.............21
2.02	Laws relating to ICT, 2006.....................................20
2.03	Judicial independence, 2006...................................36
2.04	Intellectual property protection, 2006....................17
2.05	Efficiency of legal framework, 2006.......................25
2.06	Property rights, 2006..21
2.07	Quality of competition in the ISP sector, 2006......14
2.08	No. of procedures to enforce a contract, 2006........5
2.09	Time to enforce a contract, 2006...........................27

Infrastructure environment	**2**
3.01	Telephone lines, 2004...6
3.02	Secure Internet servers, 2005.................................2
3.03	Internet hosts, 2004...1
3.04	Electricity production, 2003......................................8
3.05	Availability of scientists and engineers, 2006......18
3.06	Quality of scientific research institutions, 2006......2
3.07	Tertiary enrollment, 2004..5

Readiness component	4
Individual readiness	**19**

4.01	Quality of math and science education, 2006.......42
4.02	Quality of the educational system, 2006...............15
4.03	Quality of public schools, 2006.............................31
4.04	Internet access in schools, 2006...........................14
4.05	Buyer sophistication, 2006.....................................10
4.06	Residential telephone connection charge, 2004......4
4.07	Residential monthly telephone subscription, 2004.........33
4.08	High-speed monthly broadband subscription, 2006..............3
4.09	Lowest cost of broadband, 2006.............................2
4.10	Cost of mobile telephone call, 2005......................21

Business readiness	**4**
5.01	Extent of staff training, 2006...................................9
5.02	Local availability of research and training, 2006.....5
5.03	Quality of management schools, 2006.....................6
5.04	Company spending on R&D, 2006............................3
5.05	University-industry research collaboration, 2006....4
5.06	Business telephone connection charge, 2004.........9
5.07	Business monthly telephone subscription, 2004...37
5.08	Local supplier quality, 2006...................................10
5.09	Computer, comm., and other services imports, 2004.........44

Government readiness	**5**
6.01	Government prioritization of ICT, 2006..................24
6.02	Gov't. procurement of advanced tech products, 2006........10
6.03	Importance of ICT to gov't. vision of the future, 2006........38
6.04	E-participation index, 2005.......................................3
6.05	E-government readiness index, 2005.......................1

Usage component	16
Individual usage	**15**

7.01	Mobile telephone subscribers, 2005......................48
7.02	Personal computers, 2004...2
7.03	Broadband Internet subscribers, 2005...................15
7.04	Internet users, 2004..10
7.05	Internet bandwidth, 2004..18

Business usage	**14**
8.01	Prevalence of foreign technology licensing, 2006.......23
8.02	Firm-level technology absorption, 2006....................9
8.03	Capacity for innovation, 2006...................................9
8.04	Availability of new telephone lines, 2006..............29
8.05	Availability of mobile telephones, 2006.................54
8.06	Extent of business Internet use, 2006...................12

Government usage	**22**
9.01	Government success in ICT promotion, 2006.........28
9.02	Availability of online services, 2006.......................13
9.03	ICT use and government efficiency, 2006...............20
9.04	ICT pervasiveness, 2006...40

Uruguay

Key indicators

Population (millions), 2005...3.5
GDP (PPP) per capita (US$), 2005.....................................10,720.4
Internet users per 100 inhabitants, 2005...................................20.6
Internet bandwidth (Mbps/10,000 inhabitants), 2004.................3.1

Networked Readiness Index

Year (number of economies)	Rank
2006–2007 (122) **60**	
2005–2006 (115)...65	
2004–2005 (104)...64	

Global Competitiveness Index 2006–2007 (125)	73

Environment component	55
Market environment	**80**

1.01 Venture capital availability, 2006 100
1.02 Financial market sophistication, 2006..................................73
1.03 Technological readiness, 2006...56
1.04 State of cluster development, 2006......................................99
1.05 US utility patents, 2005 48
1.06 High-tech exports, 2004 70
1.07 Burden of government regulation, 2006 50
1.08 Extent and effect of taxation, 2006.....................................93
1.09 Time required to start a business, 2006...............................79
1.10 No. of procedures required to start a business, 2006.........61
1.11 Intensity of local competition, 2006 90
1.12 Freedom of the press, 2006...37

Political and regulatory environment	**57**

2.01 Effectiveness of law-making bodies, 2006..........................83
2.02 Laws relating to ICT, 2006...72
2.03 Judicial independence, 2006 37
2.04 Intellectual property protection, 2006..................................51
2.05 Efficiency of legal framework, 2006.....................................42
2.06 Property rights, 2006..55
2.07 Quality of competition in the ISP sector, 2006...................71
2.08 No. of procedures to enforce a contract, 2006 86
2.09 Time to enforce a contract, 2006 94

Infrastructure environment	**47**

3.01 Telephone lines, 2005..41
3.02 Secure Internet servers, 2005 42
3.03 Internet hosts, 2004 28
3.04 Electricity production, 2003 59
3.05 Availability of scientists and engineers, 2006.....................52
3.06 Quality of scientific research institutions, 2006 79
3.07 Tertiary enrollment, 2003...46

Readiness component	72
Individual readiness	**69**

4.01 Quality of math and science education, 2006 78
4.02 Quality of the educational system, 2006.............................74
4.03 Quality of public schools, 2006 61
4.04 Internet access in schools, 2006...60
4.05 Buyer sophistication, 2006 79
4.06 Residential telephone connection charge, 2005 49
4.07 Residential monthly telephone subscription, 2005 63
4.08 High-speed monthly broadband subscription, 2006 68
4.09 Lowest cost of broadband, 2006...57
4.10 Cost of mobile telephone call ....n/a

Business readiness	**69**

5.01 Extent of staff training, 2006...80
5.02 Local availability of research and training, 2006 64
5.03 Quality of management schools, 2006.................................51
5.04 Company spending on R&D, 2006 91
5.05 University-industry research collaboration, 2006................89
5.06 Business telephone connection charge, 2005 41
5.07 Business monthly telephone subscription, 2005 66
5.08 Local supplier quality, 2006 83
5.09 Computer, comm., and other services imports, 2004.........77

Government readiness	**86**

6.01 Government prioritization of ICT, 2006................................94
6.02 Gov't. procurement of advanced tech products, 2006........97
6.03 Importance of ICT to gov't. vision of the future, 2006........91
6.04 E-participation index, 2005 78
6.05 E-government readiness index, 2005...................................49

Usage component	56
Individual usage	**58**

7.01 Mobile telephone subscribers, 2005 77
7.02 Personal computers, 2004..47
7.03 Broadband Internet subscribers, 2005 49
7.04 Internet users, 2005 52
7.05 Internet bandwidth, 2004 40

Business usage	**63**

8.01 Prevalence of foreign technology licensing, 2006...............79
8.02 Firm-level technology absorption, 2006...............................96
8.03 Capacity for innovation, 2006 68
8.04 Availability of new telephone lines, 2006 26
8.05 Availability of mobile telephones, 2006...............................46
8.06 Extent of business Internet use, 2006 54

Government usage	**57**

9.01 Government success in ICT promotion, 2006.....................90
9.02 Availability of online services, 2006....................................48
9.03 ICT use and government efficiency, 2006...........................40
9.04 ICT pervasiveness, 2006 63

Venezuela

Key indicators

Population (millions), 2005..26.7
GDP (PPP) per capita (US$), 2005....................................6,186.3
Internet users per 100 inhabitants, 2005....................................12.4
Internet bandwidth (Mbps/10,000 inhabitants), 2004.................0.5

Networked Readiness Index

Year (number of economies)	Rank
2006–2007 (122) **83**	
2005–2006 (115)..81	
2004–2005 (104)..84	

Global Competitiveness Index 2006–2007 (125)   88

Environment component — 115

Market environment — 121

1.01 Venture capital availability, 2006...................................102
1.02 Financial market sophistication, 2006..............................75
1.03 Technological readiness, 2006...72
1.04 State of cluster development, 2006..................................77
1.05 US utility patents, 2005...58
1.06 High-tech exports, 2004...80
1.07 Burden of government regulation, 2006..........................122
1.08 Extent and effect of taxation, 2006..................................70
1.09 Time required to start a business, 2006..........................114
1.10 No. of procedures required to start a business, 2006.......112
1.11 Intensity of local competition, 2006................................109
1.12 Freedom of the press, 2006..120

Political and regulatory environment — 118

2.01 Effectiveness of law-making bodies, 2006......................121
2.02 Laws relating to ICT, 2006..83
2.03 Judicial independence, 2006..122
2.04 Intellectual property protection, 2006.............................115
2.05 Efficiency of legal framework, 2006................................122
2.06 Property rights, 2006..121
2.07 Quality of competition in the ISP sector, 2006..................73
2.08 No. of procedures to enforce a contract, 2006..................91
2.09 Time to enforce a contract, 2006.....................................58

Infrastructure environment — 67

3.01 Telephone lines, 2005..74
3.02 Secure Internet servers, 2005...67
3.03 Internet hosts, 2004..70
3.04 Electricity production, 2003..49
3.05 Availability of scientists and engineers, 2006..................62
3.06 Quality of scientific research institutions, 2006..............103
3.07 Tertiary enrollment, 2003...48

Readiness component — 66

Individual readiness — 71

4.01 Quality of math and science education, 2006....................97
4.02 Quality of the educational system, 2006.........................110
4.03 Quality of public schools, 2006......................................115
4.04 Internet access in schools, 2006......................................70
4.05 Buyer sophistication, 2006...80
4.06 Residential telephone connection charge, 2004................46
4.07 Residential monthly telephone subscription, 2005............62
4.08 High-speed monthly broadband subscription, 2006...........67
4.09 Lowest cost of broadband, 2006......................................56
4.10 Cost of mobile telephone call, 2005.................................65

Business readiness — 66

5.01 Extent of staff training, 2006..82
5.02 Local availability of research and training, 2006................92
5.03 Quality of management schools, 2006..............................62
5.04 Company spending on R&D, 2006....................................92
5.05 University-industry research collaboration, 2006..............70
5.06 Business telephone connection charge, 2004...................49
5.07 Business monthly telephone subscription, 2005...............57
5.08 Local supplier quality, 2006...81
5.09 Computer, comm., and other services imports, 2004.........55

Government readiness — 54

6.01 Government prioritization of ICT, 2006.............................93
6.02 Gov't. procurement of advanced tech products, 2006........72
6.03 Importance of ICT to gov't. vision of the future, 2006......104
6.04 E-participation index, 2005...22
6.05 E-government readiness index, 2005................................54

Usage component — 75

Individual usage — 66

7.01 Mobile telephone subscribers, 2005.................................62
7.02 Personal computers, 2004..60
7.03 Broadband Internet subscribers, 2005..............................52
7.04 Internet users, 2005...65
7.05 Internet bandwidth, 2004...59

Business usage — 72

8.01 Prevalence of foreign technology licensing, 2006.............55
8.02 Firm-level technology absorption, 2006............................77
8.03 Capacity for innovation, 2006...106
8.04 Availability of new telephone lines, 2006.........................73
8.05 Availability of mobile telephones, 2006............................38
8.06 Extent of business Internet use, 2006..............................79

Government usage — 84

9.01 Government success in ICT promotion, 2006...................111
9.02 Availability of online services, 2006.................................38
9.03 ICT use and government efficiency, 2006.........................67
9.04 ICT pervasiveness, 2006..104

Vietnam

Key indicators

Population (millions), 2005 .... 84.2
GDP (PPP) per capita (US$), 2005 .... 3,024.8
Internet users per 100 inhabitants, 2005 .... 12.7
Internet bandwidth (Mbps/10,000 inhabitants), 2004 .... 0.2

Networked Readiness Index

Year (number of economies)	Rank
2006–2007 (122)	82
2005–2006 (115)	75
2004–2005 (104)	68

Global Competitiveness Index 2006–2007 (125) — 77

Environment component — 82
Market environment — 79
1.01 Venture capital availability, 2006 64
1.02 Financial market sophistication, 2006 90
1.03 Technological readiness, 2006 93
1.04 State of cluster development, 2006 16
1.05 US utility patents, 2005 77
1.06 High-tech exports .... n/a
1.07 Burden of government regulation, 2006 108
1.08 Extent and effect of taxation, 2006 53
1.09 Time required to start a business, 2006 91
1.10 No. of procedures required to start a business, 2006 79
1.11 Intensity of local competition, 2006 76
1.12 Freedom of the press, 2006 110

Political and regulatory environment — 67
2.01 Effectiveness of law-making bodies, 2006 55
2.02 Laws relating to ICT, 2006 90
2.03 Judicial independence, 2006 72
2.04 Intellectual property protection, 2006 98
2.05 Efficiency of legal framework, 2006 60
2.06 Property rights, 2006 69
2.07 Quality of competition in the ISP sector, 2006 93
2.08 No. of procedures to enforce a contract, 2006 78
2.09 Time to enforce a contract, 2006 26

Infrastructure environment — 86
3.01 Telephone lines, 2005 63
3.02 Secure Internet servers, 2005 110
3.03 Internet hosts, 2004 119
3.04 Electricity production, 2003 94
3.05 Availability of scientists and engineers, 2006 58
3.06 Quality of scientific research institutions, 2006 97
3.07 Tertiary enrollment, 2004 93

Readiness component — 76
Individual readiness — 76
4.01 Quality of math and science education, 2006 65
4.02 Quality of the educational system, 2006 99
4.03 Quality of public schools, 2006 73
4.04 Internet access in schools, 2006 69
4.05 Buyer sophistication, 2006 70
4.06 Residential telephone connection charge, 2005 92
4.07 Residential monthly telephone subscription, 2005 81
4.08 High-speed monthly broadband subscription, 2006 88
4.09 Lowest cost of broadband, 2006 80
4.10 Cost of mobile telephone call, 2005 63

Business readiness — 76
5.01 Extent of staff training, 2006 71
5.02 Local availability of research and training, 2006 76
5.03 Quality of management schools, 2006 112
5.04 Company spending on R&D, 2006 79
5.05 University-industry research collaboration, 2006 80
5.06 Business telephone connection charge, 2005 82
5.07 Business monthly telephone subscription, 2005 60
5.08 Local supplier quality, 2006 91
5.09 Computer, comm., and other services imports .... n/a

Government readiness — 65
6.01 Government prioritization of ICT, 2006 42
6.02 Gov't. procurement of advanced tech products, 2006 44
6.03 Importance of ICT to gov't. vision of the future, 2006 63
6.04 E-participation index, 2005 62
6.05 E-government readiness index, 2005 86

Usage component — 84
Individual usage — 92
7.01 Mobile telephone subscribers, 2005 101
7.02 Personal computers, 2004 98
7.03 Broadband Internet subscribers, 2005 68
7.04 Internet users, 2005 64
7.05 Internet bandwidth, 2004 69

Business usage — 80
8.01 Prevalence of foreign technology licensing, 2006 115
8.02 Firm-level technology absorption, 2006 37
8.03 Capacity for innovation, 2006 36
8.04 Availability of new telephone lines, 2006 66
8.05 Availability of mobile telephones, 2006 88
8.06 Extent of business Internet use, 2006 109

Government usage — 72
9.01 Government success in ICT promotion, 2006 32
9.02 Availability of online services, 2006 96
9.03 ICT use and government efficiency, 2006 74
9.04 ICT pervasiveness, 2006 70

257

3: Country/Economy Profiles

Zambia

Key indicators

Population (millions), 2005..11.7
GDP (PPP) per capita (US$), 2005......................................930.8
Internet users per 100 inhabitants, 2004.....................................2.0
Internet bandwidth (Mbps/10,000 inhabitants), 2004................0.0

Networked Readiness Index

Year (number of economies)	Rank
2006–2007 (122) **112**	
2005–2006 (115) ....—	
2004–2005 (104)..81	

Global Competitiveness Index 2006–2007 (125)	115

Environment component	105
Market environment	**113**

1.01 Venture capital availability, 2006.......................................122
1.02 Financial market sophistication, 2006................................105
1.03 Technological readiness, 2006..114
1.04 State of cluster development, 2006 38
1.05 US utility patents, 2005 79
1.06 High-tech exports, 2004 87
1.07 Burden of government regulation, 2006 10
1.08 Extent and effect of taxation, 2006...................................119
1.09 Time required to start a business, 2006..............................65
1.10 No. of procedures required to start a business, 2006.........16
1.11 Intensity of local competition, 2006 122
1.12 Freedom of the press, 2006...118

Political and regulatory environment	**85**

2.01 Effectiveness of law-making bodies, 2006........................113
2.02 Laws relating to ICT, 2006..101
2.03 Judicial independence, 2006 94
2.04 Intellectual property protection, 2006...............................111
2.05 Efficiency of legal framework, 2006....................................55
2.06 Property rights, 2006..61
2.07 Quality of competition in the ISP sector, 2006.................119
2.08 No. of procedures to enforce a contract, 2006 16
2.09 Time to enforce a contract, 2006 50

Infrastructure environment	**113**

3.01 Telephone lines, 2005...108
3.02 Secure Internet servers, 2005..105
3.03 Internet hosts, 2004 92
3.04 Electricity production, 2003 79
3.05 Availability of scientists and engineers, 2006.....................91
3.06 Quality of scientific research institutions, 2006 100
3.07 Tertiary enrollment, 2000...113

Readiness component	110
Individual readiness	**101**

4.01 Quality of math and science education, 2006.....................92
4.02 Quality of the educational system, 2006............................58
4.03 Quality of public schools, 2006 93
4.04 Internet access in schools, 2006......................................120
4.05 Buyer sophistication, 2006 120
4.06 Residential telephone connection charge, 2004 70
4.07 Residential monthly telephone subscription, 2004 91
4.08 High-speed monthly broadband subscription charge ....n/a
4.09 Lowest cost of broadband ....n/a
4.10 Cost of mobile telephone call, 2004...................................95

Business readiness	**115**

5.01 Extent of staff training, 2006..122
5.02 Local availability of research and training, 2006 122
5.03 Quality of management schools, 2006..............................102
5.04 Company spending on R&D, 2006 121
5.05 University-industry research collaboration, 2006..............106
5.06 Business telephone connection charge, 2004 84
5.07 Business monthly telephone subscription, 2004 97
5.08 Local supplier quality, 2006 117
5.09 Computer, comm., and other services imports ....n/a

Government readiness	**116**

6.01 Government prioritization of ICT, 2006...............................69
6.02 Gov't. procurement of advanced tech products, 2006......121
6.03 Importance of ICT to gov't. vision of the future, 2006........97
6.04 E-participation index, 2005 109
6.05 E-government readiness index, 2005...............................110

Usage component	111
Individual usage	**109**

7.01 Mobile telephone subscribers, 2005 107
7.02 Personal computers, 2004..102
7.03 Broadband Internet subscribers, 2005 95
7.04 Internet users, 2004 105
7.05 Internet bandwidth, 2004 84

Business usage	**86**

8.01 Prevalence of foreign technology licensing, 2006..............64
8.02 Firm-level technology absorption, 2006..............................49
8.03 Capacity for innovation, 2006 121
8.04 Availability of new telephone lines, 2006 79
8.05 Availability of mobile telephones, 2006 92
8.06 Extent of business Internet use, 2006 105

Government usage	**120**

9.01 Government success in ICT promotion, 2006...................116
9.02 Availability of online services, 2006.................................105
9.03 ICT use and government efficiency, 2006.........................120
9.04 ICT pervasiveness, 2006 121

Zimbabwe

Key indicators

Population (millions), 2005...13.0
GDP (PPP) per capita (US$), 2005.........................2,606.7
Internet users per 100 inhabitants, 2005.....................8.4
Internet bandwidth (Mbps/10,000 inhabitants), 2004................0.0

Networked Readiness Index

Year (number of economies)	Rank
2006–2007 (122) **117**	
2005–2006 (115)...105	
2004–2005 (104)...94	

Global Competitiveness Index 2006–2007 (125)   119

Environment component	117
Market environment	**119**

1.01 Venture capital availability, 2006.............................92
1.02 Financial market sophistication, 2006....................69
1.03 Technological readiness, 2006.................................95
1.04 State of cluster development, 2006........................109
1.05 US utility patents, 2005 73
1.06 High-tech exports, 2004 81
1.07 Burden of government regulation, 2006 115
1.08 Extent and effect of taxation, 2006......................101
1.09 Time required to start a business, 2006.............109
1.10 No. of procedures required to start a business, 2006.........61
1.11 Intensity of local competition, 2006 120
1.12 Freedom of the press, 2006..................................122

Political and regulatory environment	**111**

2.01 Effectiveness of law-making bodies, 2006.........110
2.02 Laws relating to ICT, 2006.......................................98
2.03 Judicial independence, 2006 119
2.04 Intellectual property protection, 2006....................85
2.05 Efficiency of legal framework, 2006.....................118
2.06 Property rights, 2006 122
2.07 Quality of competition in the ISP sector, 2006.........102
2.08 No. of procedures to enforce a contract, 2006 66
2.09 Time to enforce a contract, 2006 52

Infrastructure environment	**103**

3.01 Telephone lines, 2005..101
3.02 Secure Internet servers, 2005 103
3.03 Internet hosts, 2004 80
3.04 Electricity production, 2003 83
3.05 Availability of scientists and engineers, 2006......89
3.06 Quality of scientific research institutions, 2006 59
3.07 Tertiary enrollment, 2003..101

Readiness component	108
Individual readiness	**110**

4.01 Quality of math and science education, 2006.....70
4.02 Quality of the educational system, 2006...............39
4.03 Quality of public schools, 2006 86
4.04 Internet access in schools, 2006..........................101
4.05 Buyer sophistication, 2006 90
4.06 Residential telephone connection charge, 2005 100
4.07 Residential monthly telephone subscription, 2005 96
4.08 High-speed monthly broadband subscription, 2006 98
4.09 Lowest cost of broadband ....n/a
4.10 Cost of mobile telephone call, 2005....................105

Business readiness	**81**

5.01 Extent of staff training, 2006.................................50
5.02 Local availability of research and training, 2006 107
5.03 Quality of management schools, 2006..................75
5.04 Company spending on R&D, 2006 68
5.05 University-industry research collaboration, 2006................65
5.06 Business telephone connection charge, 2005 92
5.07 Business monthly telephone subscription, 2005 82
5.08 Local supplier quality, 2006 82
5.09 Computer, comm., and other services imports.................n/a

Government readiness	**119**

6.01 Government prioritization of ICT, 2006...............112
6.02 Gov't. procurement of advanced tech products, 2006......116
6.03 Importance of ICT to gov't. vision of the future, 2006......121
6.04 E-participation index, 2005 109
6.05 E-government readiness index, 2005....................95

Usage component	121
Individual usage	**94**

7.01 Mobile telephone subscribers, 2005 113
7.02 Personal computers, 2004.......................................57
7.03 Broadband Internet subscribers, 2005 81
7.04 Internet users, 2005 77
7.05 Internet bandwidth, 2004 81

Business usage	**119**

8.01 Prevalence of foreign technology licensing, 2006..............70
8.02 Firm-level technology absorption, 2006.............103
8.03 Capacity for innovation, 2006 118
8.04 Availability of new telephone lines, 2006 122
8.05 Availability of mobile telephones, 2006.............122
8.06 Extent of business Internet use, 2006 100

Government usage	**121**

9.01 Government success in ICT promotion, 2006...................100
9.02 Availability of online services, 2006....................122
9.03 ICT use and government efficiency, 2006.........121
9.04 ICT pervasiveness, 2006 119

Part 4
Data Tables

How to Read the Data Tables

THIERRY GEIGER, World Economic Forum

The following pages present the data for of all the variables included in the Networked Readiness Index 2006–2007 (NRI), for all 122 economies included in *The Global Information Technology Report 2006–2007.*

The tables are organized in three sections, each composed by three pillars of the NRI, as follows:

I. Environment
1. Market environment
2. Political and regulatory environment
3. Infrastructure environment

II. Readiness
4. Individual readiness
5. Business readiness
6. Government readiness

III. Usage
7. Individual usage
8. Business usage
9. Government usage

Two types of data are used in the NRI: Survey data and hard data.

- **Survey data:** average responses in each economy to questions included in the World Economic Forum's Executive Opinion Survey, conducted in the early months of 2006.

- **Hard data:** indicators obtained from a variety of sources.

Survey data

❶ Data yielded from the World Economic Forum's Executive Opinion Survey are presented in blue-colored bar graphs. Questions asked for responses on a scale of 1 to 7, where an answer of 1 corresponds to the lowest possible score and an answer of 7 corresponds to the highest possible score. For each Survey variable, the original question and the two extreme answers are shown.

We report the average score for each economy—that is, the arithmetic mean of responses from each economy. Variable 1.02, for example, asks about the sophistication of the financial markets in the respondent's economy. On this particular variable, United Kingdom with a score of 6.76 ranks first, and therefore appears at the top. We report responses rounded to two decimal points, but use the exact figures to determine rankings. For example, in the case of variable 8.01 on the prevalence of foreign technology licensing, Sweden's average score is 5.392157 and Thailand's average score is 5.391304. These economies are therefore ranked 17th and 18th respectively, although they are both listed with the same rounded score of 5.39.

❷ A dotted line on the graph indicates the mean score across the sample of 122 economies.

❸ Standard deviations are shown next to the bars representing each economy's mean score. Standard deviation gives an indication of how closely or widely the individual responses are spread around the mean economy score. In other words, this provides information on the extent of agreement or disagreement on the question within the given economy. The smaller the standard deviation, the greater the consensus is among the respondents from

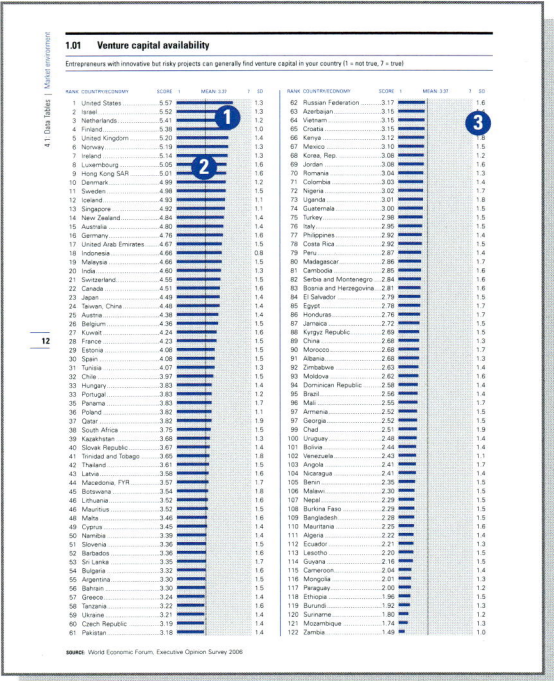

a particular economy. In the case of variable 2.06 on property rights, the standard deviation of the sample of respondents from Iceland is 0.5.

Hard data

4 While Survey data provide qualitative information, hard data are an objective measure of a quantity (for example, gross domestic product, cost of a telephone call, number of personal computers, number of procedures required to start a business, and so on). We use the latest data available from international organizations (such as the IMF, the World Bank, the International Telecommunication Union, United Nations agencies, and so on), completed, if necessary, by national sources. In the following pages, hard data variables are presented by black-shaded bar graphs. A detailed description and full source for each variable can be found in the Technical Notes and Sources section at the end of this *Report*.

When data are not available or are too old, "n/a" is used in lieu of the rank and the value.

In the case of hard data, true ties between two or more economies are possible. In such cases, shared rankings are indicated accordingly. For example, the number of days required to start a business—five—is the same in Denmark, Iceland, and the United States. Therefore, the three countries share the third place on variable 1.09.

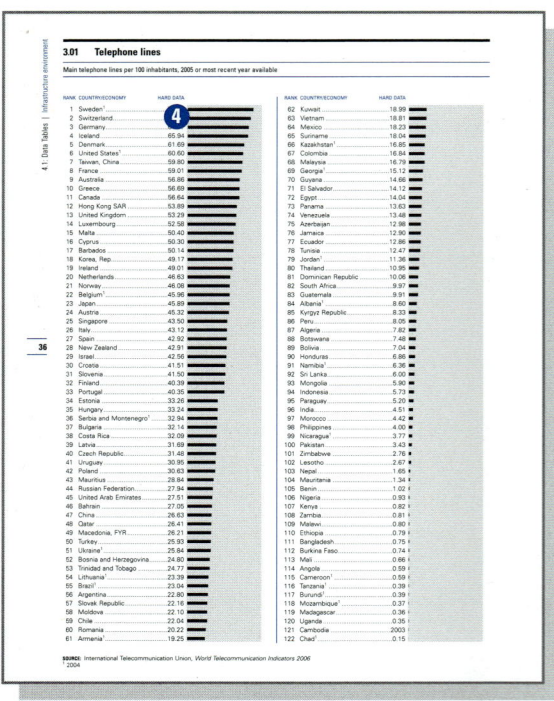

List of Data Tables

(cont'd.)

Section III Usage component

Pillar 1
Market environment

1.01 Venture capital availability

Entrepreneurs with innovative but risky projects can generally find venture capital in your country (1 = not true, 7 = true)

RANK	COUNTRY/ECONOMY	SCORE	SD
1	United States	5.57	1.3
2	Israel	5.52	1.3
3	Netherlands	5.41	1.2
4	Finland	5.38	1.0
5	United Kingdom	5.20	1.4
6	Norway	5.19	1.3
7	Ireland	5.14	1.3
8	Luxembourg	5.05	1.6
9	Hong Kong SAR	5.01	1.6
10	Denmark	4.99	1.2
11	Sweden	4.98	1.5
12	Iceland	4.93	1.1
13	Singapore	4.92	1.1
14	New Zealand	4.84	1.4
15	Australia	4.80	1.4
16	Germany	4.76	1.6
17	United Arab Emirates	4.67	1.5
18	Indonesia	4.66	0.8
19	Malaysia	4.66	1.5
20	India	4.60	1.3
21	Switzerland	4.55	1.5
22	Canada	4.51	1.6
23	Japan	4.49	1.4
24	Taiwan, China	4.48	1.4
25	Austria	4.38	1.4
26	Belgium	4.36	1.5
27	Kuwait	4.24	1.6
28	France	4.23	1.5
29	Estonia	4.08	1.5
30	Spain	4.08	1.5
31	Tunisia	4.07	1.3
32	Chile	3.97	1.5
33	Hungary	3.83	1.4
33	Portugal	3.83	1.2
35	Panama	3.83	1.7
36	Poland	3.82	1.1
37	Qatar	3.82	1.9
38	South Africa	3.75	1.5
39	Kazakhstan	3.68	1.3
40	Slovak Republic	3.67	1.4
41	Trinidad and Tobago	3.65	1.8
42	Thailand	3.61	1.5
43	Latvia	3.58	1.6
44	Macedonia, FYR	3.57	1.7
45	Botswana	3.54	1.8
46	Lithuania	3.52	1.6
46	Mauritius	3.52	1.5
48	Malta	3.46	1.6
49	Cyprus	3.45	1.4
50	Namibia	3.39	1.4
51	Slovenia	3.36	1.5
52	Barbados	3.36	1.6
53	Sri Lanka	3.35	1.7
54	Bulgaria	3.32	1.6
55	Argentina	3.30	1.5
56	Bahrain	3.30	1.5
57	Greece	3.24	1.4
58	Tanzania	3.22	1.6
59	Ukraine	3.21	1.4
60	Czech Republic	3.19	1.4
61	Pakistan	3.18	1.4
62	Russian Federation	3.17	1.6
63	Azerbaijan	3.15	1.4
64	Vietnam	3.15	1.6
65	Croatia	3.15	1.6
66	Kenya	3.12	1.8
67	Mexico	3.10	1.5
68	Korea, Rep.	3.08	1.2
69	Jordan	3.08	1.6
70	Romania	3.04	1.3
71	Colombia	3.03	1.4
72	Nigeria	3.02	1.7
73	Uganda	3.01	1.8
74	Guatemala	3.00	1.5
75	Turkey	2.98	1.5
76	Italy	2.95	1.5
77	Philippines	2.92	1.4
78	Costa Rica	2.92	1.5
79	Peru	2.87	1.4
80	Madagascar	2.86	1.7
81	Cambodia	2.85	1.6
82	Serbia and Montenegro	2.84	1.6
83	Bosnia and Herzegovina	2.81	1.6
84	El Salvador	2.79	1.5
85	Egypt	2.78	1.7
86	Honduras	2.76	1.7
87	Jamaica	2.72	1.5
88	Kyrgyz Republic	2.69	1.5
89	China	2.68	1.3
90	Morocco	2.68	1.7
91	Albania	2.68	1.3
92	Zimbabwe	2.63	1.4
93	Moldova	2.62	1.6
94	Dominican Republic	2.58	1.4
95	Brazil	2.56	1.4
96	Mali	2.55	1.7
97	Armenia	2.52	1.5
97	Georgia	2.52	1.5
99	Chad	2.51	1.9
100	Uruguay	2.48	1.4
101	Bolivia	2.44	1.4
102	Venezuela	2.43	1.1
103	Angola	2.41	1.7
104	Nicaragua	2.41	1.4
105	Benin	2.35	1.5
106	Malawi	2.30	1.5
107	Nepal	2.29	1.5
108	Burkina Faso	2.29	1.5
109	Bangladesh	2.28	1.5
110	Mauritania	2.25	1.6
111	Algeria	2.22	1.4
112	Ecuador	2.21	1.3
113	Lesotho	2.20	1.5
114	Guyana	2.16	1.5
115	Cameroon	2.04	1.4
116	Mongolia	2.01	1.3
117	Paraguay	2.00	1.2
118	Ethiopia	1.96	1.5
119	Burundi	1.92	1.3
120	Suriname	1.80	1.2
121	Mozambique	1.74	1.3
122	Zambia	1.49	1.0

MEAN: 3.37

SOURCE: World Economic Forum, Executive Opinion Survey 2006

1.02 Financial market sophistication

The level of sophistication of financial markets in your country is (1 = lower than international norms, 7 = higher than international norms)

RANK	COUNTRY/ECONOMY	SCORE	SD		RANK	COUNTRY/ECONOMY	SCORE	SD
1	United Kingdom	6.76	0.5		62	Jordan	4.01	1.3
2	Switzerland	6.63	0.8		63	Barbados	4.00	1.2
3	Luxembourg	6.45	0.6		64	Argentina	3.97	1.3
4	Hong Kong SAR	6.41	1.0		65	Kazakhstan	3.97	1.2
5	United States	6.25	1.1		66	Philippines	3.94	1.3
6	Netherlands	6.16	0.7		67	Guatemala	3.87	1.4
7	Ireland	6.11	1.0		68	Kenya	3.78	1.3
8	Sweden	6.10	0.9		69	Zimbabwe	3.77	1.4
9	Canada	6.10	0.8		70	Azerbaijan	3.74	1.5
10	Australia	6.08	0.7		71	Croatia	3.69	1.5
11	Germany	6.04	1.0		72	Sri Lanka	3.64	1.6
12	Finland	6.02	0.8		73	Uruguay	3.62	1.4
13	Singapore	5.97	0.8		74	Botswana	3.60	1.2
14	France	5.90	0.8		75	Venezuela	3.59	1.4
15	Israel	5.77	0.9		76	Egypt	3.55	1.5
16	Iceland	5.77	1.0		77	Dominican Republic	3.40	1.5
17	Denmark	5.75	0.9		78	Ecuador	3.28	1.2
18	Japan	5.73	0.9		79	Nigeria	3.25	1.5
19	South Africa	5.71	0.9		80	Nicaragua	3.24	1.3
20	Norway	5.64	0.9		81	Ukraine	3.24	1.2
21	Belgium	5.62	1.1		82	Morocco	3.22	1.4
22	Spain	5.54	1.1		83	Indonesia	3.15	0.7
23	Austria	5.54	1.2		84	Russian Federation	3.13	1.3
24	Chile	5.45	0.9		85	Romania	3.09	1.1
25	Portugal	5.44	1.0		86	Honduras	3.08	1.5
26	New Zealand	5.37	1.2		87	Macedonia, FYR	2.95	1.1
27	Panama	5.30	1.2		88	Suriname	2.95	1.3
28	Brazil	5.29	1.4		89	Tanzania	2.89	1.2
29	Estonia	5.23	1.1		90	Georgia	2.86	1.3
30	Bahrain	5.18	1.3		90	Vietnam	2.86	1.2
31	Malaysia	5.14	1.0		92	Bolivia	2.83	1.3
32	India	5.12	1.2		93	China	2.82	1.2
33	Taiwan, China	4.75	1.1		94	Bosnia and Herzegovina	2.78	1.2
34	Hungary	4.73	1.1		95	Mongolia	2.65	1.2
35	El Salvador	4.71	1.2		96	Burkina Faso	2.61	1.3
36	Turkey	4.63	1.2		97	Bangladesh	2.60	1.2
37	Greece	4.58	1.3		98	Armenia	2.58	1.3
38	Mexico	4.56	1.1		99	Moldova	2.57	1.3
39	Jamaica	4.53	1.1		100	Mali	2.53	1.4
40	Kuwait	4.49	1.4		101	Paraguay	2.52	1.1
41	Thailand	4.43	1.1		102	Benin	2.51	1.3
42	Korea, Rep.	4.41	1.3		103	Madagascar	2.50	1.4
43	Mauritius	4.37	1.0		104	Mauritania	2.49	1.3
44	Cyprus	4.36	1.3		105	Zambia	2.46	1.0
45	United Arab Emirates	4.34	1.5		106	Bulgaria	2.44	1.3
46	Italy	4.33	1.3		107	Nepal	2.44	1.3
47	Qatar	4.31	1.5		108	Serbia and Montenegro	2.43	1.2
48	Malta	4.29	1.3		109	Uganda	2.42	1.3
49	Colombia	4.28	1.2		110	Malawi	2.35	1.2
50	Peru	4.25	1.3		111	Albania	2.35	1.3
51	Czech Republic	4.23	1.1		112	Mozambique	2.33	0.9
52	Slovenia	4.22	1.2		113	Kyrgyz Republic	2.28	1.2
53	Pakistan	4.21	1.3		114	Burundi	2.23	1.4
54	Slovak Republic	4.19	1.4		115	Cambodia	2.23	1.2
55	Costa Rica	4.19	1.4		116	Angola	2.21	1.1
56	Poland	4.17	0.9		117	Lesotho	2.19	1.2
57	Trinidad and Tobago	4.11	1.4		118	Guyana	2.16	1.1
58	Lithuania	4.11	1.2		119	Ethiopia	2.13	1.2
59	Tunisia	4.07	1.1		120	Algeria	2.02	1.0
60	Namibia	4.06	1.0		121	Cameroon	1.83	0.9
61	Latvia	4.02	1.3		122	Chad	1.65	0.9

MEAN: 4.01

SOURCE: World Economic Forum, Executive Opinion Survey 2006

1.03 Technological readiness

Your country's level of technological readiness (1 = generally lags behind most other countries, 7 = is among the world leaders)

RANK	COUNTRY/ECONOMY	SCORE	1 MEAN: 3.92 7	SD
1	Finland	6.53		0.7
2	Japan	6.52		0.8
3	Sweden	6.42		0.8
4	Israel	6.41		0.8
5	Iceland	6.37		0.8
6	Germany	6.18		0.9
7	United States	6.10		1.1
8	Switzerland	6.10		1.0
9	Norway	6.08		0.8
10	Denmark	6.07		0.9
11	Singapore	5.94		0.9
12	Canada	5.83		1.1
13	United Kingdom	5.79		1.0
14	Netherlands	5.68		0.9
15	United Arab Emirates	5.62		1.2
16	Taiwan, China	5.52		0.9
17	Australia	5.51		1.1
18	Malaysia	5.49		0.9
19	France	5.48		1.1
20	Belgium	5.36		1.1
21	Korea, Rep.	5.36		0.9
22	Hong Kong SAR	5.34		1.3
23	India	5.25		0.9
24	Estonia	5.18		1.3
25	Austria	5.12		1.1
26	Chile	5.10		1.0
27	Qatar	4.89		1.3
28	Panama	4.87		1.3
29	Malta	4.84		1.1
30	Tunisia	4.81		1.2
31	New Zealand	4.78		1.2
32	Spain	4.76		1.3
33	Czech Republic	4.75		1.0
34	Barbados	4.74		1.2
35	Dominican Republic	4.67		1.4
36	Thailand	4.65		0.9
37	Ireland	4.63		1.5
38	South Africa	4.61		1.1
39	Bahrain	4.56		1.5
40	Jamaica	4.56		1.2
41	Luxembourg	4.44		1.2
42	El Salvador	4.35		1.1
43	Kuwait	4.35		1.4
44	Jordan	4.34		1.2
45	Slovenia	4.31		1.2
46	Guatemala	4.29		1.2
47	Cyprus	4.29		1.2
48	Hungary	4.27		1.0
49	Indonesia	4.27		1.1
50	Italy	4.26		1.3
51	Slovak Republic	4.21		1.2
52	Portugal	4.14		1.1
53	Costa Rica	4.12		1.3
54	Mauritius	4.11		1.1
55	Turkey	4.08		1.1
56	Uruguay	4.06		1.4
57	Mexico	4.04		1.1
58	Brazil	3.96		1.4
59	Latvia	3.96		1.4
60	Argentina	3.96		1.2
61	Peru	3.94		1.2
62	Trinidad and Tobago	3.93		1.5
63	Philippines	3.92		1.5
64	Colombia	3.81		1.2
65	Namibia	3.73		1.4
66	Egypt	3.71		1.5
67	Botswana	3.70		1.4
68	Greece	3.65		1.3
69	China	3.65		1.2
70	Lithuania	3.64		1.1
71	Morocco	3.64		1.6
72	Venezuela	3.62		1.3
73	Azerbaijan	3.60		1.6
74	Romania	3.59		1.3
75	Tanzania	3.41		1.4
76	Poland	3.40		0.9
77	Pakistan	3.37		1.2
78	Croatia	3.33		1.3
79	Sri Lanka	3.29		1.7
80	Kazakhstan	3.28		1.3
81	Ecuador	3.19		1.2
82	Kenya	3.17		1.3
83	Nicaragua	3.13		1.4
84	Russian Federation	3.12		1.5
85	Nigeria	3.04		1.6
86	Armenia	2.95		1.3
87	Uganda	2.85		1.6
88	Cambodia	2.84		1.4
89	Bulgaria	2.83		1.3
90	Madagascar	2.83		1.3
91	Honduras	2.81		1.4
92	Georgia	2.81		1.2
93	Vietnam	2.76		1.1
94	Ukraine	2.74		1.3
95	Zimbabwe	2.72		1.1
96	Bolivia	2.71		1.3
97	Albania	2.66		1.3
98	Macedonia, FYR	2.63		1.2
99	Mongolia	2.62		1.3
100	Paraguay	2.54		1.3
101	Mali	2.52		1.3
102	Algeria	2.51		1.2
103	Burkina Faso	2.51		1.2
104	Bangladesh	2.50		1.2
105	Malawi	2.37		1.0
106	Nepal	2.36		1.2
107	Cameroon	2.34		1.2
108	Benin	2.34		1.3
109	Lesotho	2.29		1.2
110	Suriname	2.28		1.1
111	Mozambique	2.28		1.2
112	Bosnia and Herzegovina	2.27		1.1
113	Guyana	2.22		1.1
114	Zambia	2.20		0.9
115	Moldova	2.19		1.1
116	Mauritania	2.17		1.4
117	Angola	2.17		1.1
118	Serbia and Montenegro	2.16		1.1
119	Ethiopia	2.09		1.1
120	Kyrgyz Republic	1.87		1.0
121	Chad	1.64		0.9
122	Burundi	1.59		0.9

SOURCE: World Economic Forum, Executive Opinion Survey 2006

1.04   State of cluster development

Strong and deep clusters are widespread throughout the economy (1 = strongly disagree, 7 = strongly agree)

RANK	COUNTRY/ECONOMY	SCORE	MEAN: 3.59	SD
1	Taiwan, China	5.52		1.2
2	United States	5.22		1.4
3	Finland	5.07		1.1
4	United Kingdom	5.06		1.0
5	Malaysia	4.96		1.3
6	Hungary	4.94		1.6
7	Singapore	4.92		1.2
8	Germany	4.90		1.4
9	United Arab Emirates	4.81		1.4
10	Romania	4.78		1.5
11	India	4.77		1.3
12	Hong Kong SAR	4.75		1.4
13	Croatia	4.72		1.6
14	Indonesia	4.68		0.8
15	Switzerland	4.67		1.4
16	Vietnam	4.62		1.6
17	Sweden	4.57		1.2
18	Netherlands	4.54		1.3
19	Belgium	4.53		1.3
20	Albania	4.51		2.1
21	Israel	4.49		1.1
22	Denmark	4.48		1.2
23	Czech Republic	4.46		1.3
24	Austria	4.42		1.4
25	Canada	4.41		1.3
26	Slovenia	4.35		1.3
27	Japan	4.33		1.8
28	Kuwait	4.32		1.7
29	France	4.27		1.6
30	Norway	4.22		1.3
31	Korea, Rep.	4.19		1.3
32	Tunisia	4.05		1.3
33	China	4.02		1.5
34	Brazil	4.02		1.3
35	Ireland	4.00		1.3
35	Spain	4.00		1.2
37	Nigeria	3.89		1.7
38	Zambia	3.86		2.0
39	South Africa	3.84		1.1
40	Thailand	3.74		1.4
41	Argentina	3.73		1.5
42	Panama	3.72		1.7
43	Mexico	3.71		1.5
44	Turkey	3.68		1.4
45	Latvia	3.68		1.7
46	Philippines	3.65		1.7
47	Iceland	3.64		1.5
47	Luxembourg	3.64		1.7
49	Chile	3.64		1.5
50	New Zealand	3.64		1.2
51	Slovak Republic	3.61		1.5
52	Cyprus	3.61		1.6
53	Australia	3.60		1.3
54	Portugal	3.59		1.4
55	Pakistan	3.59		1.3
56	Qatar	3.58		2.0
57	Guatemala	3.58		1.5
58	Cambodia	3.57		1.6
59	Egypt	3.53		1.7
60	Azerbaijan	3.53		1.8
61	Uganda	3.51		1.9
62	Bangladesh	3.51		1.5
63	Honduras	3.49		1.8
64	Peru	3.48		1.5
65	Estonia	3.45		1.5
66	Kenya	3.43		1.5
67	Colombia	3.43		1.6
68	Sri Lanka	3.42		1.6
69	El Salvador	3.38		1.2
70	Trinidad and Tobago	3.38		1.7
71	Jamaica	3.35		1.6
72	Nepal	3.35		1.6
73	Russian Federation	3.34		1.6
74	Costa Rica	3.34		1.2
75	Greece	3.33		1.6
76	Jordan	3.33		1.6
77	Venezuela	3.32		1.6
78	Mauritius	3.28		1.4
79	Poland	3.26		0.8
80	Lithuania	3.25		1.3
81	Botswana	3.23		1.6
82	Nicaragua	3.22		1.5
83	Ukraine	3.20		1.5
84	Ecuador	3.20		1.6
85	Italy	3.19		1.8
86	Dominican Republic	3.18		1.9
87	Kazakhstan	3.18		1.4
88	Morocco	3.18		1.8
89	Mauritania	3.17		2.2
90	Bahrain	3.16		1.7
91	Lesotho	3.16		1.7
92	Madagascar	3.14		1.6
93	Guyana	3.04		1.8
94	Namibia	3.04		1.6
95	Barbados	3.00		1.4
95	Ethiopia	3.00		1.9
95	Tanzania	3.00		1.7
98	Bolivia	2.99		1.6
99	Uruguay	2.94		1.6
100	Kyrgyz Republic	2.94		1.6
101	Malta	2.93		1.4
102	Armenia	2.91		1.6
103	Suriname	2.85		1.7
104	Malawi	2.79		1.6
105	Macedonia, FYR	2.74		1.6
106	Bosnia and Herzegovina	2.74		1.6
107	Paraguay	2.73		1.6
108	Mongolia	2.72		1.4
109	Zimbabwe	2.71		1.4
110	Algeria	2.65		1.4
111	Serbia and Montenegro	2.60		1.4
112	Bulgaria	2.49		1.4
113	Georgia	2.48		1.4
114	Cameroon	2.46		1.5
115	Burkina Faso	2.42		1.8
116	Moldova	2.31		1.4
117	Mozambique	2.30		1.5
118	Chad	2.22		1.4
119	Mali	2.21		1.4
120	Angola	2.20		1.2
121	Benin	2.10		1.3
122	Burundi	1.87		1.2

SOURCE: World Economic Forum, Executive Opinion Survey 2006

271

1.05 US utility patents

Number of utility patents (i.e., patents for invention) granted between January 1 and December 31, 2005, per million population

RANK	COUNTRY/ECONOMY	HARD DATA		RANK	COUNTRY/ECONOMY	HARD DATA
1	United States	250.29		62	Dominican Republic	0.22
2	Japan	236.85		63	Philippines	0.22
3	Taiwan, China	226.86		64	Colombia	0.15
4	Finland	138.46		65	Ecuador	0.15
5	Israel	137.91		66	El Salvador	0.14
6	Switzerland	136.30		67	Kazakhstan	0.14
7	Sweden	124.78		68	Peru	0.11
8	Germany	108.96		69	Tunisia	0.10
9	Korea, Rep.	91.05		70	Turkey	0.10
10	Canada	89.60		71	Egypt	0.09
11	Luxembourg	89.13		72	Guatemala	0.08
12	Singapore	80.47		73	Zimbabwe	0.08
13	Iceland	66.67		74	Sri Lanka	0.05
14	Denmark	66.30		75	Indonesia	0.04
15	Netherlands	60.92		76	Morocco	0.03
16	Austria	56.34		77	Vietnam	0.02
17	United Kingdom	52.73		78	Pakistan	0.02
18	Belgium	49.90		79	Albania	0.00
19	Norway	47.83		79	Algeria	0.00
20	France	47.37		79	Angola	0.00
21	Australia	45.10		79	Azerbaijan	0.00
22	Hong Kong SAR	40.43		79	Bahrain	0.00
23	Ireland	38.05		79	Bangladesh	0.00
24	New Zealand	30.50		79	Barbados	0.00
25	Italy	22.31		79	Benin	0.00
26	Cyprus	7.89		79	Bolivia	0.00
27	Spain	6.33		79	Bosnia and Herzegovina	0.00
28	Slovenia	6.00		79	Botswana	0.00
29	Hungary	4.55		79	Burkina Faso	0.00
30	Estonia	3.85		79	Burundi	0.00
31	Malaysia	3.48		79	Cambodia	0.00
32	Croatia	2.61		79	Cameroon	0.00
33	Malta	2.50		79	Chad	0.00
34	Czech Republic	2.45		79	Ethiopia	0.00
35	South Africa	1.84		79	Guyana	0.00
36	Greece	1.35		79	Honduras	0.00
37	Kuwait	1.11		79	Jordan	0.00
38	Russian Federation	1.03		79	Kyrgyz Republic	0.00
39	Portugal	0.95		79	Lesotho	0.00
40	Lithuania	0.88		79	Macedonia, FYR	0.00
41	Latvia	0.87		79	Madagascar	0.00
42	Mexico	0.75		79	Malawi	0.00
43	Costa Rica	0.70		79	Mali	0.00
44	Georgia	0.67		79	Mauritania	0.00
44	United Arab Emirates	0.67		79	Mauritius	0.00
46	Argentina	0.62		79	Mongolia	0.00
47	Poland	0.60		79	Mozambique	0.00
48	Uruguay	0.57		79	Namibia	0.00
49	Chile	0.55		79	Nepal	0.00
50	Brazil	0.41		79	Nicaragua	0.00
51	Bulgaria	0.39		79	Nigeria	0.00
52	Ukraine	0.39		79	Panama	0.00
53	Jamaica	0.37		79	Paraguay	0.00
54	India	0.35		79	Qatar	0.00
55	Armenia	0.33		79	Serbia and Montenegro	0.00
56	Romania	0.32		79	Slovak Republic	0.00
57	China	0.31		79	Suriname	0.00
58	Venezuela	0.30		79	Tanzania	0.00
59	Kenya	0.26		79	Trinidad and Tobago	0.00
60	Thailand	0.25		79	Uganda	0.00
61	Moldova	0.24		79	Zambia	0.00

272

SOURCE: US Patent and Trademark Office; UNFPA (March 2006)

High-tech exports

High-technology exports as percentage of total exports, 2004

RANK	COUNTRY/ECONOMY	HARD DATA
1	Singapore	48.75
2	Taiwan, China	44.60
3	Malaysia	36.86
4	Malta	34.14
5	Philippines	30.53
6	United States	26.38
7	Hong Kong SAR	25.40
8	Korea, Rep.	25.27
9	China	24.64
10	Japan	21.92
11	Hungary	21.41
12	Ireland	20.75
13	Switzerland	20.44
14	Cyprus	16.78
15	Costa Rica	15.99
16	Mexico	15.75
17	Finland	15.39
18	Netherlands	14.57
19	Israel	13.33
20	Germany	12.66
21	United Kingdom	12.26
22	France	12.20
23	Sweden	10.62
24	Czech Republic	9.99
25	Denmark	9.23
26	Canada	8.43
27	Indonesia	7.29
28	Austria	7.11
29	Estonia	6.66
30	Belgium	6.64
31	Brazil	5.45
32	Italy	5.26
33	Portugal	5.12
34	Croatia	4.66
35	New Zealand	4.29
36	Morocco	4.20
37	Georgia	4.07
38	Slovenia	4.06
39	Slovak Republic	3.86
40	Spain	3.72
41	Australia	3.60
42	Tunisia	2.94
43	Jordan	2.68
44	Norway	2.52
45	Romania	2.41
46	Greece	2.40
47	South Africa	2.28
48	Luxembourg	2.27
49	India	2.16
50	Latvia	2.15
51	Lithuania	2.13
52	Poland	2.04
53	Argentina	1.94
54	Guatemala	1.89
55	Mauritius	1.82
56	Bulgaria	1.75
57	Russian Federation	1.69
58	Colombia	1.63
59	Uganda	1.24
60	Turkey	1.22
61	Bolivia	1.03

RANK	COUNTRY/ECONOMY	HARD DATA
62	Moldova	1.00
63	Pakistan	0.98
64	El Salvador	0.85
65	Sri Lanka	0.83
66	Macedonia, FYR	0.75
67	Burkina Faso	0.73
68	Iceland	0.67
69	Ecuador	0.61
70	Uruguay	0.56
71	Paraguay	0.51
72	Chile	0.50
73	Armenia	0.45
74	Kenya	0.42
75	Nicaragua	0.39
76	Albania	0.33
77	Kazakhstan	0.32
78	Malawi	0.31
79	Peru	0.30
80	Venezuela	0.30
81	Zimbabwe	0.29
82	Bahrain	0.23
83	Burundi	0.20
84	Azerbaijan	0.18
85	Tanzania	0.16
86	Cambodia	0.13
87	Zambia	0.12
88	Qatar	0.10
89	Egypt	0.07
90	Guyana	0.06
91	Cameroon	0.06
92	Madagascar	0.05
93	Bangladesh	0.03
94	Panama	0.02
95	Algeria	0.02
n/a	Angola	n/a
n/a	Barbados	n/a
n/a	Benin	n/a
n/a	Bosnia and Herzegovina	n/a
n/a	Botswana	n/a
n/a	Chad	n/a
n/a	Dominican Republic	n/a
n/a	Ethiopia	n/a
n/a	Honduras	n/a
n/a	Jamaica	n/a
n/a	Kuwait	n/a
n/a	Kyrgyz Republic	n/a
n/a	Lesotho	n/a
n/a	Mali	n/a
n/a	Mauritania	n/a
n/a	Mongolia	n/a
n/a	Mozambique	n/a
n/a	Namibia	n/a
n/a	Nepal	n/a
n/a	Nigeria	n/a
n/a	Serbia and Montenegro	n/a
n/a	Suriname	n/a
n/a	Thailand	n/a
n/a	Trinidad and Tobago	n/a
n/a	Ukraine	n/a
n/a	United Arab Emirates	n/a
n/a	Vietnam	n/a

SOURCE: World Bank, *World Development Indicators Online Database* (December 2006); national sources

1.07 Burden of government regulation

Complying with administrative requirements (permits, regulations, reporting) issued by the government in your country is (1 = burdensome, 7 = not burdensome)

RANK	COUNTRY/ECONOMY	SCORE	SD
1	Iceland	5.34	1.2
2	Singapore	5.08	1.3
3	Finland	4.85	1.3
4	Hong Kong SAR	4.83	1.6
5	Indonesia	4.75	1.4
6	Mauritania	4.60	2.2
7	Malaysia	4.58	1.4
8	United Arab Emirates	4.33	1.6
9	Estonia	4.21	1.4
10	Zambia	4.19	1.6
11	Tunisia	4.17	1.4
12	Taiwan, China	4.12	1.2
13	Switzerland	4.07	1.5
14	Luxembourg	3.97	1.5
15	Thailand	3.93	1.3
16	Qatar	3.89	1.4
17	Ireland	3.88	1.4
18	Georgia	3.84	1.5
19	Norway	3.82	1.4
20	Chile	3.80	1.4
21	Denmark	3.74	1.5
22	Israel	3.62	1.3
23	Jordan	3.60	1.4
24	Japan	3.58	1.3
25	Burkina Faso	3.57	1.7
26	United States	3.57	1.5
27	Austria	3.56	1.3
28	Cyprus	3.54	1.3
29	Tanzania	3.52	1.5
30	New Zealand	3.48	1.4
31	Bahrain	3.42	1.7
32	Barbados	3.40	1.2
33	Malawi	3.39	1.2
34	China	3.38	1.4
35	Azerbaijan	3.37	1.6
36	Latvia	3.33	1.5
37	Canada	3.32	1.4
38	Guatemala	3.32	1.3
39	Uganda	3.29	1.5
40	United Kingdom	3.28	1.4
41	El Salvador	3.25	1.3
42	Sweden	3.24	1.6
43	Nigeria	3.23	1.7
44	Portugal	3.21	1.3
45	Netherlands	3.18	1.4
46	Morocco	3.14	1.6
47	Botswana	3.13	1.7
47	Dominican Republic	3.13	1.3
49	Korea, Rep.	3.13	1.3
50	Uruguay	3.10	1.3
51	Romania	3.10	1.4
52	Slovak Republic	3.10	1.1
53	Australia	3.09	1.3
54	Pakistan	3.07	1.2
55	Poland	3.07	1.0
56	Germany	3.06	1.4
57	Kazakhstan	3.04	1.4
58	Lithuania	3.01	1.3
59	Ethiopia	3.00	1.5
59	Mali	3.00	1.6
61	Algeria	2.99	1.5
62	South Africa	2.97	1.3
63	Turkey	2.97	1.1
64	Trinidad and Tobago	2.96	1.6
65	Nicaragua	2.94	1.5
66	India	2.94	1.4
67	Spain	2.94	1.3
68	Croatia	2.93	1.5
69	Macedonia, FYR	2.92	1.6
70	Slovenia	2.91	1.3
71	Kuwait	2.89	1.6
72	Egypt	2.86	1.6
73	Nepal	2.84	1.5
74	Panama	2.84	1.4
75	Belgium	2.82	1.3
76	Armenia	2.82	1.6
77	Hungary	2.82	1.3
78	Cambodia	2.81	1.4
79	Honduras	2.80	1.3
80	Malta	2.80	1.2
81	Paraguay	2.76	1.6
82	Mongolia	2.75	1.6
83	Kenya	2.74	1.6
84	Namibia	2.72	1.4
85	Jamaica	2.71	1.4
86	Sri Lanka	2.71	1.5
87	Greece	2.67	1.3
88	Burundi	2.65	1.6
89	France	2.65	1.4
90	Bulgaria	2.65	1.2
91	Madagascar	2.65	1.4
92	Mexico	2.62	1.2
93	Albania	2.59	1.5
94	Suriname	2.59	1.4
95	Colombia	2.57	1.3
96	Ukraine	2.55	1.4
97	Costa Rica	2.55	1.2
98	Moldova	2.53	1.1
99	Bolivia	2.53	1.2
100	Guyana	2.53	1.4
101	Lesotho	2.52	1.4
102	Argentina	2.49	1.2
103	Ecuador	2.45	1.3
104	Bangladesh	2.44	1.3
105	Benin	2.44	1.4
106	Philippines	2.43	1.2
107	Chad	2.41	1.8
108	Vietnam	2.38	1.2
109	Peru	2.35	1.2
110	Czech Republic	2.33	1.1
111	Angola	2.30	1.1
112	Mauritius	2.30	1.0
113	Russian Federation	2.23	1.3
114	Mozambique	2.21	1.2
115	Bosnia and Herzegovina	2.19	1.2
115	Zimbabwe	2.19	1.2
117	Kyrgyz Republic	2.10	1.2
118	Cameroon	2.05	1.4
119	Italy	2.04	0.9
120	Serbia and Montenegro	2.03	1.3
121	Brazil	1.87	1.1
122	Venezuela	1.83	1.0

MEAN: 3.10

SOURCE: World Economic Forum, Executive Opinion Survey 2006

1.08 Extent and effect of taxation

The level of taxes in your country (1 = significantly limits the incentives to work or invest, 7 = has little impact on the incentives to work or invest)

RANK	COUNTRY/ECONOMY	SCORE	MEAN: 3.45	SD		RANK	COUNTRY/ECONOMY	SCORE	MEAN: 3.45	SD
1	Bahrain	6.05		1.4		62	Morocco	3.20		1.6
2	Hong Kong SAR	6.01		1.0		63	Poland	3.20		1.1
3	United Arab Emirates	5.94		1.6		64	Spain	3.19		1.3
4	Kuwait	5.89		1.5		65	Canada	3.16		1.4
5	Qatar	5.77		1.6		66	Armenia	3.15		1.6
6	Iceland	5.66		1.0		67	Honduras	3.15		1.5
7	Singapore	5.64		1.0		68	Madagascar	3.12		1.6
8	Slovak Republic	5.49		1.3		69	Lesotho	3.12		1.5
9	Luxembourg	5.29		1.2		70	Venezuela	3.11		1.6
10	Ireland	5.23		1.3		71	Korea, Rep.	3.09		1.3
11	Indonesia	5.13		1.4		72	Sri Lanka	3.09		1.6
12	Malaysia	5.01		1.1		73	Mexico	3.09		1.3
13	Estonia	5.00		1.4		74	Germany	3.08		1.2
13	Mauritania	5.00		1.7		75	Greece	3.04		1.1
15	Switzerland	4.96		1.3		76	Kazakhstan	3.02		1.5
16	Botswana	4.74		1.5		77	Hungary	3.00		1.4
17	Mauritius	4.67		1.0		78	Australia	2.99		1.0
18	Tunisia	4.62		1.3		79	Ecuador	2.98		1.4
19	Cyprus	4.58		1.3		80	Peru	2.94		1.2
20	Taiwan, China	4.55		1.2		81	Turkey	2.89		1.2
21	India	4.44		1.3		82	Mali	2.89		1.7
22	Thailand	4.43		1.2		83	Croatia	2.88		1.3
23	Nigeria	4.38		1.8		84	Malawi	2.87		1.5
24	Trinidad and Tobago	4.30		1.6		85	France	2.85		1.3
25	United Kingdom	4.24		1.2		86	Macedonia, FYR	2.85		1.6
26	El Salvador	4.16		1.3		87	Czech Republic	2.83		1.2
27	South Africa	4.16		1.3		88	Albania	2.83		1.3
28	Cambodia	4.05		1.8		89	Jamaica	2.82		1.4
29	Paraguay	3.94		1.7		90	Dominican Republic	2.80		1.4
30	Chile	3.92		1.4		91	Panama	2.75		1.3
31	United States	3.92		1.4		92	Russian Federation	2.72		1.5
32	Algeria	3.90		1.8		93	Uruguay	2.72		1.3
33	Pakistan	3.87		1.4		94	Mozambique	2.66		1.5
34	Austria	3.83		1.4		95	Burundi	2.63		1.7
35	Angola	3.82		1.4		96	Bulgaria	2.57		1.6
36	Egypt	3.77		1.8		97	Serbia and Montenegro	2.57		1.5
37	Guatemala	3.77		1.4		98	Finland	2.55		1.4
38	Latvia	3.74		1.5		99	Ukraine	2.55		1.5
39	Netherlands	3.66		1.3		100	Slovenia	2.55		1.4
40	Philippines	3.64		1.3		101	Zimbabwe	2.53		1.3
41	Azerbaijan	3.64		1.7		102	Uganda	2.52		1.4
42	Namibia	3.63		1.3		103	Moldova	2.47		1.3
43	Barbados	3.63		1.2		104	Kenya	2.47		1.3
44	Nepal	3.58		1.7		105	Guyana	2.46		1.4
45	China	3.58		1.4		106	Romania	2.44		1.2
46	Costa Rica	3.57		1.4		107	Colombia	2.43		1.1
47	Bangladesh	3.56		1.6		108	Nicaragua	2.39		1.1
48	Japan	3.53		1.1		109	Chad	2.39		1.7
49	Jordan	3.53		1.6		110	Suriname	2.36		1.3
50	Georgia	3.51		1.4		111	Cameroon	2.33		1.4
51	Norway	3.46		1.1		112	Bosnia and Herzegovina	2.29		1.3
52	Tanzania	3.42		1.4		113	Mongolia	2.25		1.5
53	Vietnam	3.41		1.4		114	Denmark	2.25		1.4
54	Bolivia	3.38		1.5		115	Sweden	2.21		1.4
55	Portugal	3.37		1.2		116	Argentina	2.21		1.1
56	Burkina Faso	3.35		1.8		117	Benin	2.21		1.3
57	Israel	3.34		1.3		118	Italy	2.20		1.2
58	Ethiopia	3.32		1.7		119	Zambia	2.13		1.1
59	Malta	3.28		1.1		120	Kyrgyz Republic	2.12		1.3
60	New Zealand	3.22		0.9		121	Belgium	2.05		1.2
61	Lithuania	3.21		1.3		122	Brazil	1.53		0.9

SOURCE: World Economic Forum, Executive Opinion Survey 2006

275

1.09   Time required to start a business

Number of days required to start a business, 2006

RANK	COUNTRY/ECONOMY	HARD DATA
1	Australia	2.00
2	Canada	3.00
3	Denmark	5.00
3	Iceland	5.00
3	United States	5.00
6	Singapore	6.00
7	France	8.00
7	Jamaica	8.00
7	Portugal	8.00
10	Turkey	9.00
11	Netherlands	10.00
12	Hong Kong SAR	11.00
12	Romania	11.00
12	Tunisia	11.00
15	Morocco	12.00
15	New Zealand	12.00
17	Italy	13.00
17	Norway	13.00
19	Finland	14.00
20	Ethiopia	16.00
20	Georgia	16.00
20	Latvia	16.00
20	Sweden	16.00
24	Jordan	18.00
24	Macedonia, FYR	18.00
24	United Kingdom	18.00
27	Egypt	19.00
27	Ireland	19.00
27	Panama	19.00
30	Kazakhstan	20.00
30	Mongolia	20.00
30	Switzerland	20.00
33	Kyrgyz Republic	21.00
33	Madagascar	21.00
33	Serbia and Montenegro	21.00
36	Korea, Rep.	22.00
37	Japan	23.00
38	Algeria	24.00
38	Armenia	24.00
38	Czech Republic	24.00
38	Germany	24.00
38	Pakistan	24.00
43	Slovak Republic	25.00
44	El Salvador	26.00
44	Lithuania	26.00
46	Belgium	27.00
46	Chile	27.00
46	Mexico	27.00
49	Russian Federation	28.00
50	Austria	29.00
51	Guatemala	30.00
51	Malaysia	30.00
51	Moldova	30.00
51	Tanzania	30.00
51	Uganda	30.00
56	Benin	31.00
56	Nepal	31.00
56	Poland	31.00
59	Argentina	32.00
59	Bulgaria	32.00
61	Thailand	33.00
61	Ukraine	33.00
63	Burkina Faso	34.00
63	Israel	34.00
65	China	35.00
65	Estonia	35.00
65	India	35.00
65	Kuwait	35.00
65	South Africa	35.00
65	Zambia	35.00
71	Bangladesh	37.00
71	Cameroon	37.00
71	Malawi	37.00
74	Greece	38.00
74	Hungary	38.00
76	Albania	39.00
76	Nicaragua	39.00
78	Mali	42.00
79	Burundi	43.00
79	Nigeria	43.00
79	Trinidad and Tobago	43.00
79	Uruguay	43.00
83	Colombia	44.00
83	Honduras	44.00
85	Croatia	45.00
86	Guyana	46.00
86	Mauritius	46.00
88	Spain	47.00
89	Philippines	48.00
89	Taiwan, China	48.00
91	Bolivia	50.00
91	Vietnam	50.00
93	Azerbaijan	53.00
94	Bosnia and Herzegovina	54.00
94	Kenya	54.00
96	Slovenia	60.00
96	Sri Lanka	60.00
98	United Arab Emirates	63.00
99	Ecuador	65.00
100	Peru	72.00
101	Dominican Republic	73.00
101	Lesotho	73.00
103	Paraguay	74.00
104	Chad	75.00
105	Costa Rica	77.00
106	Mauritania	82.00
107	Cambodia	86.00
108	Namibia	95.00
109	Zimbabwe	96.00
110	Indonesia	97.00
111	Botswana	108.00
112	Mozambique	113.00
113	Angola	124.00
114	Venezuela	141.00
115	Brazil	152.00
116	Suriname	694.00
n/a	Bahrain	n/a
n/a	Barbados	n/a
n/a	Cyprus	n/a
n/a	Luxembourg	n/a
n/a	Malta	n/a
n/a	Qatar	n/a

276

SOURCE: World Bank, *Doing Business 2007: How to Reform* (2006)

1.10 Number of procedures required to start a business

Number of administrative procedures to start a business, 2006

RANK	COUNTRY/ECONOMY	HARD DATA
1	Australia	2.00
1	Canada	2.00
1	New Zealand	2.00
4	Denmark	3.00
4	Finland	3.00
4	Sweden	3.00
7	Belgium	4.00
7	Ireland	4.00
7	Norway	4.00
10	Hong Kong SAR	5.00
10	Iceland	5.00
10	Israel	5.00
10	Latvia	5.00
10	Romania	5.00
10	United States	5.00
16	Estonia	6.00
16	Hungary	6.00
16	Jamaica	6.00
16	Mauritius	6.00
16	Morocco	6.00
16	Netherlands	6.00
16	Nicaragua	6.00
16	Singapore	6.00
16	Switzerland	6.00
16	United Kingdom	6.00
16	Zambia	6.00
27	Benin	7.00
27	Ethiopia	7.00
27	France	7.00
27	Georgia	7.00
27	Kazakhstan	7.00
27	Lithuania	7.00
27	Nepal	7.00
27	Panama	7.00
27	Russian Federation	7.00
36	Bangladesh	8.00
36	Burkina Faso	8.00
36	Guyana	8.00
36	Japan	8.00
36	Kyrgyz Republic	8.00
36	Lesotho	8.00
36	Mexico	8.00
36	Mongolia	8.00
36	Portugal	8.00
36	Sri Lanka	8.00
36	Taiwan, China	8.00
36	Thailand	8.00
36	Turkey	8.00
49	Armenia	9.00
49	Austria	9.00
49	Bulgaria	9.00
49	Chile	9.00
49	Germany	9.00
49	Italy	9.00
49	Malaysia	9.00
49	Nigeria	9.00
49	Slovak Republic	9.00
49	Slovenia	9.00
49	South Africa	9.00
49	Trinidad and Tobago	9.00
61	Cambodia	10.00
61	Croatia	10.00
61	Czech Republic	10.00
61	Dominican Republic	10.00
61	Egypt	10.00
61	El Salvador	10.00
61	Macedonia, FYR	10.00
61	Madagascar	10.00
61	Malawi	10.00
61	Moldova	10.00
61	Namibia	10.00
61	Peru	10.00
61	Poland	10.00
61	Spain	10.00
61	Tunisia	10.00
61	Ukraine	10.00
61	Uruguay	10.00
61	Zimbabwe	10.00
79	Albania	11.00
79	Botswana	11.00
79	Burundi	11.00
79	Costa Rica	11.00
79	India	11.00
79	Jordan	11.00
79	Mauritania	11.00
79	Pakistan	11.00
79	Philippines	11.00
79	Vietnam	11.00
89	Bosnia and Herzegovina	12.00
89	Cameroon	12.00
89	Indonesia	12.00
89	Korea, Rep.	12.00
89	United Arab Emirates	12.00
94	Serbia and Montenegro	12.50
95	Angola	13.00
95	China	13.00
95	Colombia	13.00
95	Guatemala	13.00
95	Honduras	13.00
95	Kenya	13.00
95	Kuwait	13.00
95	Mali	13.00
95	Mozambique	13.00
95	Suriname	13.00
95	Tanzania	13.00
106	Algeria	14.00
106	Ecuador	14.00
108	Argentina	15.00
108	Azerbaijan	15.00
108	Bolivia	15.00
108	Greece	15.00
112	Venezuela	16.00
113	Brazil	17.00
113	Paraguay	17.00
113	Uganda	17.00
116	Chad	19.00
n/a	Bahrain	n/a
n/a	Barbados	n/a
n/a	Cyprus	n/a
n/a	Luxembourg	n/a
n/a	Malta	n/a
n/a	Qatar	n/a

277

SOURCE: World Bank, *Doing Business 2007: How to Reform* (2006)

1.11　Intensity of local competition

Competition in the local market is (1 = limited in most industries and price-cutting is rare, 7 = intense in most industries as market leadership changes over time)

RANK	COUNTRY/ECONOMY	SCORE	1　MEAN: 4.78　7	SD		RANK	COUNTRY/ECONOMY	SCORE	1　MEAN: 4.78　7	SD
1	Germany	6.18		0.7		62	Luxembourg	4.86		1.4
2	United Kingdom	6.07		0.9		63	Kuwait	4.85		1.5
3	Japan	6.04		0.9		64	Greece	4.73		1.3
4	India	6.00		1.0		65	Russian Federation	4.72		1.8
5	United States	5.92		1.1		66	Bangladesh	4.71		1.5
6	Hong Kong SAR	5.89		1.0		67	Qatar	4.69		1.5
7	Sweden	5.86		0.8		68	Egypt	4.67		1.7
8	Belgium	5.81		0.9		69	Sri Lanka	4.67		1.6
9	Netherlands	5.78		0.8		70	Kazakhstan	4.63		1.6
10	Canada	5.77		1.1		71	Morocco	4.61		1.6
11	Chile	5.75		1.1		72	Trinidad and Tobago	4.60		1.7
12	Australia	5.74		1.0		73	Pakistan	4.56		1.5
13	Finland	5.72		1.0		74	Italy	4.55		1.4
14	France	5.72		1.1		75	Botswana	4.54		1.4
15	Indonesia	5.70		0.7		76	Vietnam	4.50		1.5
16	Malaysia	5.70		1.0		77	Ecuador	4.48		1.5
16	New Zealand	5.70		1.1		78	Cambodia	4.47		1.5
18	Austria	5.63		1.2		79	Moldova	4.46		1.7
19	Norway	5.61		1.0		80	Ukraine	4.45		1.6
20	Ireland	5.60		1.3		81	Suriname	4.45		1.6
21	Denmark	5.57		1.2		82	Barbados	4.44		1.4
22	Taiwan, China	5.57		1.0		83	Tanzania	4.42		1.6
23	Czech Republic	5.53		1.1		84	Mauritius	4.41		1.2
24	Iceland	5.52		1.1		85	Nepal	4.40		1.6
25	Israel	5.51		1.0		86	Burkina Faso	4.34		1.6
26	Singapore	5.48		1.1		87	Mongolia	4.32		1.6
27	Turkey	5.44		1.0		88	Dominican Republic	4.32		1.4
28	United Arab Emirates	5.43		1.3		89	Macedonia, FYR	4.28		1.7
29	Malta	5.40		1.3		90	Uruguay	4.25		1.4
30	Hungary	5.39		1.1		91	Lesotho	4.25		1.6
31	Estonia	5.39		1.2		92	Bolivia	4.25		1.6
32	Cyprus	5.36		1.2		93	Serbia and Montenegro	4.25		1.6
33	Spain	5.34		1.1		94	Bosnia and Herzegovina	4.25		1.8
34	China	5.33		1.2		95	Algeria	4.24		1.6
35	Switzerland	5.28		1.4		96	Poland	4.20		1.2
36	Korea, Rep.	5.25		1.3		97	Nicaragua	4.16		1.5
37	Lithuania	5.22		1.2		98	Bulgaria	4.15		1.7
38	South Africa	5.20		1.1		99	Benin	4.13		1.4
39	El Salvador	5.20		1.4		100	Argentina	4.12		1.3
40	Brazil	5.18		1.2		101	Georgia	4.11		1.3
41	Jordan	5.18		1.2		102	Cameroon	4.11		1.5
42	Thailand	5.17		0.9		103	Nigeria	4.10		1.8
43	Tunisia	5.17		0.9		104	Guyana	4.06		1.7
44	Jamaica	5.15		1.1		105	Azerbaijan	4.04		1.6
45	Portugal	5.14		1.1		106	Paraguay	3.99		1.5
46	Slovenia	5.13		1.2		107	Armenia	3.96		1.7
47	Kenya	5.10		1.5		108	Mozambique	3.93		1.6
48	Costa Rica	5.06		1.4		109	Venezuela	3.92		1.5
49	Philippines	5.04		1.1		110	Ethiopia	3.92		1.7
50	Colombia	5.03		1.3		111	Madagascar	3.89		1.4
51	Uganda	5.00		1.5		112	Mali	3.84		1.6
52	Croatia	4.99		1.3		113	Malawi	3.81		1.6
53	Slovak Republic	4.98		1.1		114	Honduras	3.63		1.6
54	Bahrain	4.97		1.4		115	Albania	3.53		1.6
55	Latvia	4.96		1.3		116	Burundi	3.45		1.6
56	Guatemala	4.94		1.2		117	Mauritania	3.43		1.6
57	Mexico	4.93		1.2		118	Kyrgyz Republic	3.38		1.7
58	Peru	4.91		1.3		119	Chad	3.14		1.6
59	Romania	4.89		1.4		120	Zimbabwe	3.08		1.5
60	Namibia	4.89		1.4		121	Angola	2.79		1.2
61	Panama	4.86		1.3		122	Zambia	2.31		1.6

SOURCE: World Economic Forum, Executive Opinion Survey 2006

1.12 Freedom of the press

In your country, can the media publish/broadcast stories of their choosing without fear of censorship or retaliation? (1 = no, 7 = yes—whatever they want)

RANK	COUNTRY/ECONOMY	SCORE	1 MEAN: 4.93 7	SD
1	Germany	6.84		0.4
2	Denmark	6.83		0.5
3	Netherlands	6.77		0.6
4	Norway	6.67		0.7
5	Sweden	6.65		0.8
6	Switzerland	6.62		0.8
7	Finland	6.60		0.8
8	Austria	6.56		0.8
9	Portugal	6.47		0.7
10	Canada	6.40		1.1
11	Estonia	6.39		1.1
12	India	6.38		0.8
13	New Zealand	6.37		1.3
14	Belgium	6.35		1.1
15	United Kingdom	6.35		1.3
16	Peru	6.33		0.9
17	France	6.30		1.1
18	Nicaragua	6.24		1.1
19	Australia	6.20		1.1
20	Iceland	6.13		1.4
21	Israel	6.06		1.2
22	Greece	6.05		1.2
23	Slovak Republic	6.03		1.2
24	South Africa	6.03		1.1
25	Costa Rica	6.02		1.2
26	United States	5.96		1.3
27	Japan	5.88		1.3
28	Hong Kong SAR	5.87		1.6
29	Czech Republic	5.86		1.3
30	Spain	5.84		1.4
31	Chile	5.83		1.1
32	Lithuania	5.81		1.3
33	Luxembourg	5.81		1.6
34	Guatemala	5.81		1.5
35	Brazil	5.78		1.5
36	Ireland	5.66		1.7
37	Uruguay	5.64		1.5
38	Mexico	5.62		1.4
39	El Salvador	5.62		1.6
40	Hungary	5.58		1.5
41	Colombia	5.52		1.4
42	Paraguay	5.48		1.6
43	Latvia	5.47		1.6
44	Italy	5.42		1.6
45	Indonesia	5.40		0.8
46	Cyprus	5.39		1.5
47	Honduras	5.39		1.7
48	Dominican Republic	5.37		1.5
49	Bolivia	5.34		1.7
50	Barbados	5.32		1.4
51	Suriname	5.32		1.6
52	Taiwan, China	5.31		1.4
53	Malta	5.30		1.6
54	Ecuador	5.28		1.6
55	Mauritius	5.22		1.6
56	Panama	5.18		1.7
57	Benin	5.17		2.0
58	Albania	5.16		1.6
59	Jamaica	5.10		1.9
60	Trinidad and Tobago	5.09		1.9
61	Namibia	5.08		1.7
62	Turkey	5.08		1.6
63	Croatia	5.08		1.8
64	Romania	5.06		1.6
65	Korea, Rep.	5.00		1.5
66	Mali	4.98		2.0
67	Slovenia	4.89		1.8
68	Serbia and Montenegro	4.87		2.1
69	Bangladesh	4.84		1.7
70	Mauritania	4.82		2.3
71	Ukraine	4.74		1.8
72	Philippines	4.73		1.8
73	Botswana	4.73		1.7
74	Bosnia and Herzegovina	4.73		1.9
75	Malawi	4.71		1.7
76	Macedonia, FYR	4.62		1.9
77	Tanzania	4.60		1.9
78	Georgia	4.57		1.7
79	Poland	4.54		1.2
80	Lesotho	4.51		1.9
81	Cameroon	4.47		1.9
82	Azerbaijan	4.44		2.0
83	Bulgaria	4.42		1.9
84	Pakistan	4.30		1.7
85	Sri Lanka	4.28		1.8
86	Thailand	4.24		1.7
87	Mongolia	4.20		2.1
87	Mozambique	4.20		1.8
89	Algeria	4.18		1.8
90	Qatar	4.15		1.9
91	Kyrgyz Republic	4.15		1.8
92	Nigeria	4.12		2.1
93	Kuwait	4.11		2.1
94	Madagascar	4.11		1.9
95	Russian Federation	3.99		1.9
96	Guyana	3.95		2.0
97	Nepal	3.93		2.0
98	Moldova	3.92		2.0
99	Egypt	3.91		2.0
100	Burundi	3.88		2.1
101	Burkina Faso	3.79		2.1
102	United Arab Emirates	3.77		1.8
103	Tunisia	3.74		1.4
104	Uganda	3.72		2.0
105	Argentina	3.71		2.1
106	Armenia	3.65		2.0
107	Kazakhstan	3.62		2.0
108	Angola	3.61		1.9
109	Bahrain	3.59		2.0
110	Vietnam	3.55		1.8
111	Morocco	3.52		1.9
112	Malaysia	3.49		1.7
113	Cambodia	3.42		1.8
114	Kenya	3.24		1.9
115	Jordan	3.20		1.7
116	Singapore	3.01		1.6
117	China	2.81		1.8
118	Zambia	2.54		1.6
119	Chad	2.51		2.1
120	Venezuela	2.30		1.5
121	Ethiopia	2.13		1.6
122	Zimbabwe	1.56		1.1

SOURCE: World Economic Forum, Executive Opinion Survey 2006

Pillar 2
Political and regulatory environment

2.01 Effectiveness of law-making bodies

How effective is your national parliament/congress as a law-making and oversight institution? (1 = very ineffective, 7 = very effective—the best in the world)

RANK	COUNTRY/ECONOMY	SCORE	MEAN: 3.42	SD
1	Singapore	5.92		0.9
2	United Kingdom	5.54		1.3
3	Denmark	5.43		1.0
4	Australia	5.36		1.0
5	Finland	5.24		1.2
6	Iceland	5.21		0.9
7	Malaysia	5.11		1.1
8	Norway	5.01		1.1
9	Luxembourg	4.98		1.2
10	New Zealand	4.98		1.4
11	Switzerland	4.95		1.2
12	Germany	4.92		1.3
13	Netherlands	4.85		1.2
14	Canada	4.79		1.3
15	Barbados	4.73		1.2
16	Tanzania	4.64		1.5
17	South Africa	4.63		1.2
18	Tunisia	4.56		1.5
19	Sweden	4.53		1.3
20	India	4.52		1.3
21	United States	4.47		1.6
22	Ireland	4.46		1.3
23	Japan	4.40		1.2
24	Mauritius	4.33		1.3
25	Austria	4.27		1.3
26	Estonia	4.23		1.3
27	Malta	4.22		1.1
28	United Arab Emirates	4.21		1.7
29	France	4.21		1.4
30	Israel	4.21		1.3
31	Botswana	4.15		1.5
32	Qatar	4.14		1.4
33	Spain	3.99		1.4
34	Mali	3.98		1.5
35	Hong Kong SAR	3.97		1.5
36	Portugal	3.97		1.2
37	Kuwait	3.95		1.7
38	Cyprus	3.92		1.3
39	Turkey	3.91		1.3
40	Thailand	3.87		1.2
41	Greece	3.86		1.4
42	Chile	3.85		1.3
43	China	3.84		1.6
44	Kazakhstan	3.72		1.4
45	Burkina Faso	3.70		1.4
46	Jamaica	3.68		1.3
47	Belgium	3.66		1.3
48	Benin	3.63		1.6
49	Taiwan, China	3.60		1.3
50	Slovenia	3.57		1.3
51	Namibia	3.56		1.4
52	Slovak Republic	3.52		1.3
53	Latvia	3.52		1.3
54	Jordan	3.49		1.4
55	Vietnam	3.48		1.5
56	Hungary	3.45		1.4
57	Croatia	3.45		1.5
58	Cambodia	3.41		1.7
59	Pakistan	3.37		1.4
60	Uganda	3.34		1.4
61	Serbia and Montenegro	3.33		1.5
62	Morocco	3.32		1.5
63	Nigeria	3.27		1.7
64	Egypt	3.26		1.7
65	Czech Republic	3.25		1.2
65	Korea, Rep.	3.25		1.4
67	Albania	3.24		1.3
68	Lithuania	3.22		1.3
69	Algeria	3.17		1.5
70	Lesotho	3.17		1.4
71	Moldova	3.16		1.3
72	Azerbaijan	3.13		1.5
73	Kyrgyz Republic	3.05		1.3
74	Colombia	3.03		1.1
75	Sri Lanka	3.03		1.4
76	Macedonia, FYR	3.00		1.4
77	Madagascar	2.98		1.3
78	Honduras	2.95		1.3
79	Indonesia	2.94		0.7
80	Italy	2.94		1.4
81	Malawi	2.89		1.4
82	Trinidad and Tobago	2.88		1.5
83	Georgia	2.88		1.4
83	Uruguay	2.88		1.1
85	Russian Federation	2.87		1.3
86	Armenia	2.85		1.2
87	Kenya	2.84		1.4
88	Angola	2.82		1.3
89	Mozambique	2.78		1.3
90	Bahrain	2.76		1.2
91	Guyana	2.71		1.5
92	Burundi	2.69		1.3
93	Poland	2.61		1.1
94	Ukraine	2.59		1.4
95	Bangladesh	2.58		1.4
96	Mongolia	2.55		1.3
97	Nepal	2.54		1.4
98	Bulgaria	2.52		1.1
99	Romania	2.46		1.4
100	Philippines	2.43		1.4
101	Mexico	2.43		1.1
102	Ethiopia	2.41		1.5
103	Bosnia and Herzegovina	2.39		1.2
104	Panama	2.38		1.2
105	El Salvador	2.37		1.0
106	Dominican Republic	2.24		1.1
107	Mauritania	2.24		1.6
108	Costa Rica	2.16		1.5
109	Brazil	2.15		1.2
110	Zimbabwe	2.14		1.5
111	Cameroon	2.13		1.2
112	Guatemala	2.12		1.1
113	Zambia	2.08		1.3
114	Suriname	2.05		1.1
115	Bolivia	2.02		0.9
116	Argentina	1.94		0.9
117	Chad	1.92		1.2
118	Paraguay	1.82		1.0
119	Peru	1.76		0.8
120	Nicaragua	1.71		0.9
121	Venezuela	1.68		1.3
122	Ecuador	1.49		0.8

SOURCE: World Economic Forum, Executive Opinion Survey 2006

Laws relating to the use of information and communication technologies (ICT) (electronic commerce, digital signatures, consumer protection) are (1 = nonexistent, 7 = well developed and enforced)

RANK	COUNTRY/ECONOMY	SCORE	SD
1	Estonia	5.73	1.1
2	Singapore	5.71	0.8
3	Germany	5.69	1.1
4	Norway	5.62	1.0
5	United Kingdom	5.61	1.2
6	Denmark	5.58	1.1
7	Austria	5.45	1.1
8	Switzerland	5.42	1.0
9	Netherlands	5.40	0.9
10	Finland	5.40	1.0
11	Korea, Rep.	5.36	1.1
12	Malaysia	5.33	1.0
13	France	5.33	1.1
14	Sweden	5.29	1.0
15	Australia	5.27	1.0
16	Canada	5.25	1.0
17	Iceland	5.24	0.9
18	New Zealand	5.20	1.3
19	Hong Kong SAR	5.14	1.2
20	United States	5.10	1.3
21	Luxembourg	5.07	1.2
22	Israel	5.00	1.2
23	Malta	4.98	1.0
24	Chile	4.91	1.2
25	Japan	4.90	0.9
26	Taiwan, China	4.80	1.2
27	Ireland	4.78	1.2
28	South Africa	4.75	1.1
29	Slovenia	4.72	1.1
30	Belgium	4.68	1.1
31	India	4.64	1.2
31	Portugal	4.64	0.8
33	Spain	4.59	1.2
34	United Arab Emirates	4.43	1.3
35	Mauritius	4.30	1.5
36	Bulgaria	4.29	1.5
37	Thailand	4.27	1.0
38	Italy	4.25	1.4
39	Qatar	4.25	1.4
40	Hungary	4.24	1.4
41	Croatia	4.17	1.3
42	Mexico	4.16	1.3
43	Lithuania	4.15	1.4
44	Czech Republic	4.15	1.2
45	Slovak Republic	4.14	1.2
46	Colombia	4.10	1.4
47	Costa Rica	4.08	1.2
48	Brazil	4.06	1.3
49	Tunisia	4.04	1.6
50	Barbados	3.98	1.3
51	Bahrain	3.97	1.6
52	Turkey	3.95	1.3
53	Philippines	3.94	1.4
54	Greece	3.82	1.4
55	Panama	3.81	1.3
56	Dominican Republic	3.81	1.5
57	Romania	3.79	1.5
58	Azerbaijan	3.79	1.8
59	Kazakhstan	3.76	1.4
60	El Salvador	3.71	1.5
61	Cyprus	3.71	1.5

RANK	COUNTRY/ECONOMY	SCORE	SD
62	Poland	3.71	1.0
63	Latvia	3.68	1.5
64	Jordan	3.67	1.4
65	Pakistan	3.64	1.7
66	Peru	3.58	1.1
67	China	3.49	1.4
68	Jamaica	3.47	1.4
69	Nigeria	3.42	1.7
70	Serbia and Montenegro	3.38	1.5
71	Sri Lanka	3.31	1.5
72	Uruguay	3.30	1.4
73	Tanzania	3.27	1.5
74	Morocco	3.24	1.7
75	Namibia	3.20	1.4
76	Kenya	3.19	1.6
77	Argentina	3.18	1.3
78	Indonesia	3.18	0.9
79	Moldova	3.12	1.4
80	Egypt	3.10	1.6
81	Guatemala	3.10	1.3
82	Nicaragua	3.07	1.5
83	Venezuela	3.05	1.4
84	Botswana	3.05	1.4
85	Armenia	3.03	1.5
86	Russian Federation	3.00	1.3
87	Ecuador	3.00	1.3
88	Honduras	2.99	1.3
89	Kuwait	2.97	1.4
90	Vietnam	2.96	1.5
91	Ukraine	2.89	1.4
92	Uganda	2.88	1.7
93	Algeria	2.87	1.7
94	Macedonia, FYR	2.87	1.4
95	Madagascar	2.81	1.5
96	Benin	2.79	1.5
97	Trinidad and Tobago	2.73	1.4
98	Zimbabwe	2.73	1.1
99	Mongolia	2.62	1.5
100	Burkina Faso	2.60	1.6
101	Zambia	2.52	0.9
102	Malawi	2.51	1.5
103	Mali	2.48	1.6
104	Nepal	2.46	1.3
105	Cameroon	2.46	1.7
106	Lesotho	2.45	1.4
107	Bolivia	2.40	1.2
108	Ethiopia	2.37	1.4
109	Bangladesh	2.34	1.3
110	Cambodia	2.32	1.5
111	Bosnia and Herzegovina	2.31	1.1
112	Guyana	2.26	1.3
113	Mauritania	2.25	1.7
114	Mozambique	2.25	1.4
115	Paraguay	2.24	1.3
116	Georgia	2.24	1.0
117	Angola	2.22	1.6
118	Albania	2.09	1.2
118	Kyrgyz Republic	2.09	1.2
120	Burundi	2.01	1.3
121	Chad	2.01	1.4
122	Suriname	1.47	0.8

MEAN: 3.73

283

SOURCE: World Economic Forum, Executive Opinion Survey 2006

2.03 Judicial independence

Is the judiciary in your country independent from political influences of members of government, citizens, or firms? (1 = no, heavily influenced, 7 = yes, entirely independent)

RANK	COUNTRY/ECONOMY	SCORE	1 MEAN: 3.93 7	SD
1	Germany	6.51		0.9
2	Netherlands	6.43		0.9
3	Israel	6.32		0.8
4	New Zealand	6.32		1.1
5	Norway	6.31		1.0
6	Australia	6.28		1.0
7	Denmark	6.27		1.4
8	United Kingdom	6.24		1.1
9	Switzerland	6.18		1.2
10	Finland	6.16		1.3
11	Ireland	6.14		1.3
12	Iceland	6.03		1.4
13	Hong Kong SAR	5.90		1.5
14	India	5.90		1.3
15	Austria	5.89		1.4
16	Barbados	5.87		1.2
17	Sweden	5.87		1.6
18	Canada	5.75		1.6
19	Portugal	5.66		1.2
20	Qatar	5.62		1.4
21	South Africa	5.58		1.3
22	Japan	5.56		1.5
23	Luxembourg	5.46		1.5
24	Malaysia	5.34		1.2
25	Botswana	5.33		1.4
26	Malta	5.31		1.8
27	Estonia	5.27		1.6
28	Namibia	5.24		1.5
29	Singapore	5.21		1.6
30	Cyprus	5.20		1.7
31	Kuwait	5.19		1.8
32	Belgium	5.19		1.6
33	Costa Rica	5.12		1.6
34	Tunisia	5.07		1.4
35	France	5.06		1.6
36	United States	4.96		1.6
37	Uruguay	4.93		1.6
38	Jordan	4.92		1.7
39	Egypt	4.82		1.9
40	United Arab Emirates	4.78		1.7
41	Suriname	4.67		1.7
42	Mauritius	4.67		1.6
43	Malawi	4.54		1.6
44	Slovenia	4.52		1.6
45	Thailand	4.40		1.7
46	Greece	4.38		1.6
47	Jamaica	4.24		1.8
48	Hungary	4.23		1.8
49	Trinidad and Tobago	4.19		1.8
50	Turkey	4.18		1.6
51	Korea, Rep.	4.13		1.6
52	Nepal	4.10		1.7
53	Taiwan, China	4.03		1.5
54	Lesotho	4.00		1.8
55	Tanzania	3.95		1.7
56	Chile	3.86		1.6
57	Czech Republic	3.84		1.5
58	Mauritania	3.83		1.7
59	Latvia	3.79		1.7
60	Uganda	3.78		1.8
61	Benin	3.76		1.9
62	Algeria	3.74		1.7
63	Spain	3.70		1.8
64	Mali	3.66		2.0
65	Colombia	3.65		1.7
66	Mexico	3.65		1.6
67	Italy	3.58		1.7
68	Slovak Republic	3.58		1.6
69	Poland	3.51		1.1
70	Dominican Republic	3.50		1.5
71	Sri Lanka	3.49		1.6
72	Vietnam	3.48		1.8
73	Bahrain	3.44		2.0
74	Morocco	3.44		1.9
75	Lithuania	3.38		1.5
76	Philippines	3.36		1.5
77	China	3.35		1.8
78	Guatemala	3.28		1.6
79	Pakistan	3.28		1.5
80	Croatia	3.21		1.6
81	Nigeria	3.15		1.9
82	Bosnia and Herzegovina	3.12		1.4
83	Burkina Faso	3.02		1.6
84	Kenya	2.99		1.7
85	Madagascar	2.95		1.7
86	Romania	2.88		1.6
87	Angola	2.85		1.4
88	El Salvador	2.84		1.6
89	Brazil	2.82		1.5
90	Guyana	2.81		1.7
91	Indonesia	2.80		0.8
92	Azerbaijan	2.79		1.6
93	Kazakhstan	2.73		1.6
94	Zambia	2.72		1.4
95	Mongolia	2.57		1.5
96	Ukraine	2.53		1.7
97	Bulgaria	2.51		1.3
98	Bolivia	2.49		1.4
99	Bangladesh	2.49		1.4
100	Mozambique	2.47		1.2
101	Serbia and Montenegro	2.47		1.5
102	Honduras	2.46		1.3
103	Panama	2.44		1.6
104	Albania	2.43		1.3
105	Macedonia, FYR	2.42		1.3
106	Cambodia	2.31		1.5
107	Russian Federation	2.28		1.4
108	Armenia	2.27		1.4
109	Cameroon	2.24		1.4
110	Moldova	2.19		1.2
111	Ethiopia	2.18		1.5
112	Argentina	2.18		1.3
113	Ecuador	2.15		1.3
114	Georgia	2.13		1.2
115	Kyrgyz Republic	1.96		1.3
116	Peru	1.95		1.2
117	Burundi	1.84		1.1
118	Chad	1.77		1.1
119	Zimbabwe	1.66		0.9
120	Paraguay	1.65		1.3
121	Nicaragua	1.49		1.1
122	Venezuela	1.18		0.7

284

SOURCE: World Economic Forum, Executive Opinion Survey 2006

Intellectual property protection in your country is (1 = weak or nonexistent, 7 = equal to the world's most stringent)

RANK	COUNTRY/ECONOMY	SCORE	1 MEAN: 3.81 7	SD
1	Germany	6.59		0.7
2	Finland	6.41		0.8
3	Switzerland	6.38		0.9
4	Denmark	6.31		0.8
5	Netherlands	6.28		0.9
6	United Kingdom	6.20		0.9
7	Iceland	6.11		1.0
8	Sweden	6.08		1.0
9	Singapore	6.00		0.9
10	Australia	5.92		1.0
11	France	5.90		1.0
12	Japan	5.88		1.1
13	New Zealand	5.78		1.0
14	Norway	5.75		1.0
15	Luxembourg	5.68		1.1
16	Canada	5.67		1.0
17	United States	5.65		1.5
18	Austria	5.61		1.1
19	Belgium	5.58		1.3
20	Hong Kong SAR	5.49		1.4
21	Israel	5.48		1.3
22	Ireland	5.31		1.4
23	Malaysia	5.12		1.2
24	Portugal	5.08		1.2
25	South Africa	5.08		1.3
26	Taiwan, China	4.85		1.2
27	Qatar	4.84		1.4
28	United Arab Emirates	4.80		1.5
29	Spain	4.78		1.4
30	Tunisia	4.62		1.4
31	Korea, Rep.	4.59		1.3
32	Estonia	4.58		1.4
33	Hungary	4.57		1.4
34	India	4.47		1.5
35	Slovenia	4.47		1.5
36	Barbados	4.46		1.4
37	Cyprus	4.39		1.6
38	Greece	4.32		1.4
39	Mauritius	4.26		1.5
40	Namibia	4.25		1.5
41	Thailand	4.24		1.3
42	Jordan	4.21		1.5
43	Malta	4.20		1.5
44	Italy	4.17		1.5
45	Chile	4.14		1.3
46	Bahrain	4.08		1.6
47	Slovak Republic	4.00		1.4
48	Costa Rica	3.98		1.2
49	Panama	3.93		1.3
50	Burkina Faso	3.91		1.6
51	Uruguay	3.90		1.3
52	Czech Republic	3.90		1.3
53	Morocco	3.82		1.7
54	Mexico	3.76		1.5
55	Colombia	3.75		1.4
56	Croatia	3.66		1.4
57	Poland	3.62		0.9
58	Kuwait	3.62		1.7
59	Jamaica	3.61		1.4
60	El Salvador	3.56		1.4
61	Indonesia	3.55		1.0

RANK	COUNTRY/ECONOMY	SCORE	1 MEAN: 3.81 7	SD
62	Egypt	3.55		1.7
63	Brazil	3.51		1.5
64	Latvia	3.44		1.4
65	Sri Lanka	3.40		1.7
66	Mali	3.40		1.4
67	Cameroon	3.38		1.7
68	Dominican Republic	3.37		1.3
69	Madagascar	3.31		1.5
70	Lithuania	3.28		1.2
71	Turkey	3.28		1.4
72	Algeria	3.28		1.6
72	Guatemala	3.28		1.2
74	China	3.26		1.4
75	Trinidad and Tobago	3.26		1.6
76	Botswana	3.25		1.5
77	Mauritania	3.25		1.4
78	Tanzania	3.23		1.4
79	Pakistan	3.20		1.5
80	Romania	3.13		1.2
81	Kazakhstan	3.11		1.4
82	Benin	3.07		1.6
83	Philippines	3.02		1.1
84	Kenya	3.02		1.4
85	Zimbabwe	2.94		1.5
86	Argentina	2.92		1.2
87	Nigeria	2.92		1.6
88	Azerbaijan	2.90		1.4
89	Moldova	2.87		1.3
90	Nicaragua	2.81		1.3
91	Georgia	2.81		1.3
92	Armenia	2.77		1.4
93	Honduras	2.77		1.3
94	Peru	2.74		1.3
95	Malawi	2.72		1.3
96	Bulgaria	2.71		1.3
97	Ukraine	2.70		1.3
98	Vietnam	2.66		1.2
99	Ecuador	2.65		1.2
100	Kyrgyz Republic	2.57		1.5
101	Angola	2.56		1.2
102	Mozambique	2.52		1.1
103	Macedonia, FYR	2.51		1.2
104	Uganda	2.51		1.4
105	Cambodia	2.50		1.4
106	Lesotho	2.49		1.3
107	Ethiopia	2.48		1.1
108	Mongolia	2.47		1.2
109	Bosnia and Herzegovina	2.44		1.1
110	Russian Federation	2.42		1.2
111	Zambia	2.41		0.9
112	Serbia and Montenegro	2.31		1.2
113	Nepal	2.29		1.3
114	Paraguay	2.28		1.1
115	Venezuela	2.15		1.0
116	Bangladesh	2.09		1.0
117	Chad	2.08		1.3
118	Albania	2.01		1.0
119	Bolivia	1.97		0.9
120	Burundi	1.95		1.2
121	Suriname	1.89		1.0
122	Guyana	1.83		1.1

285

SOURCE: World Economic Forum, Executive Opinion Survey 2006

2.05 Efficiency of legal framework

The legal framework in your country for private businesses to settle disputes and challenge the legality of government actions and/or regulations
(1 = is inefficient and subject to manipulation, 7 = is efficient and follows a clear, neutral process)

RANK	COUNTRY/ECONOMY	SCORE	SD
1	Denmark	6.55	0.7
2	Germany	6.37	0.7
3	Iceland	6.37	0.7
4	Netherlands	6.34	0.8
5	Norway	6.29	0.7
6	Sweden	6.23	1.1
7	Switzerland	6.17	1.2
8	Finland	6.14	1.1
9	Austria	6.14	1.0
10	United Kingdom	6.03	1.1
11	Australia	5.99	1.0
12	Hong Kong SAR	5.99	1.3
13	New Zealand	5.89	1.1
14	Singapore	5.78	1.2
15	South Africa	5.57	1.3
16	Luxembourg	5.46	1.4
17	Canada	5.45	1.6
18	Malaysia	5.41	1.1
19	Japan	5.38	1.4
20	Barbados	5.29	1.2
21	Israel	5.28	1.2
22	Ireland	5.26	1.6
23	India	5.12	1.3
24	Estonia	5.10	1.5
25	United States	5.08	1.5
26	Qatar	5.08	1.2
27	France	5.07	1.5
28	Cyprus	5.06	1.4
29	Kuwait	5.06	1.7
30	Tunisia	4.96	1.3
31	Botswana	4.94	1.3
32	Costa Rica	4.91	1.3
33	Mauritius	4.85	1.7
34	United Arab Emirates	4.81	1.6
35	Belgium	4.79	1.4
36	Jordan	4.79	1.6
37	Chile	4.59	1.4
38	Namibia	4.56	1.7
39	Malta	4.56	1.5
40	Thailand	4.52	1.3
41	Taiwan, China	4.42	1.3
42	Uruguay	4.28	1.5
43	Greece	4.27	1.5
44	Hungary	4.21	1.5
45	Portugal	4.20	1.6
46	Slovenia	4.16	1.5
47	Korea, Rep.	4.16	1.4
48	Spain	4.15	1.7
49	Suriname	4.11	1.6
50	Trinidad and Tobago	4.10	1.8
51	Mauritania	4.07	1.6
52	Algeria	4.06	1.7
53	Egypt	4.05	1.9
54	Morocco	4.03	1.8
55	Zambia	4.00	1.7
56	Turkey	3.81	1.5
57	Lesotho	3.81	1.7
58	Jamaica	3.80	1.6
59	Colombia	3.78	1.5
60	Vietnam	3.76	1.3
61	Latvia	3.75	1.5
62	Tanzania	3.67	1.6
63	Malawi	3.65	1.8
64	Sri Lanka	3.63	1.8
65	Poland	3.62	1.1
66	Slovak Republic	3.62	1.3
67	Lithuania	3.58	1.3
68	Kazakhstan	3.48	1.5
69	Bahrain	3.47	1.8
70	Uganda	3.47	1.8
71	Czech Republic	3.47	1.3
72	Benin	3.44	1.5
73	Croatia	3.42	1.5
74	China	3.42	1.6
75	Indonesia	3.35	1.0
76	Mali	3.34	1.4
77	Mexico	3.31	1.5
78	Guatemala	3.29	1.3
79	Dominican Republic	3.25	1.4
80	Nigeria	3.25	1.7
81	Madagascar	3.24	1.5
82	Burkina Faso	3.20	1.5
83	Italy	3.20	1.4
84	Philippines	3.17	1.5
85	Nepal	3.14	1.4
86	Romania	3.07	1.4
87	Brazil	3.05	1.6
88	Azerbaijan	3.04	1.5
89	Pakistan	3.03	1.4
90	El Salvador	3.02	1.4
91	Kenya	3.00	1.5
92	Cambodia	2.97	1.4
93	Bosnia and Herzegovina	2.96	1.4
94	Macedonia, FYR	2.91	1.5
95	Panama	2.86	1.4
96	Ukraine	2.83	1.4
97	Honduras	2.81	1.3
98	Armenia	2.78	1.4
99	Bangladesh	2.78	1.4
100	Serbia and Montenegro	2.76	1.4
101	Angola	2.75	1.2
102	Cameroon	2.69	1.5
103	Russian Federation	2.67	1.3
104	Mozambique	2.61	1.3
105	Ethiopia	2.61	1.4
106	Albania	2.59	1.3
107	Argentina	2.57	1.2
108	Moldova	2.57	1.4
109	Peru	2.55	1.3
110	Bulgaria	2.53	1.2
111	Burundi	2.52	1.4
112	Mongolia	2.51	1.3
113	Bolivia	2.49	1.2
114	Kyrgyz Republic	2.46	1.4
115	Guyana	2.44	1.5
116	Georgia	2.43	1.3
117	Ecuador	2.30	1.2
118	Zimbabwe	2.29	1.3
119	Chad	2.18	1.2
120	Nicaragua	2.15	1.3
121	Paraguay	2.06	1.1
122	Venezuela	1.56	1.0

MEAN: 3.95

SOURCE: World Economic Forum, Executive Opinion Survey 2006

2.06   Property rights

Property rights, including over financial assets, are (1 = poorly defined and not protected by law, 7 = clearly defined and well protected by law)

RANK	COUNTRY/ECONOMY	SCORE	MEAN: 4.63	SD		RANK	COUNTRY/ECONOMY	SCORE	MEAN: 4.63	SD
1	Germany	6.76		0.5		62	Brazil	4.61		1.5
2	Iceland	6.67		0.5		63	Sri Lanka	4.61		1.6
3	Denmark	6.60		0.6		64	Czech Republic	4.60		1.4
4	Switzerland	6.57		0.7		65	Trinidad and Tobago	4.59		1.6
5	Netherlands	6.53		0.7		66	Algeria	4.48		2.0
6	United Kingdom	6.46		0.9		67	El Salvador	4.47		1.7
7	Austria	6.45		0.9		68	Burkina Faso	4.44		1.4
8	Finland	6.42		0.9		69	Vietnam	4.43		1.5
9	Ireland	6.40		0.8		70	Philippines	4.42		1.5
10	Australia	6.38		0.8		71	Armenia	4.32		1.7
11	Singapore	6.34		0.8		72	Guatemala	4.18		1.2
12	Luxembourg	6.30		1.0		73	Malawi	4.17		1.6
13	Norway	6.30		0.7		74	Dominican Republic	4.16		1.3
14	Hong Kong SAR	6.24		1.1		75	Romania	4.07		1.5
15	Japan	6.24		1.0		76	Mali	4.07		1.7
16	Sweden	6.22		1.1		77	Nepal	4.06		1.8
17	New Zealand	6.20		1.1		78	Mongolia	4.00		1.6
18	France	6.08		1.1		79	Croatia	3.99		1.5
19	Belgium	5.92		1.3		80	China	3.97		1.5
20	Canada	5.89		1.4		81	Kenya	3.94		1.6
21	United States	5.83		1.3		82	Tanzania	3.94		1.5
22	South Africa	5.78		1.0		83	Kazakhstan	3.93		1.5
23	Israel	5.77		1.0		84	Mozambique	3.89		1.6
24	Malaysia	5.76		0.9		85	Bangladesh	3.82		1.7
25	India	5.73		1.3		86	Nigeria	3.79		1.8
26	Spain	5.61		1.3		87	Serbia and Montenegro	3.77		1.7
27	Estonia	5.61		1.2		88	Indonesia	3.75		0.9
28	Mauritius	5.58		1.3		89	Bulgaria	3.75		1.5
29	Cyprus	5.52		1.4		90	Poland	3.75		1.0
30	Chile	5.48		1.3		91	Mauritania	3.74		1.5
31	Namibia	5.48		1.3		92	Suriname	3.73		1.8
32	Hungary	5.46		1.2		93	Pakistan	3.70		1.4
33	Portugal	5.46		1.4		94	Cambodia	3.65		1.8
34	Korea, Rep.	5.44		1.3		95	Benin	3.64		1.6
35	Tunisia	5.42		1.3		96	Azerbaijan	3.63		1.5
36	Barbados	5.41		1.1		97	Ethiopia	3.60		1.6
37	Greece	5.35		1.3		98	Madagascar	3.59		1.4
38	Qatar	5.33		1.5		99	Uganda	3.57		1.8
39	Taiwan, China	5.29		1.2		100	Georgia	3.54		1.4
40	Thailand	5.27		1.0		101	Moldova	3.52		1.5
41	Malta	5.15		1.5		102	Honduras	3.49		1.3
42	United Arab Emirates	5.09		1.7		103	Cameroon	3.48		1.9
43	Italy	5.08		1.5		104	Peru	3.48		1.3
44	Panama	5.02		1.4		105	Macedonia, FYR	3.47		1.7
45	Slovak Republic	5.00		1.2		106	Angola	3.47		1.5
46	Jordan	4.98		1.5		107	Nicaragua	3.43		1.6
47	Bahrain	4.97		1.7		108	Ecuador	3.43		1.3
48	Jamaica	4.96		1.4		109	Lesotho	3.37		1.8
49	Lithuania	4.91		1.4		110	Guyana	3.37		1.7
50	Slovenia	4.89		1.6		111	Ukraine	3.37		1.4
51	Latvia	4.88		1.7		112	Russian Federation	3.18		1.5
52	Botswana	4.86		1.4		113	Bolivia	3.09		1.5
53	Turkey	4.81		1.5		114	Albania	3.01		1.4
54	Kuwait	4.79		1.8		115	Paraguay	3.01		1.3
55	Uruguay	4.76		1.4		116	Kyrgyz Republic	3.00		1.4
56	Morocco	4.76		1.9		117	Burundi	2.99		1.8
57	Egypt	4.73		1.8		118	Bosnia and Herzegovina	2.96		1.2
58	Colombia	4.70		1.5		119	Argentina	2.91		1.4
59	Costa Rica	4.65		1.4		120	Chad	2.60		1.7
60	Mexico	4.63		1.5		121	Venezuela	2.36		1.3
61	Zambia	4.63		1.5		122	Zimbabwe	1.86		1.3

SOURCE: World Economic Forum, Executive Opinion Survey 2006

2.07   Quality of competition in the ISP sector

Is there sufficient competition among Internet service providers (ISPs) in your country to ensure high quality, infrequent interruptions, and low prices?
(1 = no, 7 = yes, equal to the best in the world)

RANK	COUNTRY/ECONOMY	SCORE	MEAN: 4.06	SD
1	Korea, Rep.	6.35		0.9
2	Netherlands	6.25		0.8
3	Israel	6.25		0.9
4	Germany	6.12		1.0
5	United Kingdom	5.99		1.1
6	Japan	5.90		1.1
7	Iceland	5.86		1.5
8	Hong Kong SAR	5.74		1.3
9	Finland	5.66		1.3
10	Canada	5.63		1.4
11	Austria	5.59		1.2
12	Sweden	5.58		1.4
13	Estonia	5.54		1.4
14	United States	5.54		1.6
15	Norway	5.48		1.2
16	France	5.45		1.4
17	Chile	5.43		1.2
18	Denmark	5.40		1.3
19	India	5.35		1.3
20	Taiwan, China	5.31		1.2
21	Jordan	5.25		1.2
22	Guatemala	5.22		1.3
23	Brazil	5.14		1.5
24	Switzerland	5.11		1.7
25	Belgium	5.11		1.7
26	Indonesia	5.11		1.2
27	Australia	5.10		1.4
28	Thailand	5.07		1.3
29	Singapore	5.00		1.4
30	Malta	4.98		1.4
31	Philippines	4.91		1.4
32	Malaysia	4.89		1.5
33	Lithuania	4.86		1.5
34	Panama	4.81		1.6
35	Egypt	4.74		1.6
36	Pakistan	4.72		1.3
37	El Salvador	4.63		1.4
38	Jamaica	4.46		1.5
39	Czech Republic	4.43		1.6
40	Cyprus	4.41		1.5
41	Italy	4.40		1.7
42	Slovenia	4.40		1.6
43	Portugal	4.37		1.4
44	Latvia	4.34		1.5
45	Georgia	4.33		1.7
46	Argentina	4.31		1.5
47	Dominican Republic	4.30		1.8
48	Nepal	4.27		1.6
49	Luxembourg	4.23		1.5
50	Kuwait	4.21		1.7
51	Spain	4.19		1.7
52	Slovak Republic	4.19		1.5
53	Sri Lanka	4.19		1.6
54	Turkey	4.16		1.4
55	Colombia	4.12		1.4
56	Tunisia	4.09		1.3
57	Bulgaria	4.08		1.9
58	Kenya	4.02		1.6
59	Azerbaijan	4.00		1.7
60	Greece	3.97		1.6
61	Peru	3.97		1.7
62	Ireland	3.94		1.7
63	Mali	3.91		1.6
64	Bangladesh	3.89		1.6
65	Honduras	3.89		1.7
66	Barbados	3.88		1.8
67	Nigeria	3.87		1.8
68	Croatia	3.87		1.4
69	Mexico	3.87		1.6
70	China	3.81		1.4
71	Uruguay	3.81		1.7
72	Hungary	3.77		1.6
73	Venezuela	3.71		1.7
74	Russian Federation	3.64		1.9
75	Moldova	3.63		1.7
76	Uganda	3.63		1.9
77	Bolivia	3.61		1.6
78	Malawi	3.61		1.7
79	Poland	3.58		1.2
80	Tanzania	3.57		1.5
81	Cambodia	3.55		1.7
82	Ecuador	3.54		1.4
83	Kazakhstan	3.52		1.8
84	Guyana	3.45		1.8
85	Romania	3.42		1.6
86	Botswana	3.40		1.5
87	Mongolia	3.35		1.6
88	Kyrgyz Republic	3.34		2.0
89	Serbia and Montenegro	3.34		1.6
90	South Africa	3.32		1.8
91	Madagascar	3.29		1.5
92	Mozambique	3.28		1.7
93	Vietnam	3.25		1.5
94	Burkina Faso	3.24		1.8
95	Cameroon	3.22		1.8
96	Nicaragua	3.18		1.7
97	Morocco	3.17		1.9
98	Benin	3.17		1.7
99	Namibia	3.16		1.8
100	Mauritius	3.07		1.5
101	New Zealand	3.07		1.5
102	Zimbabwe	3.06		1.7
103	Albania	3.03		1.6
104	Bosnia and Herzegovina	3.00		1.6
105	Paraguay	2.97		1.6
106	Armenia	2.96		1.8
107	Algeria	2.88		1.6
108	Ukraine	2.86		1.6
109	Angola	2.72		1.8
110	Lesotho	2.71		1.6
111	Qatar	2.64		1.9
112	Macedonia, FYR	2.61		1.7
113	Burundi	2.55		1.5
114	United Arab Emirates	2.51		1.9
115	Bahrain	2.49		1.9
116	Mauritania	2.43		1.6
117	Trinidad and Tobago	2.40		1.6
118	Costa Rica	2.09		1.5
119	Zambia	2.07		1.5
120	Chad	2.05		1.4
121	Ethiopia	1.71		1.1
122	Suriname	1.69		1.1

SOURCE: World Economic Forum, Executive Opinion Survey 2006

2.08   Number of procedures to enforce a contract

Number of administrative procedures to enforce a contract, 2006

ANK	COUNTRY/ECONOMY	HARD DATA
1	Iceland	14.00
1	Norway	14.00
3	Denmark	15.00
4	Hong Kong SAR	16.00
5	Canada	17.00
5	United States	17.00
7	Ireland	18.00
7	Jamaica	18.00
9	Australia	19.00
9	Sweden	19.00
9	Uganda	19.00
9	United Kingdom	19.00
13	Japan	20.00
13	Nicaragua	20.00
13	Sri Lanka	20.00
16	Czech Republic	21.00
16	France	21.00
16	Hungary	21.00
16	Latvia	21.00
16	Tanzania	21.00
16	Tunisia	21.00
16	Zambia	21.00
23	Croatia	22.00
23	Greece	22.00
23	Netherlands	22.00
23	Switzerland	22.00
27	Austria	23.00
27	Nigeria	23.00
27	Spain	23.00
30	Armenia	24.00
30	Georgia	24.00
30	Lithuania	24.00
30	Portugal	24.00
34	Estonia	25.00
34	Kenya	25.00
34	Philippines	25.00
34	Slovenia	25.00
38	Botswana	26.00
38	South Africa	26.00
38	Thailand	26.00
41	Azerbaijan	27.00
41	Belgium	27.00
41	Finland	27.00
41	Macedonia, FYR	27.00
41	Slovak Republic	27.00
46	Mali	28.00
46	Nepal	28.00
46	New Zealand	28.00
46	Taiwan, China	28.00
46	Ukraine	28.00
51	Dominican Republic	29.00
51	Korea, Rep.	29.00
51	Madagascar	29.00
51	Mongolia	29.00
51	Singapore	29.00
51	Suriname	29.00
57	Ethiopia	30.00
57	Germany	30.00
57	Guyana	30.00
60	Cambodia	31.00
60	China	31.00

RANK	COUNTRY/ECONOMY	HARD DATA
60	Israel	31.00
60	Malaysia	31.00
60	Namibia	31.00
60	Russian Federation	31.00
66	Argentina	33.00
66	Chile	33.00
66	Zimbabwe	33.00
69	Bulgaria	34.00
69	Costa Rica	34.00
69	Indonesia	34.00
69	Malta	34.00
69	Turkey	34.00
69	United Arab Emirates	34.00
75	Peru	35.00
76	Bosnia and Herzegovina	36.00
76	Honduras	36.00
78	Colombia	37.00
78	Kazakhstan	37.00
78	Mauritius	37.00
78	Mexico	37.00
78	Moldova	37.00
78	Trinidad and Tobago	37.00
78	Vietnam	37.00
85	Mozambique	38.00
86	Albania	39.00
86	Uruguay	39.00
88	Italy	40.00
88	Malawi	40.00
88	Mauritania	40.00
91	Burkina Faso	41.00
91	Ecuador	41.00
91	El Salvador	41.00
91	Poland	41.00
91	Serbia and Montenegro	41.00
91	Venezuela	41.00
97	Brazil	42.00
97	Morocco	42.00
99	Jordan	43.00
99	Romania	43.00
101	Kyrgyz Republic	44.00
102	Panama	45.00
103	Paraguay	46.00
104	Angola	47.00
104	Bolivia	47.00
104	Burundi	47.00
107	Algeria	49.00
107	Benin	49.00
109	Bangladesh	50.00
109	Guatemala	50.00
111	Chad	52.00
111	Kuwait	52.00
113	Egypt	55.00
113	Pakistan	55.00
115	India	56.00
116	Cameroon	58.00
116	Lesotho	58.00
n/a	Bahrain	n/a
n/a	Barbados	n/a
n/a	Cyprus	n/a
n/a	Luxembourg	n/a
n/a	Qatar	n/a

289

SOURCE: World Bank, *Doing Business 2007: How to Reform* (2006)

2.09 Time to enforce a contract

Number of days to enforce a contract, 2006

RANK	COUNTRY/ECONOMY	HARD DATA
1	New Zealand	109.00
2	Singapore	120.00
3	Kyrgyz Republic	140.00
4	Lithuania	166.00
5	Russian Federation	178.00
6	Australia	181.00
7	Kazakhstan	183.00
7	Ukraine	183.00
9	Armenia	185.00
10	Denmark	190.00
11	Sweden	208.00
12	Hong Kong SAR	211.00
13	Switzerland	215.00
14	Ireland	217.00
15	Finland	228.00
16	United Kingdom	229.00
17	Korea, Rep.	230.00
18	Latvia	240.00
19	Japan	242.00
20	Azerbaijan	267.00
21	Namibia	270.00
22	Estonia	275.00
23	Norway	277.00
24	Georgia	285.00
25	China	292.00
26	Vietnam	295.00
27	Peru	300.00
27	United States	300.00
29	Moldova	310.00
30	Mongolia	314.00
31	Belgium	328.00
32	France	331.00
33	Hungary	335.00
33	Romania	335.00
35	Malawi	337.00
36	Austria	342.00
36	Jordan	342.00
38	Canada	346.00
39	Iceland	352.00
40	Kenya	360.00
41	Macedonia, FYR	385.00
42	Albania	390.00
42	Kuwait	390.00
44	Tanzania	393.00
45	Germany	394.00
46	Algeria	397.00
47	Mauritania	400.00
48	Cambodia	401.00
49	Burundi	403.00
50	Zambia	404.00
51	Netherlands	408.00
52	Zimbabwe	410.00
53	Jamaica	415.00
53	Mexico	415.00
55	Turkey	420.00
56	Thailand	425.00
57	Malta	432.00
58	Venezuela	435.00
59	Bulgaria	440.00
60	Burkina Faso	446.00
61	Malaysia	450.00

RANK	COUNTRY/ECONOMY	HARD DATA
62	Nigeria	457.00
63	Dominican Republic	460.00
64	Paraguay	478.00
65	Chile	480.00
65	Honduras	480.00
67	Tunisia	481.00
68	Uganda	484.00
69	Nicaragua	486.00
70	Portugal	495.00
71	Ecuador	498.00
72	Botswana	501.00
73	Taiwan, China	510.00
74	Spain	515.00
75	Argentina	520.00
76	Croatia	561.00
77	Slovak Republic	565.00
78	Indonesia	570.00
79	Guatemala	583.00
80	Israel	585.00
81	Nepal	590.00
81	Serbia and Montenegro	590.00
83	Bolivia	591.00
83	Madagascar	591.00
85	Bosnia and Herzegovina	595.00
86	Philippines	600.00
86	South Africa	600.00
88	United Arab Emirates	607.00
89	Costa Rica	615.00
89	Morocco	615.00
91	Brazil	616.00
92	El Salvador	626.00
93	Mauritius	630.00
94	Uruguay	655.00
95	Guyana	661.00
96	Panama	686.00
97	Ethiopia	690.00
98	Lesotho	695.00
99	Benin	720.00
100	Greece	730.00
101	Chad	743.00
102	Cameroon	800.00
103	Czech Republic	820.00
104	Sri Lanka	837.00
105	Mali	860.00
106	Pakistan	880.00
107	Poland	980.00
108	Egypt	1,010.00
108	Mozambique	1,010.00
110	Angola	1,011.00
111	Italy	1,210.00
112	Suriname	1,290.00
113	Trinidad and Tobago	1,340.00
114	Colombia	1,346.00
115	Slovenia	1,350.00
116	India	1,420.00
117	Bangladesh	1,442.00
n/a	Bahrain	n/a
n/a	Barbados	n/a
n/a	Cyprus	n/a
n/a	Luxembourg	n/a
n/a	Qatar	n/a

SOURCE: World Bank, *Doing Business 2007: How to Reform* (2006)

Pillar 3
Infrastructure environment

3.01 Telephone lines

Main telephone lines per 100 inhabitants, 2005 or most recent year available

RANK	COUNTRY/ECONOMY	HARD DATA
1	Sweden[2]	71.54
2	Switzerland	68.66
3	Germany	66.57
4	Iceland	65.94
5	Denmark	61.69
6	United States[2]	60.60
7	Taiwan, China	59.80
8	France	59.01
9	Australia	56.86
10	Greece	56.69
11	Canada	56.64
12	Hong Kong SAR	53.89
13	United Kingdom	53.29
14	Luxembourg	52.58
15	Malta	50.40
16	Cyprus	50.30
17	Barbados	50.14
18	Korea, Rep.	49.17
19	Ireland	49.01
20	Netherlands	46.63
21	Norway	46.08
22	Belgium[2]	45.96
23	Japan	45.89
24	Austria	45.32
25	Singapore	43.50
26	Italy	43.12
27	Spain	42.92
28	New Zealand	42.91
29	Israel	42.56
30	Croatia	41.51
31	Slovenia	41.50
32	Finland	40.39
33	Portugal	40.35
34	Estonia	33.26
35	Hungary	33.24
36	Serbia and Montenegro[2]	32.94
37	Bulgaria	32.14
38	Costa Rica	32.09
39	Latvia	31.69
40	Czech Republic	31.48
41	Uruguay	30.95
42	Poland	30.63
43	Mauritius	28.84
44	Russian Federation	27.94
45	United Arab Emirates	27.51
46	Bahrain	27.05
47	China	26.63
48	Qatar	26.41
49	Macedonia, FYR	26.21
50	Turkey	25.93
51	Ukraine[2]	25.84
52	Bosnia and Herzegovina	24.80
53	Trinidad and Tobago	24.77
54	Lithuania[2]	23.39
55	Brazil[2]	23.04
56	Argentina	22.80
57	Slovak Republic	22.16
58	Moldova	22.10
59	Chile	22.04
60	Romania	20.22
61	Armenia[2]	19.25

RANK	COUNTRY/ECONOMY	HARD DATA
62	Kuwait	18.99
63	Vietnam	18.81
64	Mexico	18.23
65	Suriname	18.04
66	Kazakhstan[2]	16.85
67	Colombia	16.84
68	Malaysia	16.79
69	Georgia[2]	15.12
70	Guyana	14.66
71	El Salvador	14.12
72	Egypt	14.04
73	Panama	13.63
74	Venezuela	13.48
75	Azerbaijan	12.98
76	Jamaica	12.90
77	Ecuador	12.86
78	Tunisia	12.47
79	Jordan[2]	11.36
80	Thailand	10.95
81	Dominican Republic	10.06
82	South Africa	9.97
83	Guatemala	9.91
84	Albania[2]	8.60
85	Kyrgyz Republic	8.33
86	Peru	8.05
87	Algeria	7.82
88	Botswana	7.48
89	Bolivia	7.04
90	Honduras	6.86
91	Namibia[2]	6.36
92	Sri Lanka	6.00
93	Mongolia	5.90
94	Indonesia	5.73
95	Paraguay	5.20
96	India	4.51
97	Morocco	4.42
98	Philippines	4.00
99	Nicaragua[2]	3.77
100	Pakistan	3.43
101	Zimbabwe	2.76
102	Lesotho	2.67
103	Nepal	1.65
104	Mauritania	1.34
105	Benin	1.02
106	Nigeria	0.93
107	Kenya	0.82
108	Zambia	0.81
109	Malawi	0.80
110	Ethiopia	0.79
111	Bangladesh	0.75
112	Burkina Faso	0.74
113	Mali	0.66
114	Angola	0.59
115	Cameroon[2]	0.59
116	Tanzania[2]	0.39
117	Burundi[2]	0.39
118	Mozambique[2]	0.37
119	Madagascar	0.36
120	Uganda	0.35
121	Cambodia[1]	0.26
122	Chad[2]	0.15

SOURCE: International Telecommunication Union, *World Telecommunication Indicators 2006*
[1] 2003; [2] 2004

3.02   Secure Internet servers

Secure Internet servers per 1 million inhabitants, 2005 or most recent year available

RANK	COUNTRY/ECONOMY	HARD DATA
1	Iceland	1,009.79
2	United States	782.57
3	Canada	569.84
4	Australia	499.88
5	New Zealand	492.70
6	Luxembourg	481.71
7	Switzerland	472.54
8	United Kingdom	466.21
9	Denmark	411.38
10	Ireland	354.65
11	Malta	331.60
12	Sweden	331.45
13	Netherlands	326.96
14	Norway	309.20
15	Finland	308.29
16	Germany	274.11
17	Singapore	270.03
18	Japan	257.33
19	Austria	232.12
20	Cyprus	216.69
21	Barbados	207.75
22	Taiwan, China[2]	169.00
23	Israel	163.27
24	Hong Kong SAR	159.28
25	Belgium	117.95
26	Estonia	101.86
27	Spain	81.68
28	France	79.12
29	Slovenia	79.07
30	Costa Rica	61.70
31	Portugal	57.40
32	Panama	56.32
33	United Arab Emirates	49.19
34	Bahrain	48.17
35	Italy	45.08
36	Czech Republic	41.68
37	Croatia	39.60
38	Latvia	37.83
39	Kuwait	31.55
40	Greece	31.47
41	Hungary	30.04
42	Uruguay	25.99
43	Poland	21.96
44	Lithuania	21.67
45	Trinidad and Tobago	21.45
46	South Africa	21.29
47	Chile	21.05
48	Qatar	20.91
49	Korea, Rep.	19.96
50	Mauritius	18.43
51	Slovak Republic	18.38
52	Turkey	16.92
53	Malaysia	14.87
54	Jamaica	14.30
55	Brazil	14.15
56	Suriname	11.13
57	Argentina	10.79
58	Bulgaria	8.66
59	Mexico	8.37
60	Namibia	6.89
61	Dominican Republic	5.85

RANK	COUNTRY/ECONOMY	HARD DATA
62	Guatemala	5.56
63	El Salvador	5.38
64	Romania	5.36
65	Peru	5.18
66	Thailand	4.86
67	Venezuela	4.63
68	Georgia	4.47
69	Colombia	4.19
70	Honduras	4.16
71	Ecuador	4.08
72	Jordan	3.70
73	Moldova	3.57
74	Bosnia and Herzegovina	3.33
75	Mongolia	3.13
76	Philippines	2.53
77	Russian Federation	2.43
78	Bolivia	2.40
79	Serbia and Montenegro	2.33
80	Nicaragua	2.01
81	Sri Lanka	1.79
82	Tunisia	1.40
83	Guyana	1.33
84	Armenia	1.33
85	Ukraine	1.32
86	Paraguay	1.30
87	Kazakhstan	0.86
88	Morocco	0.73
89	India	0.60
90	Kyrgyz Republic	0.58
91	Botswana[1]	0.57
92	Egypt	0.53
93	Macedonia, FYR	0.49
94	Azerbaijan	0.48
95	Indonesia	0.47
96	Nepal	0.44
97	China	0.33
98	Mauritania	0.33
99	Albania	0.32
100	Angola	0.31
101	Pakistan	0.30
102	Kenya	0.26
103	Zimbabwe	0.23
104	Nigeria	0.18
105	Zambia	0.17
106	Madagascar	0.16
107	Malawi	0.16
108	Mozambique	0.15
109	Burkina Faso	0.15
110	Vietnam	0.15
111	Burundi	0.13
112	Cameroon	0.12
113	Benin	0.12
114	Algeria	0.09
115	Mali	0.07
116	Cambodia	0.07
117	Uganda	0.04
118	Tanzania[1]	0.03
119	Bangladesh	0.02
120	Ethiopia[1]	0.01
n/a	Chad	n/a
n/a	Lesotho	n/a

SOURCE: World Bank, *World Development Indicators Online Database* (December 2006); national sources
[1] 2004; [2] 2006

3.03   Internet hosts

Internet hosts per 10,000 inhabitants, 2004

RANK	COUNTRY/ECONOMY	HARD DATA
1	United States	6,645.16
2	Iceland	4,758.60
3	Netherlands	3,334.42
4	Denmark	2,681.94
5	Finland	2,215.16
6	Australia	1,978.27
7	Norway	1,918.44
8	Austria	1,565.75
9	New Zealand	1,504.94
10	Sweden	1,466.67
11	Taiwan, China	1,389.65
12	Japan	1,286.80
13	Singapore	1,165.93
14	Hong Kong SAR	1,132.74
15	Korea, Rep.	1,130.06
16	Luxembourg	1,125.25
17	Canada	1,110.85
18	Switzerland	1,026.68
19	Israel	789.56
20	United Kingdom	697.90
21	Portugal	552.35
22	Estonia	486.31
23	Hungary	479.17
24	Ireland	420.95
25	France	386.48
26	Czech Republic	376.84
27	Germany	366.19
28	Uruguay	333.81
29	Italy	282.03
30	Lithuania	274.25
31	Slovenia	269.67
32	Latvia	258.69
33	Greece	250.61
34	Argentina	242.42
35	Belgium	232.50
36	Slovak Republic	227.26
37	Spain	217.46
38	Brazil	192.95
39	Malta	166.73
40	Mexico	145.17
41	Chile	142.27
42	Cyprus	94.47
43	Trinidad and Tobago	93.40
44	Bulgaria	84.73
45	Croatia	78.57
46	South Africa	77.52
47	Dominican Republic	75.02
48	Poland	70.50
49	Turkey	65.56
50	United Arab Emirates	61.11
51	Russian Federation	59.24
52	Thailand	58.13
53	Malaysia	52.81
54	Colombia	42.53
55	Peru	39.69
56	Mauritius	34.41
57	Serbia and Montenegro	33.83
58	Moldova	31.21
59	Ukraine	27.03
60	Costa Rica	26.35
61	Bahrain	25.80

RANK	COUNTRY/ECONOMY	HARD DATA
62	Romania	22.64
63	Panama	21.89
64	Bosnia and Herzegovina	21.69
65	Guatemala	18.75
66	Nicaragua	17.76
67	Macedonia, FYR	17.40
68	Namibia	16.70
69	Kazakhstan	14.69
70	Venezuela	14.53
71	Paraguay	13.99
72	Georgia	12.42
73	Botswana	12.22
74	Kyrgyz Republic	11.00
75	Kuwait	10.93
76	Bolivia	9.30
77	Guyana	8.37
78	Philippines	7.91
79	Barbados	7.79
80	Zimbabwe	6.77
81	Ecuador	6.67
82	El Salvador	6.63
83	Honduras	5.67
84	Jamaica	5.44
85	Jordan	5.28
86	Indonesia	5.01
87	Armenia	4.99
88	Qatar	4.23
89	Mozambique	3.78
90	Kenya	3.09
91	Suriname	2.96
92	Zambia	2.14
93	Albania	1.65
94	Pakistan	1.65
95	Tanzania	1.57
96	Morocco	1.38
97	India	1.33
98	China	1.25
99	Benin	1.24
100	Nepal	1.15
101	Sri Lanka	1.06
102	Uganda	1.00
103	Lesotho	0.84
104	Mongolia	0.61
105	Cambodia	0.57
106	Egypt	0.50
107	Madagascar	0.49
108	Azerbaijan	0.43
109	Tunisia	0.38
110	Mali	0.33
111	Burkina Faso	0.33
112	Angola	0.30
113	Algeria	0.29
114	Cameroon	0.28
115	Burundi	0.22
116	Mauritania	0.09
117	Nigeria	0.08
118	Malawi	0.05
119	Vietnam	0.05
120	Chad	0.01
121	Ethiopia	0.01
122	Bangladesh	0.00

SOURCE: International Telecommunication Union, *World Telecommunication Indicators 2005*

3.04 Electricity production

Per capita electricity production (kWh), 2003

RANK	COUNTRY/ECONOMY	HARD DATA
1	Iceland	29,356.09
2	Norway	23,303.99
3	Canada	18,555.04
4	Kuwait	16,608.96
5	Qatar	16,382.15
6	Finland	16,157.31
7	Sweden	15,135.89
8	United States	13,941.59
9	United Arab Emirates	12,259.34
10	Australia	11,466.31
11	Bahrain	11,004.98
12	New Zealand	10,254.17
13	France	9,357.82
14	Taiwan, China	9,248.74
15	Switzerland	8,848.94
16	Paraguay	8,808.02
17	Denmark	8,587.76
18	Singapore	8,441.89
19	Japan	8,133.92
20	Czech Republic	8,117.62
21	Belgium	8,053.23
22	Austria	7,529.87
23	Estonia	7,505.61
24	Korea, Rep.	7,207.08
25	Germany	7,199.99
26	Israel	7,033.64
27	Slovenia	7,024.60
28	United Kingdom	6,644.40
29	Russian Federation	6,323.18
30	Ireland	6,226.14
31	Luxembourg	6,178.51
32	Spain	6,139.64
33	Netherlands	5,964.46
34	Slovak Republic	5,759.85
35	Malta	5,604.01
36	Lithuania	5,450.69
37	Bulgaria	5,402.15
38	Greece	5,248.98
39	Hong Kong SAR	5,219.53
40	South Africa	5,000.10
41	Trinidad and Tobago	4,961.94
42	Cyprus	4,953.39
43	Italy	4,915.72
44	Portugal	4,455.59
45	Serbia and Montenegro	4,337.95
46	Kazakhstan	4,280.57
47	Poland	3,927.35
48	Ukraine	3,768.88
49	Venezuela	3,577.32
50	Hungary	3,370.83
51	Malaysia	3,209.38
52	Chile	3,058.11
53	Bosnia and Herzegovina	2,871.23
54	Croatia	2,838.97
55	Kyrgyz Republic	2,783.51
56	Jamaica	2,716.71
57	Azerbaijan	2,585.00
58	Romania	2,536.10
59	Uruguay	2,512.16
60	Argentina	2,422.67
61	Mexico	2,164.44
62	Brazil	2,011.48
63	Turkey	1,988.08
64	Thailand	1,852.62
65	Costa Rica	1,811.62
66	Armenia	1,811.21
67	Panama	1,787.68
68	Latvia	1,711.15
69	Albania	1,690.53
70	Jordan	1,654.47
71	Dominican Republic	1,563.19
72	Georgia	1,558.93
73	China	1,480.43
74	Egypt	1,289.96
75	Tunisia	1,261.31
76	Colombia	1,064.62
77	Algeria	927.99
78	Ecuador	898.31
79	Zambia	847.92
80	Peru	843.98
81	Moldova	806.61
82	Namibia	756.81
83	Zimbabwe	684.05
84	Philippines	659.42
85	Honduras	657.19
86	Morocco	628.48
87	El Salvador	613.70
88	India	594.96
89	Mozambique	556.47
90	Guatemala	546.82
91	Pakistan	544.53
92	Indonesia	526.03
93	Nicaragua	513.86
94	Vietnam	503.29
95	Bolivia	483.29
96	Sri Lanka	395.32
97	Cameroon	233.93
98	Nigeria	160.29
99	Kenya	148.59
100	Bangladesh	144.29
101	Angola	132.59
102	Nepal	87.02
103	Tanzania	74.24
104	Ethiopia	33.48
105	Benin	9.72
n/a	Barbados	n/a
n/a	Botswana	n/a
n/a	Burkina Faso	n/a
n/a	Burundi	n/a
n/a	Cambodia	n/a
n/a	Chad	n/a
n/a	Guyana	n/a
n/a	Lesotho	n/a
n/a	Macedonia, FYR	n/a
n/a	Madagascar	n/a
n/a	Malawi	n/a
n/a	Mali	n/a
n/a	Mauritania	n/a
n/a	Mauritius	n/a
n/a	Mongolia	n/a
n/a	Suriname	n/a
n/a	Uganda	n/a

SOURCE: World Bank, *World Development Indicators Online Database* (December 2006); national sources

295

3.05 Availability of scientists and engineers

Scientists and engineers in your country are (1 = nonexistent or rare, 7 = widely available)

RANK	COUNTRY/ECONOMY	SCORE	MEAN: 4.49	SD
1	Israel	6.33		0.9
2	Japan	6.25		0.7
3	Finland	6.22		0.8
4	India	6.21		0.7
5	France	6.05		1.0
6	Switzerland	5.99		0.9
7	Czech Republic	5.93		1.0
8	Sweden	5.85		0.9
9	Canada	5.82		1.0
10	Tunisia	5.81		1.0
11	Germany	5.80		0.9
12	Denmark	5.64		1.0
13	Belgium	5.62		1.1
14	Taiwan, China	5.60		1.0
15	Singapore	5.58		0.9
16	Iceland	5.53		0.8
17	Greece	5.53		1.1
18	United States	5.52		1.2
19	Ireland	5.51		1.0
20	Morocco	5.39		1.1
21	Algeria	5.38		1.3
22	United Kingdom	5.36		1.2
23	Slovak Republic	5.35		1.2
24	Malaysia	5.33		1.2
25	Norway	5.30		0.9
26	Jordan	5.29		1.4
27	Hong Kong SAR	5.28		1.1
28	Korea, Rep.	5.25		1.0
29	Austria	5.20		1.2
30	Hungary	5.18		1.2
31	Netherlands	5.15		1.1
32	Portugal	5.14		1.0
33	Chile	5.07		1.2
34	Cyprus	5.05		1.3
35	Australia	5.05		1.1
36	Indonesia	4.98		0.6
37	Costa Rica	4.95		1.1
38	Croatia	4.93		1.3
39	Serbia and Montenegro	4.93		1.4
40	Egypt	4.88		1.5
41	Romania	4.85		1.3
42	Spain	4.83		1.3
43	New Zealand	4.80		1.2
44	Turkey	4.76		1.3
45	Thailand	4.74		1.2
46	Russian Federation	4.73		1.4
47	Italy	4.73		1.3
48	Lithuania	4.70		1.2
49	Bulgaria	4.68		1.4
50	Estonia	4.67		1.3
51	Macedonia, FYR	4.66		1.6
52	Uruguay	4.65		1.3
53	Mongolia	4.65		1.8
54	Trinidad and Tobago	4.60		1.3
55	Armenia	4.56		1.6
56	Kenya	4.55		1.4
57	Malta	4.52		1.3
58	Vietnam	4.51		1.4
59	Barbados	4.51		1.3
60	Colombia	4.51		1.3
61	Brazil	4.49		1.4
62	Venezuela	4.47		1.3
63	Kuwait	4.46		1.4
64	Benin	4.44		1.5
65	Argentina	4.44		1.3
66	Azerbaijan	4.44		1.8
67	Sri Lanka	4.42		1.5
68	Madagascar	4.41		1.4
69	Tanzania	4.35		1.4
70	Ukraine	4.31		1.5
71	Mali	4.27		1.5
72	Cameroon	4.26		1.5
73	Peru	4.26		1.4
74	Bangladesh	4.21		1.4
75	Poland	4.21		1.3
76	Luxembourg	4.21		1.4
77	Mauritania	4.18		1.8
78	Pakistan	4.17		1.4
79	Georgia	4.17		1.4
80	Qatar	4.14		1.3
80	United Arab Emirates	4.14		1.5
82	Uganda	4.10		1.6
83	Nigeria	4.10		1.7
84	Philippines	4.06		1.4
85	Mexico	4.05		1.3
86	China	4.00		1.3
86	Mauritius	4.00		1.2
88	Jamaica	3.99		1.3
89	Zimbabwe	3.92		1.5
90	Guatemala	3.88		1.1
91	Zambia	3.88		1.7
92	South Africa	3.85		1.2
93	Bosnia and Herzegovina	3.85		1.4
94	Slovenia	3.84		1.4
95	Nepal	3.79		1.4
96	Bahrain	3.79		1.6
97	Latvia	3.78		1.4
98	Moldova	3.77		1.4
99	Burundi	3.77		1.8
100	Kazakhstan	3.72		1.4
101	El Salvador	3.69		1.4
102	Panama	3.69		1.3
103	Burkina Faso	3.63		1.5
104	Nicaragua	3.61		1.5
105	Dominican Republic	3.54		1.4
106	Suriname	3.51		1.4
107	Ecuador	3.45		1.3
108	Botswana	3.43		1.3
109	Kyrgyz Republic	3.40		1.6
110	Honduras	3.35		1.3
111	Albania	3.34		1.4
112	Bolivia	3.19		1.3
113	Chad	3.10		1.7
114	Guyana	3.08		1.2
115	Lesotho	3.04		1.5
116	Ethiopia	3.04		1.3
117	Mozambique	3.00		1.3
118	Namibia	2.98		1.2
119	Malawi	2.95		1.5
120	Paraguay	2.87		1.2
121	Cambodia	2.78		1.2
122	Angola	2.41		1.0

SOURCE: World Economic Forum, Executive Opinion Survey 2006

Scientific research institutions in your country (e.g., university laboratories, government laboratories) are (1 = nonexistent, 7 = the best in their fields internationally)

RANK	COUNTRY/ECONOMY	SCORE	MEAN: 3.91	SD
1	Switzerland	6.25		0.7
2	United States	6.04		1.2
3	United Kingdom	6.03		1.0
4	Israel	6.00		0.7
5	Japan	5.85		0.7
6	Germany	5.84		0.8
7	Finland	5.73		0.7
8	Sweden	5.65		0.8
9	Belgium	5.59		0.8
10	Singapore	5.53		0.8
11	Canada	5.52		0.9
12	Netherlands	5.49		0.9
13	Denmark	5.31		0.8
14	India	5.31		1.0
15	Ireland	5.29		0.9
16	Australia	5.27		1.2
17	Malaysia	5.16		1.1
18	Norway	5.15		0.9
19	New Zealand	5.13		1.0
20	France	5.13		1.1
21	Taiwan, China	5.02		0.9
22	Korea, Rep.	5.00		1.1
23	Austria	4.89		1.0
24	Hong Kong SAR	4.82		1.0
25	South Africa	4.75		1.1
26	Hungary	4.74		1.0
27	Indonesia	4.70		0.9
28	Estonia	4.68		1.3
29	Czech Republic	4.65		1.1
30	Iceland	4.59		0.9
31	Kenya	4.47		1.4
32	Russian Federation	4.40		1.6
33	Tunisia	4.38		1.4
34	Uganda	4.31		1.4
35	Portugal	4.26		0.9
36	Brazil	4.26		1.2
37	Thailand	4.24		1.0
38	Costa Rica	4.24		1.2
39	Jamaica	4.20		1.0
40	Tanzania	4.19		1.4
41	Slovenia	4.17		1.2
42	Sri Lanka	4.12		1.5
43	Serbia and Montenegro	4.10		1.3
44	Lithuania	4.08		1.1
45	Nigeria	4.05		1.6
46	Croatia	4.04		1.3
47	Spain	4.04		1.1
48	Chile	4.02		1.1
49	Qatar	4.00		1.2
50	Barbados	3.96		1.1
51	Ukraine	3.94		1.3
52	Azerbaijan	3.93		1.7
53	Kazakhstan	3.87		1.4
54	Mexico	3.86		1.2
55	Turkey	3.86		1.1
56	Luxembourg	3.86		1.3
57	Kuwait	3.86		1.3
58	Poland	3.81		1.0
59	Zimbabwe	3.81		1.1
60	United Arab Emirates	3.79		1.4
61	Latvia	3.76		1.3
62	Pakistan	3.74		1.5
63	China	3.72		1.1
64	Greece	3.71		1.3
65	Trinidad and Tobago	3.71		1.3
66	Romania	3.70		1.3
67	Bulgaria	3.67		1.5
68	Ethiopia	3.67		1.2
69	Armenia	3.67		1.3
70	Mauritius	3.63		1.3
71	Slovak Republic	3.58		1.0
72	Argentina	3.58		1.3
73	Jordan	3.56		1.3
74	Botswana	3.55		1.4
75	Colombia	3.55		1.2
76	Burkina Faso	3.53		1.3
77	Cyprus	3.51		1.4
78	Philippines	3.51		1.5
79	Uruguay	3.51		1.1
80	Malawi	3.47		1.5
81	Mongolia	3.44		1.2
82	Morocco	3.44		1.5
83	Macedonia, FYR	3.44		1.5
84	Algeria	3.37		1.3
85	Nepal	3.37		1.4
86	Mali	3.36		1.5
87	Italy	3.35		1.5
88	Malta	3.32		1.3
89	Bangladesh	3.30		1.3
90	Panama	3.29		1.2
91	Guatemala	3.23		1.3
92	Madagascar	3.19		1.2
93	Georgia	3.18		1.2
94	Egypt	3.17		1.5
95	Mozambique	3.16		1.4
96	Suriname	3.15		1.2
97	Vietnam	3.12		1.0
98	Kyrgyz Republic	3.11		1.4
99	Guyana	3.11		1.2
100	Zambia	3.02		1.0
101	Namibia	3.02		1.2
102	Moldova	3.01		1.2
103	Venezuela	3.00		1.1
104	Bosnia and Herzegovina	2.95		1.2
105	Cambodia	2.91		1.4
106	Peru	2.88		1.0
107	Cameroon	2.86		1.4
108	Benin	2.85		1.3
109	Ecuador	2.81		1.1
110	Lesotho	2.79		1.2
111	Nicaragua	2.79		1.0
112	El Salvador	2.78		1.1
113	Dominican Republic	2.58		1.2
114	Honduras	2.58		1.2
115	Bahrain	2.55		1.2
116	Bolivia	2.54		1.1
117	Angola	2.49		1.2
118	Chad	2.41		1.4
119	Burundi	2.40		1.2
120	Albania	2.19		0.9
121	Paraguay	2.09		0.9
122	Mauritania	1.74		1.3

SOURCE: World Economic Forum, Executive Opinion Survey 2006

3.07 Tertiary enrollment

Gross tertiary enrollment rate, 2004 or most recent year available

RANK	COUNTRY/ECONOMY	HARD DATA
1	Finland	89.50
2	Korea, Rep.	88.52
3	New Zealand	85.78
4	Sweden	83.73
5	United States	82.44
6	Norway	80.47
7	Greece	79.38
8	Taiwan, China	78.60
9	Latvia	74.35
10	Denmark	73.88
11	Slovenia	73.74
12	Lithuania	73.16
13	Australia	72.21
14	Russian Federation	68.22
15	Iceland	67.69
16	Spain	65.65
17	Ukraine	65.51
18	Estonia	65.07
19	Argentina[4]	63.86
20	Italy	63.09
21	Belgium	62.54
22	Poland	61.04
23	Canada[3]	60.21
24	United Kingdom	60.13
25	Hungary	59.61
26	Netherlands	59.19
27	Ireland	58.51
28	Portugal	56.60
29	Israel	56.48
30	France	56.02
31	Japan	53.98
32	Germany	50.00
33	Austria	49.78
34	Kazakhstan	48.00
35	Singapore	47.00
36	Switzerland	46.99
37	Panama	45.30
38	Czech Republic	43.24
39	Chile	42.98
40	Georgia	41.46
41	Bulgaria	41.12
42	Thailand	40.98
43	Bolivia	40.62
44	Romania	40.18
45	Kyrgyz Republic	39.70
46	Uruguay[4]	39.34
47	Jordan	39.34
48	Venezuela[4]	39.27
49	Mongolia	38.89
50	Croatia[4]	38.71
51	Barbados[2]	37.75
52	Moldova	36.99
53	Serbia and Montenegro[2]	36.25
54	Slovak Republic	36.10
55	Cyprus	35.86
56	Bahrain	34.42
57	Peru	33.36
58	Dominican Republic	32.89
59	Egypt	32.58
60	Malaysia[4]	32.35
61	Hong Kong SAR	32.10

RANK	COUNTRY/ECONOMY	HARD DATA
62	Turkey	29.01
63	Philippines	28.80
64	Tunisia	28.60
65	Macedonia, FYR	27.99
66	Colombia	26.86
67	Malta	26.39
68	Armenia	26.20
69	Costa Rica	25.34
70	Paraguay[4]	24.42
71	Mexico	23.39
72	United Arab Emirates[4]	22.48
73	Brazil[4]	22.28
74	Kuwait	22.27
75	Algeria	19.59
76	Albania	19.28
77	Qatar	19.12
78	China	19.10
79	Jamaica[4]	18.99
80	El Salvador	18.51
81	Nicaragua[4]	17.87
82	Mauritius	17.24
83	Indonesia	16.67
84	Honduras	16.42
85	South Africa[4]	15.29
86	Azerbaijan	14.77
87	Suriname[5]	12.44
88	Luxembourg	12.39
89	Trinidad and Tobago	11.90
90	India	11.76
91	Morocco	10.62
92	Nigeria	10.18
93	Vietnam	10.16
94	Guatemala[4]	9.58
95	Guyana	9.08
96	Bangladesh[4]	6.51
97	Botswana	6.22
98	Namibia[4]	6.11
99	Nepal	5.64
100	Cameroon	5.28
101	Zimbabwe[4]	3.67
102	Mauritania	3.46
103	Uganda	3.44
104	Pakistan	3.20
105	Benin[2]	3.04
106	Sri Lanka	3.00
107	Kenya	2.91
108	Cambodia	2.91
109	Lesotho[4]	2.76
110	Madagascar	2.52
111	Ethiopia	2.49
112	Burundi	2.33
113	Zambia[1]	2.33
114	Mali	2.10
115	Burkina Faso	1.67
116	Chad	1.21
117	Tanzania	1.20
118	Mozambique	1.19
119	Angola[4]	0.85
120	Malawi	0.42
n/a	Bosnia and Herzegovina	n/a
n/a	Ecuador	n/a

SOURCE: UNESCO, Institute for Statistics (December 2006); World Bank, *World Development Indicators Online Database* (December 2006); national sources
[1] 2000; [2] 2001; [3] 2002; [4] 2003; [5] 2005

Pillar 4
Individual readiness

4.01 Quality of math and science education

Math and science education in your country's schools (1 = lag far behind that of most other countries, 7 = are among the best in the world)

RANK	COUNTRY/ECONOMY	SCORE	SD		RANK	COUNTRY/ECONOMY	SCORE	SD
1	Singapore	6.33	0.8		62	China	4.10	1.5
2	Finland	6.14	0.8		63	Georgia	4.10	1.4
3	Belgium	6.14	1.1		64	Sri Lanka	3.99	1.5
4	Switzerland	5.85	0.9		65	Vietnam	3.98	1.6
5	France	5.77	1.1		66	Kazakhstan	3.96	1.3
6	Hong Kong SAR	5.71	1.0		67	Costa Rica	3.94	1.2
7	India	5.69	1.1		68	Albania	3.94	1.4
8	Czech Republic	5.65	1.0		69	Azerbaijan	3.92	1.6
9	Tunisia	5.58	1.0		70	Zimbabwe	3.89	1.6
10	Taiwan, China	5.55	1.0		71	Armenia	3.82	1.6
11	Romania	5.52	1.2		72	Botswana	3.74	1.5
12	Malaysia	5.48	1.1		73	Algeria	3.73	1.4
13	Hungary	5.46	1.0		74	Burundi	3.72	1.8
14	Japan	5.39	1.3		75	Suriname	3.71	1.2
15	Netherlands	5.36	1.1		76	Burkina Faso	3.69	1.3
16	Ireland	5.34	1.2		77	Colombia	3.61	1.3
17	Israel	5.31	1.1		78	Uruguay	3.61	1.4
18	Estonia	5.28	1.2		79	Madagascar	3.59	1.6
19	Barbados	5.26	1.0		80	Cameroon	3.56	1.4
20	Denmark	5.24	1.2		81	Kenya	3.54	1.5
21	Slovak Republic	5.21	1.1		82	Spain	3.53	1.5
22	Canada	5.14	1.2		83	Portugal	3.51	1.3
23	Korea, Rep.	5.09	1.2		84	Kyrgyz Republic	3.51	1.5
24	Serbia and Montenegro	5.07	1.6		85	Pakistan	3.42	1.3
25	New Zealand	5.07	1.2		86	Bahrain	3.38	1.6
26	Lithuania	4.99	1.1		87	Jamaica	3.38	1.5
27	Austria	4.99	1.2		88	Argentina	3.35	1.3
28	Indonesia	4.95	0.7		89	Guyana	3.28	1.5
29	Australia	4.94	1.1		90	Nepal	3.25	1.5
30	Cyprus	4.94	1.1		91	El Salvador	3.22	1.1
31	Croatia	4.93	1.3		92	Zambia	3.19	1.2
32	Luxembourg	4.84	1.1		93	Egypt	3.15	1.5
33	Iceland	4.79	0.9		94	Mauritania	3.06	1.8
34	Germany	4.78	1.0		95	Uganda	3.02	1.5
35	Latvia	4.74	1.5		96	Mozambique	3.00	1.4
36	United Kingdom	4.71	1.3		97	Venezuela	2.97	1.3
37	Sweden	4.71	1.3		98	Brazil	2.94	1.4
38	Qatar	4.67	1.2		99	Mali	2.93	1.4
39	Slovenia	4.63	1.2		100	Chile	2.93	1.3
40	Macedonia, FYR	4.62	1.4		101	Mexico	2.91	1.2
41	United Arab Emirates	4.55	1.4		102	Nigeria	2.91	1.5
42	United States	4.55	1.4		103	Tanzania	2.90	1.2
43	Russian Federation	4.54	1.6		104	Panama	2.84	1.5
44	Malta	4.54	1.1		105	Nicaragua	2.82	1.3
45	Bosnia and Herzegovina	4.51	1.6		106	Ethiopia	2.81	1.3
46	Greece	4.50	1.3		107	Philippines	2.81	1.3
47	Thailand	4.48	1.1		108	Ecuador	2.80	1.2
48	Trinidad and Tobago	4.45	1.6		109	Guatemala	2.78	1.0
49	Morocco	4.43	1.5		110	Bangladesh	2.74	1.5
50	Ukraine	4.41	1.3		111	Lesotho	2.70	1.6
51	Bulgaria	4.40	1.6		112	Cambodia	2.67	1.3
52	Italy	4.40	1.4		113	Namibia	2.51	1.3
53	Poland	4.39	1.1		114	Honduras	2.50	1.3
54	Norway	4.37	1.2		115	South Africa	2.44	1.0
55	Benin	4.33	1.5		116	Dominican Republic	2.43	1.2
56	Jordan	4.31	1.5		117	Angola	2.42	1.0
57	Turkey	4.31	1.3		118	Malawi	2.42	1.4
58	Moldova	4.29	1.5		119	Bolivia	2.38	1.1
59	Mauritius	4.19	1.1		120	Chad	2.37	1.4
60	Mongolia	4.14	1.4		121	Paraguay	2.20	1.0
61	Kuwait	4.12	1.7		122	Peru	2.06	1.1

MEAN: 4.07

SOURCE: World Economic Forum, Executive Opinion Survey 2006

4.02   Quality of the educational system

The educational system in your country (1 = does not meet the needs of a competitive economy, 7 = meets the needs of a competitive economy)

RANK	COUNTRY/ECONOMY	SCORE	MEAN: 3.69	SD
1	Finland	6.00		1.0
2	Singapore	5.95		0.8
3	Iceland	5.87		0.9
4	Switzerland	5.82		1.0
5	Denmark	5.58		1.0
6	Ireland	5.49		1.0
7	Hong Kong SAR	5.44		1.2
8	Belgium	5.42		1.1
9	Taiwan, China	5.40		1.0
10	Malaysia	5.15		1.3
11	Tunisia	5.13		1.1
12	Australia	5.07		1.1
13	Austria	5.06		1.2
14	Canada	5.02		1.4
15	United States	5.00		1.4
16	Barbados	5.00		1.3
17	Norway	4.99		1.2
18	Netherlands	4.92		1.3
19	Japan	4.92		1.3
20	Qatar	4.85		1.4
21	New Zealand	4.80		1.4
22	Israel	4.77		1.5
23	Indonesia	4.74		0.9
24	Sweden	4.73		1.5
25	India	4.70		1.5
26	Cyprus	4.69		1.4
27	France	4.60		1.6
28	Malta	4.60		1.2
29	United Kingdom	4.55		1.5
30	Czech Republic	4.48		1.5
31	Estonia	4.42		1.4
32	United Arab Emirates	4.42		1.5
33	Germany	4.39		1.4
34	Poland	4.36		1.1
35	Latvia	4.35		1.6
36	Luxembourg	4.32		1.6
37	Kenya	4.20		1.6
38	Korea, Rep.	4.14		1.4
39	Zimbabwe	4.14		1.7
40	Costa Rica	4.12		1.4
41	Thailand	4.09		1.4
42	Hungary	4.04		1.5
43	Macedonia, FYR	4.02		1.8
44	Jordan	4.01		1.6
45	Lithuania	3.94		1.4
46	Serbia and Montenegro	3.91		2.0
47	Ukraine	3.89		1.6
48	Trinidad and Tobago	3.87		1.8
49	Slovak Republic	3.86		1.5
50	Kazakhstan	3.85		1.5
51	Romania	3.83		1.7
52	Slovenia	3.82		1.6
53	Croatia	3.77		1.5
54	Russian Federation	3.74		1.6
55	Colombia	3.68		1.5
56	Botswana	3.66		1.7
57	Portugal	3.66		1.2
58	Zambia	3.61		1.9
59	Greece	3.59		1.6
60	Philippines	3.58		1.6
61	Kuwait	3.53		1.5
62	Bosnia and Herzegovina	3.46		1.7
63	Uganda	3.42		1.9
64	Mauritius	3.41		1.4
65	El Salvador	3.40		1.3
66	Spain	3.40		1.5
67	Jamaica	3.38		1.6
68	Moldova	3.36		1.7
69	Sri Lanka	3.33		1.8
70	Nigeria	3.28		1.7
71	Italy	3.26		1.5
72	Turkey	3.25		1.5
73	Pakistan	3.24		1.5
74	Uruguay	3.24		1.4
75	Chile	3.21		1.4
76	Guyana	3.19		1.6
77	Kyrgyz Republic	3.18		1.7
78	Bahrain	3.16		1.7
79	Albania	3.13		1.5
80	Cambodia	3.12		1.5
81	Mexico	3.11		1.4
82	Bulgaria	3.11		1.5
83	Tanzania	3.07		1.4
84	Lesotho	3.06		1.7
85	Azerbaijan	3.05		1.6
86	China	3.03		1.5
87	Armenia	3.00		1.6
87	Cameroon	3.00		1.7
89	Madagascar	2.97		1.4
90	Morocco	2.94		1.5
91	Algeria	2.93		1.4
92	Benin	2.92		1.6
93	Georgia	2.92		1.3
94	Ethiopia	2.92		1.4
95	Mongolia	2.90		1.5
96	Malawi	2.87		1.8
97	South Africa	2.85		1.0
98	Argentina	2.75		1.4
99	Vietnam	2.74		1.4
100	Nepal	2.72		1.4
101	Bangladesh	2.70		1.3
102	Nicaragua	2.70		1.4
103	Egypt	2.66		1.5
104	Burundi	2.66		1.7
104	Panama	2.66		1.3
106	Namibia	2.65		1.3
107	Mali	2.60		1.5
108	Mauritania	2.60		1.2
109	Guatemala	2.59		1.2
110	Venezuela	2.59		1.3
111	Mozambique	2.57		1.2
112	Brazil	2.55		1.4
113	Burkina Faso	2.53		1.4
114	Suriname	2.47		1.2
115	Dominican Republic	2.33		1.1
116	Honduras	2.31		1.2
117	Ecuador	2.26		1.1
118	Angola	2.23		1.0
119	Chad	2.14		1.4
120	Bolivia	2.10		1.0
121	Peru	2.02		0.8
122	Paraguay	1.94		1.0

301

SOURCE: World Economic Forum, Executive Opinion Survey 2006

4.03 Quality of public schools

The public (free) schools in your country are (1 = of poor quality, 7 = equal to the best in the world)

RANK	COUNTRY/ECONOMY	SCORE	SD
1	Finland	6.51	0.8
2	Singapore	6.03	0.8
3	Switzerland	5.97	1.0
4	Belgium	5.91	1.3
5	Ireland	5.83	1.1
6	Iceland	5.73	0.9
7	Netherlands	5.59	1.1
8	Barbados	5.52	1.1
9	Canada	5.52	1.2
10	Denmark	5.46	1.2
11	Czech Republic	5.43	1.2
12	Austria	5.35	1.1
13	Taiwan, China	5.32	1.0
14	Tunisia	5.31	1.2
15	France	5.27	1.2
16	New Zealand	5.26	1.2
17	Malaysia	5.23	1.1
18	Hong Kong SAR	5.23	1.2
19	Norway	5.21	1.0
20	Japan	5.19	1.4
21	Estonia	5.12	1.2
22	Australia	5.12	1.2
23	Sweden	5.04	1.3
24	Israel	5.02	1.3
25	Germany	4.94	1.2
26	Hungary	4.87	1.2
27	Luxembourg	4.76	1.2
28	Slovak Republic	4.74	1.0
29	United Kingdom	4.72	1.4
30	Latvia	4.69	1.5
31	United States	4.66	1.4
32	Croatia	4.61	1.4
33	Slovenia	4.55	1.5
34	Cyprus	4.51	1.2
35	Malta	4.46	1.3
36	Romania	4.45	1.5
37	Serbia and Montenegro	4.39	1.9
38	Poland	4.38	1.1
39	Qatar	4.38	1.5
40	Lithuania	4.29	1.3
41	Portugal	4.25	1.3
42	Italy	4.24	1.6
43	Korea, Rep.	4.19	1.2
44	Macedonia, FYR	4.07	1.8
45	Mauritius	4.04	1.3
46	Indonesia	4.02	1.1
47	United Arab Emirates	3.88	1.6
48	Ukraine	3.86	1.5
49	Botswana	3.84	1.3
50	Costa Rica	3.73	1.3
51	Russian Federation	3.70	1.6
52	Bulgaria	3.69	1.5
53	Greece	3.67	1.2
54	China	3.65	1.5
55	Trinidad and Tobago	3.59	1.6
56	Kuwait	3.57	1.6
57	Sri Lanka	3.53	1.6
58	Thailand	3.52	1.4
59	Bosnia and Herzegovina	3.44	1.5
60	Spain	3.40	1.5
61	Uruguay	3.36	1.4
62	Kazakhstan	3.31	1.4
63	Jordan	3.25	1.5
64	Burkina Faso	3.19	1.3
65	Benin	3.16	1.4
66	Colombia	3.16	1.2
67	Morocco	3.14	1.7
68	Bahrain	3.14	1.5
69	Algeria	3.10	1.4
70	Guyana	3.05	1.4
71	Suriname	3.05	1.3
72	Albania	3.05	1.3
73	Vietnam	3.05	1.3
74	Georgia	3.03	1.3
75	Moldova	3.02	1.4
76	Turkey	3.00	1.3
77	El Salvador	2.96	1.1
78	Pakistan	2.94	1.6
79	Kyrgyz Republic	2.90	1.7
80	Jamaica	2.88	1.5
81	Kenya	2.88	1.5
82	Armenia	2.84	1.6
83	Cameroon	2.82	1.3
84	Mongolia	2.80	1.4
85	Argentina	2.75	1.4
86	Zimbabwe	2.72	1.5
87	South Africa	2.65	1.1
88	Lesotho	2.64	1.3
89	Burundi	2.63	1.6
90	Mexico	2.62	1.1
91	Philippines	2.58	1.2
92	Azerbaijan	2.58	1.4
93	Zambia	2.56	1.2
94	Panama	2.55	1.2
95	Cambodia	2.55	1.4
95	Tanzania	2.55	1.2
97	Namibia	2.55	1.3
98	Uganda	2.47	1.7
99	Madagascar	2.43	1.1
100	India	2.42	1.2
101	Ethiopia	2.39	1.3
102	Chile	2.36	1.1
103	Mozambique	2.34	1.2
104	Mauritania	2.30	1.5
105	Mali	2.27	1.2
106	Honduras	2.24	1.1
107	Nicaragua	2.24	1.0
108	Nigeria	2.22	1.4
109	Guatemala	2.22	1.0
110	Bangladesh	2.06	1.2
111	Brazil	2.05	1.2
112	Egypt	2.05	1.2
113	Chad	2.03	1.3
114	Bolivia	2.00	1.0
115	Venezuela	1.98	1.2
116	Angola	1.97	0.8
117	Ecuador	1.93	1.1
118	Paraguay	1.90	1.0
119	Malawi	1.84	1.2
120	Nepal	1.81	1.3
121	Dominican Republic	1.76	0.9
122	Peru	1.52	0.7

MEAN: 3.61

SOURCE: World Economic Forum, Executive Opinion Survey 2006

4.04 Internet access in schools

Internet access in schools is (1 = very limited, 7 = extensive—most children have frequent access)

RANK	COUNTRY/ECONOMY	SCORE	SD
1	Iceland	6.83	0.4
2	Sweden	6.46	0.7
3	Finland	6.42	0.7
4	Korea, Rep.	6.40	1.0
5	Singapore	6.29	0.8
6	Denmark	6.22	0.9
7	Hong Kong SAR	6.21	1.0
8	Netherlands	6.20	0.8
9	Austria	6.19	0.9
10	Switzerland	6.07	0.8
11	United Kingdom	6.04	0.8
12	Canada	6.04	0.8
13	Estonia	6.02	1.3
14	United States	5.87	1.2
15	Luxembourg	5.86	1.1
16	Israel	5.81	1.2
17	Australia	5.79	0.8
18	Japan	5.76	1.0
19	Slovenia	5.75	1.1
20	Taiwan, China	5.72	1.1
21	New Zealand	5.67	1.0
22	Norway	5.62	1.0
23	Belgium	5.59	1.1
24	France	5.56	1.2
25	United Arab Emirates	5.48	1.5
26	Malta	5.47	1.1
27	Czech Republic	5.34	1.2
28	Germany	5.33	1.2
29	Hungary	5.23	1.3
30	Malaysia	5.22	1.2
31	Lithuania	5.18	1.3
32	Tunisia	5.13	1.2
33	Latvia	5.10	1.4
34	Portugal	5.08	1.2
35	Ireland	4.97	1.3
36	Chile	4.86	1.3
37	Kuwait	4.85	1.7
38	Slovak Republic	4.68	1.4
39	Qatar	4.60	1.8
40	Jordan	4.59	1.6
41	Thailand	4.59	1.1
42	Cyprus	4.54	1.4
43	Bahrain	4.54	1.9
44	Spain	4.47	1.3
45	Barbados	4.46	1.4
46	Croatia	4.42	1.4
47	Kazakhstan	4.27	1.6
48	Italy	4.06	1.6
49	China	4.02	1.7
50	Poland	3.94	1.2
51	Pakistan	3.87	1.6
52	India	3.84	1.7
53	Mexico	3.83	1.5
54	Russian Federation	3.77	1.6
55	Turkey	3.76	1.4
56	Greece	3.76	1.5
57	Egypt	3.75	1.8
58	Panama	3.75	1.5
59	Peru	3.74	1.5
60	Uruguay	3.73	1.5
61	Bulgaria	3.71	1.4
62	Mauritius	3.70	1.8
63	El Salvador	3.69	1.4
64	Jamaica	3.66	1.3
65	Romania	3.59	1.6
66	Brazil	3.55	1.6
67	Argentina	3.55	1.4
68	Philippines	3.48	1.6
69	Vietnam	3.40	1.6
70	Venezuela	3.40	1.4
71	Bosnia and Herzegovina	3.36	1.6
72	Morocco	3.35	1.7
73	Colombia	3.35	1.5
74	Georgia	3.32	1.4
75	Indonesia	3.30	0.8
76	Trinidad and Tobago	3.25	1.5
77	Namibia	3.20	1.5
78	South Africa	3.18	1.2
79	Serbia and Montenegro	3.17	1.4
80	Dominican Republic	3.13	1.5
81	Costa Rica	3.08	1.5
82	Guatemala	3.07	1.1
83	Botswana	3.06	1.5
84	Macedonia, FYR	2.99	1.7
85	Azerbaijan	2.96	1.5
86	Sri Lanka	2.94	1.8
87	Moldova	2.93	1.6
88	Ukraine	2.86	1.6
89	Mongolia	2.78	1.7
90	Kyrgyz Republic	2.74	1.7
91	Nepal	2.71	1.6
92	Algeria	2.69	1.6
93	Mali	2.69	1.8
94	Ecuador	2.66	1.4
95	Nicaragua	2.62	1.6
96	Nigeria	2.59	1.6
97	Tanzania	2.49	1.4
98	Cambodia	2.47	1.5
99	Honduras	2.46	1.3
100	Armenia	2.43	1.2
101	Zimbabwe	2.36	1.3
102	Angola	2.24	1.3
103	Guyana	2.15	1.2
104	Bolivia	2.13	1.1
105	Malawi	2.13	1.4
106	Mozambique	2.05	1.4
107	Madagascar	2.03	1.3
108	Ethiopia	1.99	1.3
109	Suriname	1.99	1.0
110	Kenya	1.97	1.2
111	Bangladesh	1.96	1.1
112	Uganda	1.90	1.2
113	Benin	1.81	1.3
114	Paraguay	1.77	1.1
115	Mauritania	1.76	1.4
116	Cameroon	1.72	1.1
117	Albania	1.67	0.9
118	Lesotho	1.62	1.1
119	Burkina Faso	1.61	1.0
120	Zambia	1.60	0.7
121	Chad	1.29	0.6
122	Burundi	1.28	0.6

MEAN: 3.86

SOURCE: World Economic Forum, Executive Opinion Survey 2006

303

4.05   Buyer sophistication

Buyers in your country are (1 = unsophisticated and make choices based on lowest price, 7 = knowledgeable and demanding and buy based on superior performance attributes)

RANK	COUNTRY/ECONOMY	SCORE	MEAN: 4.05	SD
1	Japan	6.23		1.0
2	Switzerland	6.07		1.0
3	United Kingdom	5.97		0.9
4	Hong Kong SAR	5.90		1.2
5	Luxembourg	5.83		1.2
6	Netherlands	5.79		1.1
7	Belgium	5.78		1.1
8	Ireland	5.74		1.1
9	Finland	5.74		1.0
10	United States	5.73		1.3
11	France	5.72		1.2
12	Indonesia	5.70		1.1
13	Australia	5.69		1.0
14	Austria	5.69		1.2
15	Germany	5.69		1.3
16	Sweden	5.68		1.3
17	Denmark	5.68		1.2
18	Canada	5.62		1.3
19	Singapore	5.55		1.1
20	Israel	5.54		1.1
21	Korea, Rep.	5.53		1.2
22	New Zealand	5.52		1.1
23	Norway	5.46		1.3
24	Taiwan, China	5.45		0.9
25	Malaysia	5.44		1.0
26	India	5.40		1.4
27	Iceland	5.37		1.2
28	Tunisia	5.11		1.1
29	Spain	5.08		1.2
30	Chile	5.05		1.4
31	Slovenia	5.05		1.4
32	United Arab Emirates	4.84		1.5
33	Panama	4.76		1.4
34	South Africa	4.75		1.2
35	Italy	4.74		1.4
36	Costa Rica	4.61		1.5
37	Barbados	4.60		1.2
37	Estonia	4.60		1.3
39	Cyprus	4.57		1.2
40	Bahrain	4.55		1.6
41	Thailand	4.48		1.2
42	Kazakhstan	4.47		1.7
43	Greece	4.44		1.3
44	Kuwait	4.43		1.5
45	Czech Republic	4.41		1.5
46	Philippines	4.40		1.5
47	Jamaica	4.39		1.5
48	Qatar	4.35		1.5
49	Mauritius	4.33		1.4
50	Trinidad and Tobago	4.32		1.8
51	Portugal	4.31		1.0
52	Malta	4.31		1.5
53	Russian Federation	4.27		1.7
54	Turkey	4.20		1.4
55	El Salvador	4.14		1.5
56	Azerbaijan	4.14		1.7
57	Argentina	4.13		1.4
58	Brazil	4.13		1.5
59	Mexico	4.11		1.5
60	Poland	3.99		1.1
61	Sri Lanka	3.99		1.7
62	China	3.95		1.4
63	Lithuania	3.94		1.5
64	Latvia	3.91		1.6
65	Namibia	3.90		1.5
66	Croatia	3.88		1.6
67	Pakistan	3.86		1.5
68	Ukraine	3.83		1.7
69	Romania	3.82		1.7
70	Vietnam	3.82		1.5
71	Jordan	3.80		1.6
72	Dominican Republic	3.74		1.4
73	Peru	3.73		1.5
74	Cambodia	3.67		1.6
75	Colombia	3.65		1.6
76	Guatemala	3.64		1.3
77	Botswana	3.59		1.4
78	Slovak Republic	3.59		1.5
79	Uruguay	3.58		1.3
80	Venezuela	3.58		1.3
81	Tanzania	3.57		1.7
82	Kenya	3.57		1.8
83	Nigeria	3.49		1.9
84	Hungary	3.48		1.4
85	Bangladesh	3.47		1.5
86	Georgia	3.35		1.3
87	Bulgaria	3.24		1.7
88	Mongolia	3.15		1.5
89	Algeria	3.14		1.5
90	Zimbabwe	3.14		1.1
91	Egypt	3.03		1.8
92	Armenia	3.03		1.3
93	Morocco	3.01		1.6
94	Guyana	3.00		1.6
94	Nicaragua	3.00		1.6
96	Uganda	2.98		1.7
97	Albania	2.96		1.4
98	Macedonia, FYR	2.94		1.5
99	Moldova	2.93		1.4
100	Malawi	2.92		1.5
101	Bosnia and Herzegovina	2.88		1.5
102	Ecuador	2.85		1.4
103	Serbia and Montenegro	2.79		1.6
104	Kyrgyz Republic	2.78		1.6
105	Nepal	2.75		1.2
106	Honduras	2.72		1.4
107	Lesotho	2.71		1.5
108	Suriname	2.69		1.3
109	Paraguay	2.64		1.3
110	Benin	2.58		1.6
111	Mozambique	2.51		1.4
112	Ethiopia	2.47		1.3
113	Mali	2.36		1.5
114	Angola	2.34		0.9
115	Bolivia	2.30		1.1
116	Mauritania	2.29		1.5
117	Burkina Faso	2.21		1.2
118	Cameroon	2.18		1.2
119	Madagascar	2.08		1.2
120	Zambia	1.88		1.3
121	Burundi	1.79		1.1
122	Chad	1.61		1.1

304

SOURCE: World Economic Forum, Executive Opinion Survey 2006

4.06 Residential telephone connection charge

One-time residential telephone connection charge (US$) as a percentage of GDP per capita, 2005 or most recent year available

RANK	COUNTRY/ECONOMY	HARD DATA
1	Singapore	0.07
2	Luxembourg	0.09
3	Trinidad and Tobago[1]	0.11
4	United States[1]	0.11
5	New Zealand	0.12
6	Turkey[1]	0.12
7	Qatar	0.13
8	Canada	0.13
9	Greece	0.17
10	United Arab Emirates[1]	0.18
11	France	0.20
12	Germany	0.21
13	Switzerland	0.22
14	Belgium	0.22
15	Romania	0.22
16	Norway	0.24
17	Hong Kong SAR	0.24
18	Malaysia	0.27
19	Spain	0.27
20	Guyana	0.27
21	Bahrain	0.29
22	Jamaica	0.29
23	Ireland	0.30
24	Sweden	0.32
25	Finland	0.32
26	Denmark	0.32
27	Israel[1]	0.33
28	Austria	0.33
29	United Kingdom	0.35
30	Korea, Rep.	0.36
31	Cyprus[1]	0.39
32	Barbados	0.44
33	Australia	0.46
34	Slovak Republic[1]	0.47
35	Malta	0.49
36	Chile	0.50
37	Brazil	0.51
38	Tunisia	0.54
39	Slovenia[1]	0.57
40	Kuwait[1]	0.60
41	Portugal	0.60
42	Taiwan, China	0.62
43	Estonia	0.65
44	Botswana	0.66
45	Mauritius	0.69
46	Venezuela[1]	0.73
47	South Africa	0.75
48	Panama	0.83
49	Uruguay	0.85
50	Macedonia, FYR	0.88
51	Latvia	0.89
52	Japan	0.96
53	Costa Rica[1]	0.97
54	Croatia	0.97
55	Argentina	1.09
56	Ecuador	1.16
57	Namibia	1.16
58	Pakistan	1.17
59	Bulgaria	1.26
60	Algeria	1.33
61	Lithuania	1.34
62	Czech Republic[1]	1.46
63	Mexico	1.46
64	Dominican Republic	1.48
65	Hungary	1.54
66	Ukraine	1.81
67	El Salvador	2.05
68	Madagascar	2.09
69	Jordan	2.17
70	Zambia[1]	2.17
71	Honduras	2.18
72	Armenia	2.37
73	Indonesia	2.40
74	Serbia and Montenegro	2.76
75	Mauritania	2.86
76	Thailand	3.05
77	Philippines	3.07
78	India	3.30
79	Bolivia	3.75
80	Morocco[1]	4.03
81	Moldova	4.48
82	Peru[1]	4.61
83	Kyrgyz Republic[1]	4.72
84	Albania[1]	4.92
85	Colombia[1]	5.16
86	Mozambique	5.17
87	Bosnia and Herzegovina	5.19
88	Kenya	5.39
89	Azerbaijan	5.70
90	Cameroon	5.81
91	Malawi	6.15
92	Vietnam	6.15
93	Suriname[1]	6.74
94	Nepal	7.96
95	Mongolia[1]	7.97
96	Lesotho	8.21
97	Egypt	8.31
98	Mali	8.55
99	Burundi	8.82
100	Zimbabwe	9.50
101	Paraguay	9.75
102	Tanzania	10.48
103	Burkina Faso	10.72
104	Nigeria[1]	13.55
105	Nicaragua	13.72
106	Chad	14.97
107	Guatemala[1]	16.20
108	Sri Lanka	16.34
109	Benin	18.49
110	Uganda	22.90
111	Ethiopia[1]	30.05
112	Bangladesh	39.44
n/a	Angola	n/a
n/a	Cambodia	n/a
n/a	China	n/a
n/a	Georgia	n/a
n/a	Iceland	n/a
n/a	Italy	n/a
n/a	Kazakhstan	n/a
n/a	Netherlands	n/a
n/a	Poland	n/a
n/a	Russian Federation	n/a

305

SOURCE: International Telecommunication Union, *World Telecommunication Indicators 2006*; International Monetary Fund, *World Economic Outlook Online Database* (April 2006 and September 2006 editions)

[1] 2004

4.07 Residential monthly telephone subscription

Residential monthly telephone subscription to the public switched telephone network (US$) as a percentage of monthly GDP per capita, 2005 or most recent year available

RANK	COUNTRY/ECONOMY	HARD DATA		RANK	COUNTRY/ECONOMY	HARD DATA
1	Taiwan, China	0.12		62	Venezuela	1.91
2	United Arab Emirates	0.18		63	Uruguay	2.05
3	Bahrain	0.20		64	Honduras	2.13
4	Singapore	0.23		65	Armenia	2.13
5	Qatar	0.25		66	China	2.16
6	Serbia and Montenegro	0.26		67	Turkey[1]	2.26
7	Italy[1]	0.28		68	Bulgaria	2.26
8	Luxembourg	0.33		69	Mexico	2.42
9	Korea, Rep.	0.38		70	Poland[1]	2.48
10	Norway	0.46		71	Jamaica	2.65
11	Switzerland	0.46		72	Moldova	2.69
12	Finland	0.48		73	Namibia	2.83
13	Denmark	0.48		74	Ecuador[1]	2.97
14	Australia	0.48		75	South Africa	3.09
15	Sweden	0.49		76	Paraguay	3.12
16	Japan	0.51		77	Guatemala	3.17
17	Kuwait[1]	0.51		78	Indonesia	3.18
18	Trinidad and Tobago[1]	0.53		79	Jordan	3.21
19	Hong Kong SAR	0.55		80	Guyana	3.22
20	Israel[1]	0.55		81	Vietnam	3.32
21	Canada	0.56		82	Macedonia, FYR	3.40
22	Azerbaijan	0.56		83	Kyrgyz Republic[1]	3.56
23	France	0.60		84	El Salvador	3.88
24	Austria	0.62		85	Cameroon	4.07
25	Malta	0.63		86	Brazil	4.10
26	United Kingdom	0.65		87	Burundi	4.45
27	Spain	0.67		88	India	4.75
28	Suriname[1]	0.67		89	Pakistan	4.89
29	Germany	0.68		90	Sri Lanka	4.91
30	Belgium	0.70		91	Zambia[1]	5.22
31	Netherlands[1]	0.70		92	Bolivia	5.34
32	Ireland	0.72		93	Dominican Republic	5.86
33	United States[1]	0.73		94	Malawi	6.15
34	Mauritius	0.75		95	Nicaragua	6.26
35	Algeria	0.80		96	Zimbabwe	6.33
36	Slovenia[1]	0.82		97	Peru[1]	6.81
37	Greece	0.85		98	Colombia	7.60
38	Botswana	0.86		99	Bangladesh[1]	7.72
39	Tunisia	0.86		100	Morocco[1]	7.75
40	Albania[1]	0.93		101	Ethiopia[1]	8.11
41	Estonia	0.93		102	Mauritania	8.93
42	Cyprus	0.94		103	Nigeria[1]	9.03
43	Panama	1.01		104	Nepal	10.62
44	Latvia	1.08		105	Philippines	11.14
45	Thailand	1.09		106	Benin[1]	11.92
46	Bosnia and Herzegovina[1]	1.10		107	Chad	11.98
47	Argentina	1.15		108	Burkina Faso	12.87
48	Costa Rica[1]	1.16		109	Tanzania	13.37
49	New Zealand	1.23		110	Kenya	14.05
50	Portugal	1.27		111	Lesotho	15.63
51	Czech Republic	1.30		112	Uganda	22.90
52	Egypt	1.32		113	Mali	23.05
53	Romania	1.35		114	Madagascar[1]	25.62
54	Ukraine	1.36		115	Mozambique	28.70
55	Croatia	1.39		n/a	Angola	n/a
56	Chile	1.40		n/a	Cambodia	n/a
57	Slovak Republic[1]	1.42		n/a	Georgia	n/a
58	Lithuania[1]	1.48		n/a	Iceland	n/a
59	Malaysia	1.60		n/a	Kazakhstan	n/a
60	Barbados	1.84		n/a	Mongolia	n/a
61	Hungary	1.85		n/a	Russian Federation	n/a

SOURCE: International Telecommunication Union, *World Telecommunication Indicators 2006*; International Monetary Fund, *World Economic Outlook Online Database* (April 2006 and September 2006 editions)

[1] 2004

4.08 High-speed monthly broadband subscription charge

High-speed monthly broadband subscription charge (US$) as a percentage of monthly GDP per capita, 2006

RANK	COUNTRY/ECONOMY	HARD DATA
1	Malta	0.13
2	Japan	0.48
3	United States	0.57
4	Netherlands	0.87
5	Switzerland	0.93
6	Germany	1.10
7	France	1.32
8	Luxembourg	1.32
9	Norway	1.36
10	Canada	1.41
11	Italy	1.48
12	United Kingdom	1.65
13	Denmark	1.66
14	Belgium	1.67
15	Ireland	1.67
16	Sweden	1.70
17	Taiwan, China	1.79
18	Australia	1.84
19	New Zealand	1.91
20	Iceland	2.08
21	Spain	2.19
22	Austria	2.21
23	Hong Kong SAR	2.41
24	Finland	2.74
25	Korea, Rep.	2.99
26	Qatar	3.06
27	Singapore	3.27
28	Israel	3.76
29	Slovenia	3.96
30	United Arab Emirates	4.12
31	Lithuania	4.58
32	Estonia	4.85
33	Portugal	5.21
34	Cyprus	5.74
35	Malaysia	6.22
36	Greece	6.40
37	Argentina	6.65
38	Hungary	6.78
39	Czech Republic	6.84
40	Trinidad and Tobago	6.98
41	Bahrain	8.65
42	Jordan	8.66
43	Romania	8.71
44	Croatia	8.99
45	China	10.27
46	Kuwait	11.03
47	Thailand	11.20
48	Poland	11.32
49	Chile	11.66
50	Botswana	14.02
51	Bosnia and Herzegovina	19.10
52	Barbados	19.72
53	Ukraine	20.87
54	Bulgaria	21.82
55	Jamaica	22.95
56	Brazil	23.95
57	Slovak Republic	24.19
58	Dominican Republic	27.15
59	Colombia	29.07
60	Peru	29.24
61	South Africa	29.46

RANK	COUNTRY/ECONOMY	HARD DATA
62	Panama	30.02
63	Georgia	31.33
64	Serbia and Montenegro	31.39
65	India	31.81
66	Ecuador	34.72
67	Venezuela	39.69
68	Uruguay	44.51
69	Indonesia	48.10
70	Turkey	51.07
71	Mauritius	51.60
72	Latvia	54.10
73	Morocco	62.51
74	Russian Federation	64.61
75	Sri Lanka	67.13
76	Tunisia	67.41
77	Mexico	69.35
78	Suriname	75.48
79	Costa Rica	85.12
80	Guatemala	86.63
81	Philippines	93.19
82	Azerbaijan	102.41
83	Nicaragua	105.83
84	Macedonia, FYR	105.91
85	Egypt	117.89
86	Albania	218.09
87	Bolivia	241.77
88	Vietnam	291.95
89	Kyrgyz Republic	304.21
90	Algeria	344.97
91	Mongolia	487.70
92	Moldova	569.85
93	Kenya	1,016.23
94	Kazakhstan	1,367.86
95	Cameroon	1,537.45
96	Pakistan	1,822.41
97	Cambodia	1,949.22
98	Zimbabwe	3,646.23
99	Benin	3,838.87
100	Burkina Faso	4,233.13
101	Armenia	4,983.85
102	Mozambique	6,709.07
103	Bangladesh	6,727.33
104	Mauritania	12,933.07
105	Uganda	22,816.70
106	Ethiopia	36,831.12
n/a	Angola	n/a
n/a	Burundi	n/a
n/a	Chad	n/a
n/a	El Salvador	n/a
n/a	Guyana	n/a
n/a	Honduras	n/a
n/a	Lesotho	n/a
n/a	Madagascar	n/a
n/a	Malawi	n/a
n/a	Mali	n/a
n/a	Namibia	n/a
n/a	Nepal	n/a
n/a	Nigeria	n/a
n/a	Paraguay	n/a
n/a	Tanzania	n/a
n/a	Zambia	n/a

SOURCE: International Telecommunication Union, *World Information Society Report 2006*; International Monetary Fund, *World Economic Outlook Online Database* (April 2006 and September 2006 editions)

4.09   Lowest cost of broadband

Lowest sampled cost (US$) per 100 kbits/s as a percentage of monthly income (GNI), 2006

RANK	COUNTRY/ECONOMY	HARD DATA
1	Japan	0.00
2	Finland	0.01
2	France	0.01
2	Italy	0.01
2	Korea, Rep.	0.01
2	Netherlands	0.01
2	Singapore	0.01
2	Sweden	0.01
2	United States	0.01
10	Germany	0.02
10	Taiwan, China	0.02
10	United Kingdom	0.02
13	Malta	0.03
14	Canada	0.04
14	Hong Kong SAR	0.04
14	Norway	0.04
14	Switzerland	0.04
18	Belgium	0.05
18	Iceland	0.05
20	Luxembourg	0.06
21	Portugal	0.08
22	Denmark	0.10
23	Ireland	0.11
23	Slovenia	0.11
25	Austria	0.12
25	New Zealand	0.12
27	Australia	0.15
27	Lithuania	0.15
29	Czech Republic	0.18
30	Israel	0.20
31	Poland	0.24
32	Spain	0.27
33	Brazil	0.42
34	Cyprus	0.43
34	Hungary	0.43
36	Slovak Republic	0.51
37	Bosnia and Herzegovina	0.54
38	Croatia	0.58
39	Malaysia	0.66
40	Estonia	0.67
41	Greece	0.76
42	Chile	0.83
43	Argentina	0.84
44	Romania	0.88
45	Jordan	0.92
46	Mexico	1.11
47	Thailand	1.14
48	Barbados	1.15
49	China	1.33
50	Bulgaria	1.34
51	Ukraine	1.43
52	Kuwait	1.56
53	Morocco	1.72
54	Philippines	1.82
55	Costa Rica	2.06
56	Venezuela	2.17
57	Uruguay	2.38
58	Bahrain	2.50
59	Macedonia, FYR	2.62
59	Mauritius	2.62
61	Botswana	2.71

RANK	COUNTRY/ECONOMY	HARD DATA
62	Jamaica	2.83
63	Dominican Republic	2.92
64	Latvia	3.32
65	Turkey	3.37
66	El Salvador	3.49
67	Sri Lanka	3.89
68	Peru	3.91
69	Trinidad and Tobago	4.01
70	South Africa	4.05
71	Georgia	4.35
72	Serbia and Montenegro	4.90
73	Panama	5.26
74	Indonesia	5.29
75	Egypt	5.56
76	Paraguay	6.01
77	Colombia	6.43
78	Tunisia	6.69
79	India	7.07
80	Vietnam	8.01
81	Albania	8.40
82	Guyana	8.57
83	Ecuador	8.59
84	Russian Federation	9.90
85	Guatemala	10.56
86	Honduras	15.17
87	Suriname	17.28
88	Nicaragua	22.25
89	Kazakhstan	27.46
90	Azerbaijan	31.43
91	Madagascar	33.44
92	Algeria	45.60
93	Bolivia	48.83
94	Mongolia	59.39
95	Moldova	67.47
96	Kyrgyz Republic	70.31
97	Cameroon	71.49
98	Benin	209.49
99	Kenya	215.42
100	Pakistan	215.80
101	Burkina Faso	246.73
102	Armenia	247.65
103	Cambodia	255.98
104	Angola	396.39
105	Bangladesh	1,194.73
106	Mozambique	1,736.78
107	Nepal	1,754.51
108	Mauritania	1,992.63
109	Ethiopia	2,501.70
110	Uganda	5,000.00
111	Burundi	8,437.50
n/a	Chad	n/a
n/a	Lesotho	n/a
n/a	Malawi	n/a
n/a	Mali	n/a
n/a	Namibia	n/a
n/a	Nigeria	n/a
n/a	Qatar	n/a
n/a	Tanzania	n/a
n/a	United Arab Emirates	n/a
n/a	Zambia	n/a
n/a	Zimbabwe	n/a

SOURCE: International Telecommunication Union, *World Information Society Report 2006*

4.10 Cost of mobile telephone call

Cost of 3-minute local call during peak hours (US$) as a percentage of monthly GDP per capita, 2005 or most recent year available

RANK	COUNTRY/ECONOMY	HARD DATA
1	United Arab Emirates[1]	0.00
2	Hong Kong SAR[1]	0.01
3	Luxembourg[1]	0.01
4	Norway[1]	0.01
5	Latvia	0.01
6	Iceland	0.01
7	Denmark[1]	0.01
8	Cyprus[1]	0.02
9	Slovenia	0.02
10	Singapore	0.02
11	Finland	0.02
12	Qatar[1]	0.02
13	Bahrain	0.02
14	Canada	0.02
15	Switzerland[1]	0.02
16	Korea, Rep.	0.02
17	Ireland	0.03
18	Mauritius	0.03
19	Portugal[1]	0.03
20	Spain	0.03
21	United States	0.04
22	Austria	0.04
23	Guatemala[1]	0.04
24	Belgium	0.05
25	Taiwan, China[1]	0.05
26	Barbados	0.05
27	Costa Rica	0.05
28	Germany	0.05
29	France	0.06
30	Estonia[1]	0.07
31	Mexico	0.07
32	Greece	0.07
33	Trinidad and Tobago	0.07
34	Kuwait[1]	0.07
35	Italy[1]	0.07
36	Hungary[1]	0.08
37	Australia	0.08
38	Algeria	0.09
39	Croatia[1]	0.09
40	Czech Republic[1]	0.10
41	Poland	0.11
42	Malaysia[1]	0.11
43	Indonesia[1]	0.11
44	El Salvador	0.11
45	India	0.12
46	Russian Federation[1]	0.12
47	New Zealand[1]	0.13
48	Lithuania[1]	0.13
49	Panama	0.14
50	Tunisia	0.14
51	Jordan[1]	0.14
52	Malta	0.15
53	Romania	0.15
54	Argentina	0.16
55	Jamaica	0.16
56	Thailand	0.16
57	China[1]	0.18
58	Botswana	0.19
59	Serbia and Montenegro	0.20
60	Dominican Republic	0.21
61	Egypt	0.25
62	South Africa	0.29
63	Vietnam	0.29
64	Pakistan	0.29
65	Venezuela	0.32
66	Sri Lanka	0.33
67	Suriname	0.34
68	Paraguay	0.34
69	Bosnia and Herzegovina[1]	0.35
70	Brazil	0.37
71	Bulgaria	0.38
72	Ukraine	0.39
73	Georgia	0.39
74	Philippines[1]	0.40
75	Namibia	0.45
76	Nepal	0.48
77	Peru	0.49
78	Macedonia, FYR	0.59
79	Cambodia[1]	0.59
80	Nigeria[1]	0.65
81	Ecuador	0.65
82	Guyana	0.71
83	Bolivia	0.75
84	Kyrgyz Republic[1]	0.77
85	Honduras	0.79
86	Bangladesh	0.85
87	Morocco[1]	0.87
88	Lesotho[1]	0.88
89	Mauritania	0.93
90	Moldova[1]	1.01
91	Tanzania[1]	1.17
92	Mozambique[1]	1.19
93	Armenia[1]	1.22
94	Mongolia[1]	1.39
95	Zambia[1]	1.53
96	Nicaragua	1.69
97	Cameroon[1]	1.72
98	Ethiopia	1.87
99	Benin[1]	2.07
100	Burkina Faso	2.09
101	Kenya[1]	2.49
102	Uganda[1]	2.76
103	Madagascar[1]	3.13
104	Mali[1]	3.43
105	Zimbabwe	4.32
106	Malawi[1]	4.38
107	Burundi	5.72
n/a	Albania	n/a
n/a	Angola	n/a
n/a	Azerbaijan	n/a
n/a	Chad	n/a
n/a	Chile	n/a
n/a	Colombia	n/a
n/a	Israel	n/a
n/a	Japan	n/a
n/a	Kazakhstan	n/a
n/a	Netherlands	n/a
n/a	Slovak Republic	n/a
n/a	Sweden	n/a
n/a	Turkey	n/a
n/a	United Kingdom	n/a
n/a	Uruguay	n/a

SOURCE: International Telecommunication Union, *World Telecommunication Indicators 2006*; International Monetary Fund, *World Economic Outlook Online Database* (April 2006 and September 2006 editions)

[1] 2004

Pillar 5
Business readiness

5.01 Extent of staff training

The general approach of companies in your country to human resources is (1 = to invest little in training and employee development, 7 = to invest heavily to attract, train, and retain employees)

RANK	COUNTRY/ECONOMY	SCORE	MEAN: 3.82 (1 → 7)	SD
1	Switzerland	6.07		0.8
2	Denmark	5.93		0.9
3	Japan	5.88		0.9
3	Sweden	5.88		0.8
5	Austria	5.77		0.9
6	Netherlands	5.70		0.8
7	Germany	5.67		1.0
8	Finland	5.62		0.9
9	United States	5.57		1.2
10	Norway	5.52		0.9
11	Iceland	5.45		1.0
12	Singapore	5.43		1.0
13	Belgium	5.42		1.1
14	Luxembourg	5.42		1.1
15	Ireland	5.36		1.0
16	United Kingdom	5.29		1.2
17	Malaysia	5.27		1.1
18	Korea, Rep.	5.19		1.0
19	Taiwan, China	5.18		1.0
20	Australia	5.11		1.0
21	Hong Kong SAR	5.09		1.1
22	France	5.07		1.2
23	Israel	5.06		1.0
24	Canada	5.02		1.2
25	New Zealand	4.96		0.8
26	South Africa	4.90		1.1
27	Estonia	4.76		1.1
28	India	4.76		1.3
29	Czech Republic	4.72		1.1
30	Thailand	4.63		1.1
31	Costa Rica	4.54		1.2
32	Slovenia	4.43		1.2
33	Mauritius	4.41		1.0
34	Chile	4.40		1.1
35	Philippines	4.33		1.2
36	Tunisia	4.26		1.5
37	United Arab Emirates	4.26		1.6
38	Brazil	4.25		1.3
39	Turkey	4.19		1.1
40	Indonesia	4.18		0.9
41	Spain	4.04		1.2
42	Latvia	4.03		1.4
43	Lithuania	4.03		1.4
44	Malta	4.00		1.2
45	Slovak Republic	3.98		1.2
46	Kuwait	3.92		1.6
47	Mexico	3.86		1.3
48	Poland	3.84		0.9
49	Barbados	3.84		1.3
50	Zimbabwe	3.83		1.2
51	Trinidad and Tobago	3.81		1.4
52	Greece	3.81		1.2
53	Hungary	3.80		1.3
54	Jamaica	3.78		1.2
55	Portugal	3.76		0.9
56	Kenya	3.74		1.5
57	Qatar	3.72		1.7
58	Panama	3.72		1.3
59	Bahrain	3.68		1.8
60	Jordan	3.62		1.5
61	Croatia	3.60		1.3
62	Italy	3.60		1.3
63	El Salvador	3.58		1.2
64	Namibia	3.57		1.3
65	Nigeria	3.52		1.7
66	Peru	3.52		1.3
67	Colombia	3.51		1.2
68	Botswana	3.48		1.4
69	Macedonia, FYR	3.47		1.5
70	Cyprus	3.45		1.1
71	Vietnam	3.43		1.4
72	Argentina	3.43		1.1
73	Guatemala	3.43		1.2
74	Uganda	3.40		1.7
75	Sri Lanka	3.39		1.3
76	China	3.36		1.3
77	Mauritania	3.34		1.8
78	Mozambique	3.33		1.4
79	Kazakhstan	3.29		1.4
80	Uruguay	3.29		1.1
81	Romania	3.29		1.2
82	Venezuela	3.28		1.1
83	Egypt	3.27		1.7
84	Mongolia	3.23		1.6
85	Morocco	3.20		1.5
86	Malawi	3.16		1.4
87	Tanzania	3.11		1.3
88	Cambodia	3.08		1.5
89	Dominican Republic	3.07		1.3
90	Pakistan	3.06		1.5
91	Lesotho	3.05		1.4
92	Honduras	3.05		1.4
93	Azerbaijan	3.03		1.3
94	Bosnia and Herzegovina	3.00		1.5
94	Georgia	3.00		1.2
96	Guyana	2.99		1.3
97	Algeria	2.97		1.4
98	Russian Federation	2.94		1.3
99	Ukraine	2.94		1.2
100	Madagascar	2.93		1.3
101	Moldova	2.89		1.3
102	Armenia	2.81		1.3
103	Ecuador	2.81		1.1
104	Nicaragua	2.74		1.2
105	Burkina Faso	2.72		1.5
106	Angola	2.70		1.2
107	Benin	2.69		1.5
108	Suriname	2.68		1.2
109	Albania	2.67		1.3
110	Cameroon	2.66		1.4
111	Bulgaria	2.64		1.2
112	Paraguay	2.59		1.1
113	Bangladesh	2.55		1.1
114	Bolivia	2.49		1.0
115	Mali	2.49		1.3
116	Nepal	2.44		1.0
117	Ethiopia	2.40		1.0
118	Kyrgyz Republic	2.36		1.3
119	Serbia and Montenegro	2.28		1.1
120	Burundi	2.20		1.3
121	Chad	1.82		1.2
122	Zambia	1.70		1.1

5.02 Local availability of specialized research and training services

In your country, specialized research and training services are (1 = not available, 7 = available from world-class local institutions)

RANK	COUNTRY/ECONOMY	SCORE	1 MEAN: 3.94 7	SD
1	Germany	6.08		0.8
1	Japan	6.08		0.8
3	United Kingdom	6.04		1.1
4	Switzerland	6.04		0.9
5	United States	5.96		1.1
6	Finland	5.86		0.7
7	Netherlands	5.83		0.8
8	Sweden	5.78		0.7
9	Belgium	5.77		1.0
10	Israel	5.60		1.1
11	Denmark	5.58		0.9
12	France	5.57		1.1
13	Canada	5.57		1.0
14	Austria	5.26		1.0
15	Norway	5.25		0.9
16	Australia	5.19		1.2
17	Singapore	5.14		1.1
18	Hong Kong SAR	5.09		1.2
19	Iceland	5.07		1.2
20	Czech Republic	4.98		1.1
21	Taiwan, China	4.97		1.1
22	Malaysia	4.94		1.3
23	Ireland	4.91		0.9
24	Indonesia	4.83		0.8
25	New Zealand	4.82		1.1
26	Estonia	4.80		1.2
27	Italy	4.78		1.6
28	India	4.73		1.4
29	Korea, Rep.	4.72		1.1
30	South Africa	4.69		1.1
31	Chile	4.64		1.2
32	Brazil	4.63		1.4
33	Tunisia	4.62		1.4
34	Croatia	4.47		1.3
35	Spain	4.40		1.2
36	Hungary	4.39		1.2
37	Slovenia	4.38		1.2
38	Portugal	4.37		1.2
39	Poland	4.37		0.9
40	Costa Rica	4.33		1.4
41	Turkey	4.32		1.1
42	United Arab Emirates	4.31		1.6
43	Argentina	4.26		1.5
44	Romania	4.25		1.3
45	Slovak Republic	4.22		1.2
46	China	4.22		1.3
47	Mexico	4.19		1.4
48	Lithuania	4.17		1.1
49	Kenya	4.15		1.5
50	Latvia	4.14		1.3
51	Luxembourg	4.09		1.5
52	Serbia and Montenegro	4.02		1.5
53	Kuwait	3.99		1.5
54	Tanzania	3.98		1.7
55	Guatemala	3.93		1.3
56	Jamaica	3.91		1.5
57	Qatar	3.90		1.9
58	Morocco	3.89		1.7
59	Greece	3.88		1.3
60	Nigeria	3.88		1.8
61	Uganda	3.88		1.6
62	Jordan	3.86		1.5
63	Peru	3.86		1.5
64	Uruguay	3.83		1.3
65	Cyprus	3.81		1.5
66	Thailand	3.80		1.2
67	Kazakhstan	3.73		1.2
68	Azerbaijan	3.72		1.7
69	Russian Federation	3.70		1.4
70	Colombia	3.70		1.3
71	Bosnia and Herzegovina	3.64		1.4
72	Trinidad and Tobago	3.64		1.5
73	Panama	3.58		1.4
74	El Salvador	3.55		1.4
75	Philippines	3.55		1.4
76	Vietnam	3.52		1.5
77	Barbados	3.51		1.5
78	Mongolia	3.51		1.5
79	Egypt	3.49		1.5
80	Bulgaria	3.48		1.5
81	Nicaragua	3.41		1.4
82	Sri Lanka	3.41		1.5
83	Pakistan	3.38		1.5
84	Honduras	3.37		1.5
85	Mauritius	3.37		1.4
86	Ukraine	3.37		1.1
87	Macedonia, FYR	3.33		1.3
88	Malta	3.31		1.5
89	Dominican Republic	3.30		1.6
90	Ecuador	3.28		1.4
91	Burkina Faso	3.28		1.6
92	Venezuela	3.25		1.4
93	Mali	3.23		1.9
94	Moldova	3.20		1.3
95	Cambodia	3.11		1.4
96	Bahrain	3.11		1.7
97	Cameroon	3.08		1.5
98	Benin	3.06		1.6
99	Madagascar	3.06		1.4
100	Algeria	2.98		1.4
101	Botswana	2.97		1.4
102	Bolivia	2.94		1.4
103	Kyrgyz Republic	2.89		1.4
104	Armenia	2.88		1.3
105	Malawi	2.86		1.6
106	Georgia	2.80		1.3
107	Paraguay	2.75		1.4
107	Zimbabwe	2.75		1.4
109	Ethiopia	2.64		1.4
110	Mozambique	2.63		1.3
111	Guyana	2.55		1.3
112	Nepal	2.53		1.1
113	Suriname	2.52		1.3
114	Chad	2.52		1.7
115	Lesotho	2.49		1.5
116	Albania	2.46		1.3
117	Namibia	2.43		1.1
118	Bangladesh	2.40		1.4
119	Mauritania	2.22		1.8
120	Angola	2.21		1.3
121	Burundi	2.20		1.4
122	Zambia	2.15		1.2

SOURCE: World Economic Forum, Executive Opinion Survey 2006

313

5.03   Quality of management schools

Management or business schools in your country are (1 = limited or of poor quality, 7 = among the best in the world)

RANK	COUNTRY/ECONOMY	SCORE	SD
1	France	6.18	0.8
2	Switzerland	6.03	1.0
3	India	5.96	0.9
4	Canada	5.88	1.1
5	United Kingdom	5.87	1.0
6	United States	5.85	1.3
7	Belgium	5.80	1.0
8	Singapore	5.70	0.8
9	Finland	5.69	1.0
10	Denmark	5.68	0.8
11	Spain	5.58	1.2
12	Netherlands	5.57	0.9
13	Hong Kong SAR	5.56	1.1
14	Israel	5.55	1.1
15	Ireland	5.51	1.0
16	Iceland	5.47	0.9
17	Australia	5.39	1.0
18	Chile	5.38	0.9
19	South Africa	5.37	1.0
20	Tunisia	5.33	1.1
21	Sweden	5.29	0.9
22	Malaysia	5.29	1.3
23	Norway	5.25	0.9
24	Taiwan, China	5.20	0.8
25	New Zealand	5.13	1.1
26	Morocco	5.08	1.2
27	Costa Rica	5.08	1.0
28	Germany	5.00	1.2
29	Argentina	4.96	1.2
30	Estonia	4.94	1.0
31	Austria	4.94	1.3
32	Portugal	4.92	0.9
33	Hungary	4.87	1.0
34	Indonesia	4.87	0.7
35	Thailand	4.83	1.1
36	Czech Republic	4.78	1.1
37	Barbados	4.77	1.1
38	Colombia	4.72	1.1
39	Latvia	4.70	1.3
40	Qatar	4.65	1.3
41	Trinidad and Tobago	4.58	1.5
42	Slovenia	4.55	1.2
43	Mexico	4.52	1.2
44	Malta	4.50	1.2
45	Peru	4.48	1.1
46	Philippines	4.47	1.4
47	Italy	4.45	1.5
48	Jamaica	4.43	1.2
49	Cyprus	4.42	1.1
50	Lithuania	4.42	1.1
51	Uruguay	4.39	1.1
52	United Arab Emirates	4.39	1.5
53	Korea, Rep.	4.34	1.1
54	Croatia	4.31	1.5
55	Poland	4.30	1.0
56	Guatemala	4.28	1.2
57	El Salvador	4.25	1.2
58	Slovak Republic	4.23	1.1
59	Japan	4.22	1.3
60	Kuwait	4.21	1.4
61	Turkey	4.20	1.3
62	Venezuela	4.18	1.2
63	Benin	4.14	1.4
64	Brazil	4.11	1.4
65	Nicaragua	4.01	1.4
66	Sri Lanka	3.97	1.5
67	Serbia and Montenegro	3.90	1.6
68	Greece	3.90	1.5
69	Mauritius	3.89	1.3
70	Romania	3.87	1.6
71	Pakistan	3.86	1.4
72	Kenya	3.86	1.4
73	Madagascar	3.86	1.3
74	Burkina Faso	3.83	1.3
75	Zimbabwe	3.80	1.4
76	Jordan	3.74	1.5
77	Bahrain	3.74	1.7
78	Dominican Republic	3.68	1.2
79	Luxembourg	3.65	1.5
80	Bosnia and Herzegovina	3.64	1.5
81	Ecuador	3.64	1.2
82	Bulgaria	3.62	1.3
83	Panama	3.60	1.3
84	Russian Federation	3.57	1.3
85	Macedonia, FYR	3.57	1.4
86	Kazakhstan	3.57	1.2
87	Mali	3.56	1.5
88	Egypt	3.55	1.4
89	Ukraine	3.52	1.0
90	Algeria	3.47	1.3
91	Cameroon	3.43	1.4
92	China	3.42	1.3
93	Botswana	3.41	1.3
94	Nigeria	3.39	1.6
95	Suriname	3.37	1.2
96	Tanzania	3.33	1.5
97	Honduras	3.33	1.2
98	Bangladesh	3.30	1.4
99	Uganda	3.27	1.6
100	Guyana	3.26	1.3
101	Moldova	3.21	1.3
102	Zambia	3.20	1.2
103	Nepal	3.19	1.3
104	Cambodia	3.17	1.3
105	Azerbaijan	3.10	1.4
106	Georgia	3.10	1.2
107	Paraguay	3.06	1.1
108	Armenia	3.05	1.3
109	Ethiopia	3.04	1.4
110	Kyrgyz Republic	2.98	1.2
111	Albania	2.97	1.3
112	Vietnam	2.97	1.3
113	Bolivia	2.97	1.2
114	Mongolia	2.95	1.3
115	Burundi	2.78	1.4
116	Mozambique	2.78	1.1
117	Namibia	2.73	1.2
118	Malawi	2.71	1.5
119	Lesotho	2.49	1.3
120	Chad	2.44	1.4
121	Mauritania	2.40	1.4
122	Angola	2.06	0.9

MEAN: 4.19

314

SOURCE: World Economic Forum, Executive Opinion Survey 2006

5.04 Company spending on research and development

Companies in your country (1 = do not spend money on research and development, 7 = spend heavily on research and development relative to international peers)

RANK	COUNTRY/ECONOMY	SCORE	1 MEAN: 3.41 7	SD
1	Switzerland	6.17		0.8
2	Japan	6.06		1.0
3	United States	5.75		1.2
4	Germany	5.71		0.9
5	Sweden	5.67		1.0
6	Finland	5.49		0.7
7	Israel	5.33		0.9
8	Denmark	5.25		0.8
9	Korea, Rep.	5.05		1.1
10	Malaysia	4.92		1.2
11	Singapore	4.90		1.1
12	Taiwan, China	4.83		0.9
13	Netherlands	4.82		1.0
14	France	4.72		1.1
15	Ireland	4.68		1.1
16	United Kingdom	4.67		1.2
17	Belgium	4.66		1.2
18	Austria	4.63		1.0
19	Norway	4.58		1.1
20	Luxembourg	4.51		1.2
21	Iceland	4.48		1.2
22	Canada	4.45		1.2
23	Hong Kong SAR	4.37		1.3
24	South Africa	4.36		1.0
25	India	4.22		1.1
26	Indonesia	4.11		1.0
27	Slovenia	4.06		1.1
28	Australia	4.05		1.1
29	Czech Republic	4.02		1.1
30	Brazil	3.84		1.3
31	Poland	3.84		1.0
32	Estonia	3.84		1.3
33	Costa Rica	3.82		1.1
34	Kenya	3.80		1.5
35	New Zealand	3.74		1.0
36	Tunisia	3.73		1.6
37	Thailand	3.72		1.1
38	Nigeria	3.59		1.7
39	China	3.58		1.2
40	Jamaica	3.46		1.1
41	Tanzania	3.45		1.4
42	Qatar	3.40		1.4
42	United Arab Emirates	3.40		1.4
44	Russian Federation	3.40		1.4
45	Slovak Republic	3.38		1.1
46	Spain	3.38		1.0
47	Cambodia	3.32		1.6
48	Chile	3.32		1.1
49	Lithuania	3.31		1.3
50	Latvia	3.31		1.4
51	Pakistan	3.31		1.5
52	Croatia	3.30		1.2
53	Portugal	3.29		0.8
54	Azerbaijan	3.28		1.4
55	Burkina Faso	3.27		1.7
56	Philippines	3.26		1.1
57	Kazakhstan	3.26		1.5
58	Barbados	3.25		1.1
59	Hungary	3.24		1.3
60	Mexico	3.22		1.0
61	Madagascar	3.19		1.5
62	Turkey	3.19		1.1
63	Sri Lanka	3.18		1.4
64	Malta	3.17		1.2
65	Trinidad and Tobago	3.16		1.2
66	Guatemala	3.13		1.2
67	Peru	3.12		1.2
68	Zimbabwe	3.11		1.2
69	Colombia	3.10		1.2
70	Romania	3.07		1.2
71	Greece	3.05		1.2
72	Uganda	3.04		1.6
73	Mauritius	3.04		1.2
74	Panama	3.04		1.2
75	Italy	3.00		1.2
75	Morocco	3.00		1.6
77	Serbia and Montenegro	2.99		1.3
78	Cyprus	2.99		1.2
79	Vietnam	2.98		1.0
80	Namibia	2.98		1.2
81	Kuwait	2.93		1.5
82	Ukraine	2.92		1.1
83	Argentina	2.91		1.0
84	Malawi	2.89		1.3
85	Bosnia and Herzegovina	2.89		1.2
86	El Salvador	2.86		1.1
87	Botswana	2.85		1.2
88	Mongolia	2.82		1.2
89	Mali	2.80		1.6
90	Algeria	2.76		1.4
91	Uruguay	2.76		0.9
92	Venezuela	2.76		1.1
93	Ecuador	2.75		1.1
94	Jordan	2.74		1.2
95	Guyana	2.73		1.2
96	Bulgaria	2.73		1.4
97	Egypt	2.70		1.5
98	Suriname	2.69		1.1
99	Armenia	2.67		1.1
100	Macedonia, FYR	2.64		1.1
101	Nepal	2.61		1.3
102	Dominican Republic	2.59		1.2
103	Cameroon	2.59		1.2
104	Mozambique	2.56		1.3
105	Nicaragua	2.54		1.1
106	Moldova	2.49		1.2
107	Georgia	2.49		1.1
108	Angola	2.49		1.2
109	Kyrgyz Republic	2.46		1.2
110	Benin	2.46		1.4
111	Bangladesh	2.44		1.3
112	Honduras	2.43		1.1
113	Bolivia	2.43		0.9
114	Bahrain	2.26		1.2
115	Burundi	2.20		1.2
116	Ethiopia	2.20		1.0
117	Chad	2.18		1.4
118	Lesotho	2.15		1.1
119	Albania	2.14		0.9
120	Paraguay	2.11		0.9
121	Zambia	1.74		1.1
122	Mauritania	1.65		1.1

SOURCE: World Economic Forum, Executive Opinion Survey 2006

5.05 University-industry research collaboration

In its R&D activity, business collaboration with local universities is (1 = minimal or nonexistent, 7 = intensive and ongoing)

RANK	COUNTRY/ECONOMY	SCORE	1 MEAN: 3.28 7	SD
1	Switzerland	5.68		1.0
2	Sweden	5.55		0.9
3	Finland	5.53		0.9
4	United States	5.52		1.4
5	Germany	5.29		1.1
6	Israel	5.24		1.2
7	Taiwan, China	5.22		1.0
8	Singapore	5.18		1.0
9	Japan	5.17		1.1
10	United Kingdom	4.90		1.2
11	Belgium	4.89		1.1
12	Malaysia	4.86		1.6
13	Netherlands	4.86		1.2
14	Canada	4.76		1.3
15	Denmark	4.75		1.1
16	Korea, Rep.	4.62		1.3
17	Hong Kong SAR	4.61		1.1
18	Norway	4.60		1.3
19	Ireland	4.57		1.2
20	Austria	4.56		1.4
21	Iceland	4.47		1.0
22	South Africa	4.34		1.1
23	New Zealand	4.24		1.3
24	Thailand	4.16		1.3
25	Australia	4.14		1.1
26	Czech Republic	4.02		1.4
27	China	3.91		1.3
28	Estonia	3.89		1.4
29	France	3.84		1.3
30	Hungary	3.82		1.9
31	Slovak Republic	3.75		1.3
32	Tunisia	3.73		1.5
33	Portugal	3.69		1.1
34	India	3.60		1.3
35	Croatia	3.58		1.6
36	Slovenia	3.57		1.4
37	Chile	3.56		1.3
38	Poland	3.56		1.0
39	Costa Rica	3.54		1.4
40	Mexico	3.49		1.3
41	Tanzania	3.48		1.5
42	Brazil	3.48		1.5
43	Luxembourg	3.47		1.5
44	Spain	3.38		1.4
45	Colombia	3.36		1.6
46	Turkey	3.35		1.3
47	Jamaica	3.35		1.3
48	United Arab Emirates	3.33		1.6
49	Nigeria	3.28		1.7
50	Kenya	3.26		1.7
51	Latvia	3.21		1.4
52	Greece	3.19		1.4
53	Sri Lanka	3.18		1.6
54	Russian Federation	3.18		1.5
55	Lithuania	3.17		1.3
56	Trinidad and Tobago	3.16		1.3
57	Guatemala	3.15		1.4
58	Macedonia, FYR	3.14		1.5
59	Uganda	3.13		1.8
60	Qatar	3.11		1.6
61	Pakistan	3.10		1.5

RANK	COUNTRY/ECONOMY	SCORE	1 MEAN: 3.28 7	SD
62	Serbia and Montenegro	3.09		1.4
63	Kazakhstan	3.03		1.4
64	Italy	3.02		1.4
65	Zimbabwe	3.00		1.4
66	Morocco	2.98		1.7
67	Philippines	2.94		1.4
68	Mauritius	2.93		1.3
69	Ukraine	2.92		1.3
70	Venezuela	2.90		1.2
71	Mongolia	2.90		1.4
72	Azerbaijan	2.89		1.6
73	Cyprus	2.89		1.3
74	Barbados	2.89		1.2
75	Burkina Faso	2.88		1.6
76	Malta	2.87		1.3
77	Romania	2.86		1.2
78	Namibia	2.85		1.1
79	Indonesia	2.84		1.1
80	Vietnam	2.82		1.2
81	Argentina	2.81		1.3
82	Madagascar	2.78		1.4
83	Panama	2.77		1.3
84	Jordan	2.76		1.4
85	Kuwait	2.76		1.4
86	Armenia	2.67		1.4
86	Botswana	2.67		1.4
88	Cambodia	2.66		1.6
89	Uruguay	2.65		1.1
90	Ecuador	2.64		1.3
91	Honduras	2.59		1.4
92	Bosnia and Herzegovina	2.59		1.4
93	Egypt	2.57		1.5
94	Mozambique	2.54		1.5
95	Bulgaria	2.50		1.2
95	Malawi	2.50		1.3
97	Peru	2.49		1.1
98	Nicaragua	2.48		1.2
99	Dominican Republic	2.47		1.4
100	Moldova	2.47		1.4
101	Algeria	2.46		1.2
102	El Salvador	2.45		1.0
103	Suriname	2.40		1.2
104	Kyrgyz Republic	2.32		1.1
105	Angola	2.26		1.1
106	Zambia	2.24		1.0
107	Guyana	2.23		1.1
108	Georgia	2.23		0.9
109	Bolivia	2.20		1.1
110	Benin	2.19		1.3
111	Ethiopia	2.18		1.1
112	Bangladesh	2.17		1.2
113	Nepal	2.17		1.1
114	Mali	2.16		1.1
115	Burundi	2.14		1.4
116	Cameroon	2.07		1.3
117	Lesotho	2.04		1.3
118	Paraguay	1.88		1.0
119	Bahrain	1.84		1.1
120	Chad	1.70		1.0
121	Mauritania	1.60		1.3
122	Albania	1.57		0.6

SOURCE: World Economic Forum, Executive Opinion Survey 2006

5.06 Business telephone connection charge

One-time business telephone connection charge (US$) as a percentage of GDP per capita, 2005 or most recent year available

RANK	COUNTRY/ECONOMY	HARD DATA
1	Switzerland	0.07
2	Singapore	0.07
3	Luxembourg	0.09
4	Turkey[1]	0.12
5	Qatar	0.13
6	New Zealand	0.14
7	Greece	0.17
8	United Arab Emirates	0.18
9	United States[1]	0.19
10	France	0.20
11	Trinidad and Tobago[1]	0.21
12	Germany	0.21
13	Belgium	0.22
14	Romania	0.22
15	Norway	0.24
16	Hong Kong SAR	0.24
17	Malaysia	0.27
18	Spain	0.27
19	Bahrain	0.29
20	Ireland	0.30
21	Finland	0.32
22	Denmark	0.32
23	Israel[1]	0.33
24	Korea, Rep.	0.36
25	Cyprus	0.39
26	Jamaica	0.42
27	Barbados	0.44
28	Australia	0.46
29	United Kingdom	0.47
30	Slovak Republic[1]	0.47
31	Portugal	0.50
32	Chile	0.50
33	Brazil	0.51
34	Tunisia	0.54
35	Austria	0.54
36	Slovenia[1]	0.57
37	Taiwan, China	0.62
38	Estonia	0.65
39	South Africa	0.75
40	Panama	0.83
41	Uruguay	0.85
42	Macedonia, FYR	0.88
43	Latvia	0.89
44	Japan	0.96
45	Botswana	0.97
46	Costa Rica[1]	0.97
47	Croatia	0.97
48	Malta	0.97
49	Venezuela[1]	0.98
50	Argentina	1.09
51	Pakistan	1.17
52	Lithuania	1.18
53	Bulgaria	1.26
54	Kuwait[1]	1.28
55	Algeria	1.33
56	Namibia	1.33
57	Mauritius	1.39
58	Czech Republic[1]	1.46
59	Mexico	1.46
60	Dominican Republic	1.48
61	Guyana	1.61

RANK	COUNTRY/ECONOMY	HARD DATA
62	El Salvador	2.05
63	Ecuador[1]	2.39
64	Armenia	2.84
65	Mauritania	2.86
66	Thailand	3.05
67	India	3.30
68	Indonesia	3.66
69	Hungary[1]	3.71
70	Bolivia	3.75
71	Jordan	4.34
72	Moldova	4.48
73	Peru[1]	4.61
74	Honduras	5.11
75	Mozambique	5.17
76	Bosnia and Herzegovina	5.19
77	Philippines	5.37
78	Kenya	5.39
79	Serbia and Montenegro	5.51
80	Cameroon	5.81
81	Malawi	6.15
82	Vietnam	6.15
83	Colombia[1]	6.20
84	Zambia[1]	6.52
85	Suriname[1]	6.74
86	Ukraine	7.22
87	Nepal	7.96
88	Morocco[1]	8.07
89	Lesotho	8.21
90	Azerbaijan	8.55
91	Mali	8.55
92	Zimbabwe	9.50
93	Paraguay	9.75
94	Albania[1]	9.89
95	Tanzania	10.48
96	Madagascar[1]	10.59
97	Mongolia[1]	10.63
98	Burkina Faso	10.72
99	Kyrgyz Republic[1]	13.37
100	Nigeria[1]	13.55
101	Egypt	13.72
102	Chad	14.97
103	Guatemala[1]	16.20
104	Sri Lanka	16.34
105	Benin	18.49
106	Nicaragua	20.88
107	Uganda	22.90
108	Ethiopia[1]	30.05
109	Bangladesh	39.44
110	Burundi	52.94
n/a	Angola	n/a
n/a	Cambodia	n/a
n/a	Canada	n/a
n/a	China	n/a
n/a	Georgia	n/a
n/a	Iceland	n/a
n/a	Italy	n/a
n/a	Kazakhstan	n/a
n/a	Netherlands	n/a
n/a	Poland	n/a
n/a	Russian Federation	n/a
n/a	Sweden	n/a

SOURCE: International Telecommunication Union, *World Telecommunication Indicators 2006*; International Monetary Fund, *World Economic Outlook Online Database* (April 2006 and September 2006 editions)

[1] 2004

5.07 Business monthly telephone subscription

Business monthly telephone subscription to the public switched telephone network (US$) as a percentage of monthly GDP per capita, 2005 or most recent year available

RANK	COUNTRY/ECONOMY	HARD DATA		RANK	COUNTRY/ECONOMY	HARD DATA
1	United Arab Emirates	0.18		62	Bulgaria	3.55
2	Serbia and Montenegro	0.25		63	Namibia	3.61
3	Italy[1]	0.32		64	Albania[1]	3.71
4	Luxembourg	0.33		65	Panama	3.74
5	Singapore	0.35		66	Uruguay	3.88
6	Bahrain	0.37		67	Cameroon	4.07
7	Korea, Rep.	0.38		68	South Africa	4.11
8	Norway	0.46		69	Paraguay	4.37
9	Switzerland	0.46		70	Burundi	4.45
10	Finland	0.48		71	Dominican Republic	4.68
11	Denmark	0.48		72	Moldova	4.70
12	Israel[1]	0.55		73	Pakistan	4.89
13	Netherlands[1]	0.59		74	Honduras	5.32
14	France	0.60		75	Indonesia	5.63
15	Taiwan, China	0.65		76	Ecuador[1]	5.75
16	Spain	0.67		77	Macedonia, FYR	5.96
17	Suriname[1]	0.67		78	Azerbaijan	5.99
18	Hong Kong SAR	0.68		79	Barbados	6.01
19	Germany	0.68		80	Malawi	6.15
20	Belgium	0.70		81	Kyrgyz Republic[1]	6.32
21	Ireland	0.72		82	Zimbabwe	6.33
22	Austria	0.74		83	Brazil	6.37
23	United Kingdom	0.78		84	El Salvador	6.55
24	Japan	0.79		85	Jamaica	6.64
25	Algeria	0.80		86	Peru[1]	6.81
26	Slovenia[1]	0.82		87	Bosnia and Herzegovina[1]	6.88
27	Greece	0.85		88	Jordan	7.26
28	Tunisia	0.86		89	Bolivia	7.50
29	Cyprus	0.87		90	Armenia	7.67
30	Qatar	0.89		91	Bangladesh[1]	7.72
31	Australia	0.91		92	Turkey[1]	7.85
32	Estonia	0.93		93	India	7.91
33	Canada	1.01		94	Mauritania	8.93
34	Portugal	1.08		95	Nigeria[1]	9.03
35	Thailand	1.09		96	Guyana	9.65
36	Kuwait[1]	1.19		97	Zambia[1]	10.44
37	United States[1]	1.32		98	Nepal	10.62
38	Costa Rica[1]	1.35		99	Morocco[1]	10.65
39	Botswana	1.37		100	Benin[1]	11.92
40	Chile	1.40		101	Chad	11.98
41	Lithuania[1]	1.58		102	Sri Lanka	12.41
42	Croatia	1.63		103	Burkina Faso	12.87
43	Malta	1.65		104	Tanzania	13.37
44	New Zealand	1.80		105	Kenya	14.05
45	Mauritius	1.87		106	Lesotho	15.63
46	Slovak Republic[1]	1.99		107	Nicaragua	16.70
47	Egypt	2.14		108	Ethiopia[1]	17.23
48	Ukraine	2.14		109	Uganda	22.90
49	Latvia	2.15		110	Mali	23.05
50	Romania	2.20		111	Philippines	23.21
51	Czech Republic	2.33		112	Madagascar[1]	25.62
52	Hungary	2.39		113	Mozambique	28.70
53	Malaysia	2.87		n/a	Angola	n/a
54	Colombia[1]	2.98		n/a	Cambodia	n/a
55	China	3.03		n/a	Georgia	n/a
56	Mexico	3.06		n/a	Iceland	n/a
57	Venezuela	3.06		n/a	Kazakhstan	n/a
58	Trinidad and Tobago[1]	3.17		n/a	Mongolia	n/a
59	Argentina	3.27		n/a	Poland	n/a
60	Vietnam	3.32		n/a	Russian Federation	n/a
61	Guatemala[1]	3.42		n/a	Sweden	n/a

SOURCE: International Telecommunication Union, *World Telecommunication Indicators 2006*; International Monetary Fund, *World Economic Outlook Online Database* (April 2006 and September 2006 editions)

[1] 2004

5.08 Local supplier quality

The quality of local suppliers in your country is (1 = poor, as they are inefficient and have little technological capacity, 7 = very good, as they are internationally competitive and assist in new product development)

RANK	COUNTRY/ECONOMY	SCORE	MEAN: 4.37	SD
1	Germany	6.59		0.7
2	Japan	6.37		0.7
3	Switzerland	6.23		0.7
4	Austria	6.22		0.8
5	Belgium	6.11		0.8
6	Sweden	6.02		1.1
7	United Kingdom	5.97		0.9
8	Netherlands	5.97		0.7
9	Finland	5.90		0.8
10	United States	5.87		1.2
11	Denmark	5.87		0.9
12	France	5.82		1.0
13	Canada	5.81		0.8
14	Taiwan, China	5.77		0.9
15	Hong Kong SAR	5.76		1.0
16	Ireland	5.74		0.9
17	Norway	5.67		0.7
18	Australia	5.57		0.9
19	Iceland	5.57		1.1
20	New Zealand	5.57		1.0
21	Israel	5.56		1.0
22	Czech Republic	5.48		0.9
23	Chile	5.43		1.0
24	Malaysia	5.42		1.0
25	Singapore	5.41		0.9
26	Korea, Rep.	5.39		1.0
27	Luxembourg	5.36		1.0
28	India	5.28		1.2
29	South Africa	5.28		0.9
30	Italy	5.20		1.2
31	Spain	5.18		1.0
32	Estonia	5.12		1.1
33	Tunisia	5.02		0.9
34	Slovenia	4.97		1.0
35	United Arab Emirates	4.92		1.2
36	Costa Rica	4.91		1.2
37	Brazil	4.87		1.2
38	Kuwait	4.87		1.4
39	Turkey	4.83		1.0
40	Thailand	4.80		0.9
41	Portugal	4.79		1.1
42	Slovak Republic	4.78		1.0
43	Peru	4.71		1.1
44	Greece	4.69		1.1
45	Latvia	4.65		1.4
46	Mauritius	4.63		0.9
47	Colombia	4.62		1.0
48	Lithuania	4.60		1.3
49	Cyprus	4.57		1.4
50	Indonesia	4.56		0.7
51	Mexico	4.53		1.2
52	Guatemala	4.52		1.1
53	Malta	4.50		1.1
54	Panama	4.46		1.3
55	Trinidad and Tobago	4.44		1.3
56	Egypt	4.43		1.5
57	Bahrain	4.42		1.6
58	Hungary	4.41		1.2
59	Jamaica	4.39		1.1
60	Poland	4.37		0.9
61	El Salvador	4.31		1.2
62	Croatia	4.31		1.2
63	Philippines	4.27		1.0
64	Barbados	4.24		1.2
65	China	4.20		1.2
66	Pakistan	4.20		1.3
67	Kenya	4.19		1.5
68	Jordan	4.16		1.4
69	Romania	4.12		1.4
70	Kazakhstan	4.09		1.5
71	Russian Federation	4.07		1.4
72	Namibia	4.06		1.5
73	Sri Lanka	3.97		1.5
74	Argentina	3.96		1.1
75	Dominican Republic	3.91		1.2
76	Ukraine	3.89		1.3
77	Azerbaijan	3.86		1.4
78	Morocco	3.84		1.5
79	Bulgaria	3.83		1.5
80	Qatar	3.83		1.3
81	Venezuela	3.82		1.2
82	Zimbabwe	3.81		1.2
83	Uruguay	3.76		1.0
84	Bosnia and Herzegovina	3.73		1.3
85	Benin	3.72		1.5
86	Macedonia, FYR	3.69		1.3
87	Serbia and Montenegro	3.69		1.2
88	Ecuador	3.68		1.0
89	Honduras	3.63		1.2
90	Tanzania	3.63		1.2
91	Vietnam	3.60		1.3
92	Botswana	3.59		1.3
93	Nigeria	3.58		1.7
94	Bangladesh	3.54		1.3
95	Burkina Faso	3.50		1.3
96	Cambodia	3.49		1.4
97	Algeria	3.48		1.2
98	Cameroon	3.41		1.4
99	Guyana	3.41		1.3
100	Nepal	3.39		1.3
101	Nicaragua	3.39		1.2
102	Moldova	3.35		1.5
103	Mali	3.32		1.7
104	Paraguay	3.29		1.3
105	Kyrgyz Republic	3.28		1.4
106	Madagascar	3.27		1.4
107	Mauritania	3.27		1.8
108	Suriname	3.21		1.1
109	Armenia	3.21		1.4
110	Uganda	3.18		1.5
111	Mozambique	3.02		1.2
112	Malawi	2.95		1.2
113	Bolivia	2.93		1.1
114	Burundi	2.92		1.4
115	Mongolia	2.87		1.2
116	Georgia	2.86		1.3
117	Zambia	2.85		0.9
118	Chad	2.81		1.6
119	Lesotho	2.81		1.6
120	Ethiopia	2.79		1.1
121	Albania	2.70		1.1
122	Angola	2.66		1.2

SOURCE: World Economic Forum, Executive Opinion Survey 2006

319

5.09   Computer, communications, and other services imports

Computer, communications, and other services as percentage of total commercial services imports, 2004

RANK	COUNTRY/ECONOMY	HARD DATA
1	Azerbaijan	83.26
2	Angola	76.06
3	Ireland	72.42
4	Indonesia	64.73
5	Kazakhstan	64.36
6	Austria	55.77
7	Croatia	54.05
8	Netherlands	53.75
9	Sweden	51.24
10	Hungary	51.11
11	Brazil	47.42
12	Finland	47.31
13	Nigeria	47.18
14	Czech Republic	46.03
15	Spain	45.91
16	Italy	45.53
17	Macedonia, FYR	41.25
18	Suriname	41.13
19	Romania	40.85
20	Slovenia	40.72
21	Belgium	40.28
22	Poland	39.57
23	Israel	38.65
24	Korea, Rep.	38.61
25	Taiwan, China	38.40
26	France	38.33
27	Guyana	38.25
28	Canada	38.02
29	Germany	37.40
30	Switzerland	36.03
31	Portugal	35.98
32	Singapore	35.63
33	Namibia	35.10
34	Japan	35.03
35	Jamaica	35.00
36	Egypt	34.96
37	Mozambique	34.91
38	Denmark	34.84
39	Russian Federation	34.48
40	Uganda	33.82
41	Estonia	32.73
42	Ukraine	32.59
43	Pakistan	31.45
44	United States	30.87
45	China	30.23
46	United Kingdom	29.94
47	Morocco	27.59
48	Moldova	27.57
49	Luxembourg	27.47
50	Thailand	27.41
51	Argentina	26.53
52	Latvia	26.29
53	Ethiopia	26.13
54	Peru	25.89
55	Venezuela	25.85
56	Ecuador	25.73
57	New Zealand	25.25
58	Norway	25.14
59	Chile	24.84
60	Kenya	24.44
61	Kyrgyz Republic	24.33

RANK	COUNTRY/ECONOMY	HARD DATA
62	Australia	22.78
63	Costa Rica	22.59
64	Honduras	22.51
65	Tanzania	22.48
66	Cambodia	21.97
67	Mauritius	21.90
68	Turkey	21.61
69	Greece	21.33
70	Tunisia	21.22
71	Philippines	20.53
72	Iceland	20.19
73	Bulgaria	19.67
74	Bolivia	19.16
75	El Salvador	18.62
76	Colombia	18.58
77	Uruguay	18.53
78	Malta	17.80
79	Sri Lanka	17.70
80	Cyprus	17.52
81	Nicaragua	17.17
82	Nepal	17.11
83	Barbados	16.92
84	Lithuania	15.96
85	Panama	14.64
86	Armenia	13.37
87	Mongolia	12.81
88	Bosnia and Herzegovina	12.38
89	South Africa	12.29
90	Mexico	10.84
91	Georgia	9.59
92	Jordan	9.06
93	Paraguay	8.31
94	Dominican Republic	7.79
95	Bangladesh	6.13
96	Bahrain	6.03
97	Guatemala	5.83
98	Kuwait	2.21
99	Lesotho	0.10
n/a	Albania	n/a
n/a	Algeria	n/a
n/a	Benin	n/a
n/a	Botswana	n/a
n/a	Burkina Faso	n/a
n/a	Burundi	n/a
n/a	Cameroon	n/a
n/a	Chad	n/a
n/a	Hong Kong SAR	n/a
n/a	India	n/a
n/a	Madagascar	n/a
n/a	Malawi	n/a
n/a	Malaysia	n/a
n/a	Mali	n/a
n/a	Mauritania	n/a
n/a	Qatar	n/a
n/a	Serbia and Montenegro	n/a
n/a	Slovak Republic	n/a
n/a	Trinidad and Tobago	n/a
n/a	United Arab Emirates	n/a
n/a	Vietnam	n/a
n/a	Zambia	n/a
n/a	Zimbabwe	n/a

SOURCE: World Bank, *World Development Indicators Online Database* (December 2006)

Pillar 6

Government readiness

6.01   Government prioritization of ICT

Information and communication technologies (ICT) (computers, Internet, etc.) are an overall priority for the government (1 = strongly agree, 7 = strongly disagree)

RANK	COUNTRY/ECONOMY	SCORE	1 MEAN: 4.66 7	SD
1	Singapore	6.32		0.8
2	Malaysia	6.07		0.8
3	Mauritania	6.05		1.7
4	Taiwan, China	6.02		0.7
5	Portugal	5.91		0.8
6	Estonia	5.89		1.3
7	United Arab Emirates	5.88		1.2
8	Japan	5.83		0.9
9	Malta	5.83		0.9
10	Qatar	5.75		1.2
11	Denmark	5.73		1.0
11	India	5.73		1.1
13	Tunisia	5.72		1.2
14	Finland	5.64		1.1
15	Iceland	5.63		1.0
16	United Kingdom	5.59		1.2
17	Mauritius	5.52		1.4
18	Korea, Rep.	5.49		1.3
19	Thailand	5.41		1.0
20	Hong Kong SAR	5.37		1.1
21	Switzerland	5.27		1.0
22	Sweden	5.27		1.3
23	Norway	5.23		1.1
24	United States	5.20		1.3
25	Chile	5.19		1.1
26	Netherlands	5.18		1.1
27	Jordan	5.17		1.3
28	Dominican Republic	5.15		1.4
29	Tanzania	5.10		1.2
30	Canada	5.09		1.2
31	Mali	5.09		1.7
32	Mongolia	5.08		1.7
33	Azerbaijan	5.08		1.8
34	Algeria	5.06		1.8
34	El Salvador	5.06		1.5
36	Ireland	5.06		1.1
37	Bahrain	5.05		1.6
38	Luxembourg	5.05		1.2
39	Kazakhstan	5.05		1.5
40	Austria	5.05		1.2
41	Jamaica	5.01		1.2
42	Germany	5.00		1.1
42	Vietnam	5.00		1.5
44	France	4.94		1.3
45	Israel	4.90		1.3
46	Lithuania	4.90		1.5
47	South Africa	4.88		1.2
48	Barbados	4.86		1.4
49	Croatia	4.85		1.7
50	Belgium	4.82		1.3
51	New Zealand	4.82		1.3
52	Morocco	4.81		1.6
53	Sri Lanka	4.77		1.6
54	Benin	4.77		1.7
55	Australia	4.76		1.2
56	Hungary	4.76		1.3
57	Mexico	4.75		1.3
58	Spain	4.74		1.4
59	Slovenia	4.70		1.5
60	Colombia	4.69		1.2
61	Madagascar	4.69		1.6
62	Pakistan	4.63		1.6
63	Nigeria	4.63		1.8
64	Burkina Faso	4.62		1.7
65	Botswana	4.58		1.4
66	Uganda	4.56		1.9
67	Nicaragua	4.55		1.6
68	Armenia	4.52		1.7
69	Zambia	4.52		1.4
70	Nepal	4.50		1.6
71	Moldova	4.48		1.8
72	Bangladesh	4.47		1.7
73	Bosnia and Herzegovina	4.47		1.8
74	Cambodia	4.46		1.8
75	Russian Federation	4.45		1.7
76	Slovak Republic	4.41		1.2
77	Czech Republic	4.40		1.4
77	Turkey	4.40		1.4
79	Egypt	4.40		1.8
80	Guyana	4.40		1.7
81	Mozambique	4.39		1.6
82	Brazil	4.36		1.4
83	Greece	4.36		1.4
84	Peru	4.32		1.6
85	Trinidad and Tobago	4.28		1.7
86	Serbia and Montenegro	4.28		1.7
87	Kyrgyz Republic	4.25		1.8
88	Latvia	4.25		1.4
88	Malawi	4.25		1.6
90	China	4.24		1.5
91	Honduras	4.21		1.7
92	Costa Rica	4.21		1.5
93	Venezuela	4.20		1.7
94	Uruguay	4.18		1.4
95	Philippines	4.15		1.6
96	Guatemala	4.15		1.5
97	Cyprus	4.14		1.4
98	Ethiopia	4.13		1.8
99	Romania	4.11		1.5
100	Lesotho	4.09		1.6
101	Italy	4.09		1.5
102	Bolivia	4.07		1.7
103	Kenya	4.07		1.8
104	Bulgaria	4.04		1.8
105	Kuwait	4.00		1.5
106	Namibia	3.95		1.1
107	Ukraine	3.88		1.6
108	Albania	3.87		2.0
109	Macedonia, FYR	3.86		1.6
110	Panama	3.82		1.4
111	Georgia	3.78		1.3
112	Zimbabwe	3.74		1.6
113	Argentina	3.71		1.5
114	Cameroon	3.71		1.8
115	Poland	3.64		1.1
116	Angola	3.59		1.7
117	Paraguay	3.16		1.8
118	Indonesia	3.03		1.0
119	Ecuador	2.98		1.4
120	Suriname	2.95		1.9
121	Chad	2.80		2.0
122	Burundi	2.79		1.8

SOURCE: World Economic Forum, Executive Opinion Survey 2006

322

6.02 Government procurement of advanced technology products

Government purchase decisions for the procurement of advanced technology products are (1 = based solely on price, 7 = based on technological performance and innovativeness)

RANK	COUNTRY/ECONOMY	SCORE	MEAN: 3.80	SD
1	Singapore	5.50		1.1
2	Malaysia	5.20		1.2
3	Taiwan, China	5.08		0.9
4	Tunisia	5.05		1.0
5	Japan	4.98		1.0
6	Switzerland	4.91		1.0
7	Luxembourg	4.85		1.0
8	Germany	4.84		1.0
9	France	4.81		1.1
10	United States	4.76		1.4
11	Israel	4.74		1.2
12	United Arab Emirates	4.72		1.7
13	Finland	4.65		1.1
14	Korea, Rep.	4.60		1.2
15	Denmark	4.54		1.4
16	Netherlands	4.52		1.1
17	Hong Kong SAR	4.51		1.1
18	Sweden	4.50		1.4
19	Mauritania	4.46		2.3
20	Nigeria	4.45		1.6
21	China	4.44		1.3
22	Austria	4.43		1.1
23	Indonesia	4.41		1.1
24	Qatar	4.40		1.5
25	Thailand	4.39		1.1
26	Portugal	4.36		1.2
27	Ireland	4.32		1.1
28	Mali	4.23		1.3
29	Cambodia	4.22		1.6
30	Australia	4.22		1.1
31	United Kingdom	4.21		1.2
32	South Africa	4.19		1.1
33	Estonia	4.18		1.5
34	Norway	4.14		1.3
35	Algeria	4.12		1.7
36	Canada	4.11		1.3
37	Tanzania	4.10		1.8
38	Burkina Faso	4.05		1.6
39	Mauritius	4.04		1.5
40	India	4.03		1.4
41	Azerbaijan	3.99		1.3
42	Benin	3.98		1.5
43	Barbados	3.96		1.1
44	Vietnam	3.95		1.3
45	Kenya	3.93		1.4
46	Uganda	3.93		1.7
47	Iceland	3.93		1.1
47	New Zealand	3.93		1.2
47	Pakistan	3.93		1.3
50	Trinidad and Tobago	3.91		1.4
51	Serbia and Montenegro	3.90		1.6
52	Spain	3.89		1.3
53	Czech Republic	3.88		1.2
54	Chile	3.88		1.4
55	Hungary	3.87		1.2
56	Kazakhstan	3.86		1.5
57	Jamaica	3.86		1.3
58	Brazil	3.85		1.4
59	Bahrain	3.83		1.4
60	Colombia	3.80		1.4
61	Malta	3.80		1.5
62	Turkey	3.77		1.3
63	Botswana	3.76		1.4
64	Morocco	3.76		1.6
65	Madagascar	3.75		1.5
66	Sri Lanka	3.72		1.4
67	Costa Rica	3.71		1.3
68	Argentina	3.69		1.1
69	Croatia	3.69		1.6
70	El Salvador	3.67		1.4
71	Lithuania	3.65		1.2
72	Venezuela	3.65		1.4
73	Greece	3.64		1.4
74	Romania	3.64		1.7
75	Belgium	3.64		1.4
76	Poland	3.63		1.0
77	Mexico	3.63		1.5
78	Russian Federation	3.61		1.5
79	Armenia	3.60		1.5
80	Slovenia	3.60		1.4
81	Namibia	3.58		1.3
82	Egypt	3.55		1.5
83	Guatemala	3.54		1.2
84	Jordan	3.53		1.5
85	Slovak Republic	3.52		1.1
86	Cyprus	3.49		1.4
87	Malawi	3.49		1.5
88	Philippines	3.46		1.4
89	Panama	3.45		1.4
89	Ukraine	3.45		1.4
91	Dominican Republic	3.43		1.3
92	Italy	3.38		1.3
93	Latvia	3.37		1.4
94	Macedonia, FYR	3.36		1.6
95	Guyana	3.27		1.5
96	Cameroon	3.26		1.5
97	Uruguay	3.26		1.2
98	Kuwait	3.23		1.7
99	Mozambique	3.22		1.4
100	Nicaragua	3.22		1.4
101	Honduras	3.21		1.6
102	Bulgaria	3.21		1.5
103	Bangladesh	3.16		1.6
104	Peru	3.13		1.2
105	Chad	3.13		1.6
106	Georgia	3.12		1.3
107	Ethiopia	3.05		1.7
108	Ecuador	3.05		1.3
109	Bosnia and Herzegovina	3.03		1.4
110	Lesotho	3.02		1.6
111	Mongolia	2.94		1.6
112	Angola	2.93		1.5
113	Moldova	2.91		1.4
114	Suriname	2.87		1.2
115	Burundi	2.86		1.6
116	Zimbabwe	2.78		1.4
117	Paraguay	2.74		1.4
118	Nepal	2.67		1.5
119	Kyrgyz Republic	2.65		1.4
120	Bolivia	2.62		1.2
121	Zambia	2.32		1.4
122	Albania	2.13		1.3

323

SOURCE: World Economic Forum, Executive Opinion Survey 2006

6.03 Importance of ICT to government's vision of the future

The government has a clear implementation plan for utilizing ICT to improve the country's overall competitiveness (1 = strongly disagree, 7 = strongly agree)

RANK	COUNTRY/ECONOMY	SCORE	MEAN: 3.99	SD
1	Singapore	6.18		0.8
2	Malta	5.68		0.9
3	Malaysia	5.58		1.1
4	United Arab Emirates	5.53		1.2
5	Mauritania	5.49		1.8
6	Tunisia	5.49		1.2
7	Portugal	5.49		1.0
8	Iceland	5.44		1.2
9	Denmark	5.36		1.2
10	Thailand	5.33		1.1
11	Qatar	5.25		1.3
12	Estonia	5.24		1.2
13	Hong Kong SAR	5.10		1.1
14	Chile	5.09		1.1
15	Korea, Rep.	5.08		1.1
16	Finland	5.04		1.3
17	Mali	5.00		1.6
17	Taiwan, China	5.00		1.3
19	Japan	4.90		1.4
20	India	4.89		1.3
21	Norway	4.80		1.1
22	El Salvador	4.74		1.2
23	Burkina Faso	4.69		1.7
24	Sweden	4.69		1.3
25	Jordan	4.59		1.6
26	Dominican Republic	4.57		1.3
27	Luxembourg	4.55		1.5
28	United Kingdom	4.51		1.4
29	Morocco	4.51		1.5
30	Uganda	4.48		1.8
31	Austria	4.46		1.2
32	Israel	4.45		1.1
33	France	4.44		1.4
34	Mongolia	4.43		1.8
35	Jamaica	4.38		1.3
36	China	4.38		1.5
37	Kazakhstan	4.38		1.5
38	United States	4.36		1.4
39	Algeria	4.35		1.8
40	Ireland	4.34		1.7
41	Australia	4.34		1.1
42	Nigeria	4.33		1.8
43	Mexico	4.33		1.3
44	Egypt	4.33		1.7
45	Switzerland	4.32		1.1
46	Netherlands	4.30		1.1
47	Mauritius	4.30		1.7
48	Ethiopia	4.29		1.7
49	Slovenia	4.26		1.7
50	Madagascar	4.25		1.8
51	Tanzania	4.25		1.5
52	Belgium	4.25		1.3
53	Pakistan	4.19		1.4
54	Bahrain	4.17		1.6
55	Canada	4.13		1.5
56	Cambodia	4.12		1.9
57	Serbia and Montenegro	4.11		1.7
58	Hungary	4.10		1.2
59	Barbados	4.07		1.4
60	Azerbaijan	4.07		1.8
61	Guatemala	4.06		1.3
62	Slovak Republic	4.05		1.3
63	Vietnam	4.04		1.6
64	Sri Lanka	3.99		1.8
65	New Zealand	3.98		1.2
66	Botswana	3.97		1.6
67	South Africa	3.93		1.4
68	Germany	3.90		1.3
69	Turkey	3.90		1.4
70	Colombia	3.90		1.4
71	Croatia	3.89		1.7
72	Romania	3.86		1.5
73	Brazil	3.84		1.4
74	Nicaragua	3.81		1.5
75	Spain	3.80		1.5
76	Greece	3.79		1.4
77	Kenya	3.78		1.8
78	Benin	3.76		1.8
79	Lithuania	3.76		1.3
80	Cyprus	3.71		1.4
81	Italy	3.67		1.6
82	Moldova	3.66		1.8
83	Latvia	3.59		1.6
84	Mozambique	3.52		1.6
85	Bulgaria	3.51		1.7
86	Peru	3.49		1.5
87	Armenia	3.44		1.8
88	Philippines	3.43		1.6
89	Bangladesh	3.41		1.6
90	Poland	3.41		1.3
91	Uruguay	3.39		1.4
92	Czech Republic	3.38		1.4
93	Panama	3.38		1.4
94	Macedonia, FYR	3.37		1.6
95	Costa Rica	3.35		1.3
96	Trinidad and Tobago	3.26		1.8
97	Zambia	3.25		1.5
98	Nepal	3.21		1.7
99	Kuwait	3.19		1.7
100	Albania	3.17		2.0
101	Bolivia	3.11		1.6
102	Namibia	3.09		1.3
103	Angola	3.06		1.5
104	Venezuela	3.06		1.2
105	Russian Federation	3.04		1.6
106	Honduras	3.00		1.6
107	Lesotho	2.90		1.7
108	Argentina	2.90		1.5
109	Cameroon	2.88		1.8
110	Ecuador	2.84		1.4
111	Malawi	2.81		1.8
112	Indonesia	2.79		0.8
113	Ukraine	2.78		1.3
114	Guyana	2.76		1.8
115	Bosnia and Herzegovina	2.74		1.5
116	Georgia	2.69		1.5
117	Kyrgyz Republic	2.59		1.5
118	Paraguay	2.57		1.5
119	Chad	2.52		1.7
120	Burundi	2.44		1.5
121	Zimbabwe	2.03		1.3
122	Suriname	1.96		1.5

SOURCE: World Economic Forum, Executive Opinion Survey 2006

6.04 E-participation index

The e-participation index assesses the quality, relevance, usefulness, and willingness of government websites for providing online information and participatory tools and services to the people, 2005

RANK	COUNTRY/ECONOMY	HARD DATA
1	United Kingdom	1.000
2	Singapore	0.984
3	United States	0.905
4	Canada	0.873
4	Korea, Rep.	0.873
6	New Zealand	0.794
7	Denmark	0.762
7	Mexico	0.762
9	Australia	0.714
10	Netherlands	0.698
11	Estonia	0.619
12	Chile	0.587
12	Colombia	0.587
14	Sweden	0.571
15	Finland	0.556
15	Germany	0.556
17	Belgium	0.508
18	Brazil	0.492
19	Malta	0.476
19	Philippines	0.476
21	Japan	0.460
22	Switzerland	0.429
22	Venezuela	0.429
24	Austria	0.413
24	France	0.413
26	Norway	0.397
27	Hungary	0.381
28	Ukraine	0.365
29	Poland	0.349
30	Mozambique	0.333
31	Israel	0.318
31	Romania	0.318
33	South Africa	0.302
34	Indonesia	0.286
34	Turkey	0.286
36	Argentina	0.270
36	Guatemala	0.270
36	Honduras	0.270
36	Panama	0.270
36	Peru	0.270
41	Bulgaria	0.254
41	Mongolia	0.254
41	Thailand	0.254
44	Italy	0.238
45	Slovenia	0.222
46	Czech Republic	0.206
46	Kazakhstan	0.206
46	Portugal	0.206
49	China	0.191
49	Ireland	0.191
51	Cambodia	0.175
51	Croatia	0.175
51	Latvia	0.175
51	Malaysia	0.175
51	Slovak Republic	0.175
56	El Salvador	0.159
56	Greece	0.159
56	India	0.159
56	Kyrgyz Republic	0.159
60	Luxembourg	0.143
60	Russian Federation	0.143
62	Iceland	0.127
62	Macedonia, FYR	0.127
62	Mauritius	0.127
62	Pakistan	0.127
62	United Arab Emirates	0.127
62	Vietnam	0.127
68	Lithuania	0.111
68	Nicaragua	0.111
70	Jamaica	0.095
71	Bolivia	0.079
71	Cyprus	0.079
71	Egypt	0.079
71	Nepal	0.079
71	Nigeria	0.079
71	Spain	0.079
71	Trinidad and Tobago	0.079
78	Armenia	0.064
78	Dominican Republic	0.064
78	Ecuador	0.064
78	Uruguay	0.064
82	Bahrain	0.048
82	Barbados	0.048
82	Botswana	0.048
82	Costa Rica	0.048
82	Jordan	0.048
82	Madagascar	0.048
82	Qatar	0.048
82	Serbia and Montenegro	0.048
82	Sri Lanka	0.048
82	Uganda	0.048
92	Albania	0.032
92	Algeria	0.032
92	Angola	0.032
92	Cameroon	0.032
92	Guyana	0.032
92	Kenya	0.032
92	Mauritania	0.032
92	Morocco	0.032
92	Tanzania	0.032
101	Azerbaijan	0.016
101	Benin	0.016
101	Bosnia and Herzegovina	0.016
101	Burkina Faso	0.016
101	Georgia	0.016
101	Lesotho	0.016
101	Malawi	0.016
101	Paraguay	0.016
109	Bangladesh	0.000
109	Burundi	0.000
109	Chad	0.000
109	Ethiopia	0.000
109	Kuwait	0.000
109	Mali	0.000
109	Moldova	0.000
109	Namibia	0.000
109	Suriname	0.000
109	Tunisia	0.000
109	Zambia	0.000
109	Zimbabwe	0.000
n/a	Hong Kong SAR	n/a
n/a	Taiwan, China	n/a

325

SOURCE: United Nations, *Global E-Readiness Report 2005*

6.05   E-government readiness index

The e-government readiness index assesses e-government readiness based on website assessment, telecommunications infrastructure, and human resource endowment, 2005

RANK	COUNTRY/ECONOMY	HARD DATA
1	United States	0.906
2	Denmark	0.906
3	Sweden	0.898
4	United Kingdom	0.878
5	Korea, Rep.	0.873
6	Australia	0.868
7	Singapore	0.850
8	Canada	0.843
9	Finland	0.823
10	Norway	0.823
11	Germany	0.805
12	Netherlands	0.802
13	New Zealand	0.799
14	Japan	0.780
15	Iceland	0.779
16	Austria	0.760
17	Switzerland	0.755
18	Belgium	0.738
19	Estonia	0.735
20	Ireland	0.725
21	Malta	0.701
22	Chile	0.696
23	France	0.693
24	Israel	0.690
25	Italy	0.679
26	Slovenia	0.676
27	Hungary	0.654
28	Luxembourg	0.651
29	Czech Republic	0.640
30	Portugal	0.608
31	Mexico	0.606
32	Latvia	0.605
33	Brazil	0.598
34	Argentina	0.597
35	Greece	0.592
36	Slovak Republic	0.589
37	Cyprus	0.587
37	Poland	0.587
39	Spain	0.585
40	Lithuania	0.579
41	Philippines	0.572
42	United Arab Emirates	0.572
43	Malaysia	0.571
44	Romania	0.570
45	Bulgaria	0.561
46	Thailand	0.552
47	Croatia	0.548
48	Ukraine	0.546
49	Uruguay	0.539
50	Russian Federation	0.533
51	Mauritius	0.532
52	Bahrain	0.528
53	Colombia	0.522
54	Venezuela	0.516
55	Peru	0.509
56	China	0.508
57	South Africa	0.508
58	Jamaica	0.506
59	Turkey	0.496
60	Barbados	0.492
61	Qatar	0.490
62	Panama	0.482
63	Kazakhstan	0.481
64	Trinidad and Tobago	0.477
65	Jordan	0.464
66	Macedonia, FYR	0.463
67	Costa Rica	0.461
68	Kuwait	0.443
69	Kyrgyz Republic	0.442
70	El Salvador	0.423
71	Dominican Republic	0.408
72	Georgia	0.403
73	Bosnia and Herzegovina	0.402
74	Bolivia	0.402
75	India	0.400
76	Guyana	0.399
77	Botswana	0.398
78	Ecuador	0.397
79	Mongolia	0.396
80	Sri Lanka	0.395
81	Indonesia	0.382
82	Egypt	0.379
83	Guatemala	0.378
84	Azerbaijan	0.377
85	Albania	0.373
86	Vietnam	0.364
87	Armenia	0.363
88	Paraguay	0.362
89	Moldova	0.346
90	Suriname	0.345
91	Namibia	0.341
92	Nicaragua	0.338
93	Lesotho	0.337
94	Honduras	0.335
95	Zimbabwe	0.332
96	Tunisia	0.331
97	Kenya	0.330
98	Algeria	0.324
99	Uganda	0.308
100	Nepal	0.302
101	Tanzania	0.302
102	Cambodia	0.299
103	Pakistan	0.284
104	Malawi	0.279
105	Morocco	0.277
106	Nigeria	0.276
107	Madagascar	0.264
108	Cameroon	0.250
109	Mozambique	0.245
110	Zambia	0.234
111	Benin	0.231
112	Serbia and Montenegro	0.196
113	Angola	0.184
114	Bangladesh	0.176
115	Mauritania	0.172
116	Burundi	0.164
117	Chad	0.143
118	Ethiopia	0.136
119	Burkina Faso	0.133
120	Mali	0.093
n/a	Hong Kong SAR	n/a
n/a	Taiwan, China	n/a

SOURCE: United Nations, *Global E-Readiness Report 2005*

Pillar 7
Individual usage

7.01 Mobile telephone subscribers

Mobile telephone subscribers per 100 inhabitants, 2005 or most recent year available

RANK	COUNTRY/ECONOMY	HARD DATA
1	Luxembourg	154.83
2	Lithuania	127.10
3	Italy	124.28
4	Hong Kong SAR	123.47
5	Czech Republic	115.22
6	Israel	112.42
7	United Kingdom	109.77
8	Portugal	109.09
9	Estonia	108.75
10	Singapore	103.41
11	Iceland	103.40
12	Bahrain	103.04
13	Norway	102.90
14	Jamaica	101.85
15	Ireland	101.49
16	United Arab Emirates	100.86
17	Denmark	100.71
18	Austria	99.82
19	Finland	99.66
20	Taiwan, China	97.37
21	Netherlands	97.15
22	Spain	96.81
23	Germany	95.78
24	Sweden	93.31
25	Hungary	92.30
26	Qatar	92.15
27	Switzerland	91.77
28	Australia	91.39
29	Greece	90.31
30	Belgium	90.00
31	Slovenia	89.44
32	Kuwait	88.57
33	New Zealand	87.61
34	Cyprus	86.09
35	Slovak Republic	84.07
36	Russian Federation	83.62
37	Latvia	81.13
38	Bulgaria	80.83
39	Malta	80.79
40	France	79.44
41	Korea, Rep.	79.39
42	Barbados	76.65
43	Poland	75.70
44	Malaysia	75.17
45	Japan	73.97
46	South Africa	71.60
47	Chile	67.79
48	United States	67.62
49	Croatia	65.55
50	Serbia and Montenegro	63.99
51	Macedonia, FYR	62.01
52	Romania	61.51
53	Trinidad and Tobago	61.26
54	Turkey	59.58
55	Mauritius	57.29
56	Argentina	57.27
57	Tunisia	56.32
58	Suriname	51.82
59	Canada	51.44
60	Colombia	47.92
61	Ecuador	47.22
62	Venezuela	46.71
63	Botswana	46.63
64	Brazil	46.25
65	Mexico	44.34
66	Thailand[1]	42.98
67	Panama	41.88
68	Algeria	41.52
69	Philippines	41.30
70	Morocco	40.89
71	Bosnia and Herzegovina	40.81
72	Dominican Republic	40.68
73	Albania[1]	39.45
74	Guyana	37.45
75	Ukraine	37.04
76	Guatemala	35.80
77	Uruguay	35.54
78	El Salvador	35.05
79	Kazakhstan	33.42
80	Georgia	32.61
81	Paraguay	30.64
82	China	29.90
83	Jordan[1]	28.93
84	Azerbaijan	26.66
85	Bolivia	26.37
86	Moldova	25.92
87	Costa Rica	25.45
88	Namibia	24.37
89	Mauritania	24.30
90	Indonesia	21.06
91	Mongolia	21.05
92	Peru	19.96
93	Egypt	18.41
94	Honduras	17.79
95	Sri Lanka	16.21
96	Nigeria	14.13
97	Cameroon	13.84
98	Lesotho	13.65
99	Kenya	13.46
100	Nicaragua[1]	13.00
101	Vietnam	11.39
102	Armenia	10.61
103	Kyrgyz Republic	10.29
104	Benin	10.00
105	Pakistan	8.30
106	India	8.16
107	Zambia	8.11
108	Mali	7.66
109	Cambodia	7.55
110	Angola	6.86
111	Bangladesh	6.35
112	Mozambique	6.16
113	Zimbabwe	5.87
114	Uganda	5.29
115	Tanzania[1]	5.16
116	Burkina Faso	4.33
117	Malawi	3.33
118	Madagascar	2.71
119	Chad	2.15
120	Burundi	2.03
121	Nepal	0.92
122	Ethiopia	0.53

SOURCE: International Telecommunication Union, *World Telecommunication Indicators 2006*
[1] 2004

7.02 Personal computers

Personal computers per 100 inhabitants, 2004 or most recent year available

RANK	COUNTRY/ECONOMY	HARD DATA
1	Switzerland	82.33
2	United States	76.22
3	Sweden	76.14
4	Israel	73.40
5	Canada	69.82
6	Australia	68.90
7	Netherlands	68.47
8	Denmark	65.48
9	Singapore	62.20
10	Luxembourg	62.09
11	Hong Kong SAR	60.55
12	United Kingdom	60.02
13	Austria	57.63
14	Norway	57.20
15	Germany	54.54
16	Korea, Rep.	54.49
17	Japan	54.15
18	Taiwan, China	52.78
19	Ireland	49.74
20	France	49.64
21	New Zealand	48.23
22	Finland	48.22
23	Iceland	47.10
24	Estonia	46.44
25	Slovenia	35.54
26	Belgium	34.72
27	Italy	31.29
28	Cyprus	30.86
29	Slovak Republic	29.58
30	Spain	25.36
31	Czech Republic	24.00
32	Latvia	21.92
33	Costa Rica	21.89
34	United Arab Emirates	19.84
35	Malaysia	19.16
36	Poland	19.10
37	Croatia	19.07
38	Qatar	17.88
39	Kuwait	17.63
40	Bahrain	16.90
41	Mauritius	16.22
42	Lithuania	15.47
43	Malta	15.30
44	Hungary	14.62
45	Chile	13.87
46	Portugal	13.32
47	Uruguay	13.27
48	Barbados	12.55
49	Mongolia	11.86
50	Romania	11.30
51	Bosnia and Herzegovina	11.00
52	Namibia	10.94
53	Mexico	10.68
54	Brazil	10.52
55	Russian Federation	10.42
56	Greece	8.88
57	Zimbabwe	8.41
58	Argentina	8.37
59	Peru	8.29
60	Venezuela	8.19
61	South Africa	7.92
62	Trinidad and Tobago[1]	7.90
63	Macedonia, FYR	6.89
64	Armenia	6.61
65	Jamaica	6.20
66	Bulgaria	5.94
67	Paraguay	5.92
68	Thailand	5.83
69	Ecuador	5.49
70	Jordan	5.34
71	Turkey	5.13
72	Serbia and Montenegro	4.77
73	Tunisia	4.73
74	Dominican Republic	4.60
75	El Salvador	4.54
76	Botswana	4.52
77	Philippines	4.46
78	Suriname[1]	4.25
79	Georgia	4.25
80	Panama	4.10
81	China	4.08
82	Colombia	3.88
83	Guyana	3.60
84	Nicaragua	3.52
85	Egypt	3.29
86	Ukraine	2.82
87	Sri Lanka	2.72
88	Moldova	2.63
89	Bolivia	2.33
90	Morocco	2.07
91	Guatemala	1.82
92	Azerbaijan	1.78
93	Kyrgyz Republic	1.71
94	Honduras	1.57
95	Mauritania	1.41
96	Kenya	1.36
97	Indonesia	1.36
98	Vietnam	1.26
99	India	1.21
100	Bangladesh	1.19
101	Albania	1.17
102	Zambia	0.98
103	Cameroon	0.98
104	Algeria	0.90
105	Tanzania	0.74
106	Nigeria	0.68
107	Mozambique	0.59
108	Madagascar	0.50
109	Burundi	0.48
110	Nepal	0.47
111	Uganda	0.43
112	Benin	0.41
113	Mali	0.38
114	Ethiopia	0.31
115	Cambodia	0.26
116	Burkina Faso	0.21
117	Angola	0.19
118	Chad	0.17
119	Malawi	0.16
n/a	Kazakhstan	n/a
n/a	Lesotho	n/a
n/a	Pakistan	n/a

329

SOURCE: International Telecommunication Union, *World Telecommunication Indicators 2006*
[1] 2003

7.03 Broadband Internet subscribers

Total broadband Internet subscribers per 100 inhabitants, 2005 or most recent year available

RANK	COUNTRY/ECONOMY	HARD DATA
1	Iceland	26.54
2	Korea, Rep.	25.24
3	Netherlands	25.15
4	Denmark	24.87
5	Hong Kong SAR	23.56
6	Switzerland	23.13
7	Finland	22.37
8	Norway	21.46
9	Sweden	21.36
10	Canada	20.78
11	Taiwan, China	20.21
12	Belgium	19.07
13	Israel	17.82
14	Japan	17.46
15	United States	16.56
16	United Kingdom	16.53
17	Singapore	15.70
18	France	15.65
19	Luxembourg	15.08
20	Austria	14.39
21	Estonia	13.48
22	Germany	12.94
23	Barbados	11.87
24	Spain	11.70
25	Italy	11.67
26	Portugal	11.55
27	Latvia	11.30
28	Malta	11.14
29	Australia	10.43
30	Slovenia	8.64
31	New Zealand	8.22
32	Lithuania	6.83
33	Ireland	6.53
34	Hungary	6.45
35	Chile	4.54
36	Czech Republic	4.38
37	Romania	3.46
38	Qatar	3.24
39	Poland	3.23
40	Cyprus	3.20
41	United Arab Emirates	2.86
42	China	2.85
43	Slovak Republic	2.57
44	Argentina	2.18
45	Turkey	2.17
46	Mexico	2.15
47	Croatia	1.97
48	Malaysia	1.89
49	Uruguay	1.88
50	Brazil	1.77
51	Greece	1.44
52	Venezuela	1.33
53	Peru	1.25
54	Russian Federation	1.11
55	Trinidad and Tobago	0.83
56	Morocco	0.82
57	Kuwait[1]	0.78
58	Dominican Republic	0.74
59	Colombia	0.70
60	Costa Rica[1]	0.66
61	El Salvador	0.61

RANK	COUNTRY/ECONOMY	HARD DATA
62	Macedonia, FYR	0.61
63	Algeria	0.59
64	Panama	0.54
65	Bosnia and Herzegovina	0.35
66	South Africa	0.35
67	Guyana	0.27
68	Vietnam	0.25
69	Moldova	0.25
70	Suriname	0.22
71	Mauritius[1]	0.22
72	Guatemala	0.22
73	Ecuador	0.20
74	Jordan[1]	0.19
75	Tunisia	0.16
76	Egypt	0.15
77	India	0.12
78	Bolivia	0.12
79	Paraguay	0.09
80	Nicaragua[1]	0.09
81	Zimbabwe	0.09
82	Mongolia	0.07
83	Sri Lanka	0.07
84	Kyrgyz Republic	0.05
85	Armenia[1]	0.03
86	Pakistan	0.03
87	Azerbaijan	0.03
88	Bulgaria[1]	0.02
89	Kazakhstan[1]	0.01
90	Mauritania	0.01
91	Malawi	0.00
92	Burkina Faso	0.00
93	Benin	0.00
94	Lesotho	0.00
95	Zambia	0.00
96	Nigeria	0.00
97	Bangladesh[1]	0.00
97	Honduras[1]	0.00
97	Kenya[1]	0.00
97	Madagascar[1]	0.00
97	Mali	0.00
97	Namibia[1]	0.00
n/a	Albania	n/a
n/a	Angola	n/a
n/a	Bahrain	n/a
n/a	Botswana	n/a
n/a	Burundi	n/a
n/a	Cambodia	n/a
n/a	Cameroon	n/a
n/a	Chad	n/a
n/a	Ethiopia	n/a
n/a	Georgia	n/a
n/a	Indonesia	n/a
n/a	Jamaica	n/a
n/a	Mozambique	n/a
n/a	Nepal	n/a
n/a	Philippines	n/a
n/a	Serbia and Montenegro	n/a
n/a	Tanzania	n/a
n/a	Thailand	n/a
n/a	Uganda	n/a
n/a	Ukraine	n/a

SOURCE: International Telecommunication Union, *World Telecommunication Indicators 2006*
[1] 2004

7.04   Internet users

Internet users per 100 inhabitants, 2005 or most recent year available

RANK	COUNTRY/ECONOMY	HARD DATA		RANK	COUNTRY/ECONOMY	HARD DATA
1	Iceland	87.76		62	Morocco	15.18
2	Sweden	76.21		63	Mauritius[1]	14.60
3	Netherlands	73.99		64	Vietnam	12.72
4	Norway	73.59		65	Venezuela	12.39
5	Australia	70.40		66	Trinidad and Tobago[1]	12.24
6	New Zealand	68.35		67	Jordan[1]	11.22
7	Korea, Rep.	68.35		68	Thailand	11.03
8	Luxembourg	67.74		69	South Africa	10.75
9	Japan	66.59		70	Colombia	10.39
10	United States[1]	63.00		71	Mongolia	10.14
11	Barbados	59.48		72	Ukraine	9.81
12	Taiwan, China	58.01		73	Moldova[1]	9.52
13	Singapore[1]	57.87		74	Tunisia	9.46
14	Slovenia	55.41		75	El Salvador	9.26
15	Finland	53.34		76	China	8.44
16	Denmark	52.55		77	Zimbabwe	8.40
17	Canada	52.06		78	Azerbaijan	8.07
18	Estonia	51.92		79	Guatemala	7.94
19	Hong Kong SAR	50:08		80	Macedonia, FYR	7.86
20	Switzerland	49.59		81	Indonesia	7.18
21	Austria	48.93		82	Suriname	7.12
22	Italy	48.20		83	Pakistan	6.82
23	United Kingdom	47.79		84	Egypt	6.75
24	Israel[1]	46.63		85	Panama	6.39
25	Slovak Republic	46.29		86	Albania	6.01
26	Belgium	45.66		87	Algeria	5.83
27	Germany	45.35		88	Benin	5.67
28	Latvia	44.65		89	India	5.44
29	France	43.23		90	Armenia	5.34
30	Malaysia	42.37		91	Philippines[1]	5.32
31	Jamaica[1]	39.87		92	Kyrgyz Republic	5.32
32	Cyprus	39.04		93	Bolivia	5.23
33	Lithuania[1]	35.67		94	Ecuador	4.66
34	Spain	35.41		95	Georgia[1]	3.89
35	Croatia	31.88		96	Nigeria	3.80
36	Malta	31.73		97	Namibia[1]	3.73
37	United Arab Emirates	31.08		98	Honduras	3.61
38	Hungary	29.71		99	Botswana	3.40
39	Qatar	28.16		100	Paraguay	3.25
40	Portugal	28.01		101	Kenya	3.24
41	Ireland	27.64		102	Kazakhstan[1]	2.70
42	Czech Republic	26.99		103	Lesotho[1]	2.39
43	Kuwait	26.05		104	Nicaragua[1]	2.20
44	Poland	25.95		105	Zambia[1]	2.01
45	Costa Rica	25.42		106	Uganda	1.74
46	Turkey	21.86		107	Cameroon	1.53
47	Bahrain	21.33		108	Sri Lanka[1]	1.44
48	Guyana	21.30		109	Angola	1.10
49	Romania[1]	20.76		110	Tanzania[1]	0.89
50	Bosnia and Herzegovina	20.64		111	Mozambique[1]	0.73
51	Bulgaria	20.60		112	Mauritania	0.65
52	Uruguay	20.55		113	Madagascar	0.54
53	Brazil	19.50		114	Burundi	0.53
54	Serbia and Montenegro[1]	18.61		115	Mali	0.53
55	Greece	17.99		116	Burkina Faso	0.49
56	Chile	17.96		117	Nepal	0.41
57	Argentina	17.78		118	Chad	0.41
58	Mexico	17.40		119	Malawi	0.41
59	Dominican Republic	16.84		120	Cambodia[1]	0.28
60	Peru	16.45		121	Bangladesh	0.26
61	Russian Federation	15.19		122	Ethiopia	0.21

331

SOURCE: International Telecommunication Union, *World Telecommunication Indicators 2006*
[1] 2004

7.05 Internet bandwidth

International Internet bandwidth (Mbps) per 10,000 inhabitants, 2004

RANK	COUNTRY/ECONOMY	HARD DATA
1	Denmark	348.28
2	Netherlands	206.19
3	Sweden	174.93
4	United Kingdom	130.69
5	Belgium	112.52
6	Switzerland	96.38
7	Norway	94.51
8	Hong Kong SAR	70.74
9	Germany	68.61
10	Canada	67.83
11	Austria	66.55
12	Ireland	60.81
13	Singapore	57.25
14	Finland	43.36
15	Iceland	42.32
16	Estonia	35.17
17	France	33.09
18	United States	33.06
19	Luxembourg	32.40
20	Taiwan, China	31.44
21	Spain	27.89
22	Israel	24.77
23	Slovak Republic	22.94
24	Malta	19.38
25	Korea, Rep.	14.85
26	New Zealand	11.72
27	Australia	11.08
28	Japan	10.38
29	Hungary	9.90
30	Latvia	9.83
31	Portugal	8.31
32	Chile	8.24
33	Qatar	6.25
34	Greece	5.93
35	Poland	5.55
36	Cyprus	3.72
37	United Arab Emirates	3.49
38	Argentina	3.20
39	Croatia	3.19
40	Uruguay	3.09
41	Panama	2.92
42	Peru	2.03
43	Lithuania	1.93
44	Romania	1.86
45	Brazil	1.52
46	Trinidad and Tobago	1.38
47	Malaysia	1.25
48	Colombia	1.23
49	Turkey	1.23
50	Kuwait	1.12
51	Mexico	1.07
52	Suriname	1.03
53	Russian Federation	1.00
54	Serbia and Montenegro	0.87
55	Bulgaria	0.80
56	China	0.57
57	Guatemala	0.56
58	Jordan	0.55
59	Venezuela	0.51
60	Thailand	0.49
61	Bolivia	0.44
62	Tunisia	0.44
63	Moldova	0.42
64	Morocco	0.41
65	Philippines	0.39
66	Ecuador	0.37
67	Paraguay	0.26
68	Bosnia and Herzegovina	0.24
69	Vietnam	0.23
70	Egypt	0.20
71	South Africa	0.19
72	Ukraine	0.17
73	Sri Lanka	0.16
74	India	0.11
75	Indonesia	0.10
76	Armenia	0.09
77	Benin	0.06
78	Kyrgyz Republic	0.06
79	Pakistan	0.05
80	Burkina Faso	0.05
81	Zimbabwe	0.05
82	Namibia	0.04
83	Uganda	0.02
84	Zambia	0.02
85	Madagascar	0.02
86	Mali	0.02
87	Kenya	0.01
88	Bangladesh	0.00
89	Chad	0.00
90	Malawi	0.00
n/a	Albania	n/a
n/a	Algeria	n/a
n/a	Angola	n/a
n/a	Azerbaijan	n/a
n/a	Bahrain	n/a
n/a	Barbados	n/a
n/a	Botswana	n/a
n/a	Burundi	n/a
n/a	Cambodia	n/a
n/a	Cameroon	n/a
n/a	Costa Rica	n/a
n/a	Czech Republic	n/a
n/a	Dominican Republic	n/a
n/a	El Salvador	n/a
n/a	Ethiopia	n/a
n/a	Georgia	n/a
n/a	Guyana	n/a
n/a	Honduras	n/a
n/a	Italy	n/a
n/a	Jamaica	n/a
n/a	Kazakhstan	n/a
n/a	Lesotho	n/a
n/a	Macedonia, FYR	n/a
n/a	Mauritania	n/a
n/a	Mauritius	n/a
n/a	Mongolia	n/a
n/a	Mozambique	n/a
n/a	Nepal	n/a
n/a	Nicaragua	n/a
n/a	Nigeria	n/a
n/a	Slovenia	n/a
n/a	Tanzania	n/a

SOURCE: International Telecommunication Union, *World Telecommunication Indicators 2006*

Pillar 8
Business usage

8.01 Prevalence of foreign technology licensing

In your country, licensing of foreign technology is (1 = uncommon, 7 = a common means of acquiring new technology)

RANK	COUNTRY/ECONOMY	SCORE	SD
1	India	5.81	1.0
2	Singapore	5.76	1.1
3	Netherlands	5.70	1.1
4	Indonesia	5.66	1.0
5	Taiwan, China	5.66	1.1
6	Malaysia	5.60	0.8
7	South Africa	5.59	0.8
8	Australia	5.59	1.1
9	Canada	5.58	1.1
10	Japan	5.58	1.2
11	Spain	5.56	1.2
12	Hong Kong SAR	5.55	1.2
13	Portugal	5.49	1.1
14	New Zealand	5.47	1.0
15	United Arab Emirates	5.46	1.2
16	Switzerland	5.46	1.3
17	Sweden	5.39	1.2
18	Thailand	5.39	1.0
19	United Kingdom	5.38	1.0
20	Iceland	5.36	1.5
21	Qatar	5.34	1.4
22	Israel	5.34	1.3
23	United States	5.31	1.4
24	Norway	5.30	1.2
25	Jordan	5.20	1.4
26	Belgium	5.19	1.3
27	Bahrain	5.18	1.6
28	Ireland	5.18	1.1
29	Germany	5.18	1.5
30	Croatia	5.13	1.5
31	Denmark	5.08	1.4
32	Chile	5.05	1.2
33	Finland	5.04	1.3
34	Tunisia	5.02	1.3
35	Trinidad and Tobago	4.99	1.5
36	Malta	4.98	1.4
37	Czech Republic	4.95	1.1
38	Greece	4.95	1.3
39	Brazil	4.93	1.3
40	Kenya	4.89	1.5
41	Slovak Republic	4.89	1.3
42	Mexico	4.88	1.3
43	Estonia	4.87	1.4
44	Hungary	4.86	1.4
45	Egypt	4.85	1.6
46	Turkey	4.84	1.3
47	Kuwait	4.84	1.6
48	Panama	4.84	1.5
49	Korea, Rep.	4.82	1.4
50	Mauritius	4.81	1.6
51	Uganda	4.80	1.6
52	Dominican Republic	4.79	1.3
53	Jamaica	4.79	1.5
54	Austria	4.73	1.4
55	Venezuela	4.71	1.4
56	Philippines	4.66	1.5
57	Luxembourg	4.65	1.4
58	Costa Rica	4.63	1.1
59	Nigeria	4.57	1.9
60	France	4.57	1.2
61	Argentina	4.55	1.3
62	El Salvador	4.54	1.4
63	Italy	4.53	1.4
64	Zambia	4.52	1.7
65	Namibia	4.52	1.8
66	Mauritania	4.51	2.6
67	Romania	4.46	1.4
68	Botswana	4.46	1.5
69	Slovenia	4.45	1.2
70	Barbados	4.44	1.3
70	Zimbabwe	4.44	1.7
72	Lithuania	4.42	1.3
73	Cyprus	4.35	1.6
74	Sri Lanka	4.33	1.8
75	Colombia	4.28	1.5
76	Peru	4.27	1.2
77	Guatemala	4.26	1.5
78	Morocco	4.26	1.5
79	Uruguay	4.24	1.4
80	Latvia	4.23	1.5
81	Pakistan	4.17	1.6
82	Azerbaijan	4.10	1.6
83	Kazakhstan	4.06	1.5
84	Burkina Faso	4.04	1.7
85	Poland	4.00	1.3
86	Albania	3.96	1.6
87	Algeria	3.90	1.9
88	China	3.88	1.5
89	Bangladesh	3.85	1.7
90	Angola	3.84	1.8
91	Tanzania	3.83	1.6
92	Bosnia and Herzegovina	3.81	1.6
93	Macedonia, FYR	3.77	1.8
94	Malawi	3.76	1.8
95	Serbia and Montenegro	3.76	1.8
96	Nepal	3.76	1.6
97	Honduras	3.73	1.6
98	Mongolia	3.69	1.7
99	Bulgaria	3.63	1.4
100	Armenia	3.61	1.6
101	Mali	3.59	2.0
102	Ecuador	3.59	1.5
103	Cameroon	3.58	1.8
104	Lesotho	3.51	1.8
105	Russian Federation	3.50	1.5
106	Mozambique	3.49	1.6
107	Georgia	3.49	1.4
108	Madagascar	3.37	1.7
109	Cambodia	3.34	1.7
110	Benin	3.34	1.7
111	Ethiopia	3.33	1.7
112	Ukraine	3.28	1.4
113	Nicaragua	3.28	1.4
114	Guyana	3.13	1.8
115	Vietnam	3.00	1.4
116	Moldova	2.99	1.6
117	Kyrgyz Republic	2.94	1.4
118	Paraguay	2.87	1.5
119	Suriname	2.82	1.7
120	Burundi	2.68	1.9
121	Bolivia	2.54	1.2
122	Chad	2.52	1.7

MEAN: 4.46

SOURCE: World Economic Forum, Executive Opinion Survey 2006

334

8.02 Firm-level technology absorption

Companies in your country are (1 = not able to absorb new technology, 7 = aggressive in absorbing new technology)

RANK	COUNTRY/ECONOMY	SCORE	SD
1	Iceland	6.50	0.5
2	Japan	6.33	0.8
3	Sweden	6.13	0.7
4	Israel	6.13	0.8
5	Taiwan, China	6.12	0.8
6	Switzerland	6.06	0.8
7	Singapore	6.03	0.8
8	Finland	6.02	0.8
9	United States	6.01	1.1
10	Germany	5.94	0.8
11	Korea, Rep.	5.94	1.0
12	Norway	5.91	0.7
13	India	5.82	1.1
14	Denmark	5.81	0.9
15	Malaysia	5.79	0.7
16	Mauritania	5.78	1.7
17	Austria	5.72	0.9
18	Hong Kong SAR	5.70	1.0
19	Estonia	5.64	0.9
20	Australia	5.59	1.0
21	United Arab Emirates	5.58	1.1
22	Canada	5.53	1.0
23	United Kingdom	5.46	1.0
24	Ireland	5.46	1.1
25	Turkey	5.41	1.0
26	Czech Republic	5.39	0.9
27	Netherlands	5.38	0.9
28	Hungary	5.34	1.0
29	Thailand	5.33	0.8
30	South Africa	5.32	0.9
31	Slovak Republic	5.29	0.8
32	New Zealand	5.28	0.9
33	Chile	5.22	0.9
34	Luxembourg	5.22	1.0
35	Belgium	5.19	0.9
36	Tunisia	5.19	1.1
37	France	5.16	0.9
37	Vietnam	5.16	1.3
39	Kuwait	5.16	1.3
40	Malta	5.14	1.0
41	China	5.06	1.3
42	Morocco	5.00	1.5
43	Qatar	4.98	1.4
44	Armenia	4.95	1.5
45	Lithuania	4.94	1.1
46	Jamaica	4.93	1.0
47	Brazil	4.91	1.1
48	Philippines	4.90	1.3
49	Zambia	4.90	1.5
50	Panama	4.86	1.3
51	Latvia	4.84	1.2
52	Bahrain	4.82	1.4
53	Kenya	4.80	1.3
54	Trinidad and Tobago	4.78	1.3
55	Jordan	4.76	1.3
56	Spain	4.75	1.1
57	Costa Rica	4.75	1.2
58	Barbados	4.74	1.1
59	Egypt	4.73	1.5
60	El Salvador	4.73	1.1
61	Guatemala	4.72	1.0

MEAN: 4.72

RANK	COUNTRY/ECONOMY	SCORE	SD
62	Azerbaijan	4.71	1.6
63	Portugal	4.69	0.9
64	Dominican Republic	4.67	1.0
65	Kazakhstan	4.66	1.3
66	Slovenia	4.63	1.1
67	Madagascar	4.62	1.5
68	Algeria	4.61	1.8
69	Cyprus	4.59	1.2
70	Tanzania	4.58	1.4
71	Burkina Faso	4.55	1.4
72	Romania	4.55	1.2
73	Mauritius	4.52	1.0
74	Indonesia	4.50	0.9
75	Mexico	4.49	1.0
76	Peru	4.45	1.2
77	Venezuela	4.44	1.2
78	Nigeria	4.44	1.7
79	Cambodia	4.44	1.5
80	Croatia	4.43	1.5
81	Russian Federation	4.43	1.5
82	Poland	4.42	1.0
83	Greece	4.38	1.2
84	Colombia	4.37	1.1
85	Pakistan	4.36	1.1
86	Benin	4.35	1.7
87	Sri Lanka	4.34	1.6
88	Mali	4.33	1.7
89	Namibia	4.27	1.3
90	Bangladesh	4.24	1.5
91	Botswana	4.20	1.3
92	Uganda	4.18	1.7
93	Italy	4.18	1.2
94	Ukraine	4.17	1.3
95	Cameroon	4.16	1.6
96	Uruguay	4.13	1.0
97	Argentina	4.07	1.0
98	Moldova	4.02	1.5
99	Ecuador	3.98	1.0
100	Honduras	3.90	1.2
101	Nepal	3.89	1.7
102	Georgia	3.79	1.3
103	Zimbabwe	3.78	1.3
104	Serbia and Montenegro	3.76	1.4
105	Burundi	3.72	2.0
106	Albania	3.71	1.4
107	Malawi	3.70	1.4
108	Mongolia	3.69	1.4
109	Guyana	3.60	1.4
110	Suriname	3.59	1.3
111	Nicaragua	3.54	1.3
112	Lesotho	3.47	1.4
113	Bosnia and Herzegovina	3.47	1.5
114	Bulgaria	3.46	1.3
115	Macedonia, FYR	3.44	1.6
116	Mozambique	3.41	1.3
117	Chad	3.38	1.9
118	Kyrgyz Republic	3.38	1.5
119	Ethiopia	3.33	1.4
120	Angola	3.32	1.6
121	Paraguay	3.31	1.2
122	Bolivia	3.19	1.2

MEAN: 4.72

SOURCE: World Economic Forum, Executive Opinion Survey 2006

335

8.03 Capacity for innovation

Companies obtain technologies (1 = exclusively from licensing or imitating foreign companies, 7 = by conducting formal research and pioneering their own new products and processes)

RANK	COUNTRY/ECONOMY	SCORE	SD
1	Germany	6.12	0.8
2	Japan	6.00	0.8
3	Sweden	5.96	0.7
4	Finland	5.78	0.8
5	Switzerland	5.77	0.8
6	Denmark	5.77	1.0
7	France	5.69	0.8
8	Israel	5.58	0.8
9	United States	5.52	1.2
10	Austria	5.43	0.8
11	Netherlands	5.43	1.0
12	United Kingdom	5.38	1.0
13	Korea, Rep.	5.17	1.0
14	Belgium	5.15	1.0
15	Norway	4.97	1.1
16	Luxembourg	4.93	1.4
17	Taiwan, China	4.91	1.2
18	Slovenia	4.84	1.1
19	Canada	4.83	1.1
20	Italy	4.74	1.2
21	Ireland	4.60	1.2
22	Hong Kong SAR	4.59	1.4
23	Malaysia	4.56	1.5
24	Singapore	4.55	1.3
25	Iceland	4.50	1.4
26	New Zealand	4.42	1.0
27	Czech Republic	4.33	1.1
28	India	4.28	1.3
29	Brazil	4.08	1.2
30	Poland	4.08	1.0
31	Tunisia	4.07	1.3
32	Hungary	4.06	1.1
33	Costa Rica	3.97	1.2
34	Spain	3.94	1.0
35	Australia	3.93	1.1
36	Vietnam	3.73	1.2
37	South Africa	3.73	1.2
38	Pakistan	3.71	1.3
39	Estonia	3.70	1.3
40	Portugal	3.65	1.2
41	Azerbaijan	3.62	1.5
42	Latvia	3.61	1.3
43	China	3.60	1.3
44	Lithuania	3.59	1.0
45	Ukraine	3.58	1.3
46	Sri Lanka	3.58	1.4
47	Turkey	3.54	1.2
48	Slovak Republic	3.49	1.0
49	Russian Federation	3.45	1.3
50	Chile	3.45	1.1
51	Thailand	3.39	1.1
52	Kenya	3.32	1.5
53	Croatia	3.31	1.4
54	Guatemala	3.30	1.4
55	Colombia	3.29	1.2
56	Mexico	3.29	1.1
57	Peru	3.28	1.3
58	Nigeria	3.25	1.7
59	Indonesia	3.21	1.2
60	Moldova	3.21	1.4
61	Qatar	3.19	1.7
62	Kazakhstan	3.16	1.3
63	Philippines	3.15	1.0
64	Mali	3.15	1.4
65	Malta	3.11	1.2
66	Macedonia, FYR	3.11	1.4
67	El Salvador	3.08	1.1
68	Uruguay	3.07	1.0
69	Jamaica	3.04	1.4
70	Armenia	3.01	1.4
71	Benin	2.99	1.4
72	United Arab Emirates	2.99	1.5
73	Mauritius	2.96	1.2
74	Greece	2.96	1.0
75	Jordan	2.93	1.2
76	Honduras	2.92	1.4
77	Mauritania	2.91	2.2
78	Madagascar	2.90	1.4
79	Argentina	2.90	1.0
79	Bulgaria	2.90	1.2
81	Cyprus	2.90	1.1
82	Kyrgyz Republic	2.89	1.5
83	Egypt	2.88	1.4
84	Romania	2.87	1.3
85	Barbados	2.85	1.3
86	Uganda	2.83	1.7
87	Burkina Faso	2.83	1.4
88	Ecuador	2.83	1.0
89	Dominican Republic	2.78	1.3
90	Suriname	2.78	1.3
91	Panama	2.76	1.2
92	Morocco	2.74	1.4
93	Nicaragua	2.71	1.2
94	Bosnia and Herzegovina	2.70	1.2
95	Georgia	2.68	1.1
96	Mongolia	2.67	1.3
97	Mozambique	2.63	1.5
98	Tanzania	2.61	1.0
99	Guyana	2.61	1.4
100	Namibia	2.59	1.1
101	Paraguay	2.58	1.3
102	Bolivia	2.57	1.1
103	Trinidad and Tobago	2.54	1.2
104	Malawi	2.54	1.2
105	Serbia and Montenegro	2.54	1.2
106	Venezuela	2.47	0.9
107	Kuwait	2.47	1.5
108	Chad	2.46	1.4
109	Botswana	2.45	1.1
110	Cameroon	2.45	1.3
111	Nepal	2.43	1.1
112	Lesotho	2.42	1.4
113	Ethiopia	2.39	1.1
114	Cambodia	2.38	1.4
115	Bangladesh	2.37	1.2
116	Bahrain	2.36	1.2
117	Algeria	2.34	1.2
118	Zimbabwe	2.33	1.1
119	Angola	2.28	1.2
120	Burundi	2.24	1.4
121	Zambia	2.19	1.1
122	Albania	1.90	1.1

MEAN: 3.50

SOURCE: World Economic Forum, Executive Opinion Survey 2006

8.04   Availability of new telephone lines

New telephone lines for your businesses are (1 = scarce and difficult to obtain, 7 = widely available and highly reliable)

RANK	COUNTRY/ECONOMY	SCORE	MEAN: 5.40	SD		RANK	COUNTRY/ECONOMY	SCORE	MEAN: 5.40	SD
1	Iceland	6.93		0.3		62	Ireland	5.57		1.5
2	Japan	6.92		0.3		63	Argentina	5.54		1.4
3	Germany	6.90		0.3		64	Jamaica	5.52		1.3
4	Switzerland	6.86		0.3		65	Italy	5.51		1.4
5	Singapore	6.86		0.4		66	Vietnam	5.50		1.2
6	France	6.84		0.4		67	Sri Lanka	5.48		1.3
7	Hong Kong SAR	6.80		0.6		68	Qatar	5.47		1.6
8	Denmark	6.80		0.4		69	Macedonia, FYR	5.42		1.6
9	Finland	6.78		0.6		70	Philippines	5.42		1.3
10	Israel	6.77		0.4		71	China	5.31		1.3
11	Sweden	6.77		0.5		72	Bosnia and Herzegovina	5.27		1.2
12	Austria	6.76		0.5		73	Venezuela	5.20		1.3
13	Netherlands	6.74		0.5		74	Uganda	5.20		1.7
14	Norway	6.70		0.5		75	Namibia	5.19		1.4
15	United Kingdom	6.68		0.8		76	Tanzania	5.17		1.6
16	Canada	6.65		0.6		77	South Africa	5.12		1.3
17	Belgium	6.62		0.8		78	Georgia	5.11		1.2
18	United Arab Emirates	6.58		0.7		79	Zambia	5.09		1.4
19	Jordan	6.56		0.7		80	Bulgaria	5.06		1.4
20	Chile	6.56		0.6		81	Bolivia	5.06		1.4
21	Slovak Republic	6.52		0.6		82	Algeria	5.04		1.6
22	Korea, Rep.	6.49		0.9		83	Moldova	5.04		1.3
23	El Salvador	6.44		0.8		84	Mozambique	5.03		1.4
24	Cyprus	6.42		0.8		85	Kazakhstan	4.93		1.4
25	Luxembourg	6.41		0.7		86	Indonesia	4.93		1.0
26	Uruguay	6.36		0.9		87	Mali	4.91		1.9
27	Estonia	6.36		0.8		88	Romania	4.87		1.2
28	Czech Republic	6.35		0.7		89	Pakistan	4.86		1.4
29	United States	6.35		1.2		90	Russian Federation	4.85		1.5
30	India	6.31		0.7		91	Nigeria	4.70		1.8
31	Hungary	6.30		1.0		92	Botswana	4.54		1.4
32	Tunisia	6.29		0.8		93	Burkina Faso	4.48		1.6
33	Taiwan, China	6.26		0.8		94	Poland	4.46		1.4
34	Dominican Republic	6.26		1.2		95	Serbia and Montenegro	4.39		1.7
35	Bahrain	6.23		1.0		96	Ukraine	4.33		1.7
36	Guatemala	6.20		1.1		97	Trinidad and Tobago	4.30		1.5
37	Australia	6.19		0.8		98	Mongolia	4.28		1.7
38	Portugal	6.17		1.2		99	Kyrgyz Republic	4.24		1.6
39	New Zealand	6.15		1.2		100	Cambodia	4.22		1.6
40	Malta	6.14		0.8		101	Nepal	4.19		1.5
41	Peru	6.14		0.7		102	Costa Rica	4.18		1.7
42	Slovenia	6.13		1.0		103	Nicaragua	4.08		1.6
43	Egypt	6.12		1.1		104	Ecuador	4.08		1.6
44	Morocco	6.04		1.3		105	Burundi	4.05		2.1
45	Kuwait	6.02		1.1		106	Ethiopia	4.00		1.6
46	Colombia	6.01		0.8		107	Armenia	3.94		1.8
47	Malaysia	6.01		0.9		108	Madagascar	3.93		1.7
48	Croatia	6.01		1.1		109	Paraguay	3.87		1.7
49	Brazil	5.99		1.0		110	Honduras	3.77		1.9
50	Mauritius	5.96		1.1		111	Kenya	3.62		1.7
51	Mexico	5.94		1.1		112	Cameroon	3.61		1.7
52	Lithuania	5.94		1.0		113	Angola	3.57		1.4
53	Greece	5.94		1.0		114	Lesotho	3.56		1.8
54	Spain	5.92		1.1		115	Albania	3.37		1.8
55	Panama	5.89		1.0		116	Malawi	3.29		1.7
56	Thailand	5.89		0.9		117	Chad	3.26		2.1
57	Latvia	5.84		1.2		118	Guyana	3.25		1.6
58	Turkey	5.82		1.1		119	Suriname	2.99		1.4
59	Azerbaijan	5.77		1.4		120	Benin	2.93		1.6
60	Mauritania	5.77		1.6		121	Bangladesh	2.55		1.3
61	Barbados	5.65		1.1		122	Zimbabwe	2.22		1.1

SOURCE: World Economic Forum, Executive Opinion Survey 2006

337

8.05 Availability of mobile telephones

Mobile or cellular telephones for your business are (1 = not available, 7 = as accessible and affordable as in the world's most technologically advanced countries)

RANK	COUNTRY/ECONOMY	SCORE	SD
1	Israel	6.94	0.2
2	Sweden	6.92	0.3
3	Germany	6.92	0.3
4	Iceland	6.90	0.3
5	Japan	6.87	0.4
6	Finland	6.86	0.5
7	Netherlands	6.85	0.4
8	Singapore	6.84	0.4
9	Hong Kong SAR	6.83	0.4
10	Austria	6.82	0.5
11	France	6.80	0.5
12	Denmark	6.78	0.4
13	Estonia	6.78	0.8
14	United Kingdom	6.76	0.5
15	Jordan	6.76	0.5
16	Switzerland	6.74	0.5
17	Czech Republic	6.74	0.4
18	Chile	6.74	0.6
19	Kuwait	6.73	0.7
20	Dominican Republic	6.71	0.6
21	Hungary	6.70	0.6
22	Norway	6.70	0.6
23	Guatemala	6.70	0.5
24	Belgium	6.66	0.5
25	India	6.64	0.6
26	Slovak Republic	6.63	0.6
27	El Salvador	6.63	0.5
28	Korea, Rep.	6.63	0.9
29	United Arab Emirates	6.61	0.7
30	Greece	6.60	0.5
31	Bahrain	6.58	0.6
32	Jamaica	6.57	0.6
33	Luxembourg	6.56	0.7
34	Spain	6.56	0.7
35	Slovenia	6.55	0.7
36	Philippines	6.55	0.7
37	Egypt	6.53	0.7
38	Venezuela	6.53	0.7
39	Algeria	6.53	0.7
40	Canada	6.48	0.8
41	Cyprus	6.48	0.8
42	Lithuania	6.48	0.7
43	Portugal	6.47	1.1
44	Italy	6.42	0.8
45	Barbados	6.40	0.7
46	Uruguay	6.40	0.8
47	Taiwan, China	6.40	0.9
48	Colombia	6.39	0.7
49	Malaysia	6.38	0.7
50	Turkey	6.38	0.8
51	Mauritius	6.37	0.7
52	Georgia	6.36	0.8
53	Brazil	6.34	0.9
54	United States	6.34	1.0
55	Croatia	6.33	1.0
56	Panama	6.31	1.0
57	Australia	6.31	0.8
58	Latvia	6.31	1.0
59	Paraguay	6.30	1.1
60	Tunisia	6.29	0.8
61	Azerbaijan	6.26	1.0

RANK	COUNTRY/ECONOMY	SCORE	SD
62	Ireland	6.26	0.9
63	Mexico	6.24	0.8
64	Honduras	6.22	1.0
65	Argentina	6.22	0.9
66	Peru	6.21	1.0
67	Malta	6.14	0.7
68	Ecuador	6.13	1.0
69	Morocco	6.12	1.2
70	Russian Federation	6.09	1.1
71	Sri Lanka	6.09	1.0
72	Mauritania	6.07	1.5
73	Indonesia	6.06	0.8
74	Bangladesh	6.03	0.8
75	Bolivia	6.02	1.1
76	Nicaragua	6.00	1.2
76	South Africa	6.00	0.9
76	Thailand	6.00	0.9
79	Macedonia, FYR	5.98	1.2
80	Kazakhstan	5.95	1.1
81	Tanzania	5.94	0.9
82	Namibia	5.94	0.9
83	Romania	5.89	1.7
84	New Zealand	5.87	0.7
85	Bulgaria	5.85	1.3
86	Mozambique	5.82	1.4
87	Uganda	5.81	1.2
88	Vietnam	5.74	1.0
89	Moldova	5.72	1.1
90	Kenya	5.69	1.3
91	Mali	5.65	1.7
92	Zambia	5.65	0.9
93	Serbia and Montenegro	5.64	1.4
94	Botswana	5.63	1.0
95	Nigeria	5.59	1.5
96	Ukraine	5.53	1.3
97	Cameroon	5.50	1.3
98	Bosnia and Herzegovina	5.49	1.2
99	Qatar	5.47	1.5
100	Burkina Faso	5.45	1.4
101	China	5.44	1.3
102	Pakistan	5.43	1.4
103	Malawi	5.34	0.9
104	Trinidad and Tobago	5.27	1.2
105	Madagascar	5.23	1.2
106	Guyana	5.20	1.3
107	Kyrgyz Republic	5.19	1.4
108	Mongolia	5.17	1.6
109	Cambodia	5.16	1.4
110	Lesotho	5.13	1.3
111	Albania	5.13	1.5
112	Angola	5.09	1.5
113	Chad	5.06	2.0
114	Poland	4.97	1.5
115	Suriname	4.96	1.4
116	Benin	4.88	1.5
117	Ethiopia	4.83	1.4
118	Nepal	4.78	1.3
119	Costa Rica	4.56	1.6
120	Armenia	4.53	1.4
121	Burundi	4.52	2.0
122	Zimbabwe	4.31	1.3

MEAN: 6.08

SOURCE: World Economic Forum, Executive Opinion Survey 2006

8.06 Extent of business Internet use

In your country, companies use the Internet extensively for buying/selling goods and services and for interaction with customers (1 = strongly disagree, 7 = strongly agree)

RANK	COUNTRY/ECONOMY	SCORE	MEAN: 3.90	SD
1	Korea, Rep.	6.14		1.0
2	United Kingdom	6.07		0.9
3	Estonia	6.05		1.0
4	Sweden	5.67		1.4
5	Netherlands	5.67		1.0
6	Germany	5.65		0.9
7	Switzerland	5.61		1.0
8	Iceland	5.60		1.3
9	Canada	5.59		1.2
10	Denmark	5.56		1.0
11	Norway	5.48		1.1
12	United States	5.47		1.5
13	Japan	5.42		1.3
14	Israel	5.42		1.4
15	Finland	5.40		1.3
16	Australia	5.23		1.3
17	Austria	5.20		1.1
18	Taiwan, China	5.12		1.2
19	France	5.05		1.2
20	Czech Republic	5.05		1.2
21	Singapore	5.00		1.1
22	New Zealand	4.98		1.4
23	Brazil	4.97		1.3
24	Hong Kong SAR	4.93		1.4
25	Ireland	4.89		1.5
26	Chile	4.82		1.2
27	Malaysia	4.75		1.2
28	Thailand	4.67		1.4
29	Lithuania	4.65		1.4
30	Slovenia	4.63		1.4
31	India	4.61		1.3
32	Belgium	4.59		1.2
33	Luxembourg	4.48		1.4
34	Malta	4.41		1.3
35	Slovak Republic	4.32		1.3
36	Latvia	4.29		1.6
37	Portugal	4.29		1.1
38	Panama	4.28		1.7
39	Guatemala	4.25		1.5
40	Italy	4.19		1.5
41	Poland	4.17		1.1
42	Russian Federation	4.10		1.7
43	Hungary	4.08		1.5
44	Spain	4.08		1.4
45	United Arab Emirates	4.06		1.5
46	Peru	4.02		1.3
47	Cyprus	4.01		1.4
48	Indonesia	4.01		1.1
49	South Africa	3.99		1.4
50	Pakistan	3.95		1.4
51	Philippines	3.94		1.4
52	Jamaica	3.93		1.6
53	El Salvador	3.92		1.3
54	Uruguay	3.92		1.5
55	Argentina	3.91		1.6
56	Mexico	3.89		1.4
57	Azerbaijan	3.88		1.9
58	Croatia	3.87		1.5
59	Tunisia	3.83		1.6
60	Barbados	3.79		1.6
61	Turkey	3.78		1.4
62	Bosnia and Herzegovina	3.77		1.8
63	Costa Rica	3.76		1.5
64	Trinidad and Tobago	3.75		1.8
65	Egypt	3.72		1.6
66	Qatar	3.71		1.6
67	Jordan	3.68		1.7
68	Kuwait	3.64		1.7
69	Kenya	3.63		1.8
70	Kazakhstan	3.62		1.9
71	Colombia	3.60		1.5
72	Mauritania	3.57		2.5
73	Namibia	3.56		1.8
74	Uganda	3.53		1.9
75	China	3.51		1.5
76	Dominican Republic	3.50		1.7
77	Sri Lanka	3.49		1.6
78	Honduras	3.44		1.5
79	Venezuela	3.44		1.4
80	Romania	3.41		1.5
81	Nigeria	3.38		1.8
82	Bahrain	3.34		1.6
83	Mali	3.29		1.9
84	Nepal	3.29		1.6
85	Bulgaria	3.29		1.5
86	Tanzania	3.21		1.6
87	Greece	3.21		1.3
88	Ukraine	3.19		1.5
89	Ecuador	3.17		1.2
90	Armenia	3.16		1.6
91	Guyana	3.16		1.6
92	Burkina Faso	3.15		1.9
93	Cambodia	3.12		2.0
94	Serbia and Montenegro	3.11		1.9
95	Nicaragua	3.10		1.5
96	Bangladesh	3.09		1.7
97	Mongolia	3.06		1.6
98	Georgia	3.01		1.7
99	Madagascar	2.98		1.6
100	Zimbabwe	2.97		1.4
101	Mauritius	2.96		1.4
102	Malawi	2.95		1.5
103	Benin	2.92		1.7
104	Bolivia	2.87		1.3
105	Zambia	2.86		1.1
106	Suriname	2.85		1.7
107	Morocco	2.83		1.6
108	Botswana	2.83		1.4
109	Vietnam	2.81		1.6
110	Ethiopia	2.75		1.7
111	Paraguay	2.75		1.3
112	Mozambique	2.72		1.5
113	Lesotho	2.69		1.6
114	Kyrgyz Republic	2.67		1.5
115	Macedonia, FYR	2.65		1.5
116	Angola	2.61		1.7
117	Albania	2.61		1.5
118	Moldova	2.52		1.5
119	Cameroon	2.42		1.6
120	Burundi	2.34		1.3
121	Algeria	2.17		1.5
122	Chad	1.94		1.4

SOURCE: World Economic Forum, Executive Opinion Survey 2006

339

Pillar 9
Government usage

9.01 Government success in ICT promotion

Government programs promoting the use of ICT are (1 = not very successful, 7 = highly successful)

RANK	COUNTRY/ECONOMY	SCORE	1 MEAN: 4.09 7	SD
1	Singapore	5.90		0.7
2	Malta	5.56		1.0
3	Malaysia	5.56		1.0
4	Tunisia	5.54		1.0
5	Mauritania	5.52		1.6
6	Taiwan, China	5.49		0.9
7	United Arab Emirates	5.49		1.1
8	Estonia	5.43		1.1
9	Iceland	5.31		1.0
10	Qatar	5.21		1.1
11	Mali	5.13		1.4
12	Japan	5.12		1.3
13	Burkina Faso	5.06		1.2
14	Denmark	5.03		1.2
15	Thailand	5.02		1.0
16	Finland	4.94		1.1
17	India	4.88		1.1
18	Sweden	4.84		1.1
19	Portugal	4.83		0.7
20	Hong Kong SAR	4.83		1.3
21	France	4.81		1.3
22	Korea, Rep.	4.78		1.2
23	Israel	4.74		1.1
24	Austria	4.71		1.2
25	Norway	4.70		1.0
26	Madagascar	4.70		1.4
27	Mauritius	4.67		1.1
28	United States	4.66		1.4
29	Tanzania	4.65		1.3
30	Switzerland	4.62		1.1
31	Jordan	4.61		1.5
32	Vietnam	4.60		1.3
33	Bahrain	4.59		1.4
34	Netherlands	4.58		1.2
35	Luxembourg	4.57		1.3
36	Algeria	4.56		1.6
37	Chile	4.53		1.2
38	Canada	4.51		1.0
39	Azerbaijan	4.47		1.5
40	Uganda	4.46		1.6
41	Morocco	4.45		1.5
42	Ireland	4.43		1.4
43	Barbados	4.42		1.2
44	Benin	4.42		1.7
45	Australia	4.42		1.0
46	Germany	4.40		1.1
46	Pakistan	4.40		1.4
48	Nigeria	4.37		1.8
49	Egypt	4.37		1.6
50	Kazakhstan	4.34		1.3
51	Jamaica	4.34		1.2
52	El Salvador	4.32		1.3
53	Slovenia	4.29		1.3
54	Hungary	4.25		1.2
55	Lithuania	4.24		1.3
56	United Kingdom	4.24		1.6
57	Brazil	4.23		1.4
58	Mongolia	4.21		1.6
59	South Africa	4.20		1.0
60	Botswana	4.18		1.3
61	Croatia	4.16		1.3
62	Cameroon	4.16		1.7
63	Romania	4.12		1.5
64	China	4.06		1.3
65	Dominican Republic	4.05		1.5
66	Belgium	4.03		1.3
67	Mexico	4.03		1.2
68	Philippines	3.96		1.4
69	Sri Lanka	3.94		1.5
70	Turkey	3.91		1.3
71	Cyprus	3.90		1.2
72	New Zealand	3.89		1.2
73	Colombia	3.88		1.2
74	Kenya	3.87		1.5
75	Trinidad and Tobago	3.87		1.4
76	Mozambique	3.80		1.4
77	Slovak Republic	3.77		1.2
78	Italy	3.74		1.5
79	Ethiopia	3.72		1.5
80	Burundi	3.71		2.0
81	Moldova	3.69		1.7
82	Costa Rica	3.62		1.3
83	Greece	3.62		1.3
84	Nepal	3.61		1.6
85	Serbia and Montenegro	3.61		1.3
86	Kuwait	3.60		1.5
86	Spain	3.60		1.4
88	Czech Republic	3.59		1.4
89	Cambodia	3.59		1.6
90	Uruguay	3.59		1.2
91	Angola	3.55		1.3
92	Bosnia and Herzegovina	3.53		1.5
93	Armenia	3.53		1.6
94	Nicaragua	3.51		1.4
95	Guyana	3.51		1.4
96	Latvia	3.50		1.3
97	Russian Federation	3.48		1.5
98	Malawi	3.47		1.3
99	Guatemala	3.45		1.3
100	Zimbabwe	3.35		1.5
101	Peru	3.32		1.1
102	Bulgaria	3.32		1.5
103	Macedonia, FYR	3.31		1.6
104	Georgia	3.28		1.3
105	Namibia	3.27		1.3
106	Lesotho	3.26		1.5
107	Bangladesh	3.23		1.5
108	Poland	3.22		1.0
109	Argentina	3.20		1.3
110	Honduras	3.18		1.4
111	Venezuela	3.16		1.3
112	Panama	3.15		1.4
113	Ukraine	3.14		1.2
114	Bolivia	3.00		1.3
115	Indonesia	2.87		0.9
116	Zambia	2.83		1.2
117	Kyrgyz Republic	2.79		1.4
118	Albania	2.65		1.4
119	Chad	2.62		1.5
120	Ecuador	2.54		1.1
121	Suriname	2.53		1.3
122	Paraguay	2.44		1.3

SOURCE: World Economic Forum, Executive Opinion Survey 2006

9.02 Availability of online services

In your country, online government services such as personal tax, car registrations, passport applications, business permits, and e-procurement are (1 = not available, 7 = extensively available)

RANK	COUNTRY/ECONOMY	SCORE	MEAN: 3.63	SD
1	Estonia	6.45		0.7
2	Singapore	6.26		0.9
3	Ireland	5.89		1.1
4	Denmark	5.78		1.1
5	Malta	5.76		0.9
6	Iceland	5.70		1.0
7	Sweden	5.65		1.5
8	United Kingdom	5.63		1.1
9	Chile	5.61		1.1
10	Hong Kong SAR	5.59		1.1
11	Norway	5.56		1.0
12	Austria	5.54		1.2
13	United States	5.45		1.3
14	Canada	5.45		1.2
15	Australia	5.40		1.4
16	Malaysia	5.33		1.2
17	Taiwan, China	5.29		1.2
18	New Zealand	5.20		1.4
19	Netherlands	5.19		1.2
20	Korea, Rep.	5.15		1.3
21	Finland	5.10		1.4
22	France	5.07		1.2
23	Brazil	5.07		1.3
24	Qatar	5.03		1.3
25	United Arab Emirates	5.01		1.4
26	Israel	5.00		1.4
27	Switzerland	4.99		1.3
28	Thailand	4.83		0.9
29	Germany	4.57		1.3
30	Portugal	4.50		1.1
31	Lithuania	4.45		1.5
32	El Salvador	4.44		1.3
33	Mexico	4.41		1.2
34	Belgium	4.34		1.3
35	Luxembourg	4.32		1.6
36	Spain	4.32		1.4
37	Slovenia	4.21		1.3
38	Venezuela	4.09		1.3
39	Tunisia	4.09		1.7
40	India	4.06		1.4
41	Japan	4.06		1.5
42	Peru	4.05		1.4
43	Dominican Republic	4.04		1.3
44	Kazakhstan	3.99		1.8
45	Hungary	3.97		1.5
46	Guatemala	3.97		1.5
47	China	3.97		1.6
48	Uruguay	3.89		1.5
49	Uganda	3.82		2.4
50	Cyprus	3.81		1.6
51	Jamaica	3.80		1.4
52	Turkey	3.78		1.5
53	Argentina	3.76		1.5
54	Bolivia	3.69		1.4
55	Nigeria	3.68		1.9
56	South Africa	3.67		1.6
57	Mauritania	3.66		2.7
58	Italy	3.64		1.6
59	Pakistan	3.62		1.6
60	Colombia	3.59		1.4
61	Egypt	3.59		1.6
62	Bahrain	3.54		1.6
63	Greece	3.51		1.3
64	Croatia	3.51		1.7
65	Poland	3.49		1.4
66	Romania	3.47		1.6
67	Panama	3.43		1.5
68	Philippines	3.36		1.6
69	Bulgaria	3.36		1.4
70	Ecuador	3.35		1.3
71	Costa Rica	3.25		1.3
72	Azerbaijan	3.23		2.2
73	Latvia	3.19		1.7
74	Slovak Republic	3.16		1.3
75	Mauritius	3.15		1.4
76	Nicaragua	3.07		1.4
77	Kenya	3.05		2.0
78	Czech Republic	3.03		1.3
79	Jordan	2.99		1.6
80	Honduras	2.88		1.5
81	Morocco	2.86		1.7
82	Sri Lanka	2.86		2.0
83	Botswana	2.82		2.0
84	Russian Federation	2.82		1.6
85	Burkina Faso	2.81		1.9
86	Bosnia and Herzegovina	2.78		1.7
87	Indonesia	2.78		0.8
88	Ethiopia	2.77		2.1
89	Mali	2.77		2.0
90	Ukraine	2.77		1.7
91	Macedonia, FYR	2.68		1.7
92	Serbia and Montenegro	2.63		1.5
93	Malawi	2.61		2.0
94	Tanzania	2.55		1.7
95	Barbados	2.53		1.7
96	Vietnam	2.51		1.6
97	Namibia	2.50		1.7
98	Chad	2.46		1.9
99	Moldova	2.40		1.7
100	Algeria	2.37		1.6
101	Benin	2.36		1.7
102	Madagascar	2.29		1.5
103	Lesotho	2.29		2.0
104	Mongolia	2.27		1.6
105	Zambia	2.25		1.4
106	Kuwait	2.23		1.5
107	Paraguay	2.17		1.2
108	Nepal	2.17		1.7
109	Burundi	2.15		1.7
110	Angola	2.12		1.7
111	Kyrgyz Republic	2.05		1.3
112	Guyana	1.98		1.9
113	Cambodia	1.93		1.7
114	Cameroon	1.88		1.5
115	Trinidad and Tobago	1.85		1.3
116	Armenia	1.82		1.1
117	Mozambique	1.77		1.3
118	Georgia	1.72		1.2
119	Bangladesh	1.62		1.1
120	Suriname	1.56		1.2
121	Albania	1.54		0.9
122	Zimbabwe	1.53		1.1

SOURCE: World Economic Forum, Executive Opinion Survey 2006

9.03   ICT use and government efficiency

In your view, ICT use by the government has improved the efficiency of government services and has facilitated interaction with business and civil society (1 = strongly disagree, 7 = strongly agree)

RANK	COUNTRY/ECONOMY	SCORE	SD
1	Estonia	6.12	0.9
2	Singapore	6.01	0.9
3	Iceland	5.90	0.9
4	Chile	5.81	1.0
5	Denmark	5.78	1.2
6	Mauritania	5.71	2.0
7	United Arab Emirates	5.65	1.4
8	Malta	5.61	1.1
9	Hong Kong SAR	5.60	1.1
10	Sweden	5.59	1.0
11	Malaysia	5.58	1.0
12	Italy	5.57	1.7
13	Qatar	5.51	1.1
14	Thailand	5.48	0.9
15	Portugal	5.44	1.1
16	Ireland	5.40	1.4
17	Taiwan, China	5.38	1.1
18	Finland	5.38	1.0
19	Brazil	5.28	1.4
20	United States	5.26	1.2
21	Norway	5.26	1.1
22	Korea, Rep.	5.23	1.0
23	Israel	5.22	1.2
24	Netherlands	5.22	1.1
25	Tunisia	5.19	1.5
26	Austria	5.16	1.1
27	India	5.10	1.3
28	Canada	5.08	1.3
29	France	5.08	1.4
30	Switzerland	5.03	1.2
31	Mexico	4.96	1.2
32	Dominican Republic	4.94	1.3
33	United Kingdom	4.90	1.4
34	Australia	4.87	1.3
35	El Salvador	4.86	1.0
36	Spain	4.85	1.4
37	Cambodia	4.77	2.0
38	Germany	4.73	1.2
39	Luxembourg	4.71	1.5
40	Uruguay	4.60	1.4
41	Guatemala	4.60	1.5
42	Turkey	4.59	1.5
43	Morocco	4.57	1.8
44	China	4.56	1.6
45	Jamaica	4.53	1.2
46	Peru	4.53	1.2
47	Mali	4.51	1.9
47	New Zealand	4.51	1.2
49	Cyprus	4.46	1.5
50	Azerbaijan	4.45	1.6
51	Philippines	4.40	1.6
52	Madagascar	4.38	1.7
53	Belgium	4.38	1.3
54	Colombia	4.38	1.6
55	Kazakhstan	4.37	1.6
56	Lithuania	4.33	1.5
57	Japan	4.31	1.5
58	Argentina	4.30	1.5
59	Bolivia	4.29	1.6
60	Burkina Faso	4.29	2.1
61	Nigeria	4.29	1.8
62	Greece	4.28	1.6
63	Pakistan	4.25	1.5
64	Bahrain	4.22	1.6
65	Egypt	4.19	1.7
66	Slovenia	4.18	1.5
67	Venezuela	4.17	1.6
68	Tanzania	4.14	1.7
69	Hungary	4.10	1.5
70	Latvia	4.10	1.6
71	Slovak Republic	4.10	1.6
72	Algeria	4.08	2.0
73	Uganda	4.08	2.0
74	Vietnam	4.07	1.8
75	South Africa	4.07	1.5
76	Ethiopia	4.06	1.9
77	Jordan	4.04	1.6
78	Romania	4.03	1.8
79	Serbia and Montenegro	4.01	2.0
80	Croatia	3.99	1.7
81	Ecuador	3.99	1.5
82	Nicaragua	3.97	1.4
83	Kenya	3.96	2.0
84	Benin	3.93	2.0
85	Nepal	3.93	2.0
86	Panama	3.93	1.6
87	Russian Federation	3.92	1.9
88	Sri Lanka	3.88	1.7
89	Georgia	3.76	1.7
90	Costa Rica	3.73	1.5
91	Poland	3.71	1.4
92	Ukraine	3.69	1.6
93	Barbados	3.66	1.6
94	Honduras	3.61	1.6
95	Mauritius	3.59	1.8
96	Botswana	3.55	1.7
97	Czech Republic	3.51	1.4
98	Kuwait	3.49	1.8
99	Bangladesh	3.48	1.8
100	Mongolia	3.43	1.9
101	Bulgaria	3.39	1.6
102	Kyrgyz Republic	3.35	2.1
103	Armenia	3.33	1.7
104	Bosnia and Herzegovina	3.29	1.7
105	Moldova	3.26	1.7
106	Namibia	3.20	1.6
107	Cameroon	3.20	2.0
108	Paraguay	3.16	1.9
109	Macedonia, FYR	3.15	1.6
110	Angola	3.09	1.8
111	Burundi	3.08	2.0
112	Chad	3.06	1.9
113	Malawi	3.05	1.9
114	Mozambique	3.02	1.6
115	Indonesia	2.95	1.1
116	Suriname	2.89	2.0
117	Guyana	2.89	1.9
118	Trinidad and Tobago	2.83	1.6
119	Lesotho	2.81	1.8
120	Zambia	2.32	1.2
121	Zimbabwe	2.17	1.4
122	Albania	2.03	1.1

MEAN: 4.29

SOURCE: World Economic Forum, Executive Opinion Survey 2006

344

The presence of ICT in government offices in your country is (1 = very rare, 7 = commonplace and pervasive)

RANK	COUNTRY/ECONOMY	SCORE	SD		RANK	COUNTRY/ECONOMY	SCORE	SD
1	Singapore	6.42	0.7		62	Turkey	4.41	1.4
2	Estonia	6.29	0.9		63	Uruguay	4.39	1.6
3	Korea, Rep.	6.16	1.2		64	South Africa	4.39	1.3
4	Switzerland	5.91	1.1		65	Albania	4.32	2.1
5	Finland	5.82	1.0		66	Belgium	4.32	1.3
6	Netherlands	5.75	1.1		67	Botswana	4.30	1.3
7	Iceland	5.75	1.0		68	Guatemala	4.28	1.4
8	Austria	5.71	1.2		69	Kazakhstan	4.25	1.5
9	Sweden	5.68	1.3		70	Vietnam	4.24	1.6
10	Denmark	5.62	0.9		71	Nicaragua	4.18	1.5
11	Mauritania	5.62	1.6		72	Uganda	4.16	1.7
12	Tunisia	5.57	1.3		73	Mauritius	4.15	1.5
13	Taiwan, China	5.57	1.1		74	Barbados	4.11	1.3
14	Norway	5.55	1.0		75	Czech Republic	4.10	1.3
15	Chile	5.50	1.1		76	Colombia	4.10	1.4
16	Malta	5.49	0.9		76	Dominican Republic	4.10	1.4
17	Malaysia	5.48	0.9		78	Armenia	4.04	1.5
18	Hong Kong SAR	5.46	1.1		79	Tanzania	3.98	1.6
19	Luxembourg	5.42	1.3		80	Greece	3.97	1.3
20	United Kingdom	5.40	1.1		81	Kuwait	3.97	1.7
21	Australia	5.38	1.2		82	Egypt	3.87	1.5
22	Ireland	5.32	1.0		83	Peru	3.86	1.4
23	Slovak Republic	5.32	1.1		84	Pakistan	3.86	1.5
24	United Arab Emirates	5.30	1.2		85	Cyprus	3.83	1.3
25	Slovenia	5.30	1.2		86	Madagascar	3.74	1.7
26	Germany	5.27	1.3		87	Poland	3.71	1.2
27	Algeria	5.26	1.4		88	Namibia	3.68	1.5
28	Canada	5.25	1.3		89	Argentina	3.63	1.5
29	Spain	5.24	1.5		90	Panama	3.59	1.4
30	Thailand	5.22	1.1		91	Nigeria	3.53	1.9
31	New Zealand	5.20	1.2		92	Bosnia and Herzegovina	3.48	1.5
32	Croatia	5.19	1.3		93	Macedonia, FYR	3.48	1.5
33	Serbia and Montenegro	5.12	1.4		94	Russian Federation	3.47	1.6
34	Qatar	5.09	1.1		95	Guyana	3.46	1.5
35	Japan	5.08	1.5		96	Philippines	3.43	1.5
36	Bulgaria	5.08	1.4		97	Bolivia	3.42	1.5
37	Romania	5.06	1.2		98	Sri Lanka	3.41	1.6
38	Latvia	5.05	1.5		99	Honduras	3.41	1.6
39	France	5.03	1.4		100	Ukraine	3.39	1.5
40	United States	5.02	1.3		101	Trinidad and Tobago	3.34	1.5
41	Lithuania	5.02	1.2		102	Ethiopia	3.34	1.6
42	El Salvador	4.96	1.1		103	Georgia	3.31	1.4
43	Portugal	4.89	0.8		104	Venezuela	3.28	1.4
44	Israel	4.85	1.0		105	Cambodia	3.27	1.6
45	Benin	4.83	1.6		106	Cameroon	3.27	1.7
46	Azerbaijan	4.79	1.7		107	Malawi	3.24	1.6
47	Mexico	4.70	1.3		108	Burundi	3.20	1.8
48	Jordan	4.64	1.4		109	Ecuador	3.15	1.3
49	Mali	4.64	2.0		110	Angola	3.15	1.5
50	Morocco	4.64	1.7		111	Chad	3.12	1.9
51	Mongolia	4.63	1.8		112	Mozambique	3.12	1.5
52	Hungary	4.62	1.4		113	Kenya	3.10	1.5
53	Jamaica	4.59	1.2		114	Paraguay	3.05	1.5
54	Burkina Faso	4.57	1.9		115	Lesotho	2.93	1.5
55	Brazil	4.54	1.4		116	Nepal	2.90	1.5
56	Bahrain	4.50	1.5		117	Indonesia	2.81	0.6
57	Moldova	4.47	1.6		118	Kyrgyz Republic	2.79	1.4
58	India	4.45	1.3		119	Zimbabwe	2.72	1.4
59	Italy	4.45	1.5		120	Bangladesh	2.64	1.3
60	China	4.43	1.5		121	Zambia	2.40	1.2
61	Costa Rica	4.42	1.5		122	Suriname	2.21	1.1

MEAN: 4.38

SOURCE: World Economic Forum, Executive Opinion Survey 2006

Technical Notes and Sources

The data used in this *Report* represent the best available estimates from various national authorities, international agencies, and private sources at the time the *Report* was prepared. It is possible that some data will have been revised or updated by national sources after publication. Throughout the statistical tables in this publication, "n/a" denotes that the value is not available, or that available data are unreasonably outdated or do not come from a reliable source.

The following section provides additional information and definitions for the hard data indicators that enter the composition of the Networked Readiness Index and are presented in the Data Tables section of this *Report*.

Pillar 1: Market environment

1.05   US utility patents

Number of utility patents (i.e., patents for invention) granted between January 1 and December 31, 2005, per million population

Utility patents are recorded such that the origin of the patent is determined by the first named inventor at the time of the grant. Patents per million population are calculated by dividing the number of patents granted to a country in 2005 by that country's population in the same year.

Source: U.S. Patent and Trademark Office (March 2006); United Nations Population Fund, *State of World Population 2005*

1.06   High-tech exports

High-technology exports as percentage of total exports, 2004

The value of high-technology exports is expressed as a percentage of the total value of goods and services exports. According to the World Bank, high-technology exports are products with high R&D intensity, as in aerospace, computers, pharmaceuticals, and scientific instruments.

Source: World Bank, *World Development Indicators Online Database* (December 2006); Economist Intelligence Unit, *CountryData Database* (2006); national sources

1.09   Time required to start a business

Number of days required to start a business, 2006

Source: World Bank, *Doing Business 2007: How to Reform* (2006)

1.10   Number of procedures required to start a business

Number of administrative procedures required to start a business, 2006

Source: World Bank, *Doing Business 2007: How to Reform* (2006)

Pillar 2: Political and regulatory environment

2.08   Number of procedures to enforce a contract

Number of procedures from the moment the plaintiff files a lawsuit in court until the moment of payment, 2006

Source: World Bank, *Doing Business 2007: How to Reform* (2006)

2.09   Time to enforce a contract

Number of days required to resolve a dispute as of 2006.

Source: World Bank, *Doing Business 2007: How to Reform* (2006)

Pillar 3: Infrastructure environment

3.01   Telephone lines

Main telephone lines per 100 inhabitants, 2005 or most recent year available

A *main telephone line* is a telephone line connecting the subscriber's terminal equipment to the public switched telephone network and that has a dedicated port in the telephone exchange equipment.

Source: International Telecommunication Union, *World Telecommunication Indicators 2006*

3.02   Secure Internet servers

Secure Internet servers per 1 million inhabitants, 2005 or most recent year available

Secure Internet servers are servers using encryption technology in Internet transactions.

Source: World Bank, *World Development Indicators Online Database* (December 2006); national sources

3.03   Internet hosts

Internet hosts per 10,000 inhabitants, 2004

Internet hosts refers to the number of computers in an economy that are directly linked to the worldwide Internet network. This statistic is based on the country code in the host address and thus may not correspond with the actual physical location.

Source: International Telecommunication Union, *World Telecommunication Indicators 2005*

3.04   Electricity production

Per capita electricity production (kWh), 2003

Electricity production is measured at the terminals of all alternator sets in a station. In addition to hydropower, coal, oil, gas, and nuclear power generation, it covers generation by geothermal, solar, wind, and tide and wave energy as well as that from combustible renewables and waste. Production includes the output of electricity plants designed to produce electricity only, as well as that of combined heat and power plants.

Source: World Bank, *World Development Indicators Online Database* (December 2006); national sources

347

3.07 Tertiary enrollment

Gross tertiary enrollment rate, 2004 or most recent year available
According to the World Bank's World Development Indicators, the *gross tertiary enrollment rate* corresponds to the ratio of total enrollment, regardless of age, to the population of the age group that officially corresponds to the tertiary education level. Tertiary education, whether or not leading to an advanced research qualification, normally requires, as a minimum condition of admission, the successful completion of education at the secondary level.

Source: UNESCO, Institute for Statistics (December 2006); World Bank, *World Development Indicators Online Database* (December 2006); national sources

Pillar 4: Individual readiness

4.06 Residential telephone connection charge

One-time residential telephone connection charge (US$) as a percentage of GDP per capita, 2005 or most recent year available
This measure refers to the one-time charge involved in applying for basic telephone service for residential purposes.

Source: International Telecommunication Union, *World Telecommunication Indicators 2006*; International Monetary Fund, *World Economic Outlook Database* (April 2006 and September 2006 editions)

4.07 Residential monthly telephone subscription

Residential monthly telephone subscription to PSTN (US$) as a percentage of monthly GDP per capita, 2005 or most recent year available
Residential monthly telephone subscription refers to the recurring fixed charge for a residential subscriber to the public switched telephone network. The charge should cover the rental of the line but not the rental of the terminal (for example, the telephone set) where the terminal equipment market is liberalized. In some cases, the rental charge includes an allowance for free or reduced-rate call units. If there are different charges for different exchange areas, the largest urban area is used.

Source: International Telecommunication Union, *World Telecommunication Indicators 2006*; International Monetary Fund, *World Economic Outlook Database* (April 2006 and September 2006 editions)

4.08 High-speed monthly broadband subscription charge

High-speed monthly broadband subscription charge (US$) as a percentage of monthly GDP per capita, 2006
The International Telecommunication Union considers *broadband* to be any dedicated connection to the Internet of 256 kilobits per second (kbits/s) or faster, in both directions. The monthly charge reflects the Internet service provider charge for one month of service. It does not include installation fees or modem rental charges if they are charged separately. Speed expressed in kbits/s represents the advertised maximum theoretical download speed and not speeds guaranteed to users. High-speed monthly charge refers to a faster and typically more expensive offer available in the economy.

Source: International Telecommunication Union, *World Information Society Report 2006*; International Monetary Fund, *World Economic Outlook Database* (April 2006 and September 2006 editions)

4.09 Lowest cost of broadband

Lowest sampled cost (US$) per 100 kbits/s as a percentage of monthly income (GNI), 2006
The lowest sampled cost in US dollars per 100 kilobits per second (kbits/s) gives the most cost-effective subscription based on criteria of least cost per 100 kbits/s. The International Telecommunication Union calculates it by dividing the monthly subscription charge in US dollars by the theoretical download speed, and then multiplying by 100. The lowest cost per 100 kbits/s across all Internet service providers is used to compute the lowest sampled cost as a percentage of monthly income (GNI).

Source: International Telecommunication Union, *World Information Society Report 2006*

4.10 Cost of mobile telephone call

Cost of 3-minute local call during peak hours (US$) as a percentage of monthly GDP per capita, 2005 or most recent year available

Source: International Telecommunication Union, *World Telecommunication Indicators 2006*; International Monetary Fund, *World Economic Outlook Database* (April 2006 and September 2006 editions)

Pillar 5: Business readiness

5.06 Business telephone connection charge

One-time business telephone connection charge (US$) as a percentage of GDP per capita, 2005 or most recent year available
This measure refers to the one-time charge involved in applying for basic telephone service for business purposes.

Source: International Telecommunication Union, *World Telecommunication Indicators 2006*; International Monetary Fund, *World Economic Outlook Database* (April 2006 and September 2006 editions)

5.07 Business monthly telephone subscription

Business monthly telephone subscription to the PSTN (US$) as a percentage of monthly GDP per capita, 2005 or most recent year available
Business monthly telephone subscription refers to the recurring fixed charge for a business subscriber to the public switched telephone network.

Source: International Telecommunication Union, *World Telecommunication Indicators 2006*; International Monetary Fund, *World Economic Outlook Database* (April 2006 and September 2006 editions)

5.09 Computer, communications, and other services imports

Computer, communications, and other services as percentage of total commercial services imports, 2004
Computer, communications, and other services include such activities as international telecommunications; postal and courier services; computer data; news-related service transactions between residents and nonresidents; construction services; royalties and license fees; miscellaneous business, professional, and technical services; and personal, cultural, and recreational services. The total volume of computer, communications, and other services imports is divided by the total volume of commercial service imports, defined as the total service imports minus imports of government services not included elsewhere.

Source: World Bank, *World Development Indicators Online Database* (December 2006)

Pillar 6: Government readiness

6.04   E-participation index

The e-participation index assesses the quality, relevance, usefulness and the willingness of government websites for providing online information and participatory tools and services to the people, 2005

Source: United Nations, *Global E-Readiness Report 2005*. The report is available at www.unpan.org/egovkb.

6.05   E-government readiness index

The e-government readiness index assesses e-government readiness based on website assessment, telecommunications infrastructure, and human resource endowment, 2005

Source: United Nations, *Global E-Readiness Report 2005*. The report is available at www.unpan.org/egovkb.

Pillar 7: Individual usage

7.01   Mobile telephone subscribers

Mobile telephone subscribers per 100 inhabitants, 2005 or most recent year available

The term subscribers refers to users of mobile telephones subscribing to an automatic public mobile telephone service that provides access to the public switched telephone network using cellular technology. This can include analogue and digital cellular systems but should not include noncellular systems. Subscribers to fixed wireless, public mobile data services, or radio paging services are not included.

Source: International Telecommunication Union, *World Telecommunication Indicators 2006*

7.02   Personal computers

Personal computers per 100 inhabitants, 2004 or most recent year available

According to the World Bank, *personal computers* are self-contained computers designed to be used by a single individual.

Source: International Telecommunication Union, *World Telecommunication Indicators 2006*

7.03   Broadband Internet subscribers

Total broadband Internet subscribers per 100 inhabitants, 2005 or most recent year available

The International Telecommunication Union considers *broadband* to be any dedicated connection to the Internet of 256 kilobits per second or faster, in both directions. *Broadband subscribers* refers to the sum of DSL, cable modem, and other broadband (for example, fiber optic, fixed wireless, apartment LANs, satellite connections) subscribers.

Source: International Telecommunication Union, *World Telecommunication Indicators 2006*

7.04   Internet users

Internet users per 100 inhabitants, 2005 or most recent year available

Internet users are people with access to the worldwide network.

Source: International Telecommunication Union, *World Telecommunication Indicators 2006*

7.05   Internet bandwidth

International Internet bandwidth (Mbps) per 10,000 inhabitants, 2004

This measure shows the total capacity of international Internet bandwidth in megabits per second.

Source: International Telecommunication Union, *World Telecommunication Indicators 2006*

349

About the Authors

Scott C. Beardsley

Scott Beardsley is a Director in McKinsey & Company's Brussels Office. Since joining the firm in 1989, he has been particularly active in helping clients around the world on a range of strategy, regulation, reputation and stakeholder management, performance transformation, and sales & marketing topics in the telecommunications, technology, and media sectors, and has recently led a variety of internal research initiatives. Over the past decade he has served many fixed and mobile telephone companies in emerging economies in the Middle East and Africa, Eastern Europe, Latin America, and Asia, in addition to numerous telecommunications firms and a leading global equipment provider in the West. He is now a global leader of McKinsey's telecommunications practice, and formally leads McKinsey's Strategy Practice in Europe, the Middle East, and Africa. A frequent author and public speaker, he has written extensively on a variety of telecommunications, broadband, media, and strategy topics. He has co-authored chapters in several of the World Economic Forum's *Global Information Technology Report*s and delivered presentations on digital readiness and telecommunications sector reform, as well as the future of telecommunications regulation. Prior to joining McKinsey, Mr Beardsley was Editor and Marketing Manager at the *MIT Sloan Management Review* and has worked in the strategic sales and product marketing functions for Advanced Micro Devices and Analog Devices of the semiconductor industry. Mr Beardsley was a Henry S. Dupont III Scholar (highest honors) for outstanding academic performance at the MIT Sloan School of Management, where he graduated with an MBA in Corporate Strategy and Marketing. He holds a Bachelor of Science degree in Electrical Engineering magna cum laude from Tufts.

Duarte Braga

Duarte Braga is an Associate Principal in the Lisbon Office of McKinsey & Company. Since joining the firm in 2000, he has mainly served clients in the Telecommunications and Financial Institutions sectors. He is a leader in McKinsey's Regulatory Management Practice and has been working with major telecommunications groups in Europe, South America, Asia, and the Middle East, supporting companies in defining their regulatory strategies, including the outlining and management of several regulation topics as well as interacting with different authorities on a broad range of issues such as new investments, retail product regulation, wholesale obligations and tariffs, interconnection agreements, and overall industry structure. He has also worked on developing competitive and pricing strategies for major mobile and wireline companies and on cost optimization "lean" efforts in mobile companies in Europe and South America. For another European telecommunications

company, he evaluated the opportunity of entering the electronic mobile payments market in parallel with the development of an m-commerce strategy through a joint venture with a major financial institution. Prior to joining McKinsey, he worked for Compta, a Portuguese IT/Telecoms services and equipment integrator, as a Senior Pre-Sales Engineer, where he was responsible for coordinating sales and engineering as well as for developing multiservice telecommunications networks. For Siemens, he worked as a Software Engineer and was responsible for a group in charge of the design and implementation of a test environment for a new generation of communications equipment. And for INESC, a Portuguese electronics R&D agency, he worked as a Researcher, having developed a DECT wireless telephone prototype and a customer database support system. Mr Braga holds a degree in Electrical and Computer Engineering from Instituto Superior Técnico in Portugal and an MBA from INSEAD in France.

Soumitra Dutta

Soumitra Dutta is the Roland Berger Chaired Professor of Business and Technology, and Dean of External Relations at INSEAD. Professor Dutta obtained his PhD in Computer Science and his MSc in business administration from the University of California at Berkeley. His current research is on technology strategy and innovation at both corporate and national policy levels. His latest books are *The Global Information Technology Report 2005–2006: Leveraging ICT for Development* (Palgrave Macmillan, March 2006) and *The Information Society in an Enlarged Europe* (Springer, February 2006). He has authored seven other books, including *The Bright Stuff* (Financial Times/Prentice Hall 2002) and *Embracing the Net* (Financial Times 2001). He has won several awards for research and pedagogy, including awards for the European Case of the Year from the European Case Clearing House in 1995, 1997, 1998, 2000, and 2002. His research has been showcased in the international media and he has taught in and consulted with international corporations across the world. He is a fellow of the World Economic Forum.

Luis Enriquez

Luis Enriquez is a Principal in McKinsey & Company's Brussels Office, where he has worked primarily in the areas of corporate finance, strategy, and telecommunications. He has had extensive experience in telecommunications, focusing on corporate finance, strategy, operations and regulation. Prior to joining McKinsey, Dr Enriquez worked extensively on telecommunications liberalization and regulation issues. In 1994 he assisted the Czech Ministry of Finance in developing price regulations to support the privatization of Cesky Telecom (then SPT Telecom), and taught courses and seminars for the Ministry staff and other industry stakeholders. He has participated in proceedings on liberalization and privatization in Mexico, Argentina, Poland, and other Eastern European and Latin American countries. He assisted the Chief Economist of the Federal Communications Commission in areas including interconnection, universal service subsidies, and developing dispute-resolution mechanisms, and has worked with US incumbents and new entrants on various regulatory topics. Dr Enriquez has a BA in Economics from Harvard University and a PhD in Economics from the University of California at Berkeley, where he focused on the economic dynamics of interconnection among telecommunications networks.

Roger Farnsworth

As a Senior Manager on Cisco's Executive Thought Leadership Team, Roger Farnsworth guides Cisco executive-level research on the cultural, technological, and economic trends that offer imminent promise as the network becomes the platform for life's experiences. In more than 10 years with Cisco, Mr Farnsworth has held several key marketing roles, including Director of Voice Product and Technology Marketing, Director of the Service Provider Solutions Marketing Team, Director of Marketing of the Optical Networking Group, and leadership of the Cisco Security Product Marketing Team. Mr Farnsworth has been in the communications industry since 1980 and holds a BS degree in Information Systems Management from the University of San Francisco.

Koichi Fujinuma

Koichi Fujinuma is Assistant Director of the International Policy Division, Telecommunications Bureau, Ministry of Internal Affairs and Communications (MIC), Japan. Mr Fujinuma is currently responsible for making international ICT policies and strategies for MIC. After joining the Ministry in 2000, he held positions in the International Frequency Policy Office and the Mobile Satellite Communication Division. He attended several ITU-R meetings as a head of the Japanese delegation. He has an MA degree from Stanford University and a Master of International Affairs degree from Columbia University.

Thierry Geiger

Thierry Geiger is an Economist with the Global Competitiveness Network at the World Economic Forum. His responsibilities include construction and computation of a range of indexes as well as data analysis for various projects and studies. His main areas of expertise are econometrics and international trade. Mr Geiger holds a BA in Economics from the University of Geneva with a specialization in monetary and financial economics, and an MA in Economics from the University of British Columbia, in Vancouver. During his studies, he was a Member of the Board of Junior Entreprise Genève. He is also Co-founder of Procab Studio S.A., an IT company based in Geneva.

352

Lionel Gibbons

As a member of Cisco's Executive Thought Leadership Team, Lionel Gibbons does research and analysis into the effects of networking technology, providing insight to Cisco executives, customers, and partners. He was previously the Director of Technology Policy, formulating and communicating positions on a variety of networking technologies with a focus on broadband access and content-protection technologies. Mr Gibbons also served as the Director of Strategic Technology Communications in the Office of the Chief Strategy Officer, where he was responsible for articulating Cisco's strategic technology direction and policies. Additionally, Mr Gibbons was the Director of Strategic Marketing for the IOS Technologies Division, where he helped define the technology strategy for the industry's leading core networking software—Cisco IOS. Prior to joining Cisco, Mr Gibbons was a Marketing Director at 3Com, where his responsibilities included policy-based networking technology and dial access concentrator products. Previously, he directed marketing at Eicon Technology. Before making the switch to marketing, Mr Gibbons was a systems software designer. He holds a BSc degree in Physics from Concordia University in Montreal.

Mehmet Güvendi

Mehmet Güvendi is a Principal at McKinsey & Company's Istanbul Office. Since joining the firm in March 1999, he has worked in strategy, regulation, operations, and IT topics in many different sectors. In particular, he has extensive experience in regulation in the telecommunications sector, where he has helped clients in Europe, Asia, and the Middle East. Before joining McKinsey, Mr Güvendi was an IT Group Manager at Procter & Gamble Company. He worked as an IT Manager for six years in Western Europe, North America, and Turkey. He led multifunctional global process design teams for planning and managed several major global pilot projects around the world. He also managed a data center and a multinational communications networks, and was in charge of IT systems and operations at several manufacturing sites. Mr Güvendi is a member of the Prime Ministry's Telecom Special Expertise Committee for the development of the Turkish National Five-Year Development Plan. He sits on the Advisory Council of Bilkent University Industrial Engineering Department. Mr Güvendi holds a BS degree with high honors in Industrial Engineering from Bilkent University in Ankara.

Markus Haacker

Markus Haacker is an Economist at the African Department of the International Monetary Fund. He has worked extensively on issues related to economic development, including numerous publications on economic aspects of HIV/AIDS, health and development, and the economic role of modern information technologies in low-income countries. For the IMF, he edited and contributed to *The Macroeconomics of HIV/AIDS*, a major study of the economic and fiscal consequences of the epidemic, published in 2004. His work on the economic role of information and communication technologies has been featured in the *Global Information Technology Report 2004–2005*, the IMF's World Economic Outlook, and the *Wall Street Journal*.

Bruno Lanvin

Bruno Lanvin is the World Bank's regional coordinator (Europe and Central Asia) for Telecommunications, IT, and E-strategies issues. He is also the Head of Client Capacity Building for the same issues. From June 2001 to December 2003, he was the Manager of the Information for Development Program (infoDev). In 2000 Dr Lanvin was appointed Executive Secretary of the G8 DOT Force. Until then he was Head of Electronic Commerce in the United Nations Conference on Trade and Development (UNCTAD) in Geneva, and occupied various senior positions including Chief of Cabinet of the Director General of the United Nations in New York, and Head of Strategic Planning and later Chief of the SME Trade Competitiveness Unit of UNCTAD/SITE. He was the main drafter, team leader, and editor of *Building Confidence: Electronic Commerce and Development*, published in January 2000. In 2003 and in 2004 he co-edited *The Global Information Technology Report* (INSEAD, The World Economic Forum, infoDev). He holds a BA in Mathematics and Physics from the University of Valenciennes (France), an MBA from Ecole des Hautes Etudes Commerciales (HEC) in Paris, and a PhD in Economics from the University of Paris I (La Sorbonne) in France.

Anat Lewin

Anat Lewin is an Operations Analyst for the World Bank's Global Information and Communication Technologies Department (GICT). Her main responsibilities consist of research, analysis, and project supervision in the areas of ICT policy, e-government, ICT applications, statistical capacity building, and monitoring and evaluation for results—primarily in Middle Eastern and North African (MNA) as well as Eastern European and Central Asian (ECA) countries. She has co-authored a number of papers on ICT and development, particularly on telecommunications policy. At the World Bank, Ms Lewin also worked in statistics as a Data Advisory Services Consultant for the Development Data Group. Prior to joining the World Bank in 2000, Ms Lewin worked for the United Nations Economic Commission for Africa in Ethiopia on ICT, science, and technology issues in Africa. She holds a Master of International Affairs degree from Columbia University in New York and a Bachelor of Arts degree in International Relations from the University of Toronto in Canada.

Tracey Lewis

Tracey Lewis is a Content Manager for the Executive Thought Leadership and Corporate Positioning Teams at Cisco. In this capacity she contributes to Cisco's business and technical research agenda, most recently analyzing the impact and opportunity for ICT in emerging countries around the world in their quest to transform their infrastructures and economies to take maximum advantage of rising globalization. Previously she led a variety of marketing strategy, communications, and operations teams for a service provider customer segment team and for several advanced technology areas, including security and optical networking. Prior to joining Cisco in 1998, Ms Lewis held senior marketing positions at Sun Microsystems in the technical graphics workstations and server and storage product groups. She also drove market development initiatives for Sun in the areas of electronic commerce and Internet security. Ms Lewis earned a BS degree in Business Administration, with emphases in Marketing and Finance, from the University of California, Berkeley. She has also completed several graduate courses through UC Berkeley and USC, and holds a Cisco Certified Networking Associate (CCNA) certification.

Irene Mia

Irene Mia is a Senior Economist with the Global Competitiveness Network at the World Economic Forum. Her responsibilities include researching competitiveness issues. She is also an editor of *The Global Information Technology Report* and is responsible for competitiveness research in Latin America and Hiberia at the World Economic Forum. Before joining the Forum, she worked at the Headquarters of Sudameris Bank in Paris for a number of years, holding various positions in the International Affairs and International Trade Divisions. Her main research interests are in the field of development, international trade and economic integration (with special reference to the Latin American region), and competitiveness. She has written and spoken extensively on issues related to national competitiveness. Dr Mia holds an MA in Latin American Studies from the Institute of Latin American Studies, London University, and a PhD in International Economic and Trade Law from Bocconi University in Italy.

Kuniko Ogawa

Kuniko Ogawa is currently Deputy Director of the International Affairs Division, Telecommunications Bureau, Ministry of Internal Affairs and Communications (MIC), Japan. She is in charge of US–Japan relations in the ICT field, including the US-Japan Regulatory Reform and Competition Policy Initiative under the US-Japan Economic Partnership for Growth. Before undertaking this position, she served as a Senior Officer for International Public Relations, revised Asia broadband programs, and attended the IGF as a member of delegation. After joining the Ministry of Posts and Telecommunications in 1995, she worked in the Telecommunications Business Department, the Radio Department, and graduated from the Fletcher School of Law and Diplomacy, Tufts University, Massachusetts, in 1998.

John G. Palfrey, Jr.

John Palfrey is Clinical Professor of Law and Executive Director of the Berkman Center for Internet & Society at Harvard Law School. His research and teaching is focused on the impact of the Internet on democracy. Prof. Palfrey teaches courses with a focus on Internet law, intellectual property, e-commerce, and digital democracy. He is a principal investigator of the OpenNet Initiative, a collaborative research project with the University of Cambridge, the University of Oxford, and the University of Toronto, which involves the study of the way that countries block their citizens' access to the Internet. He has published a number of scholarly papers related to the Internet's relationship to intellectual property, international governance, and democracy. Along with Professors Jonathan Zittrain and William Fisher, he co-authored an amicus brief to the United States Supreme Court in *MGM v. Grokster*. He is a graduate of Harvard College, the University of Cambridge, and Harvard Law School. He writes a blog at http://blogs.law.harvard.edu/palfrey/.

Marsha Powell

Marsha Powell is a senior member of Cisco's Executive Thought Leadership Team, which is responsible for defining business and technology strategies for global and regional businesses and organizations in all major segments of the economy including manufacturing, transportation, retail, health care, finance, and government. Ms Powell began at Cisco as a Senior Manager in the Internet Business Solutions Group, where she designed business strategies and processes for Cisco's enterprise sales organization that aligned business and technology architectures with the vision and objectives of customers. Prior to joining Cisco, Ms Powell worked for Cap Gemini Ernst & Young, Clear Channel Communications, UCLA, British Petroleum, and the IBM Corporation, and served as an executive adviser to leaders of some of the world's most innovative companies, assisting them in areas related to strategy, enterprise transformation, advanced technology investments, and innovative solutions to enhance customer loyalty and drive new revenue.

Sergio Sandoval

Sergio Sandoval is a Knowledge Associate in McKinsey & Company's Brussels Office. Since joining the firm in 2001, he has been serving clients in Europe, the Middle East, and Asia on strategy, regulation, and stakeholder management topics in the telecommunications, banking, and electricity sectors. He is also a member of McKinsey's Strategy Practice, where he focuses on developing knowledge around key regulatory topics. Prior to joining McKinsey, Mr Sandoval was employed as an Advisor to the Colombian Minister of Finance on macroeconomic policy matters. Additionally, he worked as a macroeconomic Advisor to the President of the Republic of Colombia. Mr Sandoval holds a Bachelor of Science degree in Economics (highest honors) and a Master of Science degree in Macroeconomics (highest honors) from Los Andes University in Colombia. He also holds an MBA (highest honors) from Solvay Business School in Belgium.

Hideo Shimizu

Hideo Shimizu has been Vice Minister for Policy Coordination (International Affairs), Ministry of Internal Affairs and Communications (MIC), Japan since 2006. As Vice Minister, he is currently head of the MIC's International Team and deals principally with regulatory harmonization, bilateral dialogue, ODA, and other matters of international coordination in the ICT area. Before assuming this position, he was Director General for Policy Planning, MIC, responsible for ICT strategic policy planning, broadcasting policy, and regulations. He made policies and plans to promote digital broadcasting toward full digitalization of broadcast media in early 2010. He also promoted R&D toward the realization of a ubiquitous network society and standardization in ICT fields. He has held important positions in the MIC, such as Director-General of the Postal Services Policy Planning Bureau, Director-General of the Finance Department. He graduated from the University of Tokyo with a Bachelor's degree in Law.

Wim Torfs

Wim Torfs is an Engagement Manager in McKinsey & Company's Brussels Office, where he has worked primarily in the areas of telecommunications, strategy, and finance. He has had extensive experience in telecommunications, mainly focusing on corporate strategy and regulation. Prior to joining McKinsey, Mr Torfs worked extensively on telecommunications and postal services liberalization and regulation issues in Europe, both for regulators and for incumbent operators. Mr Torfs has a BA degree in Commercial Engineering from the University of Antwerp, Belgium, where he focused on the impact of technological convergence on the EU regulatory framework. He also has a Master's degree in Management from the Vlerick Leuven Ghent Management School, Belgium.

Graham Vickery

Graham Vickery is Head of the Information Economy Group in the Information, Computer and Communications Policy Division at the Organisation for Economic Co-operation and Development (OECD), covering information technology, e-business, and the ICT industry. He has authored numerous OECD publications on the information economy, technology strategies, and government policies. He holds a BA in Economics from the University of Melbourne and a PhD in Chemistry from the University of Adelaide in Australia.

Sacha Wunsch-Vincent

Sacha Wunsch-Vincent is an Economist in the Information, Computer and Communications Policy Division at the Organisation for Economic Co-operation and Development (OECD) in Paris. He has recently authored an OECD study about the role of China in the ICT industry and a series of OECD studies on digital broadband content. He holds a Master's degree in International Economics from the University of Maastricht and a PhD in Economics from the University of St Gallen (Switzerland). He was Visiting Fellow at the Institute for International Economics and teaches International Economics at SciencesPo (Paris) and the World Bank Institute.

Partner Institutes

Albania
Institute for Contemporary Studies (ISB)
Artan Hoxha, President
Ilir Ciko, Researcher
Julia Dhimitri, Researcher

Algeria
Centre de Recherche en Economie Appliquée pour le
    Développement (CREAD)
Professor Yassine Ferfera, Director
Youcef Benabdallah, Assistant Professor

Angola
Serviços de Organização e Finanças (SOF)
Marcolino Meireles, Manager

Argentina
IAE—Universidad Austral
Marcelo Paladino, Vice Dean
Ariel A. Casarin, Assistant Professor

Armenia
Economy and Values Research Center
Manuk Hergnyan, Chairman
Sevak Hovhannisyan, Senior Research Associate
Anna Makaryan, Research Analyst

Australia
Australian Industry Group
Heather Ridout, Chief Executive
Tony Pensabene, Associate Director, Economics & Research

Austria
Austrian Institute of Economic Research (WIFO)
Professor Karl Aiginger, Director
Gerhard Schwarz, Coordinator, Survey Department

Azerbaijan
Azerbaijan Marketing Society
Sanar Mammadov, Executive Director
Ashraf Hajiyev, Project Coordinator
Saida Mammadova, Consultant

Bahrain
Bahrain Competitiveness Council
Sulaf Zakharia, Secretary-General

Barbados
Arthur Lewis Institute for Social and Economic Studies,
    University of West Indies (UWI)
Andrew Downes

Bangladesh
Centre for Policy Dialogue (CPD)
Dr Debapriya Bhattacharya, Executive Director
Professor Mustafizur Rahman, Research Director
Dr Khondaker Golam Moazzem, Research Fellow

Belgium
Vlerick Leuven Gent Management School
Professor Dr Lutgart Van den Berghe, Executive Director; Chairman,
    Competence Centre Entrepreneurship, Governance and Strategy
Professor Dr Harry P. Bowen, Economics and International Business

Benin
Micro Impacts of Macroeconomic Adjustment Policies
    (MIMAP) Benin
Epiphane Adjovi, Business Coordinator
Maria-Odile Attanasso, Deputy Coordinator
Cosme Vodounou, Responsible Axe Thémathique
Damien Mededji, Researcher

Bosnia and Herzegovina
MIT Center, the Faculty of Economics, Sarajevo University
Professor Zlatko Lagumdzija
Dr Fikret Causevic
Dr Zeljko Sain

Botswana
Botswana Institute for Development Policy Analysis (BIDPA)
Dr N.H. Fidzani, Executive Director
Kedikilwe P. Maroba, Program Coordinator

Brazil
Fundação Dom Cabral
Professor Carlos Arruda, Associate Dean for Research
    and Development
Rafael Tello, Researcher
Diogo Lara, Research Assistant

Movimento Brasil Competitivo (MBC)
José Fernando Mattos, President
Claudio Leite Gastal, Director
Jorge H. S. Lima, Project Coordinator

Bulgaria
Center for Economic Development
Anelia Damianova, Senior Expert

Burkina Faso
Societe d'Etudes et de Recherche Formation pour le
    Developpement (SERF)
Abdoulaye Tarnagda, Director General

Burundi
Center of Scientific Research in Economics (CURDES)
Pascal Rutake, Dean of Economics
Ferdinand Bararuzunza, Professor of Economics

Cambodia
Economic Institute of Cambodia
Sok Hach, Director
Chan Vuthy, Researcher
Tuy Chak Riya, Research Associate

Cameroon
Comité de Compétitivité (Competitiveness Committee)
Lucien Sanzouango, Permanent Secretary

355

Canada

Institute for Competitiveness and Prosperity
Roger Martin, Dean of the Rotman School of Management,
   University of Toronto, and Chairman of the Institute for
   Competitiveness and Prosperity
James Milway, Executive Director of the Institute for
   Competitiveness and Prosperity

Chad

Groupe de Recherches Alternatives et de Monitoring du Projet
   Pétrole-Tchad-Cameroun (GRAMP-TC)
Professor Gilbert Maoundonodji, Director
Lydie Beassemda, Program Officer
Yode Miangotar, Researcher

Chile

Universidad Adolfo Ibáñez
Andres Allamand, Dean, School of Government
Catalina Mertz, Director, Institute of Political Economics
Sergio Selman, Project Coordinator

China

Institute of Economic System and Management
National Development and Reform Commission
Dr Zhou Haichun, Deputy Director and Professor
Dong Ying, Professor
Chen Wei, Research Fellow

Colombia

National Planning Department
Santiago Montenegro Trujillo, General Director
Orlando Gracia Fajardo, Entrepreneurial Development Director
Victor Manuel Nieto, Economist

Croatia

National Competitiveness Council
Mira Lenardic, Secretary General
Martina Hatlak, Research Assistant

Cyprus

Center of Applied Research, Cyprus College
Dr Bambos Papageorgiou

The Cyprus Development Bank
Maria Markidou-Georgiadou, Manager, International Banking
   Services Unit and Business Development

Czech Republic

CMC Graduate School of Business
Dr Dagmar Glueckaufova, Interim President & Academic Dean
Jarmila Krupickova, Coordinator
Daniela Sedlackova, Executive Assistant to the President

Denmark

Copenhagen Business School
Department of International Economics and Management
Lars Håkanson, Head of Department
Anne Sluhan, Administrative Director

Ecuador

Escuela Superior Politécnica del Litoral (ESPOL)
Escuela de Postgrado en Administración de Empresas (ESPAE)
Virginia Lasio, Acting Director
Sara Wong, Professor
Lorena Carlo, Project Assistant

Egypt

The Egyptian Center for Economic Studies
Dr Hanaa Kheir-El-Din, Executive Director and Director of Research
Amal Refaat, Economist

Estonia

Estonian Chamber of Commerce and Industry
Siim Raie, Director General

Ethiopia

Ethiopian Economic Association/Ethiopian Economic Policy
   Research Institute
Assefa Admassie, Director
Kibre Moges, Senior Researcher
Worku Gebeyehu, Researcher

Finland

ETLA—The Research Institute of the Finnish Economy
Petri Rouvinen, Research Director
Pasi Sorjonen, Head of the Forecasting Group
Pekka Ylä-Anttila, Managing Director

France

HEC School of Management—Paris
Bernard Ramanantsoa, Professor, Dean of HEC School of
   Management
Bertrand Moingeon, Professor, Associate Dean for Executive
   Education

Georgia

Business Initiative for Reforms in Georgia
Irakli Burdiladze, Executive Director
Mamuka Tsereteli, Founding Member of the Board of Directors
Giga Makharadze, Founding Member of the Board of Directors

Germany

WHU—Otto Beisheim School of Management
Professor Michael Frenkel, Chair, Macroeconomics and
   International Economics

Greece

Federation of Greek Industries
Thanasis Printsipas, Economist, Research and Analysis
Antonis Tortopidis, Coordinator, Research and Analysis

Guyana

Institute of Development Studies, University of Guyana
Clive Thomas, Director
Karen Pratt, Research Associate

Hong Kong SAR

The Hong Kong General Chamber of Commerce
David O'Rear, Chief Economist

Federation of Hong Kong Industries
Alexandra Poon, Director

Hungary

Kopint-Datorg, Economic Research
Dr Éva Palócz, Deputy General Director
Ágnes Nagy, Project Manager

Iceland

IceTec
Hallgrímur Jónasson, General Director
Eydís Arnviðardóttir, Information Manager, Innovation Centre
Hallfríður Benediktsdóttir, Information Manager, Innovation Centre

India

Confederation of Indian Industry
Tarun Das, Chief Mentor
Kavita Choudhry, Deputy Director

Indonesia

LP3E-Kadin Indonesia
M.S. Hidayat, Chairman
Tulus Tambunan, Director

Ireland
Competitiveness Survey Group, Department of Economics,
   University College Cork
Dr Eleanor Doyle
Rosemary Kelleher
Niall O'Sullivan
Dr Bernadette Power

Israel
Manufacturers' Association of Israel (MAI)
Shraga Brosh, President
Yoram Blizovsky, Managing Director
Dan Catarivas, Director, Foreign Trade and International Relations
   Division

Italy
SDA Bocconi
Secchi Carlo, Full Professor of Economic Policy, Bocconi University
Paola Dubini, Associate Professor, Strategic and Entrepreneurial
   Management Department
Olga E. Annushkina, Assistant Professor, Strategic and
   Entrepreneurial Management Department

Jamaica
The Private Sector Organisation of Jamaica (PSOJ)
Lola Fong Wright, Chief Executive Officer
Stephanie Logan, Administrative Officer

Mona School of Business (MSB), University of the West Indies
Professor Neville Ying, Executive Director (Acting)
Michelle Tomlinson, Survey Coordinator
Patricia Douce, Survey Coordinator

Japan
Hitotsubashi University
Graduate School of International Corporate Strategy (ICS)
Yoko Ishikura, Professor and Associate Dean

Jordan
Ministry of Planning & International Cooperation
Jordan National Competitiveness Team
Amjad Attar, Director

Kazakhstan
Kazyna Development Fund
Prasad Bhamre, Vice-Chairman

Kenya
Institute for Development Studies, University of Nairobi
Professor Dorothy McCormick, Director
Walter Odhiambo, Research Fellow
Paul Kamau, Research Fellow

Korea
Graduate Institute of Management , Seoul School of Integrated
   Science and Technologies (aSSIST)
Dean Cheol Ho Shin, Professor of Strategy and International Business
Shin Hyo Kim, Senior Researcher
So Young Lee, Researcher

Kuwait
Economics Department, Kuwait University
Dr Reyadh Faras, Assistant Professor
Dr Mohammed El-Sakka, Professor
Dr Mohammad Ali Alomar, Assistant Professor
Dr Abdullah Al Salman, Assistant Professor

Kyrgyz Republic
Economic Policy Institute "Bishkek Consensus"
Marat Tazabekov, Chairman
Lola Abduhametova, Program Coordinator

Latvia
Institute of Economics, Latvian Academy of Sciences
Dr Raita Karnite, Director

Lesotho
Sechaba Consultants
Barbara Nkoala, Associate Consultant

Lithuania
Statistikos Tyrimai—Statistical Surveys, Vilnius
Benonas Miksas, Director

Luxembourg
Chamber of Commerce of Luxembourg
Carlo Thelen, Member of the Managing Board
Jean-Christophe Burkel, Attaché, Economics Department

Macedonia, FYR
National Entrepreneurship and Competitiveness Council (NECC)
Mirjana Apostolova, President of the Assembly
Minco Jordanov, President of the Managing Board
Saso Trajkoski, Executive Director

Madagascar
Centre of Economic Studies, University of Antananarivo
Pépé Andrianomanana, Director

Malawi
Malawi Investment Promotion Agency
Alick C. E. Sukasuka, Acting Deputy General Manager

Malaysia
Institute of Strategic and International Studies (ISIS)
Dato' Mohamed Jawhar Hassan, Chairman and Chief Executive
   Officer

National Productivity Corporation (NPC)
Dato' Nik Zainiah Nik Abdul Rahman, Director General
Chan Kum Siew, Senior Manager

Mali
Groupe de Recherche en Economie Appliquée et Théorique (GREAT)
Massa Coulibaly, Coordinator

Malta
Competitive Malta—Foundation for National Competitiveness
Dr John C. Grech, President
Margrith Lutschg-Emmenegger, Vice President
Adrian Said, Chief Coordinator

Mauritius
Joint Economic Council of Mauritius
Raj Makoond, Director

Mauritania
Centre d'Information Mauritanien pour le Développement
   Economique et Technique (CIMDET/CCIAM)
Moustapha Sidibé, Director
Chekroud Ould Bouhake
Aminata Niang

Mexico
Ministry of the Economy
Dr Eduardo J. Solis Sanchez, Chief of the Office for Investment
   promotion
Lic. Veronica Orendain De Los Santos, Director of Promotion,
   Office for Investment Promotion

Center for Intellectual Capital and Competitiveness
Dr Rene Villarreal, President
René Alejandro Villarreal, Vice-President

Moldova
Center for Strategic Territorial Development
Ruslan Codreanu, Executive Director
Andrei Smic, Expert, Regional Economic Development

357

358

Mongolia
Open Society Forum (OSF)
Munkhsoyol Baatarjav, Manager of Economic Policy
Erdenejargal Perenlei, Executive Director

Morocco
Université Hassan II
Fouzi Mourji, Professor of Economics

Mozambique
EconPolicy Research Group
Dr Peter Coughlin, Partner
Professor Dr Paulo N. Mole, Partner

Namibia
Namibian Economic Policy Research Unit (NEPRU)
Dr Christoph Stork, Senior Researcher

Nepal
Centre for Economic Development and Administration (CEDA)
Dr Ramesh Chandra Chitrakar, Executive Director
Santosh Kumar Upadhyaya, Researcher
Menaka Rajbhandari Shrestha, Researcher

Netherlands
Erasmus Strategic Renewal Center, Erasmus University Rotterdam
Professor Frans A. J. Van den Bosch
Professor Henk W. Volberda

New Zealand
Business New Zealand
Phil O'Reilly, Chief Executive
Marcia Dunnett, Manager, Business Services

Nigeria
Nigerian Economic Summit Group (NESG)
Dr Felix Ogbera, Associate Director, Research
Chris Okpoko, Senior Consultant, Research

Norway
BI Norwegian School of Management
Professor Torger Reve
Eskil Goldeng, Researcher

Pakistan
Pakistan Institute of Development Economics
Nadeem Ul Haque, Director
Faheem Jehangir, Research Economist

Paraguay
Centro de Analisis y Difusion de Economia Paraguaya (CADEP)
Dionisio Borda, Director
Fernando Masi, Research Member
Jaime Escobar, Research Member

Peru
Centro de Desarrollo Industrial (CDI), Sociedad Nacional de Industrias
Luis Tenorio, Executive Director
Néstor Asto, Project Director

Philippines
Makati Business Club
Guillermo M. Luz, Executive Director
Marc P. Opulencia, Deputy Director
Michael B. Mundo, Chief Economist

Poland
Warsaw School of Economics
Professor Bogdan Radomski, Associate Professor

Portugal
PROFORUM, Associação para o Desenvolvimento da Engenharia
Ilídio António de Ayala Serôdio, Vice President of the Board of Directors

Qatar
Qatari Businessmen Association (QBA)
Issa Abdul Salam Abu Issa, Secretary-General
Bassam Ramzi Massouh, General Manager
Ahmed El-Shaffee, Economist

Romania
Group of Applied Economics (GEA)
Dragos Pislaru, Executive Director
Dr Liviu Voinea, Research Director
Anca Rusu, Program Coordinator

Russian Federation
Academy of National Economy, Bauman Innovation
Dr Alexei Prazdnitchnykh, Principal, Associate Professor

Institute for Private Sector Development and Socio-Economic Analysis (IPSSA)
Irina Evseyeva

Stockholm School of Economics, Russia
Professor Carl F. Fey, Associate Dean of Research
Dr Igor Dukeov, Research Fellow

Serbia and Montenegro
Jefferson Institute
Aaron Presnall, Director of Studies

Singapore
Economic Development Board
Tan Choon Shian, Director, Planning and Policy, Marcom and Client Services
Chua Kia Chee, Head, Research and Statistics Unit

Slovak Republic
Business Alliance of Slovakia (PAS)
Robert Kicina, Executive Director
Gabriel Machlica, Project Manager

Institute for Economic and Social Reforms (INEKO)
Eugen Jurzyca, Director

Slovenia
Institute for Economic Research
Dr Art Kovacic
Professor Peter Stanovnik
Dr Mateja Drnovšek
Professor Aleš Vahcic

South Africa
Business Unity South Africa (BUSA)
Jerry Vilakazi, Chief Executive Officer
Friede Dowie, Chief Officer Strategic Services

Spain
Anselmo Rubiralta Center for Globalization and Strategy, IESE Business School
Professor Eduardo Ballarín
María Luisa Blázquez, Research Associate

Sri Lanka
Institute of Policy Studies
Indika Siriwardena, Database Manager

Suriname
Institute for Development Oriented Studies (IDOS)
John R.P. Krishnadath, President
Ashok Hirschfeld

Sweden
Center for Strategy and Competitiveness, Stockholm School of Economics
Professor Örjan Sölvell
Dr Christian Ketels

Switzerland
University of St. Gallen
Professor Dr Franz Jaeger, Director, Research Institute for
  Empirical Economics and Economic Policy

Taiwan, China
Council for Economic Planning and Development
Dr Sheng Cheng Hu, Chairman
J. B. Hung, Director, Economic Research Department
Chung Chung Shieh, Researcher, Economic Research Department

Tanzania
Economic and Social Research Foundation
Professor Haidari Amani, Executive Director
James Kajuna, Research Assistant, Commissioned Studies
  Department

Thailand
National Economic and Social Development Board
Dr Ampon Kittiampon, Secretary-General
Arkhom Termpittayapaisith, Deputy Secretary-General

Trinidad and Tobago
Arthur Lok Jack Graduate School of Business
Dr Rolph Balgobin, Executive Director
Deryck Omar, Managing Consultant, Centre for Strategy &
  Competitiveness
Narisha Khan, Research Analyst

Tunisia
Institut Arabe des Chefs d'Entreprises
Faycal Lakhoua, Conseiller

Turkey
TUSIAD Sabanci University Competitiveness Forum
Professor Dr A. Gunduz Ulusoy, Director
Hande Yegenoglu, Project Specialist

Uganda
Makerere Institute of Social Research, Makerere University
Delius Asiimwe, Senior Research Fellow
Wilson Asiimwe, Graduate Fellow
Robert Apunyo, Research Associate

Ukraine
CASE Ukraine, Center for Social and Economic Research
Vladimir Dubrovskiy, Leading Researcher
Oleksandr Rohozynsky, Executive Director

United Arab Emirates
Economic & Policy Research Unit, Zayed University
Dr Kenneth Wilson, Director

United Kingdom
London Business School
Dr Rebecca Harding, Executive Director, Global Entrepreneurship
  Monitor

United States
US Chamber of Commerce
David Hirschmann, Senior Vice President
John C. Clark, Associate Director, Information Resources
Julie Morris

Uruguay
Universidad ORT
Professor Isidoro Hodara

Venezuela
CONAPRI—National Council for Investment Promotion
Patricia Wallis, Consulting Manager
Giuseppe Rionero, Junior Consultant, Special Projects

Vietnam
Central Institute for Economic Management (CIEM)
Dr Dinh Van An, President
Phan Thanh Ha, Deputy Director, Department of Macroeconomic
  Management
Pham Hoang Ha, Senior Researcher, Department of Macroeconomic
  Management

Institute for Economic Research of HCMC
Tran Du Lich, Director
Du Phuoc Tan, Head of the Research Management and International
  Cooperation Department
Doan Nguyen Ngoc Quynh, Researcher of the Research
  Management and International Cooperation Department

Zambia
Institute of Economic and Social Research (INESOR),
  University of Zambia
Dr Mutumba M. Bull, Director
Dr Inyambo Mwanawina, Assistant Director and Coordinator,
  Economics and Business Research Program
Kabombo Fenete, Research Affiliate

Zimbabwe
Graduate School of Management, University of Zimbabwe
Professor A.M. Hawkins

**Bolivia, Costa Rica, Dominican Republic, Ecuador, El Salvador,
Guatemala, Honduras, Nicaragua, Panama**
INCAE Business School Latin American Center for Competitiveness
  and Sustainable Development
Roberto Artavia, Rector
Arturo Condo, Dean
Marlene de Estrella, Director of External Relations

Estonia, Latvia, Lithuania
Stockholm School of Economics in Riga
Dr Anders Paalzow, Rector
Dr Karlis Kreslins, Associate Professor

The World Economic Forum would like to thank Cisco Systems, Inc., for their invaluable support of this *Report*.

Cisco Systems, Inc., is the worldwide leader in networking for the Internet. Today networks are an essential part of business, education, government, and home communications, and Cisco Internet protocol–based (IP) networking solutions are the foundation of these networks. Cisco hardware, software, and service offerings are used to create Internet solutions that allow individuals, companies, and countries to increase productivity, improve customer satisfaction, and strengthen competitive advantage. The Cisco name has become synonymous with the Internet, as well as with the productivity improvements that Internet business solutions provide. At Cisco, our vision is to change the way people work, live, play, and learn. Cisco is interested in developing a deeper understanding of the relationship between information and communication technologies and organizational benefits such as innovation, improved productivity, and competitive advantage. Research projects such as *The Global Information Technology Report 2006–2007* help to identify issues and gain insight into these relationships, and apply them to improve productivity and standard of living. Cisco is deeply committed to the mission and values of the World Economic Forum, and is proud to be the sponsor of this *Report*.

Positions in the articles and papers included in this *Report* are in no way endorsed by Cisco Systems, Inc. To view Cisco's positions on public policy matters, please visit the Government Affairs homepage: http://www.cisco.com/gov.